LIFE AND DEATH IN ASIA MINOR

Studies in Funerary Archaeology Vol. 10

LIFE AND DEATH IN ASIA MINOR IN HELLENISTIC, ROMAN, AND BYZANTINE TIMES

STUDIES IN ARCHAEOLOGY AND BIOARCHAEOLOGY

Edited by

J. RASMUS BRANDT, ERIKA HAGELBERG,
GRO BJØRNSTAD, AND SVEN AHRENS

 OXBOW | books

Oxford & Philadelphia

Published in the United Kingdom in 2017 by
OXBOW BOOKS
The Old Music Hall, 106–108 Cowley Road, Oxford OX4 1JE

and in the United States by
OXBOW BOOKS
1950 Lawrence Road, Havertown, PA 19083

Hardback Edition: ISBN 978-1-78570-359-1
Digital Edition: ISBN 978-1-78570-360-7 (epub)

A CIP record for this book is available from the British Library

Library of Congress Cataloging-in-Publication Data

Names: Brandt, J. Rasmus, editor of compilation. | Hagelberg, Erika, editor
   of compilation. | Bjørnstad, Gro, editor of compilation. | Ahrens, Sven,
   editor of compilation.
Title: Life and death in Asia Minor in Hellenistic, Roman, and Byzantine
   times : studies in archaeology and bioarchaeology / edited by J. Rasmus
   Brandt, Erika Hagelberg, Gro Bjørnstad, and Sven Ahrens.
Description: Oxford ; Philadelphia : Oxbow Books, 2016. | Series: Studies in
   funerary archaeology ; vol. 10 | Includes bibliographical references and
   index.
Identifiers: LCCN 2016036111 (print) | LCCN 2016054635 (ebook) | ISBN
   9781785703591 (hardback) | ISBN 9781785703607 (digital) | ISBN
   9781785703607 (epub) | ISBN 9781785703614 (mobi) | ISBN 9781785703621
   (pdf)
Subjects: LCSH: Turkey--Antiquities. | Human remains (Archaeology)--Turkey. |
   Excavations (Archaeology)--Turkey. | Death--Social
   aspects--Turkey--History--To 1500. | Social archaeology--Turkey. |
   Turkey--History--To 1453
Classification: LCC DR431 .L55 2016 (print) | LCC DR431 (ebook) | DDC
   939/.200909--dc23
LC record available at https://lccn.loc.gov/2016036111

Printed in Malta by Melita Press Ltd
Typeset in India by Lapiz Digital Services, Chennai

For a complete list of Oxbow titles, please contact:

UNITED KINGDOM
Oxbow Books
Telephone (01865) 241249, Fax (01865) 794449
Email: oxbow@oxbowbooks.com
www.oxbowbooks.com

UNITED STATES OF AMERICA
Oxbow Books
Telephone (800) 791-9354, Fax (610) 853-9146
Email: queries@casemateacademic.com
www.casemateacademic.com/oxbow

Oxbow Books is part of the Casemate Group

Front cover: Hierapolis, North Necropolis, Tomb A18, seen from the north-east. © *Italian Archaeological Mission
at Hierapolis in Phrygia; Oslo University excavations; photo: J. Rasmus Brandt.*
Back cover: Hierapolis, North-East Necropolis, Tomb C92 (Eutyches' tomb), inside burials, seen from the south-west.
© *Italian Archaeological Mission at Hierapolis in Phrygia; Oslo University excavations.*

Contents

PART II: FROM DEATH TO LIFE: DEMOGRAPHY, HEALTH, AND LIVING CONDITIONS

Acknowledgements

In 2007 the Department of Archaeology, Conservation, and History at the University of Oslo, on the generous invitation of Prof. Francesco D'Andria, the director of the *Italian Archaeological Mission at Hierapolis in Phrygia*, started archaeological investigations of the North-East Necropolis at Hierapolis. The excavations and accompanying surveys, lasting till 2014, were primarily financed by a mixture of public and private funds, while a research project built up around the fieldwork, *Thanatos: Dead bodies – Live data. A study of funerary data from the Hellenistic-Roman-Byzantine town Hierapolis in Phrygia, Turkey*, was financed by the Norwegian Research Council (2010-2013).

The present publication is the result of an international conference held at Oslo and Fredrikstad, in Norway, Oct. 7-10, 2013, a conference which tried to sum up the results of three preceding annual workshops on Anatolian funerary archaeology in historical times, held in respectively Rome, Istanbul, and Lecce, all financed by the *Thanatos* project. To the conference were invited both scholars who had participated at one or more of the preceding workshops, but also new participants to expand on some of the previous arguments. The publication contains both papers, in elaborated form, presented at the conference, but also some developed from arguments presented at the one or more of the preceding annual workshops. In addition, two articles (by respectively Giuseppe Scardozzi and Michael Schultz and Tyede. H. Schmidt-Schultz) were put forward in the preparative stages of the present publication. They both add important information to the volume.

Without generous financial support from various public and private institutions our excavations at Hierapolis would never have taken place. For these important contributions we are very grateful to the University of Oslo ('Småforsk'), Institutt for sammenliknende kulturforskning, Statskraft, Rainpower ASA, Sigval Bergesen d.y. og hustru Nankis stiftelse til almennyttige formål, Stiftelsen Thomas Fearnley, Nils og Heddy Astrup, and Hans Rasmus Astrup.

Without the excavations the workshops and concluding conference would never have taken place. For this the Norwegian Research Council, through the *Thanatos* project, is to be thanked. The Norwegian Research Council has also offered, for which we are most grateful, a generous financial contribution towards the publication of the present volume

When it comes to the arrangement of the three three-day international workshops we are very grateful to the technical and administrative support from the respective persons and institutions:

- 2010, in Rome: The Norwegian Institute in Rome, by its then director prof. Turid Karlsen Seim, the administration, Ms. Anne Nicolaysen and Ms. Mona Elisabeth Johansen, and the caretaker couple Ms. Anna Muzi and Mr. Nicola Quinzi. Our thanks also go to Dr Maria Cataldi Dini of the then Soprintendenza per i beni archeologici dell'Etruria Meridionale for help in obtaining permission to visit Etruscan tombs at Tarquinia otherwise closed to the public, during the last day's excursion.
- 2011, in Istanbul: The Swedish Research Institute, by its then deputy director Dr Marianne Boqvist and administrative staff, Ms. Helin Şemmikanlı and Ms. Birgitta Kurultay; and also to Institut français d'Études Anatoliennes and its 'pensionnaire scientifique' for archaeology, Dr. Olivier Henry, who hosted the workshop for one day. Mr. Muhsin Baser at Pakero Travel gave the participants an interesting tour of Istanbul the final day of the workshop.
- 2012, in Lecce: Scuola di Specializzazione in Archeologia at Università del Salento, by its director prof. Francesco D'Andria and dott. Corrado Notario, and the mayor of Cavallino, Mr. Michele Lombardi, for opening the workshop at the Convento dei Domenicani. For practical matters like lodging in bed and breakfast places, dinners, and local transport much valuable help was offered by the Lecce Information office, in particular Mr. Marco Bianchi and the chauffeur Mr. Dario Cava. Ms. Emanuela D'Andria gave us a much appreciated evening tour of the town and prof. D'Andria gave the group a tour to some of his important Messapian excavations in the Salento

peninsula: Cavallino, Vaste, Chiesa di S. Stefano, and Castro, while the last afternoon visit to Otranto was left for the participants to enjoy for themselves.

For the arrangement of the conference at Oslo and Fredrikstad in 2013 our sincere thanks go to Anne Siri Wathne at the Department of Archaeology, Conservation and History at the University of Oslo, for practical support regarding the arrangement of the part of the conference which started in Oslo; to Christine Lande at the Museums of the Østfold County, who went beyond the call of duty to make the arrangements for the part of the conference which was held at Isegran, part of the old defence works of the 16th-century town Fredrikstad, a success; to Ms. Camilla Cecilie Wenn who, without asking, took upon herself the practical task of administering the breakfast and lunches at Isegran and making certain everyone was comfortable; to Ms. Liv Skjelbred, at the information office at Fredrikstad, who arranged an afternoon boat trip with freshly cooked shrimps in the old ferry boat 'Skjærhalden' to the archipelago outside Fredrikstad, at the inlet to the Oslofjord; and to Mr. Wiggo Andersen, who connected us to the right people in Fredrikstad and gave us a tour of the old town and arranged for a dinner serving wine produced in his own vineyard in South Africa; and to prof. Jan Bill, the director of the Viking Ship Museum (part of the University of Oslo), for offering a most interesting tour of the museum the last day of the conference.

Regarding the publication of the present volume our sincere thanks go to Ms Priscilla Field, who diligently copy-edited all the papers written by the non-native English speakers. For any mistakes which remain the editors take the blame.

Sincere thanks go to the staff of Oxbow Books for their professional handling of this publication from our first preliminary request to the final product. These expressions of thanks are also extended to the anonymous peer reviewers, who made very valuable suggestions to improve the quality of the publication. In particular, we should like to thank Ms Clare Litt, who patiently has supported this publication project from its inception.

Last, but not least, we should like to thank all the participants (no one mentioned, no one forgotten), both those from the workshops and those from the concluding conference, who all contributed to making the workshops and the conference successful events, and in particular to those who found the time and energy to rewrite their papers into what makes up this book. We were very touched by their many expressions of thanks afterwards.

J. Rasmus Brandt, Erika Hagelberg,
Gro Bjørnstad, and Sven Ahrens
Oslo, November 2015

Abbreviations

Since the present publication covers contributions from archaeologists and natural scientists working in many different fields of research, the editors have decided not to use abbreviations of journals, nor of ancient authors and their works, but to have them all written in full.

List of Contributors

AHRENS, SVEN, completed his studies of classical archaeology and art history in 2001 at the Humboldt-University, Berlin, with the doctoral thesis *Die Architekturdekoration von Italica, Santiponce (Sevilla)*. In 2002/2003 he received the travel scholarship of the German Archaeological Institute. Ahrens has worked on excavations in Germany, Spain, Norway, and on the excavation in Hierapolis by the University of Oslo, 2007–2013 (in the years 2010–2013 with a post-doctoral scholarship). Since 2013, he has been employed as archaeologist and researcher at the Norwegian Maritime Museum, Oslo.

BJØRNSTAD, GRO, is a zoologist and specialist in molecular evolution and ancient DNA studies. She obtained her doctorate at the Norwegian School of Veterinary Science (NSVS), for a project on genetic diversity of horse breeds, and has afterwards worked at the International Livestock Research Institute in Nairobi, the Natural History Museum in Oslo, NSVS and at the University of Oslo. She is interested in population genetics of past and present organisms, and has written and co-authored publications on the genetic history of reindeer in Scandinavia and chicken in Africa, on conservation genetics of African ungulates and salmon, and on DNA variations of the Viking Age population of Norway. She performed the DNA analysis on the human remains from Hierapolis in the *Thanatos* project (see Introduction).

BLAIZOT, FRÉDÉRIQUE is a researcher at the French National Institute for Preventive Archaeological Research (Lyon) and is associated with the University of Bordeaux (PACEA, UMR 5199). Amongst her many excavations, she excavated the Late Antique cemetery of Porsuk (Turkey) and a collective grave for children in the necropolis of Alexandria (Egypt). She has published an important synthesis on funeral practices in Roman Gaul: *Pratiques et espaces funéraires de la Gaule durant l'Antiquité* (Gallia 66.1/2009).

BRANDT, J. RASMUS, Professor emeritus in Classical Archaeology, University of Oslo, graduated with a DPhil degree from Oxford University. He was director of the Norwegian Institute in Rome (1996–2002) and at present he is director of the *Thanatos* research project at Hierapolis, Turkey (see Introduction). He has written widely on Greek and Roman archaeology and has recently edited two anthologies: *Greek and Roman Festivals. Content, Meaning, and Practice* (Oxford University Press 2012) with J. W. Iddeng, and *Death and Changing Rituals. Function and Meaning in Ancient Funerary Practices* (Oxbow Books 2014) with M. Prusac and H. Roland.

CASTEX, DOMINIQUE is director of research in Anthropology at the French Center for Scientific Research (CNRS). Between 2010 and 2015, she was the director of the laboratory of anthropology of the University of Bordeaux (PACEA, UMR 5199). Her main research is on the mortality crisis in historic periods, from funeral practices to demographic studies. She was director of the excavations of the central sector of the catacomb of Saints Peter and Marcellinus (Rome). She is in 2016 resident scholar at the French School of Rome.

D'ANDRIA, FRANCESCO, Professor emeritus in Archaeology and History of Greek and Roman Art and formerly head of the School of Specialization in Archaeology, both at the Salento University, Lecce, graduated from the 'Cattolica' University in Milan. On the Salento peninsula he has set up archaeological parks at Cavallino, Acquarica, Vaste, and Castro. In 2000 he founded IBAM, the Institute for Archaeological and Monumental Heritage of the Italian National Research Council, and functioned as its head till 2010. Since 2000 he has been the Director of the *Missione Archeologica Italiana a Hierapolis di Frigia* where he has worked for more than thirty years. He is the author of numerous scholarly books and articles.

DEMİREL, F. ARZU, Assistant Professor in Palaeoanthropology, Mehmet Akif Ersoy University, has a PhD from Ankara University. She has been directing a survey on Neogene and Pleistocene sites in the Burdur region since 2010 and is also involved with the analysis of human remains from some Hellenistic, Roman, and Byzantine sites in Turkey including Alanya (Antalya), Amorium (Afyonkarahisar), Herakleia-Perinthos (Tekirdağ), Miletus (Aydın) and Pisidian Antioch (Isparta). Recently she has been focusing on Byzantine human remains and some of her recent articles are 'Two weapon-related traumas from the Enclosure, 2008' (*Amorium Reports 3,* Ege Yayınları, 2012) and 'Analysis of human remains from Herakleia Perinthos' (*Istanbuler Forschungen* 55, Wasmuth-Verlag, 2016).

GOLDMAN, ANDREW L. is a professor in the History Department and chair of the Classical Civilizations Department at Gonzaga University in Spokane, WA, where he has taught since 2002. The main focus of his research has been the investigation of the Roman-period settlement and cemeteries at Gordion, where he has been an active team member since 1992. Between 2004 and 2006, he directed new fieldwork in the Roman-period strata that provided the first conclusive evidence that the site functioned as an auxiliary military base from the mid-1st to early 2nd century AD. He is currently publishing the final results of the 1950-2006 Roman excavations, including recent articles on the epigraphic finds, military equipment, and Roman imperial gemstones.

GRUPE, GISELA, Prof. Dr. in biology, did her education at the University at Göttingen, where she took her diploma degree in 1982 and a PhD in 1986 on trace element and stable isotope analysis of archaeological skeletal remains. Until 1989 she worked there as assistant lecturer in physical anthropology. In 1990 she was a recipient of the Heisenberg scholarship (German Science Foundation) and did her habilitation and *venia legendi* on *Physical Anthropology and Environmental History*. Since 1991 she has been professor in Physical Anthropology and Environmental History at the Ludwig-Maximilians-University of Munich and in the period 1991–2015 she was Director of the Bavarian State Collection for Anthropology. Her main research focuses on man/environmental relations in the Holocene and archaeometry.

HAGELBERG, ERIKA, has been Professor of Evolutionary Biology at the University of Oslo, Norway, since 2002. She holds a PhD in biochemistry from Cambridge University, and a master in history and philosophy of science from University College London. She pioneered the earliest applications of bone DNA typing in archaeology and forensic identification, and has held teaching positions at the University of Cambridge, UK, and the University of Otago, New Zealand. With Professor J. R. Brandt, she was co-investigator in the *Thanatos* project funded by the Research Council of Norway.

IŞIN, GÜL, is professor in Mediterranean art and archaeology at the Department of Archaeology, Akdeniz University, Antalya, Turkey. Her current projects concentrate on Tlos. Her former projects involved Arykanda, Sagalassos, Caunos, Pednelissos, and Patara. Her research areas are Lycian archaeology and culture, pottery, Pisidia, Mediterranean art and archaeology, sculpture, terracotta figurines, and identity problems in ancient cultures.

JAOUEN, KLERVIA, a Post-doc at the Department of Human Evolution, Max Planck Institute for Evolutionary Anthropology in Leipzig, Germany, is interested in the development of new isotopic tools for archeological purposes. Her research focuses on the link between the isotopic variability of metals (Fe, Cu, Zn and Sr) and parameters such as diet, mobility or metabolism (menopause, diseases). Currently, she is exploring the isotopic variation of different elements within food webs to verify whether new isotopic tracers could shed light on ancient diets. In 2014 she published with V. Balter, 'Menopause effect on blood Fe and Cu isotope compositions', *American Journal of Physical Anthropology* 153, 280–5; and in 2013 with M. L Pons and V. Balter, 'Iron, copper and zinc isotopic fractionation up mammal trophic chains', *Earth and Planetary Science Letters* 374, 164–72.

KIESEWETTER, HENRIKE is a physician and a physical anthropologist and has a doctorate in Medicine and a PhD in palaeoanthropology from Tübingen University. For her dissertation in palaeoanthropology she examined skeletons from a Neolithic graveyard in the Emirate of Sharjah (United Arab Emirates) with a focus on palaeopathology and on traces of violence. She was excavating and analysing the human remains from Troy and Kumtepe in Turkey for many years for the Troy Archaeological Project (Universities of Tübingen, Germany and Cincinnati, USA). Since 2012 she has been involved in the *Thanatos* research project of the Oslo University at Hierapolis.

KORKUT, TANER is professor in the Department of Archaeology at the Faculty of Literature, Akdeniz University in Antalya (Turkey). He has formerly been involved in excavations and surveys at Kyzikos, Patara, Tlos, and Sillyon. Since 2009 he has been director of the Tlos excavations and has conducted the extensive surveys around Tlos. He has written widely, both books and articles, on his major research fields: the traditions of burials and grave monuments, iconography, pottery, sculpture, and architecture. He also does research on the meaning of urbanism in Lycia.

LAFOREST, CAROLINE has recently been awarded her Ph.D on the use of the Roman and Proto-byzantine collective tomb 163d, in the North Necropolis of Hierapolis, focusing in particular on funeral gestures. While at Hierapolis she also took part in excavations of different funerary structures run by the Italian Archeological Mission. She works at the University of Bordeaux (PACEA, UMR 5199), and is associated with the French Institute of Anatolian Studies in Istanbul. For the academic year 2015–2016 she holds a research scholarship at the Koç University at Istanbul.

LIGHTFOOT, CHRISTOPHER S. is the Curator of Roman Art in the Department of Greek and Roman Art, The Metropolitan Museum of Art, New York. He was the director of the Amorium Excavations Project between 1993 and 2013 and continues to publish the results of his excavations there. He is about to publish *Amorium Reports 5: Roman and Byzantine Inscriptions, Seals, Stamps, and Graffiti*, which will include many funerary inscriptions from Amorium and its territory.

MÜLDNER, GUNDULA is a Lecturer in Bioarchaeology at the Department of Archaeology, University of Reading, UK, where she specializes in the reconstruction of diet and mobility of humans and animals by isotope analysis of skeletal remains. She has worked and published extensively on Roman and Medieval populations from Britain and abroad.

NAUMANN, ELISE is an archaeologist working at the Department of Archaeology, Conservation and History at the University of Oslo, Norway. Her research focuses on isotope analyses on human remains, with an emphasis on dietary practices and mobility patterns. In her PhD project she conducted analyses on human remains from Iron Age Norway. In 2014 she published two articles, one with M. Krzewinska, A. Götherström, and G. Eriksson, 'Changes in dietary practices and social organization during the pivotal Late Iron Age period in Norway (AD 550–1030): Isotope analyses of Merovingian and Viking Age human remains', *American Journal of Physical Anthropology* 155.3, 322–31; the other with T. D. Price and M. P. Richards, 'Slaves as burial gifts in Viking Age Norway? Evidence from stable isotope and ancient DNA analyses', *Journal of Archaeological Science* 41, 533–40.

NEHLICH, OLAF is an anthropologist and is specialized in dietary and mobility reconstruction of archaeological remains by stable isotope analysis. He mainly focuses on the analysis of stable sulphur isotopes as a dietary marker to infer the possible input of freshwater and marine diets. He is affiliated with the Department of Human Evolution at the Max-Planck Institute for Evolutionary Anthropology and the Department of Anthropology at The University of British Columbia. He is currently working in an isotope company for commercial isotope analyses.

NOVÁČEK, JAN, physical anthropologist and anatomist, is a tenured scientist and curator of the anthropological collection at the Thuringia State Service for Cultural Heritage and Archaeology in Weimar. Furthermore, he is a lecturer at the Institute of Anatomy and Cell Biology at the University Medical School Göttingen. He is specialized in human anatomy and its variations, palaeopathology, cremations, and microscopic investigation on human skeletal remains. His PhD thesis dealt with new methods of microscopic investigation of cremated human remains for which he was given the Hubert Walter award for outstanding performance in the field of physical anthropology from the German Anthropological Society in 2013. Currently, he is involved in several projects in the eastern Mediterranean and Central Europe. The results of his investigation of late Roman skeletons from the Harbour Necropolis in Ephesus will soon be published.

ÖĞÜŞ, ESEN is assistant director of the Aphrodisias excavations, and works as Assistant Professor of Ancient Mediterranean Art at Texas Tech University. She is currently an Alexander von Humboldt Foundation research scholar in Munich, Germany. Her research interests include Roman sarcophagi, funerary art, Roman sculpture and treatment of images in late Antiquity. She is currently working on the manuscript of a book on the 'columnar sarcophagi from Aphrodisias', where she examines the relationship between sarcophagus production and social status in Asia Minor. She has published on the social history of Asia Minor as attested by funerary inscriptions and archaeology.

PROPSTMEIER, JOHANNA, graduated in summer 2012 from Ludwig-Maximilians-University Munich in biology (physical anthropology) with the MSc-dissertation 'Die Lebensbedingungen in Pergamon: Nahrungsrekonstruktion mit Hilfe stabiler Stickstoff- und Kohlenstoffisotope einer römischen und spätbyzantinischen Nekropole', which has formed the basis for her presentation in the present volume. She has been an intern at the Archaeology Isotope Lab group at the University of British Columbia.

RICHARDS, MICHAEL is a Professor in the Department of Archaeology at Simon Fraser University, Canada, and is a Fellow of the Society of Antiquaries, and the Royal Society of Canada. His research is in the field of archaeological science, with a focus on isotope analysis, and Dr. Richards' main research interest is in the range and variety of modern human diets, and how our diets have evolved and changed over time. Among his many publications, note two works from 2009, one with J.-J. Hublin (eds.) *The Evolution of Hominin Diets: Integrating approaches to the study of Palaeolithic subsistence*, Dordrecht, Springer Netherlands; the other with E. Trinkaus, 'Isotopic evidence for the diets of European Neandertals and early modern humans,' *Proceedings of the National Academy of Sciences, USA* 106, 16034–16039.

RONCHETTA, DONATELLA is professor of History of Ancient Architecture at the Polytechnic of Turin (Italy), where she has been teaching this topic for over thirty years. Since 1965, that is almost 50 years, she has taken an active part in the Italian Archaeological Mission at Hierapolis in Phrygia, being responsible for the scientific investigations of the funerary architecture of the town's necropoleis. Among her publications, note a specific work on tumuli: Significance of the tumulus burial among the funeral buildings of Hierapolis of Phrygia, in O. Henry and U. Kelp (eds.) *Tumulus as sema: Space, politics, culture and religion in the first millennium BC*, Boston and Berlin, De Gruyter 2016.

SCARDOZZI, GIUSEPPE is a Ph.D. researcher at the Institute for the Archaeological and Monumental Heritage of the National Research Council of Italy (CNR-IBAM). He is responsible for the CNR-IBAM office in Lecce and for the Ancient Topography, Archaeology and Remote Sensing Laboratory. He has been coordinator of numerous research projects in Italy and in the Near East, focusing mainly on ancient topography, landscape archaeology, aerial archaeology, satellite remote sensing for archaeology, GIS, and webGIS for cultural heritage. He works in the Italian Archaeological Mission at Hierapolis in Phrygia since 2003, where he is coordinator of the archaeological surveys in the urban area and territory.

SCHEELEN, KRISTINA is a physical anthropologist and prehistorical archaeologist. She works at the Institute of Anatomy and Embryology at the University Medical School Göttingen and is a member of the Palaeopathology Work Group. She focuses on palaeopathology, bioarchaeology, human ecology, cremations, and forensic facial reconstruction. Currently, she is writing her PhD thesis at the Institute of Biology and Chemistry at the University of Hildesheim, dealing with traces of violence in Early Bronze Age skeletons from Slovak Republic. She investigates human skeletal remains for several archaeological projects in the eastern Mediterranean and Central Europe. The results of her investigation of early Ottoman skeletons from Ephesus/Ayasuluk were recently published (K. Scheelen, J. Nováček, and M. Schultz (2015) Anthropologische und paläopathologische Untersuchung menschlicher Skeletüberreste aus dem Friedhof um die Türbe. In S. Ladstätter (ed.) *Die Türbe im Artemision* (Sonderschriften ÖAI 53), 377-487). Further 17th century burials from Limyra (Lycia) will soon be issued.

SCHMIDT-SCHULTZ, TYEDE H., a biochemist and physical anthropologist at the Institute of Anatomy and Embryology at the University Medical School Göttingen, holds a doctorate in biochemistry. Her work focuses on proteomic in ancient and recent human bone, nature, aetiology and epidemiology of diseases in past populations, molecular neurobiology and special factors for myelination of oligodendrocytes in the Göttingen minipig. She has carried out fieldwork in southeast Europe, the Americas, the Near East and China. For publications, see, for example, T. H. Schmidt-Schultz and M. Schultz (2015) Ag-85, a major secretion protein of *Mycobacterium tuberculosis*, can be identified in ancient bone. *Tuberculosis* 95, 87–92; and T. H. Schmidt-Schultz and M. Schultz (2005) Intact growth factors are conserved in the extracellular matrix of ancient human bone and teeth: a storehouse for the study of human evolution in health and disease. *Biological Chemistry* 386,767–76.

SCHULTZ, MICHAEL, Professor emeritus in anatomy and director of the Palaeopathology Work Group at the Institute of Anatomy and Embryology at the University Medical School Göttingen (UMG) and co-opted professor at the Institute of Biology and Chemistry at the University of Hildesheim holds a doctorate in both medicine and in physical anthropology. His work focuses on palaeopathology of children, aetiology and epidemiology of diseases, palaeoanthropology, comparative and functional anatomy, and forensic taphonomy. He was president of the Palaeopathology Association (2001–2003) and of the German Anthropological Society (1997–1998, 1999–2000) and is the holder of the Medal of Honour from the Charles University of Prague for his work on microscopic palaeopathology. He has been visiting professor (e.g. at Basel, Bradford, Cairo, Mexico City, Kiev, Tucson, and Vienna) and has carried out research in many countries around the world (e.g. in central, south, and east Europe including Russia, the Americas, the Near East including Turkey, Egypt and Iraq, and China).

STESKAL, MARTIN studied Classical Archaeology and Ancient History at the University of Vienna. He is currently a tenured scholar at the Austrian Archaeological Institute in Vienna where he directs archaeological field projects and works as a lecturer at the Department of Classical Archaeology at the University of Vienna. Since 2015 he has been the assistant director of the excavations of Ephesus. His research interests include: ancient mortuary landscapes and practice, the built environment, ancient settlement patterns, functional analysis of material culture, Greek and Roman cultural and social history, the Roman East.

TEEGEN, WOLF-RÜDIGER, was educated at the Universities of Göttingen and Rome (La Sapienza) in European Prehistory, Physical Anthropology, Medical History, and Archaeology of the Near East. He took a PhD 1996 at Göttingen University, and his Habilitation degree in 2006 on animal palaeopathology at Leipzig University. Since 2009 he works at the Institute of Prehistoric Archaeology and Archaeology of the Roman Provinces, and the

ArchaeoBioCenter, Ludwig-Maximilians-University Munich. As an osteoarchaeologist he has participated in excavations in Pergamon, Priene, Kyme, Elaia, and Aigai in Turkey. His main research interests are funeral archaeology, palaeopathology, skeletal biology, and social archaeology.

UYGUN, ÇILEM, is assistant professor in the Archaeology Department at Mustafa Kemal University in Hatay (Turkey). She formerly took part in the excavations at Patara and the surveys of Tlos in the Lycia region. In addition, she has directed the excavations of rural settlements in the territory of Antioch. She has worked as a science team member at excavations in Tlos since 2005. Her areas of specialization are Hellenistic and Roman pottery, ancient jewellery, and sculpture.

WENN, CAMILLA CECILIE is an archaeologist with a cand.philol. degree in Classical Archaeology from the University of Oslo, analysing Central-Italian pottery. She is at present working for the Museum of Cultural History (University of Oslo) as excavation leader, conducting various projects primarily from the Norwegian Iron Age, including settlements, graves

and production sites. She has been part of the Hierapolis excavation project since 2009, and has previously excavated in Italy and Denmark. Her most recent research has focused on graves, grave disturbances, reuse of graves, and trade relations as evidenced in graves, in Roman-Byzantine Hierapolis and Iron Age southern Norway.

WONG, MEGAN is a PhD student in the Department of Archaeology at Simon Fraser University, Canada. She is a part of the Archaeology Isotope Lab group and is an archaeologist specializing in the reconstruction of prehistoric diet and mobility patterns through the use of isotopic analysis of human skeletal remains.

YILDIZ, M. ERTAN is assistant professor in Classics and epigraphy at the Faculty of Literature, Department of Ancient Languages and Cultures, Akdeniz University in Antalya. He formerly took part in the surveys of Keleinai/ Apameia Kibotos in the Phrygia region. He has published on Greek inscriptions, as, for example the recent article 'Kelainai/Apameia Kibotos'tan Dört Yeni Yazıt (Vier neue Inschriften aus Kelainai/Apameia Kibotos), *Olba* 22, 291–306.

ADDRESSES

Dr Sven Ahrens
Norwegian Maritime Museum, Bygdøynesveien 37, N-0286 Oslo; Norway
E-mail: sven.ahrens@marmuseum.no

Dr Gro Bjørnstad
Division of Forensic Sciences, Norwegian Institute of Public Health, PO Box 4404 Nydalen, N-0403 Oslo; Norway
E-mail: gro.bjornstad@gmail.com

Dr Frédérique Blaizot
INRAP (French National Institute for Preventive Archaeological Research), 11 rue d'Annonay F-69 675 Bron Cedex; France
E-mail: frederique.blaizot@inrap.fr

Prof. Dr J. Rasmus Brandt
Istituto di Norvegia, Viale Trenta Aprile 33, I-00153 Roma; Italy
E-mail: j.r.brandt@iakh.uio.no

Prof. Dr Dominique Castex
Anthropologie des Populations Passées et Présentes, PACEA - UMR 5199, Université de Bordeaux, Bâtiment B8, Allée Geoffroy Saint Hilaire, CS 50023, F-33615 Pessac Cedex; France
E-mail: dominique.castex@u-bordeaux.fr

Prof. Dr Francesco D'Andria
Dipartimento Beni Culturali, Università del Salento, Via Birago 64, I-73100 Lecce; Italy
E-mail: francesco.dandria@unisalento.it

Assistant Professor Dr F. Arzu Demirel
Mehmet Akif Ersoy University, Faculty of Arts and Sciences, Department of Anthropology, TR-15030 Burdur; Turkey
E-mail: arzudemirel@mehmetakif.edu.tr

Dr Andrew L. Goldman
Gonzaga University, 502 E. Boone Ave, Spokane, Washington 99258-0035; USA
E-mail: goldman@gonzaga.edu

Prof. Dr. Gisela Grupe
Ludwig-Maximilians-University, Biocenter Martinsried, Grosshaderner Str. 2, D-82152 Martinsried; Germany
E-mail: g.grupe@lrz.uni-muenchen.de

Prof. Dr Erika Hagelberg
Department of Biosciences, University of Oslo, PO Box 1066 Blindern, 0316 Oslo; Norway
E-mail: erika.hagelberg@ibv.uio.no

Prof. Dr Gül Işın
Akdeniz University, Faculty of Literature Department of Classical Archaeology, TR-07058 Antalya; Turkey
E-mail: gulisin@akdeniz.edu.tr

Dr Klervia Jaouen
Department of Human Evolution, Max-Planck Institute for Evolutionary Anthropology, D-04103 Leipzig; Germany
E-mail: klervia_jaouen@eva.mpg.de

Dr Dr Henrike Kiesewetter
Via Pier Lombardo 5, I-20135 Milano; Italy
E-mail: henrike.kiesewetter@web.de

Prof. Dr Taner Korkut
Akdeniz University, Faculty of Literature, Department of Archaeology, TR-07058 Antalya; Turkey
E-mail: tkorkut@akdeniz.edu.tr

Dr Caroline Laforest
Anthropologie des Populations Passées et Présentes, PACEA - UMR 5199, Université de Bordeaux, Bâtiment B8, Allée Geoffroy Saint Hilaire, CS 50023, F-33615 Pessac Cedex; France
E-mail: caroline.laforest@u-bordeaux.fr; laforest.caro@gmail.com

Dr Chris Lightfoot
Department of Greek and Roman Art, The Metropolitan Museum of Art, 1000 Fifth Avenue New York, New York 10028-0198; USA
E-mail: Christopher.Lightfoot@metmuseum.org

Dr Gundula H. Müldner
Department of Archaeology, University of Reading, Whiteknights, PO Box 227, Reading RG6 6AB, UK
E-mail: g.h.mueldner@reading.ac.uk

Dr Elise Naumann
Department of Archaeology, Conservation, and History, University of Oslo, P.O. Box 1008 Blindern, N-0315 Oslo; Norway
E-mail: elise.naumann@iakh.uio.no; elisena@gmail.com

Dr Olaf Nehlich
Department of Anthropology, University of British Columbia, 3124-6303 N.W. Marine Drive Vancouver, British Columbia V6T 1Z1; Canada
and
Department of Human Evolution, Max-Planck Institute for Evolutionary Anthropology D-04103 Leipzig; Germany
E-mail: olaf@nehlich.com

Dr Jan Nováček,
Thuringia State Service for Cultural Heritage and Archaeology, Humboldtstrasse 11, D-99423 Weimar; Germany
and
Institute of Anatomy and Cell Biology, University Medical School Göttingen, Kreuzbergring 36, D-37075 Göttingen; Germany
E-mail: jan.novacek@med.uni-goettingen.de

Dr Esen Öğüş
Institut für Klassische Archäologie, Ludwig-Maximilians-Universität, Katharina-von-Bora Str. 10, D-80333 Munich, Germany
E-mail: ogus@post.harvard.edu

Johanna Propstmeier
Am alten Sportplatz 3, D-85646 Anzing; Germany
E-mail: jo_hanna@gmx.de

Prof. Dr Michael P. Richards
Department of Archaeology, Simon Fraser University, 888 University Dr., Burnaby, British Columbia, V5A 1S6; Canada
and
Department of Human Evolution, Max-Planck Institute for Evolutionary Anthropology, D-04103 Leipzig; Germany
E-mail: michael_richards@sfu.ca

Prof. Dr Donatella Ronchetta
Interuniversity Department of Regional and Urban Studies and Planning, Politecnico di Torino, Viale Mattioli 39, I-10125 Torino; Italy
E-mail: donatella.ronchetta@polito.it

Giuseppe Scardozzi
Consiglio Nazionale delle Ricerche, Istituto per i Beni Archeologici e Monumentali (CNR-IBAM), Strada per Monteroni, Campus Ecotekne (Università del Salento), I-73100 Lecce; Italy
E-mail: g.scardozzi@ibam.cnr.it

Kristina Scheelen M.A.
Institute of Anatomy and Embryology, University Medical School Göttingen, Kreuzbergring 36 D-37075 Göttingen; Germany
and
Institute of Biology and Chemistry, University of Hildesheim, Universitätsplatz 1, D-31141 Hildesheim; Germany
E-mail: scheelen@uni-hildesheim.de

Dr Tyede H. Schmidt-Schultz
Institute of Anatomy and Embryology, University Medical School Göttingen, Kreuzbergring 36 D-37075 Göttingen; Germany
Email: tschmidt-schultz@web.de

Prof. Dr Michael Schultz
Institute of Anatomy and Embryology, University Medical School Göttingen, Kreuzbergring 36 D-37075 Göttingen; Germany
and
Institute of Biology and Chemistry, University of Hildesheim, Universitätsplatz 1, D-31141 Hildesheim, Germany
E-mail: mschult1@gwdg.de

Dr Martin Steskal
Austrian Archaeological Institute, Franz Klein-Gasse 1,
A-1190 Wien; Austria
E-mail: martin.steskal@oeai.at

Prof. Dr Wolf-Rüdiger Teegen
ArchaeoBioCenter & Institute for Pre- and Protohistoric
Archaeology and Archaeology of the Roman Provinces,
Ludwig-Maximilians-University, Geschwister-Scholl-Platz
1, D-80539 München; Germany
E-mail: w.teegen@lmu.de

Asist. Prof. Dr. Çilem Uygun
Mustafa Kemal University, Department of Archaeology,
TR-31100 Hatay; Turkey
E-mail: cilemuygun@hotmail.com

Camilla Cecilie Wenn
Museum of Cultural History, University of Oslo, P.O.Box
6762 St. Olavs plass, N-0130 Oslo; Norway
E-mail: c.c.wenn@khm.uio.no

Megan Wong
Department of Archaeology, Simon Fraser University,
888 University Dr., Burnaby, British Columbia, V5A
1S6; Canada
E-mail: wong.meganb@gmail.com

Dr Ertan Yıldız
Akdeniz University, Faculty of Literature, Department
of Ancient Languages and Cultures TR-07058 Antalya;
Turkey
E-mail: ertanyy@hotmail.com

Introduction
Dead bodies – Live data: Some reflections from the sideline

J. Rasmus Brandt

Dead bodies – Live data

In the summer of 2007 the Department of Archaeology, Conservation, and History at the University of Oslo, started archaeological investigations of the North-East Necropolis at Hierapolis. The investigations, covering a) surveys of the whole necropolis area to achieve a full record of visible tombs and sarcophagi and their typologies as well as the topographical extension and organization, and b) focused excavations of selected tombs and tomb areas to achieve a chronology of use and of changing funerary practices, have continued for 4–6 weeks every year till 2014.

In 2010 the Norwegian Research Council gave considerable financial support to a four-year research project: *Thanatos: Dead bodies – Live data. A study of funerary data from the Hellenistic-Roman-Byzantine town Hierapolis in Phrygia, Turkey*, which emerged from the surveys and excavations of the North-East Necropolis. The primary object of the research project was to investigate, in a social setting, an urban population in detail over a long period of time, including studies on funerary architecture and landscape perception, organization and entrepreneurship, practices and rituals, mortuary behaviour, genetic relationships and origins, palaeodemography, health and disease, behaviour, and diets and individual mobility patterns. The aim of the project was, in addition to the archaeological research, to make a wide use of radiocarbon dates and osteological, ancient and mitochondrial DNA, and isotope analyses. In the present publication this perspective has been widened to include contributions from other sites in Asia Minor.

The most important sites discussed are entered on a map of Asia Minor (Fig. 0.1), but the articles refer to many more than those signalled. This limitation in space is done on purpose. By pulling together funerary information from a rather limited, defined region the aim has been to reduce the presence of many regional differences found in ancient Anatolia (Ousterhout 2010, 87; Moore 2013, 84–5) with the hope that we shall be able to present data that can more

easily be compared and transformed into more generic cultural and historical overviews than if we had collected funerary results from larger areas within the Anatolian and/ or Mediterranean areas. The reader will quickly discover that Hierapolis plays a central part in many of the contributions in the present publication, a situation difficult to avoid, due to the quantity of data now being generated through just concluded surveys and excavations.

The history of Hierapolis, founded around 200 BC, destroyed and abandoned due to an earthquake in the mid-7th century AD, but reborn to a new and different life some generations later, before its final abandonment in the 14th century AD, also forms the chronological frame for the present publication. It is a recurring habit in historical and archaeological publications to go by defined time periods; from recent years can be mentioned many interesting and well-reviewed books on Roman funerary customs, as, for example, Cormack 2004; Flämig 2007; Brink, Green, and Green (eds.) 2008; Carrol and Rempel (eds.) 2011; Hope and Huskinson (eds.) 2011. By stretching the timeline to cover more periods the intent has been to more easily pick up changes in funerary practices and life conditions, significant and insignificant, not only within periods, but between periods. Therefore, many of the articles cover the important, but at times turbulent, transition from a pagan to a Christian society, and the traumatic change when the Roman urban centres disappeared in the 6th/7th centuries and a new society, both socially, politically, and urbanistic, a few generations later both literally and metaphorically grew out of the ruins of the old ones.

Unfortunately, in all sites discussed in the present volume, there is a clear cultural and historical break with the Turkish gradual take-over of the area, when a completely new communication and settlement infrastructure was built up. In places which show some Turkish activity, like at Hierapolis (Arthur 2012, 297–9) and Pergamon (Schultz and Schmidt-Schultz, this publication), burial grounds have not been detected.

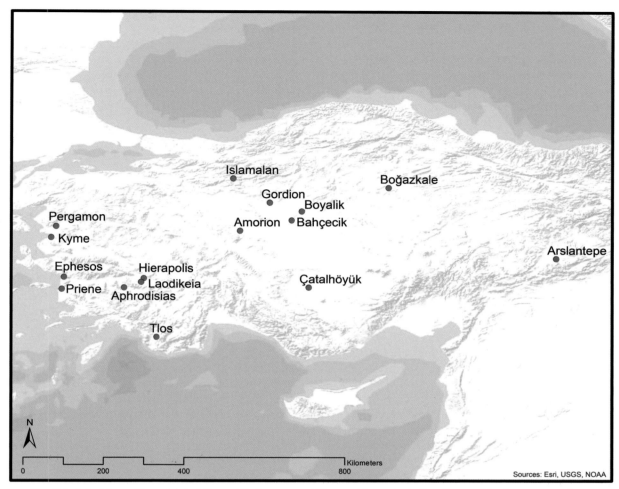

Fig. 0.1. Asia Minor. Map of Asia Minor with the names of the most important sites discussed in the present volume.

Ancient towns, in addition to their role as living places and trading centres for people, also functioned as communities where experiences, stories, thoughts, and memories were collected, revised, and adapted, and where the past and the future met in activities of the present (Casey 1996, 24–26). And within the territory of the town it was the religious place which was 'the most rooted in the past and collective memories to impact upon the present' (Lock *et al.* 2005, 151). The nomadic Seljuks and later Turkoman populations had no political, cultural, or religious roots in the Classical Anatolian landscape, for that reason many Turkish towns grew up in areas without a Greek, Roman, and/or Byzantine urban past. Denizli in the Lykos valley is a good example, even if it had some Byzantine roots (Arthur 2012, 299); in the Middle Ages it gradually came to take over the functions of the previous urban centres in the valley, like Colossae, Laodikeia, Tripolis, and Hierapolis. To my limited knowledge, places in which there are preserved cemeteries from both Roman, Byzantine, and Seljuk and later Turcoman times, are few, like, for example, Çatalhöyük, Ephesos,

Istanbul, and Corinth; among these Çatalhöyük, about which we shall certainly know more in the future, may be the only one where pagan Romans, Christian Byzantines, and Muslim Turks were buried in the same necropolis area.

In order to give an overview of the wide range of funerary studies under development in Asia Minor, this publication tries to give an insight into a cross-section of this research ranging from basic to applied research. Basic research in archaeology has, over the years, lost its hegemony to theoretical questions, but without the one, the other has no data to work with. In a long future perspective it is my personal experience that the basic research and presentation of data will have a longer scholarly lifetime than the theoretical results; though this does not turn theoretical studies into unimportant short-lived ephemeral research; the questions posed and interpretations made are part of the total picture of knowledge, which lives in a constant flux in its attempts at achieving a wider and in-depth understanding of how people lived in a distant, complex past – seen in both a synchronic as well as in a diachronic perspective.

Even if theoretical issues are not specifically addressed in the present publication the majority of the articles, if not all, in one way or the other can be included under the wide theoretical umbrella of body theory and materiality.

A concern of both the present publication and the *Thanatos* project has been to find a tighter historical and cultural dialogue between the results in the archaeological field of study with those in the scientific fields. Still the results address different aspects of the life and death problems. Articles from the first field of study are concerned with the archaeology and history of death in a social context, or how life in a material context is translated into death, death standing out, along with birth, as the most important transition event in the life of man. Articles in the second field of study try, through osteological, DNA and isotope analyses, to bring life back into the dead bodies in order to better understand the general living conditions of man in different periods and locations. This is the reason why the two kinds of studies have been presented here under different headings: Part I: *From life to death. Death and the social and funerary setting;* Part II: *From death to life. Man and the ancient life conditions.* In the second part of this introduction I shall try to indicate some areas in the study of man in the past in which the two fields of study may meet.

Chapters 14–20 by, respectively, Wong *et al.*, Propstmeier *et al.*, Teegen, Kiesewetter, Schultz and Scmidt-Schultz, Demirel, and Novacek *et al.*, should be read together, since they all contain information on diets and pathologies which both support and overlap each other and give an interesting, rather coherent picture of the life conditions in Asia Minor, both in the Roman/early Byzantine and in the mid-/late Byzantine periods. Some cross-referencing has been done between the articles, including the articles in the archaeological section, but not in a comprehensive way. See also below for some further observations. However, first a few words shall be said about body theory and materiality.

Body theory and materiality

The Hellenistic-Roman-Byzantine society at Hierapolis, as all the other contemporary societies in Asia Minor discussed here, was a multifaceted living organism, as expressed through its many preserved monuments, artefacts, and inscriptions. The tombs add literally a human factor to the urban organism. In the interplay between dead bodies and their material contexts we possess an important tool to study social meanings and relationships, not only as singular and synchronic, but also as constantly changing diachronic events in which the meanings are continuously attacked and defended, defined and redefined. The aim of the *Thanatos* project, the final results of which are under preparation for publication in the Italian series *Hierapolis di Frigia*, has been to unveil the complexity of the ancient society, not its uniformity. The wide variety of empirical data collected has required the application of, within a social setting, theoretical approaches that are both polysemic and diverse. The studies, as implied by the title of the project, have therefore taken their point of departure in body theory and materiality, two theoretical issues, as already stated, which are also central to the present publication.

Body theory purports the study of both the body *as* material (i.e. the dead bodies as preserved skeletons, the object of osteological studies, DNA and isotope analyses, and C14 dates) and the body *and* material (i.e. the dead body in its many-faceted contexts, as, for example, its relation to the urban settlement and the landscape, the shape of the tomb, the grave goods and the inscriptions) (see in particular, Sofaer 2006, 62–88).

Materiality concerns archaeological material in its widest sense, i.e. all kinds of material found in archaeological excavations, including dead bodies. It is not a theory in itself, but an umbrella concept for a wide range of theories placing the archaeological material in a social context (Glørstad and Hedeager 2008, 27–8). Body theory and materiality are thus two approaches which have much in common. The present publication has, for that reason, pulled together data from three different fields of study: genetic, osteological, and archaeological studies, the accessed data of which have been analyzed from different methodological and theoretical standpoints.

Under the concept body *as* material the more important issues investigated and discussed are, for example, questions like:

- Demography: mortality rates and life expectancy among age groups, sexes, and social groups (Ahrens, Demirel, Kiesewetter, Schultz and Schmidt-Schultz, Teegen).
- Life standards/living conditions seen both in a synchronic and a diachronic perspective, in short osteobiographies (Demirel, Kiesewetter, Nováček *et al.*, Propstmeier *et al.*, Schultz and Schmidt-Schultz, Teegen).
- Pathology, or various kinds of sicknesses, disease, and nutritional deficiencies, which leave traces in bones and/or teeth (Demirel, Kiesewetter, Nováček *et al.*, Schultz and Schmidt-Schultz, Teegen).
- Body actions, or the study of wear and tear, lesions, and injuries on the bones due to activities connected with crafts or physical expressions of various sorts, including combat (Kiesewetter, Nováček *et al.*, Teegen).
- Genetic variations, related to ethnic populations and families (Hagelberg and Bjørnstad).
- Mobility patterns and diet (Propstmeier *et al*; Wong *et al.*).
- Post-mortem life of the skeletons (archaeothanatology) (Laforest *et al.*) (on the use of this concept, see Duday 2009).

Under the concept body *and* material, the issues discussed are:

- Necropoleis (urban, rural) and topography (Ahrens, D'Andria, Goldman, Işin and Yıldız, Korkut and Uygun, Lightfoot, Öğüş, Ronchetta, Scardozzi, Steskal, Wenn *et al.*).
- Death, territory, urban transformations and lines of communication (D'Andria, Scardozzi).
- City border, liminality, pollution and purification (Steskal, Wenn *et al.*).
- Landscape perception and visibility (Ronchetta, Scardozzi, Steskal, Wenn *et al.*).
- Funerary architecture (body containers), material, construction techniques, production, and costs (Ahrens, D'Andria, Goldman, Işin and Yıldız, Korkut and Uygun, Lightfoot, Öğüş, Ronchetta, Scardozzi, Steskal).
- Funerary epigraphy, grave markers: ownership, social and family organizations, gender roles, status, symbols, and craft affiliations (Ahrens, D'Andria, Işin and Yıldız, Korkut and Uygun, Laforest *et al.*, Lightfoot, Öğüş, Ronchetta, Scardozzi, Steskal, Wenn *et al.*).
- Grave goods: symbols and significance (Goldman, Korkut and Uygun, Laforest *et al.*, Wenn *et al.*).
- Funerary rights, rituals, and practices: taboos, orientation, inhumation and cremation, locals and non-locals, pagans, Jews, and Christians (D'Andria, Goldman, Işin and Yıldız, Korkut and Uygun, Laforest *et al.*, Lightfoot, Steskal, Wenn *et al.*).
- Use and reuse of tombs/graves (continuity and change) related to practical and/or symbolic conditions/reasons; burials of saints (D'Andria, Goldman, Korkut and Uygun, Laforest *et al.*, Lightfoot, Ronchetta, Steskal, Wenn *et al.*).
- Memory, identity, and mental changes (D'Andria, Goldman, Öğüş, Scardozzi, Steskal, Wenn *et al.*).

Some reflections from the sideline

Being the chief editor and having had the gratifying privilege to read through all the articles more than once, I have made some observations I should like to bring forward, in a meagre attempt in this introduction to open up some discussions I hope will be carried further. Not all contributions will be mentioned here, only a highly personal selection, which must not be read as a priority list with regard to their contents, only as a reflection of some personal research interests. I use the word reflection on purpose in order to underline that the observations and questions asked are not the result of profound studies, but are rather questions of curiosity which may or may not be worth approaching more seriously. And since I am an archaeologist by profession, and myself a contributor to the volume, the articles that have inspired my observations are mainly those in the bioarchaeological section (Part II), which

have brought new information to my limited knowledge of the ancient micro-Asian societies, the underlying aim being to approach historical and cultural questions in which the present archaeological and scientific studies can meet and develop further. I have three observations to make: the first on the living conditions and the stability of the Roman societies both in Asia Minor and beyond, the second on nutrition, and the third on health.

Living conditions

Despite the high number of Roman towns, Roman society was basically rural – a large majority of the population lived in the countryside. Their daily work and preoccupations were tied to the fields and their cultivation of agricultural products. They were tightly bound to the rhythms of the nature, the changing seasons, the weather, and the soil fertility. It was a society, to use the concepts of the German historian Reinhart Koselleck (1977; see also 2002, 218–35), built on a large space of experience in agriculture, but with a low future horizon of expectations beyond cultivation, i.e. there was a continuity, but low tension between the two concepts. The same may be said for the artisans in the towns – their occupation was also built on high experience in their crafts, but with low future expectations beyond their professions (Koselleck 1977, 198). It was a stable society held together by a web of strict laws and norms, with low social mobility, and where the prospect for the future was mainly not to have it worse than at the present moment. The ancients certainly had the idea of the future as a temporal space, but this space was not left for social ideologies or the bureaucracy to prognosticate, rather for masters in divination, prophecies, and the reading of signs sent mankind by the gods, and for the examiners of oracles and the Sibylline books to interpret and take appropriate action in order to prevent present and future instability (cf. Santangelo 2013, 4). The future in Antiquity, so to speak, was taken care of by religious institutions and persons whose thinking was rooted in the past and the collective memory, a safe guarantee for stability.

This claim for stability is also reflected in the necropoleis, even if these shifted in their locations and their programmatic exposures over time. At Hierapolis, for example, the Republican tradition of placing the tombs along the roads in a dialogue between the dead and the living was in the Imperial period exchanged for tombs withdrawn from the hectic life along the roads to the peaceful hills behind with a beautiful view of the town and the surrounding landscape (Wenn *et al.*, this volume). At the same time, however, especially in the Roman Imperial period, a high degree of standardization (a sign of social stability) was reached both in the choice of type and size of the tombs and sarcophagi. Even the tomb inscriptions were highly standardized giving the name of the dead and family relations limiting the use of the tomb to family members and imposing heavy fines on those misusing or desecrating it. This standard formula lasted, with small variations, for centuries.

With the Christians some social ideologies of the pagan society were broken and new spaces of experiences introduced, but these were experiences aimed at the expectations of life beyond death, not at the creation of new expectation horizons in the present life. But since the other-worldly expectations raised from worldly experiences could never be controlled, no conflict would arise between experience and expectations (Koselleck 1977, 198–9).

If the society was hit by natural disasters, drought, flooding, fire, earthquakes, or epidemics, the main concern was to bring the society back into balance, and regain the past rhythm of life. With the exception of single, strong earthquakes it is difficult to read such temporary set-backs in the archaeological records, tombs included. However, reading the social archaeological and bioarchaeological data over long periods, some differences in the living conditions may appear also carrying information in what these differences consisted.

The watershed event at Hierapolis, distinguishing between two ways of life, is linked to the earthquake in the middle of the 7th century AD, when the classical Roman city was abandoned and a new one grew out of the ruins of the old some 150–200 years later: smaller, with a completely new urban infrastructure, and without the many public buildings of the classical town. In the 6th and 7th centuries many other pre-Roman and Roman towns were for various reasons abandoned, reduced in size, or moved to a much smaller area nearby due to earthquakes (Laodikeia, Tripolis, and Sagalassos: Kumsar *et al.* 2015; Sintubin *et al.* 2003), foreign assaults (Sardis captured by the Persians in AD 616: Foss 1975a; 1975b; 1976, 53–66; Rautman 2011, 24–6), because they simply imploded when the countryside could not produce enough products for the town to which it belonged to survive, and the town became too small to consume and distribute the products produced in the countryside (as I imagine may have been the case with Aphrodisias in the 7th century AD, though see also Ratté 2001, 144–5; Ratté 2012; Ratté and De Staebler 2011, 123–4), or for other combinations of reasons (Ephesos: Ladstätter and Pülz 2007; Pergamon and its harbour town Elaia: Radt 2001, 52–3; Pirson 2010, 197–200).

It was not the first time these cities had been hit by natural disasters or other tragic events, but in the 6th century the Byzantine Empire had lived through a dramatic series of natural disasters, which may have caused a significant demographic reduction and thus reduced human strength and energy to withstand and restore the fragile society. The events in the 6th century AD can be summarized as follows (Brandt 2016). First of all, a severe earthquake storm (i.e. not a single earthquake, but a series of earthquakes within a short period of time; within two generations (AD 513–572) the Empire was hit by 39 registered strong earthquakes, equal to nearly a quarter of all the 169 earthquakes registered in the seven centuries AD 300–1000 (Guidoboni *et al.* 1994; Ambraseys 2009). Secondly, the outbreak of the Justinian bubonic plague in AD 541, a plague which up till AD 750 returned 17 times with an average of 12 years between each outbreak (Little ed. 2007). And thirdly, in AD 536–537 a dust veil, caused by a volcanic eruption in the northern hemisphere, shaded the sun for 18 months – and which according to the latest research was followed up by another dust veil, but of shorter duration, in AD 540 (Stigl *et al.* 2015; see also Hodges 2010; Gräslund and Price 2012). These two dust veil events of global extent caused climatic disturbances, which had a strong impact on agriculture and must have caused bad crops for many years thereafter resulting in starvation and deaths. The events necessitated the acquisition of new experiences in short periods of time, but the many tragic moments which hit the Byzantine families may have created doubts and disillusioned expectations for the future adding despair to misery.

Unfortunately, the skeletal material treated in this publication is in many cases too badly dated and too limited to be able to read the effects on the health of individuals living in this period, but it is wise to keep this in mind when bone material from before and after the watershed mark of the 7th century AD is compared. If it had been possible to separate this early Byzantine bone material of the 6th and 7th centuries from the earlier Roman and the immediately succeeding Byzantine one, it would, perhaps, have been possible to see interesting differences between these three data sets, and between these three separately and the bone material from the later mid-Byzantine period. As the data available at present shows, both Kiesewetter and Teegen, supported by Demirel and by Schultz and Schmidt-Schultz (all this volume), conclude in their respective presentations that the living conditions of the mid-Byzantine populations in Asia Minor were noticeably worse than those of the Romans/early Byzantines. This is a picture, as mentioned by the authors, which compares well with that of other analyses outside Asia Minor. Agriculture was still the main occupation of the population, but the community lacked, perhaps in addition to some of the medical and nutritional knowledge of the past, the organized sanitation infrastructure of the Roman towns. When it comes to the living conditions in the mid-Byzantine period it appears that the loss of the past space of experiences (and perhaps also economic means) did not cause views to be revised for the future horizon of life expectations.

This period of physical transition overlaps in Anatolia with a period of spiritual transition (from pagan to Christian) and there is, of course, no causal relation between them. In order to better understand the underlying processes of material culture preservation versus changes, it would, however, be of interest to see if possible changes in the contemporary daily-life material culture were more conditioned by external physical conditions than by individual spiritual transitions, and to see if a change of living conditions can also be detected in the material data, in what sense and to what extent.

Nutrition

Living conditions are, to a great extent, determined by the level of housing and sanitation, nutrition and health. The two articles on isotope analysis on bones from respectively Pergamon (Propstmeier *et al.*) and Hierapolis (Wong *et al.*) demonstrate that in both places the main diet was composed of C_3 plants and animal meat and/or milk. The same observation has earlier also been obtained at Sagalassos (Fuller *et al.* 2012). The result is not surprising for the inland towns of Hierapolis and Sagalassos, but more so for Pergamon (and also according to unpublished preliminary results for Ephesos), two towns lying close to or on the coast. However, the result harmonizes well with other recent studies of regions bordering on Asia Minor as, for example, Greece, where it appears that seafood was never an important element in the diet of the Greeks from the Bronze Age to the Middle Ages (Roberts *et al.* 2005, 48).

Why is this so? Seafood definitely made up part of the menu for the Romans in Rome and along coastal areas of Italy. Ample evidence for this can be found in ancient literature (cf. Rowan 2014, 61, n. 5), in mosaics, in the so-called *asaroton* motifs, the unswept floor, where fish-bones appear regularly (see, for example, Perpignani and Fiori 2012), and in the many villas (*villae maritimae*) with fish farms along the coasts of the Italian peninsula (but also inland) (Lafon 2001)

This has now been confirmed through the analysis of human skeletons (see from Ostia, for example, Prowse 2011). Recent investigations from Herculaneum demonstrate that the basic seafood was coastal (shallow water, estuaries), not deep sea fish (Rowan 2014), the catch being made with line and hook and various types of catching nets. One popular way of fishing was by the use of seine nets for which two teams were needed, one to hold on to the one end of the net, while the other was rowed out in a wide semi-circle back to the shore further along the coast. The catch was made by slowly pulling the net towards the shore. This kind of fishing is dependent on a smooth shoreline, best done along long low-water beaches, which abound on the Italian peninsula. They are much more seldom along the rocky Aegean coasts of Greece and Asia Minor. Could this be one reason for the low presence or even absence of seafood in diets of the people living in this area? For the same reason villas for fish farming in the Aegean area are very rare, if they exist at all, and if none have been discovered, is this due

to the fact that they did not exist, or that they have they not been looked for (cf. Lafon 2001, 164, 212–13)? And is there a relation between preserved fishing gear and areas of where fish is eaten and where it is not? Fish, it shall be remembered, in Roman and early Byzantine times, was also a commodity of overseas commerce (see, for example, Arndt *et al.* 2003; Theodoropoulou 2014; Čechová 2014).

The absence, or low use, of fish in the diet of people bordering on the Aegean Sea has recently, for the period from the Neolithic to Classical times (5th century BC), been questioned (Vika and Theodoropoulou 2012). The two authors have, for the Aegean, demonstrated that freshwater and marine fish are often indistinguishable in their $\delta^{13}C$- and $\delta^{15}N$-values (see also Propstmeier *et al.*, this volume), and that the $\delta^{15}N$-values are lower than samples reported from the Atlantic. They continue that '$\delta^{13}C$ could be a better indicator of fish consumption, however the enrichment will be moderate and the possibility of C_4 plant consumption needs to be accounted for'. But since 'up to present research suggests that the ancient Greek terrain was dominated by C_3 plants', they conclude that 'fish consumption may have been much more frequent in Greek antiquity than previously thought' (Vika and Theodoropoulou 2012, 1625).

Here lies a challenge hidden both for the isotope analysts and for archaeologists and historians. Vika and Theodoropoulou (p. 1619) make references, in addition to fish remains, to literary sources, the material culture, and iconography, all of which are abundant in the area. However, a thorough study of the sources may be necessary to analyze the context and reasons for their use: do they all refer to the eating of fish, or were they used as metaphors or pure decoration, and how do these sources compare with similar ones from areas where noticeable fish consumption has been documented through isotope analyses. Another question, already touched upon, is also to better understand to what extent the geological formation of the coastline and coastal waters had an influence on fishing techniques and possibilities.

We should also ask if the lack of (or reduced) intake of seafood in certain areas was connected to social questions, like prices and status, or if the avoidance of fish consumption could be explained by the presence of dietary prohibitions/ taboos or ethical ideas. Purcell (1995, 132; cf. also Osborne 1990, 26–28) claims that fish is the only animal consumed by man that eats human flesh and puts this forward as one possible explanation for humans in some areas not eating fish. The fish you eat could have fed on the relative or friend you lost at sea. This is a folk belief, which has existed at all times; a good modern European example is the mackerel, which was not eaten for that reason (Pontoppidan 1752, 219).[1] In late classic medical thought, in the explanatory system of sicknesses, referred to as humoral pathology, seafood is classified as 'cold and moist', in contrast to meat being considered 'warm and dry' (cf. Prowse 2011, 427–8).

Could this system of thoughts have had its roots in much older thinking in which 'cold and moist' food was not considered beneficial for the health? This brings us to the last observation to be made.

Health

There is a close connection between nutrition and health – it is not enough to feel full after a meal. The food must be varied and in most Mediterranean areas it would have been possible to have had a balanced diet based on local products, though, perhaps, with some limitations in the winter months. The situation with seafood above, however, may be a good indication that the ancients did not exploit their latent food resources fully. In fact, there is good reason to think that the daily diet of the ancients was rather badly balanced. Health problems connected to malnutrition is a reoccurring discovery in the analysis of ancient human bones.

According to Foxhall and Forbes (1982) 70% of the daily intake of calories in Rome came from wheat. Abrasion of the teeth (the molars) and tooth decay caused by caries suggests that carbohydrates in the form of C_3-plants (as wheat) formed the basis of the diet also in the towns in Asia Minor analyzed in the present volume (see Propstmeier *et al.*; Wong *et al.*; Kiesewetter; Teegen; Nováček *et al.*). This indicates an unbalanced diet with low protein- and calcium-content. With the exception of thiamin and Vitamin E, cereal products are rather lacking in other vitamins – and the lack of Vitamins A, C, and D can result in serious health and injury consequences. Food with starch free of Vitamin D, leads to the development of rickets, or a weakening of the bones, which could be helped by being out in the sun and/or eating egg, milk, fish and meat. (Brothwell 1969, 179–82; Rickman 1980, 7; Garnsey 1998, 246–9). With a diet so highly based on cereals the risk for rickets is strong and even more so if the person is protected from the sun. According to Soranus (*Gynaeceia* 2.44), this was a sickness in particular occurring in Rome; the analysis made of the skeletons in the present volume demonstrates that rickets was not a serious problem for any of the respective populations at any time period. Only one case at Amurium (Demirel) and one at Arslantepe (Schultz and Schmidt-Schultz), both children, have been registered. On the other hand, scurvy, due to the lack of Vitamin C, seems to have been a more serious problem than perhaps envisaged, but which could be easily cured by, for example, eating fresh fruit.

The lack of Vitamin A afflicts the eyes leading to night-blindness, to xerophthalmia progressing to karatomalacia, and even to blindness (Garnsey 1998, 233). Eye sickness was a recurring phenomenon in Antiquity, not traceable in the skeletons, but appearing repeatedly in inscriptions from sanctuaries of Asklepios, as, for example, from that of Epidauros (Martzavou 2012, 199-203: A4, A9, A11, A18, A20, B2, B12). The strong and bright sunlight of the Mediterranean apparently had a beneficiary effect on potential cases of rickets, but did certainly not have a positive effect on the eyes. Is there any possibility to compare the contrary effects the sun had on people living off a diet with deficiencies of both Vitamins A and D to see if there is any reciprocity between the two deficiency sicknesses and the sun? And archaeologically speaking, did the Greeks and Romans do anything to reduce the effects of blinding sunlight? One way could have been to use paint colours on houses, temples, and marble statues, as was done, but was this a concurrent reason, or only a side-effect?

A problem not unusual in small, isolated communities, is inbreeding between humans, in many cases leading to mental disturbances and physical deformations among individuals, some resembling a pathological situation, like rickets, others being rather genetically conditioned, like teeth with only one root instead of the normal two to three roots (d'Ercole and Pellegrini 1991, 75). In none of the cases presented in this volume does inbreeding appear to have been a problem, even if some of the areas studied for some periods may have been small and isolated. On the contrary, the results of the strontium isotope analyses from mid-Byzantine Hierapolis demonstrates that the small Medieval settlement born out of the abandoned Roman city may have witnessed a strong, if not, perhaps, a relatively stronger presence of foreign people of various origins than its Roman predecessor (see Wong *et al.*; Wenn *et al.*, both this volume). Is this a situation also experienced by other settlements in mid-Byzantine Asia Minor?

Concluding remarks

Many more reflections could have been presented and some of the observations made and questions asked may not even have answered the over-ruling question as to how to establish a tighter historical and cultural dialogue between the results in the archaeological field of study with those in the scientific fields. Still, the amount of available data is small, but the presentations which follow demonstrate the many potential clue hidden in the data and what they can reveal of both life and death in ancient communities. Each new set of data in both fields of study will open up new questions and bring corrections and new information to former answers, methods, and theories. Everything hangs together with everything, the challenge is to see the funerary data in a wider context, knowing that the ancient society, as well as the present, was far more complex than our sources can reveal.

Note

1   Erik Pontoppidan the Younger (1698–1764), a Danish-Norwegian bishop; the quoted text, in my own English translation, is as follows: the mackerel 'readily eats human flesh and seeks the one who swims naked, so that he in all haste is devoured, if he falls into a flock or shoal of mackerels.'

Bibliography

Ambraseys, N. (2009) *Earthquakes in the Mediterranean and Middle East. A multidisciplinary study of seismicity up to 1900.* Athens and Cambridge, Academy of Athens and Cambridge University Press.

Arndt, A., Van Neer, W., Hellemans, B., Robben, J., Volckaert, F., and Waelkens, M. (2003) Roman trade relationships at Sagalassos (Turkey) elucidated by ancient DNA of fish remains. *Journal of Archaeological Science* 30.9, 1095–1105.

Arthur, P. (2012) Hierapolis of Phrygia: The drawn-out demise of an Anatolian city. In N. Christie and A. Augenti (eds.) *Vrbes Extinctae. Archaeologies of abandoned Classical towns*, 275–305. Farnham and Burlington (VT), Ashgate.

Botte, E. and Leitch, V. (eds.) *Fish & ships. Production and commerce of salsamenta during Antiquity. Production et commerce des salsamenta durant l'Antiquité. Actes de l'atelier doctoral, Rome 18–22 juin 2012* (Bibliothèque d'Archéologie Méditerranéenne et Africaine 17). Arles and Aix-en-Provence, Éditions Errance and Centre Camille Jullian.

Brandt, J. R. (2016) Bysantinsk skjebnetid. Om jordskjelv, pest og klimatiske endringer i Anatolia i tidlig bysantinsk tid (300–800 e.Kr.). *Nicolay arkeologisk tidsskrift* 127, 40–6 (expanded English version forthcoming).

Brink, L., Green, O. P. and Green, D. (2008) *Commemorating the dead. Texts and artifacts in context.* Berlin and New York, Walter de Gruyter.

Brothwell, D. R. and Brothwell, P. (1969) *Food in Antiquity.* London, Thames and Hudson.

Carrol, M. and Rempel, J. (eds.) (2011) *Living through the dead. Burial and commemoration in the Classical world.* Oxford, Oxbow Books

Casey, E. S. (1996) How to get from space to space in a fairly short stretch of time: Phenomenological prolegomena. In K. Feld and K. H. Basso (eds.) *Senses of place*, 13–52. Santa Fe, School of American Research Press.

Čechová, M. (2014) Fish products and their trade in Tauric Chersonesos/Byzantine Cherson: The development of a traditional craft from Antiquity to the Middle Ages. In Botte and Leitch (eds.), 229-36.

Cormack, S. H. (2004) *The space of death in Roman Asia Minor* (Wiener Forschungen zur Archäologie 6). Vienna, Phoibus.

Dally, O. and Ratté, C. (eds.) (2011) *Archaeology and the cities of Asia Minor in late Antiquity* (Kelsey Museum Publication 6). Ann Arbor (MI).

Demirel, F. A. (this volume) Infant and child skeletons from the Lower City Church at Byzantine Amorium, 306–17.

Duday, H. 2009 (reprint 2011): *The archaeology of the dead. Lectures in archaeothanatology.* Oxford, Oxbow Books.

d'Ercole, V. and Pellegrini, W. (1991) *Il museo archeologico di Campli.* Campli, Sopritendenza archeologica dell'Abruzzo, Chieti, and Comune di Campli.

Flämig, C. (2007) *Grabarchitektur der römischen Kaiserzeit in Griechenland.* Rahden/Westfalen, Verlag Marie Leidorf.

Foss, C. (1975a) The Persians in Asia Minor and the end of Antiquity. *English Historical Review* 90, 721–47.

Foss, C. (1975b) The fall of Sardis in 616 and the value of evidence. *Jahrbuch der Österreichischen Byzantinistik* 24, 11–22.

Foss, C. (1976) *Byzantine and Turkish Sardis.* Cambridge (Mass.), Harvard University Press.

Foxhall, L. and Forbes, H. A. (1982) Sitometria: The role of the grain as a staple food in Classical Antiquity. *Chiron* 12, 41–90.

Fuller, B. T., Cupere, B., de Marinova, E., Van Neer, W., Waelkens, M., and Richards, M. P. (2012) Isotopic reconstruction of human diet and animal husbandry practices during the Classical-Hellenistic, Imperial, and Byzantine periods at Sagalassos, Turkey. *American Journal of Physical Anthropology* 149.2, 157–71.

Garnsey, P. (1998) Mass diet and nutrition in the city of Rome. In P. Garnsey (W. Scheidel (ed.)) *Cities, peasants and food in Classical Antiquity*, 226–52. Cambridge, Cambridge University Press.

Glørstad, H. and Hedeager, L. (eds.) (2008) *Materiality. Six essays on the materiality of society and culture.* Bricoleur Press: Lindome (Sweden).

Gräslund, B. and Price, N. (2012) Twilight of the gods? The 'dust veil event' of AD 536 in a critical perspective. *Antiquity* 86, 428–43.

Guidoboni E., Comastri, A., and Traina, G. (1994) *Catalogue of ancient earthquakes in the Mediterranean area up to the 10th century.* Rome, Istituto nazionale di geofisica.

Hodges, R. (2010) AD 536: The year Merlin (supposedly) died. In A. Bruce Mainwaring, R. Giegengack, and C. Vita-Finzi (eds.) *Climate crises in human history*, 73–84. Philadelphia, American Philosophical Society (Lightning Rod Press, vol. 6).

Hope, V. M. and Huskinson, J. (2011) *Memory and mourning. Studies on Roman death.* Oxford, Oxbow Books.

Kiesewetter, H. (this volume) Toothache, back pain, and fatal injuries: What skeletons reveal about life and death at Roman and Byzantine Hierapolis, 268–85.

Koselleck, R. (1977) 'Erfahrungsraum' und 'Erwartungshorizont' – zwei historische Kategorien. In G. Patzig, E. Scheibe, and W. Wieland (eds.) *Logik, Ethik, Theorie der Geisteswissenschaften, XI. Deutscher Kongress für Philosophie*, 191–208. Hamburg, Felix Meiner Verlag. (Reprinted from U. Engelhardt, V. Sellin, and H. Stuke (eds.) (1976) *Soziale Bewegung und politische Verfassung. Beiträge zur Geschichte der modernen Welt* (Industrielle Welt, Sonderband Werner Conze zum 31. Dezember 1975), 13–33. Stuttgart, Klert Verlag.)

Koselleck, R. (2002) *The practice of conceptual history. Timing history, spacing concepts.* Stanford (CA), Stanford University Press.

Kumsar, H., Aydan, Ö., Şimşek, C., and D'Andria, F. (2015) Historical earthquakes that damaged Hierapolis and Laodikeia antique cities and their implications for earthquake potential of Denizli basin in western Turkey. *Bulletin of Engineering Geology and the Environment* (doi: 10.1007/s10064-015-0791-0).

Ladstätter, S. and Pülz, A. (2007) Ephesos in the Late Roman and Early Byzantine period: Changes in its urban character from the third to the seventh century AD. In A. G. Poulter (ed.) *The transition to Late Antiquity: On the Danube and beyond*, 391–433. Oxford, Oxford University Press.

Lafon, X. (2001) *Villa maritima: recherches sur les villas littorals de l'Italie Romaine (IIIe siècle av. J.-C./IIIe siècle ap. J.-C.).* Paris, Boccadr, École Française de Rome.

Little, L. K. (ed.) (2007) *Plague and the end of Antiquity. The pandemic of 541–750.* Cambridge, Cambridge University Press (containing many articles on the subject).

Lock, G., Gosden, C., and Daly, P. (2005) *Segsbury camp: Excavations in 1996 and 1997 at an Iron Age hillfort on the Oxfordshire Ridgeway* (Oxford University School of Archaeology, Monograph 61). Oxford, Oxford University School of Archaeology.

Martzavou, P. (2012) Dream, narrative, and the construction of hope in the 'Healing Miracles' of Epidauros. In A. Chaniotis (ed.) *Unveiling emotions. Sources and methods for the study of emotions in the Greek world* (Habes: Heidelberger Althistorische Beiträge und Epigraphische Studien, Band 51), 177–204. Stuttgart, Franz Steiner Verlag.

Moore, S. V. (2013) *A relational approach to mortuary practices within Medieval Byzantine Anatolia.* PhD-diss., Newcastle University.

Nováček, J., Scheelen, K., and Schultz, M. (this volume) The wrestler from Ephesus: Osteobiography of a man from the Roman period based on his anthropological and palaeopathological record, 318–38.

Osborne, C. (1990) Boundaries in nature: Eating with animals in the 5th century BC. *Bulletin of the Institute of Classical Studies* 37, 15–29.

Ousterhout, R. (2010) Remembering the dead in Byzantine Cappadocia: The architectural settings for commemoration. In *Transactions of the State Hermitage Museum LIII: Architecture of Byzantium and Kievan Rus from the 9th to the 12th centuries. Materials of the International Seminar November 17–21, 2009*, 87–98. St. Petersburg, The State Hermitage Publishers.

Parrish, D. (ed.) (2001) *Urbanism in Western Asia Minor. New studies on Aphrodisias, Ephesos, Hierapolis, Pergamon, Perge and Xanthos* (Journal of Roman Archaeology, Suppl. 45). Portsmouth, Rhode Island.

Perpignani, P. and Fiori, C. (2012) *Il mosaico 'non spazzato': studio e restauro all'asaroton di Aquileia.* Ravenna, Edizioni del Girasole.

Pirson F. (2010) Pergamon – Bericht über die Arbeiten in der Kampagne 2009. *Archäologischer Anzeiger* 2010.2, 139–236.

Pontoppidan, E. (1752) *Det første Forsøg paa Norges naturlige Historie, forestillende dette Kongeriges Luft, Grund, Fielde, Vande, Væxter, Metaller, Mineralier, Steen-Arter, Dyr, Fugle, Fiske og omsider Indbyggernes Naturel, samt Sædvaner og Levemaade*, 2 vols. Copenhagen (translated into German in 1753 (Copenhagen) and English: *The natural history of Norway: Containing a particular and accurate account of the temperature of the air, the different soils, waters, vegetables, metals, minerals, stones, beasts, birds, fishes...*, London 1755). See https://archive.org/stream/detfrsteforsgpaa00pont#page/n7/mode/2up (consulted Nov. 2, 2015)

Propstmeier, J., Nehlich, O., Richards, M. P., Grupe, G., Müldner, G. H., and Teegen, W.-R. (this volume) Diet in Roman Pergamon: Preliminary results using stable isotope (C, N, S), osteoarchaeological and historical data, 237–49.

Prowse (2011) Diet and dental health through the life course in Roman Italy. In S. C. Agarwal and B. A. Glencross (eds.) *Social bioarchaeology.* Chichester, Wiley-Blackwell.

Purcell, N. (1995) Eating fish: The paradoxes of seafood. In J. Wilkins, M. Dobson, and D. Harvey (eds.) *Food in Antiquity: Studies in ancient society and culture*, 132–49. Exeter, University of Exeter Press.

Radt, W. (2001) The urban development of Pergamon. In Parrish (ed.), 43–56.

Ratté, C. (2001) New research on the urban development of Aphrodisias in Late Antiquity. In Parrish (ed.), 116–47.

Ratte, C. (2012) Introduction. In C. Ratté and P. D. Staebler (eds.) *The Aphrodisias regional survey* (Aphrodisias V), 1–38. Leiden and Boston, Brill.

Ratté, C. and De Staebler, P. D. (2011) Survey evidence for Late Antique settlement in the region around Aphrodisias. In Dally and Ratté (eds.), 123–36.

Rautman, M. (2011) Sardis in Late Antiquity. In Dally and Ratté (eds.), 1–26.

Rickman, G. (1980) *The corn supply of ancient Rome.* Oxford, Clarendon Press.

Roberts, C., Bourbou, C., Lagia, A., Triantaphyllou, S., and Tsaliki, A. (2005) Health and disease in Greece. Past, present and future. In H. King (ed.) *Health in Antiquity*, 32–58. London and New York, Routledge.

Rowan, E. (2014) The fish remains from the Cardo V sewer: New insights into consumption and the fishing economy of Herculaneum. In Botte and Leitch (eds.), 61–73.

Santangelo, F. (2013) *Divination, prediction and the end of the Roman Republic.* Cambridge, Cambridge University Press.

Schultz, M. and Schmidt-Schultz, T. H. (this volume) Health and disease of infants and children in Byzantine Anatolia between AD 600 and 1350, 286–305.

Sintubin, M., Muchez, P., Similox-Tohon, D., Verhaert, G., Paulissen, E., and Waelkens, M. (2003) Seismic catastrophes at the ancient city of Sagalassos (SW Turkey) and their implications for the seismotectonics in the Burdur-Isparta area. In *Geological Journal* 38, 359–74.

Sofaer, J. (2006) *The body as material culture.* Cambridge University Press, Cambridge.

Stigl, M., Winstrup, M., McConnell, J. R., *et al.* (2015) Timing and climate forcing of volcanic eruptions for the past 2,500 years. In *Nature*: doi:10.1038/nature14565.

Teegen, W.-R. (this volume) Pergamon – Kyme – Priene: Health and disease from the Roman to the Late Byzantine period in different locations of Asia Minor, 250–67.

Theodoropoulou, T. (2014) *Salting the East*: Evidence for salted fish and fish products from the Aegean Sea in Roman times. In Botte and Leitch (eds.) 218–28.

Vika, E. and Theodoropoulou, T. (2012) Re-investigating fish consumption in Greek antiquity: results from $\delta^{13}C$ and $\delta^{15}N$ analysis from fish bone collagen. In *Journal of Archaeological Science* 39, 1618–1627.

Wenn, C. C., Ahrens, S., and Brandt, J. R. (this volume) Romans, Christians, and pilgrims at Hierapolis in Phrygia. Changes in funerary practices and mental processes, 196–216.

Wong, M., Naumann, E., Jaouen, K., and Richards, M. (this volume) Isotopic investigations of human diet and mobility at the site of Hierapolis, Turkey, 228–36.

Part I

From life to death: Death and the social and funerary setting

The Sanctuary of St Philip in Hierapolis and the tombs of saints in Anatolian cities

Francesco D'Andria

Abstract

The age of Constantine, when Christian worship began to be freely practised, saw the start of a process that would cause a radical transformation of the cities of the Empire. The construction of churches, which became the new centres of social aggregation, had a profound impact on the ancient cityscapes. In this context the cities of Anatolia provide some very significant examples. Many necropoleis, whose main function during the Imperial period had been the social visualization of family groups, now took on new functions connected to the veneration of the tombs of the saints. Large sanctuaries grew up around these tombs, and within the urban layout, new lines of communication were opened up. In Anatolia the examples of Hierapolis (the tomb and sanctuary of St Philip), Ephesus (the Seven Sleepers and St John) and Meriamlik (the tomb of St Thecla) demonstrate these dynamics in an exemplary way.

Keywords: Anatolia, Ephesus, Hierapolis in Phrygia, St John, St Philip, St Thecla, Seven Sleepers, Silifke, tombs of saints, urban landscapes.

Introduction

A recent book by Ann Marie Yasin (2009), dedicated to saints and church spaces, tackles a theme of great interest for understanding the development of cities in late Antiquity. The author investigates the role that the cult of the saints – including the building of churches in their honour and the veneration of their tombs – played in the construction of those spaces where the new social and political order of cities was manifested in the 4th to 6th centuries AD. From the reign of Constantine onwards, with Christians now enjoying freedom of worship, churches became the new centres of aggregation. They replaced the pagan sanctuaries, which (especially in Asia Minor) were completely destroyed and erased from urban and suburban landscapes, as in the case of the famous temple of Artemis in Ephesus.

In this framework the necropoleis, particularly those where the remains of saints and martyrs were venerated, took on new functions. There was a shift away from the self-representation of family groups by means of reliefs on burial monuments and complex messages, sometimes of a juridical nature, expressed in inscriptions, to a more 'public' function, particularly in those parts of the ancient necropoleis that were linked to the presence of *sancta corpora*. The tombs attributed to saints thus became the central point of complex and dynamic urban development, with a shift in the fulcrum of the city's layout, modifying the road network and creating new concentrations of monuments in areas outside the main settlements, even those with only minor necropoleis.

Exemplary in this regard is the case of Rome and the Old St Peter's Basilica, built by Constantine over the tomb of St Peter. However, recent excavations in Milan, conducted in the courtyards of the Università Cattolica, in a suburban area dated to the Republican and Imperial eras, have revealed new data regarding the urban transformations associated with the cult of other martyrs' burial sites.[1] As well as gardens and peasants' houses, there are necropoleis with tombs belonging to figures of not particularly high social rank, except for the sarcophagus with the deposition of a woman accompanied by a rich set of grave goods. It was in this area – following a dream – that Bishop Ambrose decided to conduct excavations in order to find the bodies of the Milanese martyrs Gervasius and Protasius. Those 'archaeological' investigations led to the discovery of the saints' bodies, which were transferred to the basilica 'ad martyres', built by order of the bishop in this area outside the walls beyond the Imperial circus (Cagiano de Azevedo 1968; Lusuardi Siena 1990, 124). A monumental complex, corresponding to the current Basilica of Sant'Ambrogio, was thus created that transformed a suburban necropolis area into one of the most extraordinary religious centres in northern Italy.

In late Roman cities the public role of the necropoleis where the tombs of the saints are located is amply attested elsewhere in the territories of the Empire and is cited in an extensive corpus of literature. Around these locations, dynamics of self-representation were activated, in which the saint took on the identifying characteristics of the city; an emblematic case of the role played by the venerated tombs is that of St Demetrius of Thessaloniki (Fig. 1.1) (Yasin 2009, 171–5). The large five-aisle basilica built over his tomb is characterized by a complex functional articulation of spaces, starting with the tomb itself, positioned on the left-hand side of the building's entrance (Fig. 1.2). In the middle of the central nave is the *ciborium*, a hexagonal structure like an ancient mausoleum, inside which the saint manifests himself and dialogues with the faithful. Lastly, in the apse, an underground chamber is used for the miracle of the *myron*, the perfumed oil emitted by the bones of the Saint; a similar miracle is attributed to St Nicholas of Myra, both in his tomb in Lycia and after the transfer of his relics to Bari in Italy.

That a strategy for representing the city's identity was activated around the tomb of St Demetrius is clear from the

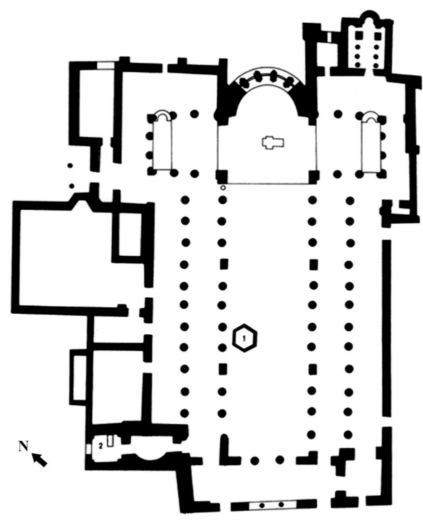

Fig. 1.1. Thessaloniki. Basilica of St Demetrius; plan: 1. Ciborium; 2. Tomb of St Demetrius (adapted from Bakirtzis 2002, fig. 1; by courtesy of the author).

large corpus of literary texts dedicated to the miracles of the martyr (Lemerle 1979–1981, 2–110). Inscribed in the famous church dedicated to the saint, next to the mosaic representing a distinguished cleric of Thessaloniki to whom St Demetrius frequently appeared, is a prayer that clearly

Fig. 1.2. Thessaloniki. Basilica of St Demetrius, Tomb of the Saint (after Bakirtzis 2002, fig. 9; by courtesy of the author).

links the saint to the city: 'Most happy martyr of Christ, you who love the city (φιλοπολις) take care of both citizens and strangers (φροντιδα τιθη καὶ πολιτῶν καὶ ξένων)'.[2] Indeed, one of the miracles tells of the apparition of the saint to a pilgrim. The same connection with the city is also seen inside the *ciborium*, where there was a gold throne with St Demetrius sitting on it and a silver throne on which sat the Lady Eutaxia, the personification of the Tyche of Thessaloniki (Pallas 1979; Bakirtzis 2002, 179).

Ephesus and Saint Thecla in Seleucia

This civic aspect of the saints' tombs is indispensable for an understanding of the new urban landscapes that characterized the cities of proto-Byzantine Anatolia, and the case of Ephesus encapsulates many of the issues linked to these themes (Fig. 1.3). Recent investigations conducted by Norbert Zimmermann at the Cemetery of the Seven Sleepers (Zimmermann 2011) have clarified many aspects of its chronology and provided original data for understanding the role of the necropoleis within the urban system (Fig. 1.4). The new stratigraphic and stylistic interpretation of the paintings in the houses of Ephesus has enabled the decorations to be dated to the 2nd or 3rd centuries AD, rather than the 4th to 5th centuries posited by previous authoritative studies. This late dating had also been attributed to the paintings of the *arcosolia* of the Cemetery of the Seven Sleepers, which had also therefore

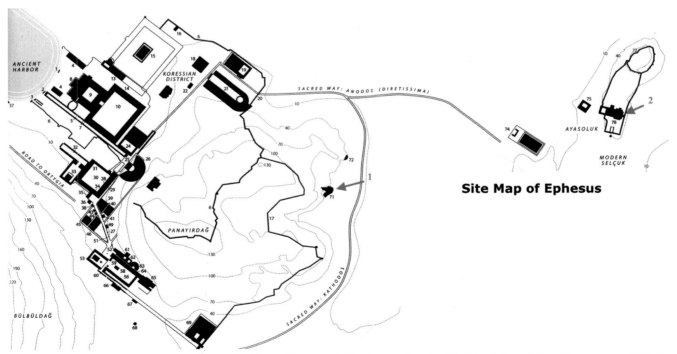

Fig. 1.3. Ephesus. Plan of the city: 1. Cemetery of Seven Sleepers; 2. Basilica and tomb of St John (adapted from Zimmermann 2011; by courtesy of the author).

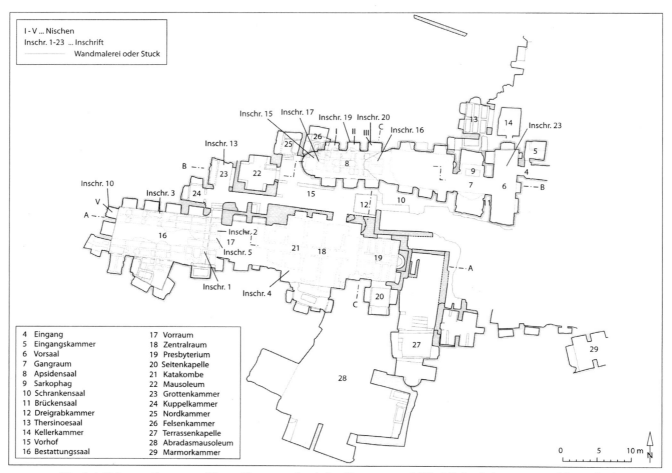

Fig. 1.4. Ephesus. Cemetery of Seven Sleepers, plan (after Zimmermann 2011, 368 fig. 3; by courtesy of the author).

been considered to be proto-Byzantine, whereas the new research dates the origins of the complex to no later than the 3rd century. This makes it possible to consider in a new light the marble epigraphs, described by Miltner in 1937 (Miltner 1937, 201–11; Zimmermann 2011, 393–402), in which the names of the deceased are accompanied by the formula χαῖρε ἐν Θεῷ, clearly indicating a Christian context (Fig. 1.5). The Imperial-era cemetery was situated outside the city on the road that leads to the hill of St John after running alongside the Sanctuary of Artemis. It was a collective burial site, comparable to the Roman catacombs in Rome, and represents a highly distinctive example for Asia, belonging to the Christian community of Ephesus, managed directly by the bishop. In the reign of Theodosius, this led to the development of the legend, linked to the complex, of the reawakening-resurrection of the seven boys who had fallen asleep during the persecution of Decius (Fig. 1.6). The site was expanded with the construction of churches, chapels and new burials, and began to attract pilgrims, not only Christians but also Muslims, since the miracle of the Seven Sleepers is cited in one of the suras of the Koran.

The pattern of settlement of Ephesus was transformed in the proto-Byzantine period and the necropolis began to play an essential role in relation to the hill of Ayasoluk, where another burial was at the origin of a new settlement. Around the tomb of the Apostle John a *sacellum* was built, followed in the reign of Justinian by the majestic basilica that houses, in the area of the presbytery, the cavity where the saint started to breathe again on the day of his panegyris (Fig. 1.7). With the emergence of the two cemeteries, the topography of the ancient provincial capital changed radically, giving way to a new settlement that still lies in the shadow of the hill of Ayasoluk, whose name conserves the memory of the miraculous tomb (in Turkish it means 'holy breath').

Another important example is found in Seleucia (Silifke) in Isauria in eastern Anatolia, which was the centre of the cult of the *thalamos* (tomb) of St Thecla. The tomb became the site of a large pilgrimage complex that was also visited, between 381 and 383, by Egeria, on her way to the Holy Land. Unfortunately the complex, known as Meriamlik, has not yet been the object of systematic research and our main sources of information are the Byzantine texts

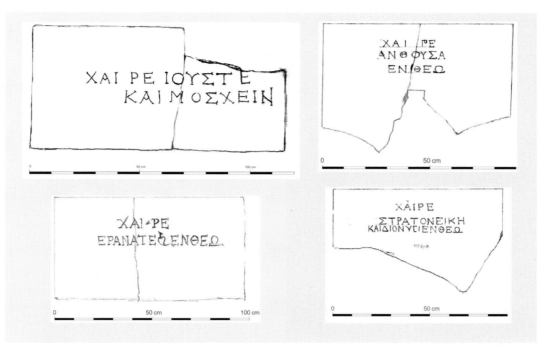

Fig. 1.5. Ephesus. Cemetery of Seven Sleepers, marble epigraphs (Roman period) (after Zimmermann 2011, 394, 396, 398; figs. 33, 37, 39, 40; by courtesy of the author).

recounting the saint's miracles (*logoi*). This sanctuary also lies outside the city and the miracles refer to a path through the fields and an uphill stretch of road that the pilgrims had to follow. As in other Christian shrines, healing took place by means of incubation, lamp oil, and water from a spring (πηγή) (Fig. 1.8).[3]

The cult of the tombs of saints continues today in the Muslim world (Chambert-Loir and Guillot 1995), which displays certain interesting forms of interaction with the Byzantine world. The sanctuary of Haci Bektaş Veli in central Anatolia developed around the *türbe* (mausoleum) built in 1367 in the village of Sulucukara Höyük. The site is the focus of worship by the Alevi community, but it is also frequented by Christians, who believe that there was once an ancient monastery there and that the venerated tomb is that of St Charalambos (or St Eustathios) (Zarcone 1995, 314). A complex ritual is practised in the sanctuary, involving an itinerary passing through seven doors, worshippers moving 'on all fours' at the entrance, propitiatory offerings, and healing practices around the *dilek agacı* (the tree of desires), to which the faithful tie strips of cloth (Zarcone 1995).

The cult of Saint Philip in Hierapolis of Phrygia

I shall conclude with the example of Hierapolis, presenting the most recent research, which enabled us, in 2011, to identify the tomb and the sanctuary of the Apostle Philip.

The transformation of the classical towns of Asia Minor in Byzantine times was dramatic, with evident effects on urban layout and disruption caused by wars, natural catastrophes, and the dynamics of human history (Arthur 2012, 278–9).

In Hierapolis, the extraordinary conservation of the town has enabled us to learn much about the urban transformation following a strong earthquake that took place in the second half of the 4th century; the town's appearance changed completely, with the abandonment of monumental buildings such as the northern agora and the nearby theatre, which then provided building material for the new city walls and marble for the lime kilns (D'Andria *et al.* 2005–2006, 349–59; Scardozzi 2008, 40–3). In this northern area of the town, outside the fortifications, the imposing building now interpreted as a Roman basilica (Scardozzi (ed.) 2015, 112–3), was transformed, probably as early as the 5th century, into a church, with lateral spans and a central aisle covered with three cross-vaults. The building presents strong signs of the seismic events affecting this region and constitutes an extraordinary example of the reuse, in the proto-Byzantine period, of imposing buildings of the Roman era. This church can also be linked to the cult of the martyrs' tombs and to the creation of new religious centres of urban aggregation in the 5th and 6th centuries AD in the peri-urban areas. Indeed, the church was built in an area surrounded by the tombs of the North Necropolis and, according to an interesting hypothesis by Paolo Verzone, it might

Fig. 1.6. Russian icon with representation of Seven Sleepers (Wikimedia Commons, https://commons.wikimedia.org/wiki/File: Seven_Sleepers_icon.jpeg).

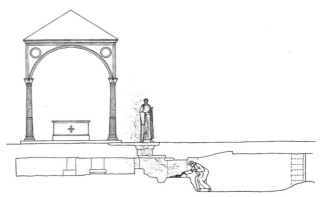

Fig. 1.7. Ephesus. Basilica of St John, the cavity under the altar (after Pülz 2010, 81 fig. 6; by courtesy of the author).

be connected to the veneration of the tombs of Kyriakos and Klaudianos, two local martyrs cited in the Syriac martyrology (Verzone 1956, 40–5; Huttner 2013, 341).

The fortifications reduced the size of the city from 72 to 60 hectares, creating an architectural structure that

radically redefined the new urban landscape. The churches now formed the nodes of the new *Forma Urbis*, replacing the ancient sanctuaries, which were systematically destroyed. The sanctuary of Apollo, the religious centre of the city, was demolished during the 5th and 6th centuries, and the sacred area was turned into a dump for old building materials. Only the oracle (Temple A) appears to have escaped destruction, perhaps because the well that provided thermal waters was saved due to its healing qualities (Semeraro 2012, 298–302).

Maintenance of the *cura aquarum* is also well attested in the nearby sanctuary of Hades, brought to light during the most recent excavation campaigns. Starting in Hellenistic times, a complex system of cult buildings

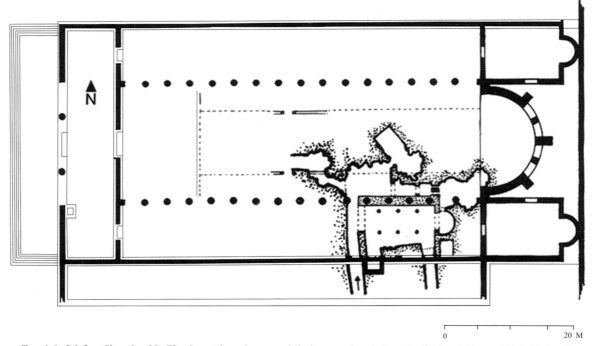

Fig. 1.8. Silifke. Church of St Thecla, with underground thalamos, plan (after Herzfeld and Guyer 1930, 38 fig. 1).

was erected around the *Ploutonion*, where, during the 2012 campaign, an inscription dedicated to the god of the Underworld and to his wife Kore was discovered by the entrance to the cave. This entrance was purposely obliterated in the 6th century by an enormous dump of stones and architectural fragments arising from the demolition of the sanctuary, and a large wall was built in front of the ancient *theatron*, partly with blocks from the destruction of its travertine seating. However, the thermal waters continued to be used for therapeutic purposes and two large basins were thus built in front of the wall (D'Andria 2013b).

It is clear that this period saw the deactivation of the distinctive characteristics of the town's religion. Specifically, the practices linked to the oracle (Sanctuary of Apollo) and health (the *Ploutonion* and the cult of Hades in the hypostasis of Serapis) were absorbed by Christian Hierapolis through the figure of the Apostle Philip. The cult of the saint appears to be closely linked to prophecy: the literary sources speak of his three daughter prophetesses, two of whom were buried next to their father while the body of the third was venerated in Ephesus. In the *Acta Philippi* there is also a clear element of Montanism, a heresy that had actually

originated in the nearby city of Pepuza.[4] In the 5th to 6th centuries the religious activities linked to prophecy and healing were transferred from the centre of the city, where the Sanctuary of Apollo and the *Ploutonion* were located, to the north-eastern hill of the settlement. Here stood the large pilgrimage sanctuary dedicated to the Apostle Philip, the complex layout of which has been brought to light during recent excavations (D'Andria 2013a).

The publication of the Atlas of Hierapolis (D'Andria, Scardozzi, and Spanò 2008, 95–7 folio 21) has helped us to understand how the octagonal *Martyrion* on top of the hill, excavated by Verzone in the late 1950s, was part of a much larger architectural complex situated immediately outside the eastern town walls (Figs. 1.9–10).

The excavations of 2011–2012 brought to light a church that has its nucleus in tomb C127, one of the 1st-century monumental tombs built in the Roman necropolis (Fig. 1.11).[5] The tomb, sited within a burial plot that developed until the beginning of the 4th century, appears to have been marked by distinctive elements from as early as Roman Imperial times. On the side wall of the tomb, which would later be incorporated or obliterated by the late ancient building, was a rectangular hollow (0.50×0.62 m) with holes that

Fig. 1.9. Hierapolis. Aerial photograph with the hill of St Philip (in circle) and the Roman theatre in the middle (photo A. Gandolfi; with kind courtesy).

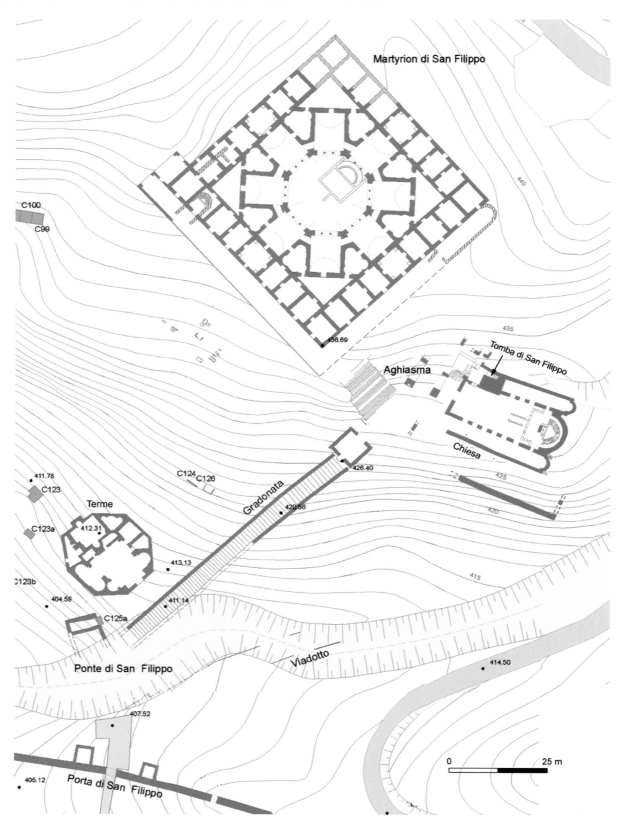

Fig. 1.10. Hierapolis. Plan of the hill with octagonal Martyrion and Church of St Philip (archive of the Italian Archaeological Mission at Hierapolis).

Fig. 1.11. Hierapolis. Tomb C127 in the Roman necropolis, 2nd–3rd centuries AD (3D reconstruction by Max Limoncelli; archive of the Italian Archaeological Mission at Hierapolis).

may have served to lodge a metal sheet with an inscription or a bas-relief. This recess can be attributed to the period preceding the construction of the church and may provide a clue to the particular importance attributed to the tomb as early as the Roman or late ancient period. Also considered a clue to the importance attributed to tomb C127 in the mid-Imperial period is the name *Apolleinarios,* in the nominative case, carved in 2nd-century characters on the wall to the left of the door. In the necropolis of Hierapolis, inscriptions referring to the tomb's owner are usually in the genitive case, followed by the term *bomos* or *soros*. The reference to the name of the celebrated bishop of Hierapolis who lived during the 2nd century, the author of an apologia of Christianity that was also sent to the emperor Marcus Aurelius, might not be coincidental (Ritti 2011–2012, 53–5 fig. 1).

Furthermore, the tradition of the presence of the tomb of the Apostle Philip in Hierapolis dates back to the 2nd century, being cited in important documents such as the letter of Polycrates, bishop of Ephesus, to Pope Victor, cited by Eusebius (*Historia Ecclesiastica* III, 31, 2–4):

καὶ γὰρ καὶ κατὰ τὴν Ἀσίαν μεγάλα στοιχεῖα κεκοίμηται (...), Φίλιππον τὸν τῶν δώδεκα ἀποστόλων, ὃς κεκοίμηται ἐν Ἱεραπόλει,...
and indeed even in Asia great stars rest (…); Philip, one of the twelve apostles, who rests in Hierapolis...

Together with the dialogue between the Roman priest Gaius and Proculus, a representative of Asian Montanism, these literary testimonies assert the nobility of the Church of Asia with respect to Rome, which holds the *tropaia* of Peter and Paul, and refer to the presence of the tombs of the apostles John in Ephesus and Philip in Hierapolis.[6] In addition, an important epigraphical document, unfortunately now lost, attests to the presence in the Phrygian city of the church of the Apostle Philip; the inscription is carved on a sarcophagus dated to the 5th century, and cites Eugenios, archdeacon, responsible for the church of the saint, apostle and theologian Philip.[7]

In the course of the 4th century the area of the tomb was separated from the rest of the necropolis and enclosed by a rectangular building in which the characteristics of the later cult appear (Fig. 1.12). Thus it may be considered a *memoria* relating to the cult of the tomb (D'Andria 2011–2012, 25–8 fig. 18) and represents the beginning of an uninterrupted process which saw the monumentalization of the eastern hill. The interior of this space is characterized by the presence of two large basins, about two metres deep and completely revetted with large marble slabs, which were used for practices of ritual immersion. Due to the depth of the basins the faithful could immerse themselves only with the help of two assistants, and this suggests that, as early as the 4th century, around the venerated tomb, an

Fig. 1.12. Hierapolis. The tomb of St Philip inside the 4th-century memoria (3D reconstruction by Max Limoncelli; archive of the Italian Archaeological Mission at Hierapolis).

organization had been established that enabled the execution of complex ritual activities. The building had a geometric mosaic pavement and a bench on its southern side which, based on evidence of what happened in the later phases, could have been used for incubation practices following immersion in the pools. The use of water appears to have already been a characteristic of the cult related to the tomb and, as noted above, shows a continuity with the cults of Roman Hierapolis, particularly the *manteion* of Apollo and the thermal springs of the *Ploutonion*, in which water played an important role in oracular activities and healing practices.

In front of the church stood the *aghiasma*, a fountain that marks the point of arrival of the aqueduct that brought water from springs on the plateau and fed the basins alongside tomb C127. The *aghiasma* was characterized by a shell-like marble canopy that perhaps deliberately recalls the analogous placement above the thermal spring of the oracle of Apollo.

In the late 4th and early 5th centuries, with the new urban layout marked by the fortification walls, the area around the tomb and the pilgrimage route leading from the northern gate were subject to large-scale monumentalization. On the eastern side of the city walls, a new gate flanked by two towers was opened and a bridge was built over the ravine that separates the north-eastern hill from the city; during the

2012 campaign we created a structure with modern materials that replicates the function of the ancient bridge, allowing visitors to reach the entrance to the sanctuary more easily. A stairway led from here to the summit of the hill, forming the axis around which the main monuments of this phase, the baths and the *Martyrion,* are oriented. Both these buildings are based on an octagonal plan. That the complex was part of a single project is also shown by the dimensions of the baths building, which correspond to the central body of the *Martyrion* (Fig. 1.13).[8]

Whereas the *Martyrion* complex was abandoned after the earthquake of the 7th century, ritual activity in the church built around tomb C127 continued into the 12th century. The church was repaired after the damage of the earthquake and had rich marble decoration in the mid-Byzantine era (9th to 10th centuries). This aspect also shows the importance that the local Christian community attributed to the place and the tomb.

The construction of the three-aisled church that obliterated the primitive *memoria* has been dated to the 5th to 6th centuries. It differs from the structure of the *Martyrion* in terms of its poorer construction technique and other aspects. In the church of the sepulchre the walls were built using reused materials, many of which came from the Roman tombs that covered the whole area. They were demolished in

Fig. 1.13. Hierapolis. The hill with the Bath Building (to the left); on the top the Martyrion and the Church of St Philip (3D reconstruction by Max Limoncelli; archive of the Italian Archaeological Mission at Hierapolis).

order to create the flat surface on which the church was built. The structures built around tomb C127 contained, as infill material, entire marble and travertine sarcophagi; among the collapsed walls, fragments of decorated sarcophagi were also discovered, clearly having been reused in the structure. In contrast, the building technique used for the walls of the *Martyrion* is precise and shows no use of reused materials: the structures of the central octagon were built with specially cut blocks from quarries and confirm the presence of skilled masons who were familiar with Imperial-era techniques used for cutting and laying travertine blocks.[9] Thus the two buildings were built by different groups of craftsmen in chronologically distinct periods. Nonetheless, the church that incorporates the tomb is large (length 35 m; width 21.50 m), particularly in the development of the nave (width 10 m), which is separated from the aisles by two rows of piers with arches. The rubble of the building has revealed the presence of women's galleries (*matronei*), with marble columns and Ionic capitals supporting impost blocks decorated with crosses flanked by vegetal motifs. Furthermore, the two aisles terminate in two chapels (*parakklesia*) provided with separate entrances. In this phase, there was a marble staircase within the *narthex* enabling access to a platform surrounding tomb C127. This must have created a compulsory path for pilgrims with an exit stair behind the tomb, where there was a landing decorated with a mosaic with braid motives and tondos with figures of fish and birds, datable to the 5th to 6th centuries.

A marble *templon* with eight columns, Pergamene capitals and an *epistylion,* bearing a dedicatory inscription and decorated with geometric motifs and crosses, was built within the *bema*. At this time the relics were kept in a barrel-vaulted room below the altar supported by a monolithic marble block; a terracotta pipe linked the level of the altar to the underlying room and was probably predisposed for offering *myron* and introducing *brandea*, strips of cloth which, on touching the venerated relics, became relics themselves which the pilgrims took with them as phylacteries (D'Andria 2011–2012, 28–33 figs. 23–4). Indeed, the inscription found on the marble altar table in the northern *parekklesion* mentions Dorotheus, indicated as *myrodotes,* and refers to the *myron*. The slab above the main altar also bears a dedicatory inscription mentioning the metropolitan bishop Theodosius. The formula of the dedication uses the expression from the Gospel in which the penitent thief says to Jesus, '...μνήσθητί μου ὅταν ἔλθῃς εἰς τὴν βασιλείαν σου' (Luke 23, 42) 'remember me, when you come into your kingdom' (Ritti 2011–2012, 54–6 figs. 2–3). The same bishop is probably indicated in the monogram of a smaller *epistilion* that was found near the tomb; it belongs to a *ciborium* or baldaquin that was built over the platform above the tomb, indicating to the pilgrims the sacred place where burning oil lamps were probably placed. The structure of the martyrial altar of Hierapolis recalls that of the altar above the tomb of the Apostle in the Basilica of St John in Ephesus. Here too the altar is positioned above an underground cavity in which the sacred relics were kept, connected by a manhole to the level of the presbytery; it is here that the miraculous powder issued from the tomb on the day of the panegyris of St John,[10] when the Apostle started to breathe again.

That the link between the sanctuary and the bishops of Hierapolis was very strong is also indicated by an exceptional document: a list of bishops painted in red on a marble slab that must have been located in the presbytery. Among the names is Auxanon, already known for having participated in the fifth council of Constantinople in 553 (Ritti 2011–2012, 58–61 fig. 6).

In the new church the system of basins associated with healing practices that were sited near the tomb was enlarged. Alongside the two large basins two further basins were added for individual immersion, furnished with a marble hatch that permitted the water to flow into the larger basins. There was a further basin, circular in shape, in the centre of the nave. The space between the tomb and the *templon* was thus entirely occupied by this hydraulic system.[11]

The complexity of these proto-Byzantine buildings is also confirmed by the recent investigations conducted in Laodicea by Celal Şimşek. In the centre of the city, in a dominant position in the valley of the Lykos, a short distance from Hierapolis, the excavations have brought to light a monumental church with three naves. Given its position in the city and the presence of a large and sumptuous baptistery, with an extensive residential area, probably the bishop's palace, it can be interpreted as the cathedral. Inside the *bema,* precisely below the altar, Turkish archaeologists discovered terracotta piping that enabled the input and drainage of water in connection with a basin that was clearly associated, as in the church of St Philip, with liturgical practices involving water and with its symbolic value.[12]

In the light of the new finds, Paolo Verzone's observation on the sanctuary of St Philip gains even greater significance. He attributed the 28 rooms that surrounded the *Martyrion* to hospitality for pilgrims and posited that incubation was practised there, as in other famous pilgrimage sanctuaries in the East, including the Cosmidion of Constantinople.[13] In the complex at Hierapolis, the inclusion of ancient healing practises within the Christian cult, particularly those linked to Asclepius and Serapis, is thus clear.[14] The therapeutic dimension of the cult of Hades-Serapis emerges most clearly in the extraordinary complex of the *Ploutonion*, where the destructive forces of the underworld, including the emission of fatally poisonous gases, were transformed via the beneficial properties of the thermal waters, which were able to heal and to promote fertility by irrigating the fields (Vitruvius, *De architectura*, VIII, 3).

Together with textual sources, the extraordinary complexity of the buildings and ritual practices now identified on the eastern hill of Hierapolis allows us to attribute the sanctuary and the tomb at the centre of the church to Philip the Apostle. This site in Hierapolis can now be listed among the most famous Eastern pilgrimage sanctuaries, such as St John in Ephesus (Büyükkolancı 2001; Pülz 2012, 80–4) or St Thecla (Meriemlik) in Seleucia (Herzfeld und Guyer 1930, 38ff.; Maraval 2004, 224–5). To confirm this interpretation we present a bronze bread stamp for pilgrims found in Hierapolis and now kept in the Virginia Museum of Art, Richmond (Gonosova and Kondoleon 1994, n. 94, 270–3). At the centre of the representation is the image of St Philip, indicated by the inscription, flanked by two stairways that lead to two different buildings, the one to the left covered by a dome, and the one on the right by a gabled roof, which effectively refer to the buildings actually discovered in the excavations.

New research in the Lykos valley

The new research conducted in the valley of the Lykos, particularly in Laodicea and Tripolis, has also provided a substantial body of new data that help to understand how the urban layouts were radically transformed in the proto-Byzantine period, in connection with the rise of numerous Christian places of worship. In Laodicea for example, the discovery of the large cathedral church, in the heart of the city, constitutes a fundamental element for our knowledge, thanks both to the exceptional state of conservation and to the serious conservation measures that have been implemented.[15] More complex is the interpretation of the other Christian religious buildings present in the city and their dedication to the bishops, confessors, and martyrs who were eminent figures in the local church. Among these was the figure of Sagaris, a bishop martyred in the 2nd century AD, who is also cited in the famous letter of Polycrates to Pope Victor concerning the correct date of Easter as one of the important figures of the Church of Asia who supported his position. This document, recorded by Eusebius (*Historia Ecclesiastica* V, 24, 5), also refers to the burial of the martyr in Laodicea (ὃ ἐν Λαοδικεία κεκοίμηται).[16] It can reasonably be hypothesized that after Sagaris' death – as with St Philip in Hierapolis – a cult developed around his tomb, which would subsequently become the nucleus of an important sanctuary in the proto-Byzantine period.

In the reconstruction of the topography of Laodicea, a process currently benefiting from the most recent investigations, one building in particular, not yet excavated, presents very interesting features. It lies to the east of the proto-Byzantine fortifications, in an extra-urban area partly occupied by the Eastern Necropolis. Already highlighted by Traversari,[17] the building is recognizable from the structures still visible on the surface as an octagon with apses on four sides, positioned within a circular rotunda with a diameter of about 34 m (Fig. 1.14). The monument's extra-urban position, adjoining the Roman necropolis, and its octagonal layout recall some features of the *Martyrion* of St Philip in nearby Hierapolis, which, however, was larger; indeed, the square-plan building that incorporates the octagon is 60 m long on one side. The octagonal church in Laodicea can thus also be interpreted as an extra-urban martyrial complex, and systematic excavations may be expected to highlight – as with the sanctuary of St Philip – the full extent of the affected area and the other structures within it (Fig. 1.15). These elements suggest a strong connection between the building and the Christian memories of Laodicea; in the distinctive characteristics of this building I believe that it is possible to recognize the central nucleus of a more extensive martyrial complex built in the proto-Byzantine period over the tomb of the saint, bishop and martyr Sagaris. This is for the time being merely a working hypothesis, but it would certainly be consistent with the dynamics of religious competition which, in the Christian era as in the previous Imperial period, is believed to have raged between the two main cities of the valley of the Lykos. It is no surprise furthermore that the competition between the two cities was based on figures of saints of unequal importance such as Philip, one of the twelve apostles, and bishop Sagaris. Indeed, a similar situation is attested by an inscription from Ephesus, discovered in the cathedral of St Mary (*Die Inschriften von Ephesos* Ia, 45). The inscription reproduces the text of an Imperial letter, perhaps by Justinian, which makes reference to a controversy between the cities of Ephesus and Smyrna,

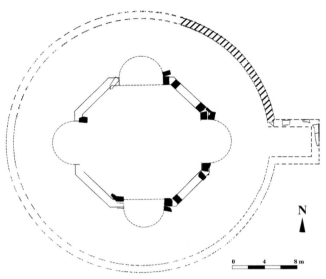

Fig. 1.14. Laodikeia. Octogonal Martyrion in the Eastern Necropolis (after Şimşek 2007, 282 fig. 99a; by courtesy of the author).

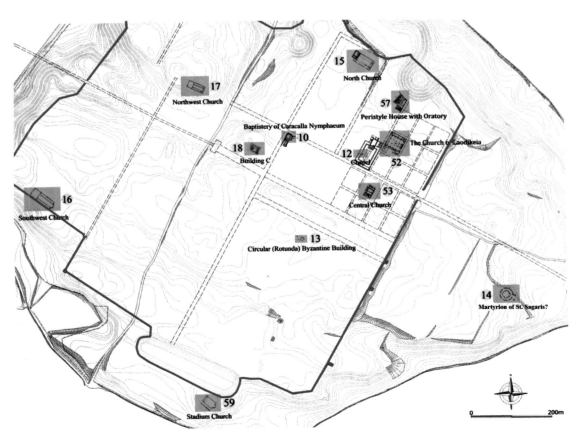

Fig. 1.15. Laodikeia. Plan of the city in the 6th century (adapted from Şimşek 2015; 20 fig. 15; by courtesy of the author).

whose bishop demanded to be made Metropolitan of Asia, emphasizing the antiquity of his church, founded by the martyr Policarp. However, the document stresses the primacy of the Metropolitan See of Ephesus, based on the authority of the Apostle John, recognizing, however, the autocephaly of Smyrna, precisely in view of the prestige deriving from the presence of the martyred saint (Amelotti and Migliardi Zingale 1985, 125 n. 15; Feissel 1999, 127–8 n. 29).

In Hierapolis, the 7th century marked a turning point for the entire city: the *Martyrion* was abandoned, following the burning of its wooden dome, which left many traces on the travertine piers of the octagonal room; in mid-Byzantine times only a small chapel on the western side of the destroyed proto-Byzantine building recalled the importance of the monument.

Conversely, in the 9th and 10th centuries, thanks to the presence of the venerated tomb, the church of St Philip saw further development, in a widely attested phase in the city in which mid-Byzantine contexts, characterized by a class of lead-glazed pottery known in Italian as 'ceramica a vetrina pesante', are frequent.[18] Although the church was reconstructed, the whole system of pools appears to have been filled in, with an *opus sectile* pavement laid over it throughout the nave.

This phase saw a change in religious practice, with the centrality of the Eucharistic liturgy and the martyr cult linked to the presence of the tomb.[19]

A further interruption was probably the result of a new earthquake in the 10th or 11th century, also attested in other parts of the city. The roof of the church is assumed to have collapsed and, in the 11th to 12th centuries, a cemetery was established in the nave, clearly demonstrating a desire to be buried next to the tomb, which continued to be an object of veneration.[20] That the memory of the apostle was still alive is also attested by the visit of Frederick Barbarossa's army in 1190: *Sequenti die in Lytania majori transivimus c dirutam civitatem Jerapolis, ubi S. Philippus Apostolus passus est* (*Historia de Expeditione Friderici Imperatoris* IV, 25–7).

The cemetery ceased to be used in the 13th and 14th centuries and the entire area was occupied by a Seljuk settlement that saw the tomb as living space, reusing the ancient stone beds (D'Andria 2011–2012, 40–1).

That the site was still visited by pilgrims is attested by the burial of a French pilgrim in a Roman tomb immediately to the west of the *Martyrion*. The tomb contained five lead *signa peregrinorum*, indicating the stations of the journey: from the sanctuary of St Leonard of Noblat, to St Mary of Rocamadour, Mary Magdalene at Saint-Maximin La Sainte-Baume, Rome and, after a trip by sea, the coast of Asia Minor to reach the ancient Persian Royal road to the Holy Land, which passed through Hierapolis (Ahrens 2011–2012).

The sanctuary of St Philip in Hierapolis and the tomb were the subject, in various forms, of veneration over the course of many centuries; now, thanks to the recent stratigraphic investigations of the complex, quite rare in terms of the wealth of data and the systematic application of multidisciplinary techniques, the complex is proving highly useful for understanding the interactions between urban landscapes and the cult of saints in Anatolia.

Notes

1 Information on the excavations in the area can be found in Rossignani and Lusuardi Siena 1990, 23–48; Lusuardi Siena, Rossignani, and Sannazzaro 2011.

2 Bakirtzis 2002, 176, focal point of the pilgrim cult of St Demetrios was the splendid silver-plated wooden *ciborium*, described by the Archbishop John of Thessalonike in the book of the *Miracula*; Yasin 2009, 237; Brubaker 2004, 69ff.

3 The only research into the site and monuments of Meriamlik dates back to the 1930s: Herzfeld and Guyer 1930. The literary traditions regarding the saint are referred to in a recent work by Davis (2001), unfortunately so full of bibliographical errors as to make consultation rather difficult.

4 Amsler 1996, 30–2; for the *Acta Philippi*, in the light of the archaeological documentation, see D'Andria 2011, 34–6. For a systematic investigation of Montanism, see Tabbernee 1997, for Hierapolis, 91–5, 497–508.

5 A preview of the results of these excavations appears in D'Andria 2011–2012, 3–52.

6 For a recently published and very helpful summary focusing on the early centuries of Christianity in the valley of the Lykos, see Huttner 2013.

7 Εὐγένιος ὁ ελάχιστος ἀρχιδιακ(ονος) κὲ ἐφεστ(ὼς) τοὺ ἁγίου κ' ἐνδόξου αποστόλου κὲ θεολόγου Φιλίππου. The inscription copied by Cockerell at the beginning of the 19th century was described by Gardner (1885, 346 no. 71) and examined by scholars such as Ramsay (1895–1897, 419) and Tabbernee (1997, 502ff. no. 83).

8 For the octagonal baths building, see D'Andria 2010, 55–67; Caggia 2012, 165–87.

9 See Verzone 1960, which still provides the most valid summary of the monument today.

10 The miracle is described by the Catalan chronicler Ramon Muntaner (*Cronica* 234), who visited Ephesus at the beginning of the 14th century: Foss 1979, 127.

11 For healings performed by means of water in Byzantine sanctuaries, see Ousterhout 2015, 65–77.

12 For an in-depth presentation of the building, with its rich decoration including mosaics and wall paintings, see Simsek 2015, for the plumbing network below the altar, 37–42 figs. 33, 44–8.

13 Verzone 1978, 1060–1. The incubation was practiced not only in the eastern Mediterranean but also in Rome, in one chapel of S. Maria Antiqua, connected to a Greek community in the early eighth century: Knipp 2002, 11–16.

14 On pilgrimage to healing Byzantine shrines, see Talbot 2002. On themes linked to the continuity of pagan cults in Christian contexts, see Dal Covolo and Sfameni Gasparro 2008.

15 For the project to restore and cover the building, see Şimşek 2015, 86–95.

16 See Huttner 2013, 334–5, on the martyrdom of Sagaris, bishop and martyr of Laodicea under the governor Servilius Paulus.

17 Traversari 2000, 94–5; the octagonal building is interpreted as a *martyrion*, inspired by the model of the *Martyrion* of St Philip in Hierapolis. See also Şimşek 2007, 281–2.

18 Arthur 1997. The presence of ceramics and mid-Byzantine contexts throughout the settlement was extensively demonstrated by the most recent excavations: Polito 2012; Semeraro 2012; Cottica 2012; Arthur, Bruno, Leo Imperiale, and Tinelli 2012; Caggia and Caldarola 2012.

19 We know that some of the relics had already reached Rome via Constantinople in the second half of the 6th century; Pope John III (561–574; *Liber Pontificalis* I, 305) is believed to have built the *Ecclesia apostolorum Philippi et Jacobi,* today the Basilica of the Holy Apostles, on the site where the relics were housed: Cecchelli 1999, 84–6.

20 Caggia 2014, 146–52, presentation of the mid-Byzantine burials inside the Church of St Philip; cf. also Wenn *et al.,* this volume.

Bibliography

Ahrens, S. (2011–2012) A set of Western European pilgrim badges from Hierapolis of Phrygia, in D'Andria, 67–74.

Amelotti, M. and Migliardi Zingale, L. (eds.) (1985) *Le Costituzioni giustinianee nei papiri e nelle epigrafi.* Milan, A. Giuffrè.

Amsler, F. (1996) *Actes de l'apôtre Philippe.* Turnhout, Brepols Publishers.

Arthur, P. (1997) Un gruppo di ceramiche alto medievale da Hierapolis (Pamukkale, Denizli), Turchia Occidentale. *Archeologia Medievale* 24, 531–40.

Arthur, P. (2012) Hierapolis of Phrygia: The drawn-out demise of an Anatolian city. In N. Christie and A. Augenti (eds.) Vrbes Extinctae. *Archaeologies of abandoned Classical towns,* 275–305. Farnham and Burlington (VT), Ashgate.

Arthur, P., Bruno, B., Leo Imperiale, M., and Tinelli, M. (2012) Hierapolis bizantina e turca. In D'Andria, Caggia, and Ismaelli (eds.) 565–83.

Atlante Hierapolis (2008) = F. D'Andria, G. Scardozzi, and A. Spanò (eds.) *Hierapolis di Frigia II. Atlante di Hierapolis di Frigia.* Istanbul, Ege Yayınları.

Bakirtzis, C. (2002) Pilgrimage to Thessalonike: The tomb of St Demetrios. *Dumbarton Oaks Papers* 56, 175–92.

Brubaker, L. (2004) Elites and patronage in Early Byzantium: The evidence from Hagios Demetrios at Thessalonike. In J. Haldon, L. I. Conrad, and A. Cameron (eds.) *Elites old and new in the Byzantine and Early Islamic Near East: Papers of the sixth workshop on Late Antiquity and Early Islam,* 63–90. Princeton (NJ), Darwin Press.

Büyükkolancı, M. (2001) *St Jean. Hayatı ve Anıtı.* Selçuk, Efes 2000 Vakfı.

Caggia, M. P. (2012) Recenti ricerche sulla collina di San Filippo a Hierapolis di Frigia. L'edificio termale di età bizantina. In R. D'Andria and K. Mannino (eds.) *Gli allievi raccontano. Atti dell'incontro di studio per i trent'anni della Scuola di Specializzazione in Beni Archeologici, Università del Salento, Cavallino (Le), Convento dei Domenicani, 29–30 gennaio 2010* (Università del Salento. Scuola di Specializzazione in Beni Culturali 'Dinu Adamesteanu'. Archeologia e storia 10.1), 165–87. Galatina, Congedo.

Caggia, M. P. (2014) La collina di S. Filippo a Hierapolis di Frigia: osservazioni sulle fasi di occupazione bizantina e selgiuchide (IX–XIV sec.). *Scienze dell'Antichità,* 20.2, 143–61.

Caggia, M. P. and Caldarola, R. (2012) La collina e il ponte di San Filippo. In D'Andria, Caggia, and Ismaelli (eds.), 601–36.

Cagiano de Azevedo, M. (1968) Appunti sulla relazione di S. Ambrogio e lo scavo del sepolcro dei SS. Gervasio e Protasio. In *Notizie dal Chiostro del Monastero Maggiore* 1–2, 1–9.

Cecchelli, M. (1999) SS. Philippus et Iacobus, basilica (SS. Apostoli). In *Lexicon topographicum Urbis Romae* IV, 84–6. Rome, Quasar.

Chambert-Loir, H. and Guillot, C. (eds.) (1995) *Le culte des saints dans le monde musulman.* Paris, École française d'Extrême-Orient.

Cottica, D. (2012) Nuovi dati sulle produzioni ceramiche dall'*insula* 104: i contesti della Casa dell'Iscrizione Dipinta (VII–IX sec. d. C.). In D'Andria, Caggia, and Ismaelli (eds.), 453–68.

Dal Covolo, E. and Sfameni Gasparro, G. (eds.) (2008) *Cristo e Asclepio. Culti terapeutici e taumaturgici nel mondo mediterraneo antico fra cristiani e pagani. Atti del Convegno Internazionale, Accademia di Studi Mediterranei, Agrigento, 20–21 novembre 2006.* Rome, LAS.

D'Andria, F. (2010) Peregrinorum utilitate. Le terme di San Filippo a Hierapolis nel V sec. d. C. In R. D'Amora and E. S. Pagani (eds.) *Hammam. Le terme nell'Islam. Convegno Internazionale di Studi, Santa Cesarea Terme, 15–16 maggio 2008,* 55–67. Florence, L. S. Olschi.

D'Andria, F. (2011) Conversion, Crucifixion and Celebration. St Philip's Martyrium at Hierapolis. *Biblical Archaeology Review* 37, 34–46.

D'Andria, F. (2011–2012) Il Santuario e la Tomba dell'apostolo Filippo a Hierapolis di Frigia. *Rendiconti della Pontificia Accademia Romana di Archeologia* 84, 1–75 (including three appendices).

D'Andria, F. (2013a) La reine des Nymphes. Eaux et paysages urbains à Hiérapolis de Phrygie. *Revue Archéologique,* 115–24.

D'Andria, F. (2013b) Il Ploutonion a Hierapolis di Frigia. *Istanbuler Mitteilungen* 63, 157–217.

D'Andria, F., Scardozzi, G., and Spanò, A. (eds.) (2008) *Atlante di Hierapolis di Frigia* (Hierapolis di Frigia II). Istanbul, Ege Yayınları.

D'Andria, F., Caggia, M. P., and Ismaelli, T. (eds.) (2012) *Hierapolis di Frigia V. Le attività delle campagne di scavo e restauro 2004–2006.* Istanbul, Ege Yayınları.

D'Andria, F., Zaccaria Ruggiu, A., and Ritti, T. (2005–2006) L'iscrizione dipinta con la 'Preghiera di Manasse' a Hierapolis di Frigia (Turchia). *Rendiconti della Pontificia Accademia Romana di Archeologia* 78, 349–449.

Davis, S. J. (2001) *The cult of St Thecla. A tradition of women's piety in Late Antiquity.* Oxford, Oxford University Press.

Feissel, D. (1999) Epigraphie administrative et topographie urbaine: l'emplacement des Actes inscrits dans l'Ephèse protobyzantine (IV–VI s.). In R. Pillinger, O. Kresten, F. Krinzinger, and E. Russo (eds.) *Efeso paleocristiana e bizantina* (Archäologische Forschungen 3), 121–32. Vienna, Verlag der Österreichischen Akademie der Wissenschaften.

Foss C. (1979) *Ephesus after Antiquity. A Late Antique, Byzantine and Turkish city.* Cambridge and New York, Cambridge University Press.

Gardner, E. A. (1885) Inscriptions copied by Cockerell in Greece. *Journal of Hellenic Studies* 6, 340–63.

Gonosovà, A. and Kondoleon, C. (1994) *Art of Late Rome and Byzantium in the Virginia Museum of Fine Arts.* Richmond, The Museum.

Herzfeld, E. and Guyer, S. (1930) *Meriamlik und Korykos. Zwei christliche Ruinenstätten des rauhen Kilikiens* (Monumenta Asiae Minoris antiqua, 2). Manchester, Manchester University Press.

Huttner, U. (2013) *Early Christianity in the Lycus Valley (Early Christianity in Asia Minor* (ECAM), vol. 1). Leiden, Brill

Knipp, D. (2002) The Chapel of Physicians at Santa Maria Antiqua. *Dumbarton Oaks Papers* 56, 1–23.

Lemerle, P. (1979–1981) *Les plus anciens recueils des miracles de Saint Démetrius et la pénétration des Slaves dans les Balkans.* Paris, Éditions du Centre National de la Recherche Scientifique.

Lusuardi Siena, S. (1990) *Il cimitero ad Martyres.* In *Milano capitale dell'impero romano, 286–402,* 124. Milan, Silvana.

Lusuardi Siena, S., Rossignani, M. P., and Sannazaro, M. (2011) *L'abitato, la necropoli, il monastero. Evoluzione di un comparto del suburbio milanese nei cortili dell'Università Cattolica.* Milan, Vita e Pensiero.

Maraval, P. (2004) *Lieux saints et pèlerinages d'Orient. Historie et geographie. Des origines à la conquéte arabe.* Münster, Cerf.

Miltner, F. (1937) in C. Praschniker, F. Miltner, and H. Gerstinger, *Das Coemeterium der Sieben Schlaefer* (Forschungen in Ephesos 4.2). Baden, Rohrer.

Ousterhout, R. G. (2015) Water and healing in Constantinople. Reading the architectural remains. In *Life is short, art long. The art of healing in Byzantium. Catalogue of exhibition in Suna ve Inan Kirac Pera Muzesi* (in English and Turkish), 65–77. Istanbul, Pera Müzesi.

Pallas, D. I. (1979) Le ciborium hexagonal de Saint-Démetrios de Thessalonique. *Zograf* 10, 44–58.

Polito, C. (2012) *Saggi di scavo nell'area del teatro.* In D'Andria, Caggia, and Ismaelli (eds.), 177–205.

Pülz, A. (2010) Ephesos als christliches Pilgerzentrum. *Mitteilungen Christlichen Archäologie* 16, 80–4.

Ramsay, W. M. (1895–1897) *The cities and bishoprics of Phrygia.* Oxford, The Clarendon Press.

Ritti, T. (2011–2012) *Alcune iscrizioni rinvenute nella chiesa di S. Filippo.* In D'Andria, 53–61.

Rossignani, M. P. and Lusuardi Siena S. (1990) La storia del sito alla luce delle indagini archeologiche. In M. L. Gatti Perer (ed.), *Dal monastero di S. Ambrogio all'Università Cattolica.* Milan, Vita e Pensiero.

Scardozzi, G. (2008) *Le fasi di trasformazione dell'impianto urbano.* In *Atlante di Hierapolis,* 31–48.

Scardozzi G. (ed.) (2015) *Hierapolis di Frigia VII. Nuovo atlante di Hierapolis di Frigia. Cartografia archeologica della città e delle necropoli.* Istanbul, Ege Yayınları.

Semeraro, G. (2012) *Ricerche nel Santuario di Apollo.* In D'Andria, Caggia, and Ismaelli (eds.), 293–324.

Şimşek, C. (2007) *Laodikeia (Laodicea ad Lycum).* Istanbul, Ege Yayınları.

Şimşek, C. (2015) *Church of Laodikeia. Christianity in the Lykos valley.* Denizli, Denizli Metropolitan Municipality and Prof. Dr Celal Şimşek.

Tabbernee, W. (1997) *Montanist inscriptions and testimonia. Epigraphic sources illustrating the history of Montanism* (Patristic monograph series, 16). Macon, Mercer University Press.

Talbot, A.-M. (2002) Pilgrimage to healing shrines: The evidence of miracle accounts. *Dumbarton Oaks Papers* 56, 153–73.

Traversari, G. (2000) *Laodicea di Frigia* (Rivista di Archeologia, Suppl. 24). Rome, Giorgio Bretschneider.

Verzone, P. (1956) Le chiese di Hierapolis in Asia Minore. *Cahiers Archéologiques* 8, 37–61.

Verzone, P. (1960) Il Martyrium ottagono a Hierapolis di Frigia. Relazione preliminare. *Palladio. Rivista di storia dell'architettura e restauro* 10, 1–20.

Verzone, P. (1978) Le primitive disposizioni del Martyrium di Hierapolis. In *Proceedings Xth International Congress of Classical Archaeology, Ankara-Izmir 1973,* 1057–62. Ankara, Türk Tarih Kurumu.

Wenn, C. C., Ahrens, S., and Brandt, J. R. (this volume) Romans, Christians, and pilgrims at Hierapolis in Phrygia: Changes in funerary practices and mental processes, 196–216.

Yasin, A. M. (2009) *Saints and church spaces in the Late Antique Mediterranean. Architecture, cult and community.* Cambridge, Cambridge University Press.

Zarcone, T. (1995) *Le mausolèe de Haci Bektash Veli en Anatolie central (Turquie).* In Chambert-Loir and Guillot (eds.), 309–19.

Zimmermann, N. (2011) Das Sieben-Schläfer-Zoemeterium in Ephesos. Neue Forschungen zu Baugeschichte und Ausstattung eines ungewöhnlichen Bestattungskomplexes. *Jahreshefte des Österreichischen Archäologischen Institutes in Wien* 80, 365–407.

Necropoleis from the territory of Hierapolis in Phrygia: New data from archaeological surveys

Giuseppe Scardozzi

Abstract

This paper presents some results of research activities conducted during the 2005–2008 fieldwork campaigns in the ancient territory of Hierapolis in Phrygia (Denizli Province, south-western Turkey). The research aimed at reconstructing the ancient topography and settlement patterns of the territory (previously not studied) from the Prehistoric times to the Ottoman age. During the archaeological surveys in the study area (the north-eastern sector of the Lykos valley and the Uzunpınar plateau), some ancient villages of different dimensions in use between the Hellenistic age and the Roman Imperial period (some also in the Byzantine period) were identified and studied. In some cases, even their necropoleis were identified and documented, recording the types of tombs (tumuli, chambers, sarcophagi, chamosoria, *pit graves) and their location in relation to the settlement area and the topography of the site. The collected funerary evidence is important new documentation to add to our knowledge of the necropoleis in ancient rural villages of southern Phrygia. It is also very interesting to note the presence, in some necropoleis, of isolated Hellenistic tumuli that could have belonged to the dominant classes of the rural communities, which had adopted the funerary typologies of the urban aristocracy.*

Keywords: ancient villages, archaeological surveys, chamber tombs, *chamosoria*, Hierapolis in Phrygia, pit graves, sarcophagi, territory, tumuli.

Introduction

The systematic archaeological surveys, carried out in the ancient territory of Hierapolis in Phrygia (Denizli Province, south-western Turkey) between 2005 and 2008, provide a lot of data to aid in the reconstruction of the ancient topography of the area and its settlement patterns from the Prehistoric period to the Ottoman age. The research was carried out by the Laboratory of Ancient Topography, Archaeology and Remote Sensing of the Institute for Archaeological and Monumental Heritage (National Research Council of Italy) under the auspices of the Italian Archaeological Mission directed by Professor F. D'Andria. The surveys aimed at studying a wide area, which is believed, in antiquity, to have been largely under the control of Hierapolis. The investigated territory had not been previously studied; before this research, only few data were available thanks to the unsystematic archaeological surveys conducted by W. M. Ramsay at the end of the 19th century and the information gleaned from epigraphic documents recovered during the 20th century (Ritti 2002, with previous bibliography).

The research activities enabled the discovery of numerous necropoleis of the Hellenistic and Roman period in the north-eastern part of the Çürüksu (ancient Lykos) river valley, where Hierapolis lies and on the Uzunpınar plateau, to the north of the city; these are both large cemeteries linked to ancient villages and small cemeteries or isolated tombs linked to small settlements or farms (Fig. 2.1).[1] The archaeological remains were documented using GPS systems; they were positioned on topographic maps (with medium or large scales) and high-resolution and ortho-rectified satellite images. The necropoleis were studied and surveyed considering the types of tombs (tumuli, chambers,

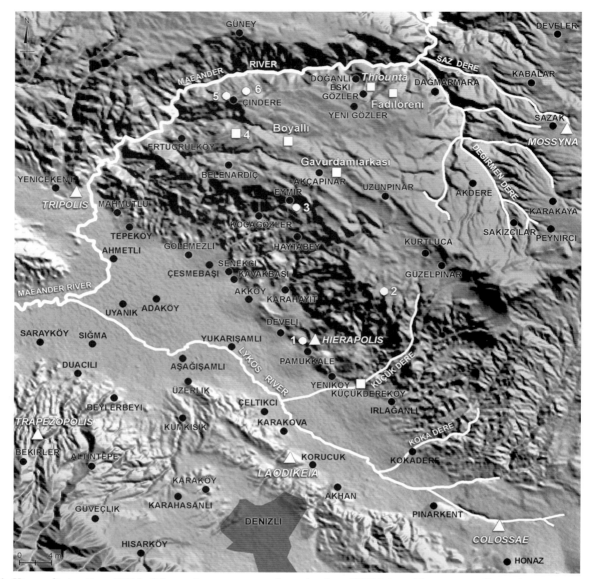

Fig. 2.1. Hierapolis territory. DEM of the study area processed starting from SRTM data: The ancient villages (squares) and the sites (circles) of the territory of Hierapolis where necropoleis or isolated tombs were identified during surveys.

sarcophagi, *chamosoria*, and pit graves) and their location in relation to the settlement area and the topography of the site.

In all the cases, the investigated necropoleis and funerary monuments had been subject to illegal digs and for the most part interred; so, no precise dating elements could be retrieved and often it was possible to suggest only a general chronology, based on comparison with the evidence from Hierapolis and the dating of the contexts where the tombs lie.

All the data presented in this paper comes from archaeological surveys and no regular stratigraphic excavations were conducted in the necropoleis of the study area; so, further research and archaeological excavations in

some test sites may result in new interesting data that will complement the present study.

The Lykos valley

The research conducted in the north-eastern sector of the Lykos valley (about 200–300 m above mean sea level), to the north of the river, enabled the documentation of many archaeological remains in the territory nearby Hierapolis, mainly dating from the Hellenistic age to the early Byzantine period, in particular aqueducts, roads, farms, and marble and travertine quarries. At a greater distance from the city, along the terraced slopes that descend from the Uzunpınar plateau, other farms and some medium-large ancient villages, with their necropoleis, were also identified. Two isolated tumuli were found in the Lykos valley, in the plain of Çukurbag, located just west of the travertine terrace where Hierapolis lies (Figs. 2.1.1; 2.2); they were sited along an ancient road leading from the city in the direction of the modern village of Develi and the Lykos river. The two tumuli are located on the last terrace facing the Lykos plain from the north-east, characterized by the presence of several ancient alabaster and travertine quarries. The building characteristics and measurements of the two funerary monuments are the same as the Hellenistic tumuli in the necropoleis that surround

the urban area of Hierapolis (D'Andria 2003, 48–62, 66–9, 87–8, 191–2, 205–6; Ronchetta and Mighetto 2012; D'Andria, Scardozzi, and Spanò 2008, *passim*; Ronchetta 2012; Scardozzi 2015, *passim*). They can be connected to urban aristocratic families who might have owned properties and residences in the plain below the terrace of the city, and who built their funerary monuments close to the main road that gave access to Hierapolis from the west.

The eastern tumulus (Figs. 2.2.A; 2.3) is located about 400 m west of the urban area. It is 9.50 m in diameter and has an almost completely preserved *crepidoma* made of squared blocks, dry-stone set and very well interlocked.

Fig. 2.2. Hierapolis territory. High-resolution satellite image (Pleiades taken on November 2013) showing the location of the two tumuli in the Çukurbag area: Above, general view of the area between Hierapolis and the modern villages of Develi and Pamukkale; below, a detail of the Çukurbag area.

Fig. 2.3. Hierapolis territory. The eastern tumulus in the Çukurbag plain: A) general view; B) dromos *and entrance side; C–D) two details of chamber (in C, the door).*

The width of the *crepidoma* wall corresponds to one row of blocks (about 0.56 m), while the height consists of two rows (max. height preserved above the interment: 0.80 m). The vaulted chamber is square in plan (2.74 m per side) and is interred up to the door's lintel; therefore, the benches, which must have been arranged on three sides, are not visible. The entrance to the chamber, which opened to the south-east and is 0.73 m wide, was accessed through a *dromos* (3.20 m long and 1.40 m wide), now buried and originally covered with slabs, now missing. The walls of the chamber are built in well-interlocked travertine slabs, dry-stone set without metal bolts or hinges; over the perfectly vertical walls the barrel vault is set, supported, on the entrance and opposite walls, by two lunettes that functioned as a centring. The partially collapsed vault was formed by six monolithic slabs set as voussoirs (length *c.* 3 m, width 0.70 m, thickness 0.36 m). The inner surface of the walls and of the vault's intrados were chiselled, while the extrados, which was covered in earth, was left roughly hewn.

The second tumulus (Figs. 2.2.B; 2.4) is located about 300 m west of the first, in a position dominating a wide stretch of the Lykos valley. The tumulus is poorly preserved and disturbed by recent activities by looters. Only a few squared blocks of the *crepidoma* are visible and a diameter of about 10.50 m can be measured. The funerary chamber, partially buried, has a square plan (2.20×2.20 m) and its walls, on which traces of plaster can be distinguished, are preserved up to the height at which the barrel vault begins; the entrance door (0.70 m wide) on the south-eastern side could be reached via a *dromos*, now buried. Inside the chamber, the funerary benches are visible (0.28 m thick and 0.65 m wide), arranged on three sides: they are located about 0.75 m from the chamber floor and were supported by small cylindrical pilasters with a rectangular base.

Again in the Lykos valley, some necropoleis were found around the modern village of Küçükdereköy, about 4 km south-east of Hierapolis (Figs. 2.1; 2.5). These funerary areas belonged to an ancient village which was located on the same site of the modern one, where several blocks and architectural elements in marble and travertine (reliefs, columns, capitals, etc.), mainly dating to the late Hellenistic age and Imperial Roman period, have been re-used; among them is a fragment of a funerary inscription dating to the 3rd century AD (Ritti and Scardozzi 2016, 812-3). The ancient settlement, that seems to have survived into Byzantine times, was probably under the control of

Fig. 2.4. Hierapolis territory. The western tumulus in the Çukurbag plain: A) general view; B) chamber; C) detail of funerary benches.

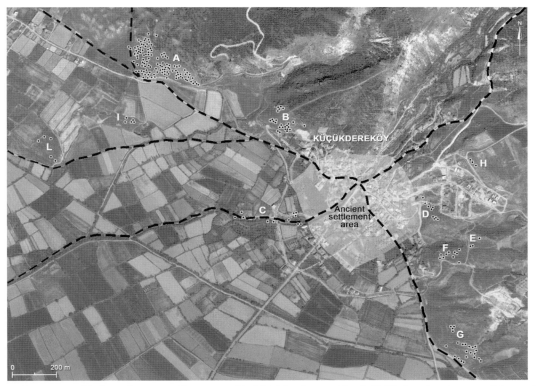

Fig. 2.5. Hierapolis territory. High resolution satellite image (QuikBird-2 taken on March 2005) showing the area of the modern village of Küçükdereköy: The site of the ancient settlement and the surrounding necropoleis are highlighted; the hypothetic ancient roads are added.

Hierapolis and situated on the slopes of the hills that climb steeply toward the south-eastern sector of the Uzunpınar plateau. The site controlled a natural road (the closed valley of the 'Küçük Dere' stream) from the Lykos valley to the plateau. It is also possible that the watercourse once marked the southern limit of the Hierapolitan territory in the direction of Laodikeia and Colossae.

The necropoleis located on the hills immediately to the west and south-east of Küçükdereköy are characterized by the presence of pit graves (generally 0.80–0.81 m wide, 1.80–1.82 m long and 0.70–0.90 m deep), cut directly into the bedrock plateau and originally covered by irregularly shaped slabs (generally two or three for each grave). The largest and most important one is the necropolis located about 500 m north-west of modern village, along the ancient road coming from Hierapolis, today a country path (Fig. 2.5.A). The cemetery extends on a slope facing south and is characterized by the presence of several clandestine digs. There are about 150 tombs that were identified during surveys and located using a DGPS system in RTK mode on

a topographic map (Fig. 2.6) obtained from the processing of an ortho-rectified Ikonos-2 satellite image taken in 2004 (Di Giacomo, Ditaranto, and Scardozzi 2011; Castrianni, Di Giacomo, and Ditaranto 2010–2011). About 80% of the tombs are pit graves, cut directly into the bedrock, which were covered by irregularly shaped slabs (Fig. 2.7.A). The remaining tombs consist of *chamosoria* (Fig. 2.7.B) and travertine sarcophagi laid directly on the rock plateau or on simple basements made with limestone blocks (Fig. 2.7.C); in only a few cases are the sarcophagi placed on more elaborate *hyposoria* (Fig. 2.7.D), quadrangular in plan and furnished with moulded benches (Fig. 2.8.A), which are mainly near the route of the road to Hierapolis. Many tombs are also along a less important road, with an east/west direction.

These sarcophagi, which generally lie near the main roads within the necropolis or in a prominent position, are simple, with a socle at the bottom, as in the II type of the North Necropolis at Hierapolis (Vanhaverbeke and Waelkens 2002, 120–1, tab. 1); in a few cases, they have bosses on

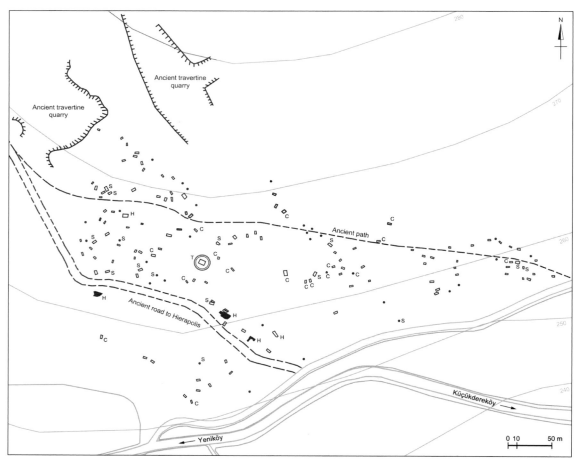

Fig. 2.6. Hierapolis territory. General plan of the main necropolis west of Küçükdereköy. About 80% of tombs are pit graves, while the other ones are chamosoria *(C) and sarcophagi (S), also positioned on elaborated* hyposoria *(H); in the western sector there is also one* tumulus *(T).*

Fig. 2.7. Hierapolis territory. The main necropolis west of Küçükdereköy: A) example of pit grave; B) chamosorion; C) sarcophagus; D) a collapsed basement for sarcophagi.

Fig. 2.8. Hierapolis territory. The main necropolis west of Küçükdereköy: A) bench decorated with lion's paws; B) general view of the tumulus with the remains of crepidoma indicated by arrow; C) funerary chamber of the tumulus; D) marble phallus re-used in the north-western periphery of the village.

the short sides, but the caskets are not decorated. The lids are very simple: there are examples with or without corner acroteria and moulded edges, such as in the types I, II, III of the North Necropolis at Hierapolis (Vanhaverbeke and Waelkens 2002, 123–4, tab. 8); some have bosses on the short sides or on all four sides.

The most important tomb is a tumulus (Figs. 2.8.B–C) located in the central-western sector of the necropolis, in a dominating spot, not far from the ancient road to Hierapolis; it surely belonged to the most important family of the ancient village. It exerted a centralizing function on the surrounding tombs, conditioning their orientation and arrangement. The funerary chamber, opening to the south-east, has a rectangular plan (3×2.62 m) and it is buried for the most part; the walls are of travertine slabs, dry-stone set and chiselled on the inner faces. The *crepidoma* is preserved in only a few squared travertine blocks, which measure about 9 m in diameter. On the back wall, the lunette is visible, made of two slabs, which must have supported the now collapsed barrel vault. Towards the entrance of the chamber was either a *dromos* or an antechamber, not preserved. Probably the top of the earth cone of this tumulus was the original location of the large marble phallus, which today has been re-used

on the north-western periphery of the village (Fig. 2.8.D), just 800 m south-east of the tomb.

Moreover, immediately to the north of the necropolis there are some ancient travertine quarries that were used for the extraction of sarcophagi (the bosses facilitated their transport) and building materials (blocks and architectural elements) for the funerary monuments. The quarries were also used extensively for the buildings of the ancient village at the same site of Küçükdereköy.

The other necropoleis close to Küçükdereköy (Fig. 2.9), another sepulchral area, with about 30 tombs consisting of rectangular graves, is in the north-western periphery of the village (Fig. 2.5.B), and three other small necropoleis are on the hills at the south-eastern periphery, where in total there are about 25 pit graves (Fig. 2.5.D-E-F). Another small necropolis is about 380 m towards the south (Fig. 2.5.G), consisting of about 25 pit graves of the same type, widely excavated by looters; it is near the ancient route to Colossae, which partially corresponds to the modern road between Küçükdereköy and Denizli. Moreover, some pit graves are at the north-eastern periphery of the village (Fig. 2.5.H), not far from the ancient route to the Uzunpınar plateau, while some architectural materials that belonged to destroyed

Fig. 2.9. Hierapolis territory. Four examples of pit graves from the necropoleis surrounding Küçükdereköy.

funerary monuments and fragments of sarcophagi are close to the ancient route, today traced by a country path, which flanked the Küçük Dere at the southern periphery of Küçükdereköy (Fig. 2.5.C) and linked the village to the southern sector of the Lykos plain and Laodikeia. Finally, ten pit graves are situated to the south and south-west of the main western necropolis (Fig. 2.5.L-I), excavated on two rock outcrops sited along an ancient route from the ancient village to the central sector of the Lykos valley, nowadays a country path; these tombs could also belong to ancient farms, whose remains (blocks, tiles, bricks, and pottery) are scattered in the nearby fields.

The Uzunpınar plateau

The broad plateau north of Hierapolis, on whose western slopes the city lies, has at its centre the large modern village of Uzunpınar; the plateau is between 1150–1250 m above mean sea level at its southern part and between 850–950 m above mean sea level at its northern part. The epigraphic documentation of Imperial Roman and early Byzantine times seems to indicate that it belonged to the territory of Hierapolis, but it was probably under the control of the city

since the Hellenistic age. The archaeological surveys in this territory allowed three large ancient villages ascribable to Hellenized indigenous populations to be identified, at the sites of Gavurdamıarkası Tepe, Boyallı, and Eski Gözler, the last one has been identified as Thiounta thanks to some epigraphic documents. Other small or medium-large villages are present, together with some farms of a medium or large size (Ritti, 2002; Ritti and Scardozzi 2016, 814-46). Moreover, some rural sanctuaries were identified: in particular, the one of Apollo Karios, east of Güzelpınar, and the sacred area of the deities of the Motaleis community, south of Dağmarmara (Ritti, Miranda, and Scardozzi, 2012).

An isolated tumulus was found about 5 km north-east of Hierapolis, near the southern limit of the Uzunpınar plateau, about 3.5 km south of the modern village of Kurtluca. The funerary chamber (Figs. 2.1.2; 2.10.A.1 and B), damaged by clandestine digs, has a square plan (2.40×2.40 m) and is built of travertine ashlar blocks, dry-stone set, finished with chisels on the inner face and left rough on the outside. The entrance was on the south-eastern side and no traces of the *crepidoma* were found. Another two chamber tombs are about 250 m southwards, on the edge of the plateau (Fig. 2.10.A.2); they have the funerary chamber cut into

Fig. 2.10. Hierapolis territory. Tombs in the area south-west of the modern village of Kurtluca: A) location of tumulus (1) and two chamber tombs (2) on a QuickBird-2 satellite image taken on March 2005; B) remains of the tumulus; C–D) remains of the two chamber tombs.

the bedrock and the roofing built with large limestone slabs. Both were damaged by illegal excavations and are partially collapsed: the first (Fig. 2.10.C), which opened eastwards, has a rectangular plan (1.10×2.05 m) and is 1.20 m high. The second (Fig. 2.10.D), with more monumental traits, is about 10 m south of the first and has a 2.70×3.30 m chamber, 1.50 m high and is partially filled with the collapsed roof, made of large travertine slabs (2.55×0.85×0.45 m). Its façade, facing southwards, has a 0.67 m wide door, 1.50 m high, and was completed by an abutting frame. The tombs are near an ancient route (nowadays a country path) that descends from the plateau to Hierapolis and the Lykos valley. No remains of a large ancient settlement were found in the surroundings, but only a few large farms of the Hellenistic and Roman period, to which the tumulus and the two chamber tombs could have belonged.

A small necropolis is situated at the south-western end of a plain located south of the modern village of Eymir, immediately south of the edge of the Uzunpınar plateau, about 10 km north of Hierapolis (Fig. 2.1.3). This cemetery is just on the southern margins of an ancient small settlement (Fig. 2.11.1), extending over a surface of about 1 ha and inhabited at least from the late Hellenistic up to the Byzantine period. In the necropolis are some graves (dug in the ground and encased with limestone slabs), partially destroyed by agricultural works, and one tumulus (named Toptaş Tepe) damaged by illegal excavations, which belonged to the most important family of the community. It has a rectangular chamber (2.20×1.66 m) built using large travertine ashlar blocks (Figs. 2.12.A–B), well refined by chisel on the inner façade and only roughly hewn outwards; the walls are 0.63 m thick and preserved to a maximum elevation of 1 m. The chamber opened northwards, where a walled structure built in small irregular blocks acts as an access corridor (length 5.5 m); it was probably added in a second phase, maybe when the tumulus was re-used in a non-funerary function. No traces of the *crepidoma* are preserved.

Other graves, dug in the ground and encased with limestone slabs (some of which have also bosses for lifting purposes: Fig. 2.12.C) and damaged by illegal excavations, are about 800 m to the north-west (Fig. 2.11.2–3), close to the western limit of the ancient settlement and the edge of the small plateau of Eymir. Other tombs of the same typology (Fig. 2.12.D) are between 1.5 and 2 km to the south-west (Fig. 2.11.4–7), along an ancient route, nowadays preserved as a country path that descends in the direction of the Lykos

Fig. 2.11. Hierapolis territory. Ancient tombs in the area of the modern village Eymir georeferenced on a QuickBird-2 satellite image taken on March 2010; the route of the 'horse road' mentioned by W. M. Ramsay highlighted.

Fig. 2.12. Hierapolis territory. Ancient tombs in the area of Eymir: A–B) tumulus; C) slab from a grave; D) example of grave.

valley; they are damaged by illegal excavations and probably belong to nearby ancient farms. Finally, it is important to remember the presence of several caves (Fig. 2.11.8) in the rock face on the top of the steep slope that is immediately to the north of Eymir. They are excavated on two or three levels and have irregular shapes and different dimensions (for example, 3.50×7 m, height 1.90 m; 6×7.60 m, height 1.80 m). They were used as houses during the Ottoman period and there are no elements that suggest a previous usage as chamber tombs.

Another small necropolis was found about 6 km north-west of the modern village of Belenardiç (17 km north of Hierapolis), at the north-western limit of the Uzunpınar plateau (Fig. 2.1.4). It is located about 230 m eastwards of a small settlement extending over a surface of about 2 ha (Fig. 2.13.A.1), inhabited at least in late Hellenistic and Roman Imperial periods and was located near an ancient route that descended to the Maeander valley. In the necropolis are some graves (dug in the ground and encased with limestone slabs) partially destroyed by agricultural works, and at least one tumulus, partially destroyed by illegal digs (Fig. 2.13.B). It has a rectangular chamber (2.55×2.10 m), built using quite well-refined limestone blocks and

characterized by two lunettes on the short sides that must have supported the now collapsed barrel vault; the chamber has the entrance southwards 0.60 m wide and is partially interred (therefore, no funerary benches are visible). No traces of *crepidoma* are preserved. Moreover, other remains probably identifiable with two tumuli, extensively damaged by illegal excavations and mostly buried, are respectively 30 m north and 150 m west of the first one (Fig. 2.13.A.2–3).

Other necropoleis were identified and documented close to the three main ancient villages of the Uzunpınar plateau: the Gavurdamıarkası Tepe, at the site named Boyallı and in the Eski Gözler-Thiounta area. The Gavurdamıarkası Tepe, the site of a large ancient settlement located in the central sector of the plateau, lies just 500 m east of the modern village of Akçapınar, 13 km north of Hierapolis (Fig. 2.1). The remains of the settlement cover a surface area of about 17 ha, along the southern slope of the hill and in the valley below (Scardozzi 2011, 119–20; 2012, 138–9). The finds, which are spread around on the ground and piled up along the field boundaries, testify to its use from at least the late Hellenistic age to the Early Byzantine period. It had an important location that controlled the central part of the Uzunpınar plateau and the main road that rose from

Fig. 2.13. Hierapolis territory. Ancient settlement north of the modern village of Belenardiç: A) location of the remains georeferenced on a QuickBird-2 satellite image; B) funerary chamber of a tumulus.

the northern side of the Lykos valley and crossed the area in a north-east direction towards the Çal plateau. This route, nowadays partially preserved as country roads, was used in 1883 by Sir William Mitchell Ramsay (who called it 'horse road') during his exploration of southern Phrygia (Ramsay 1883, 376–7; 1895, 122–4). It was also included on the geological map of western Turkey surveyed by Alfred Philippson in 1901 (Philippson 1914, map at the end of the book). Ramsay visited the remains of the ancient village on the Gavurdamıarkası Tepe, calling it 'Geuzlar-kahve' and highlighted the presence of several vaulted tombs that he dated to the Roman period (Ramsay 1895, 124). The scholar

emphasized the importance of the site and identified it as the ancient city of Mossyna (Ramsay 1887, 350; 1895, 122–4), mentioned by literary and epigraphic sources, today located near Sazak, in the Çal plateau (Ritti 2002, 43 and 47; Castrianni and Scardozzi 2010).

The main necropolis of the settlement is on the flat top of the hill (Fig. 2.14.1), near an ancient road heading north, nowadays a country path. Here, in 2006–2007 the remains of at least 13 chamber tombs were found preserved, although damaged by illegal excavations and mostly collapsed and buried (Fig. 2.14.A). They were irregularly grouped over about 0.5 ha located on the

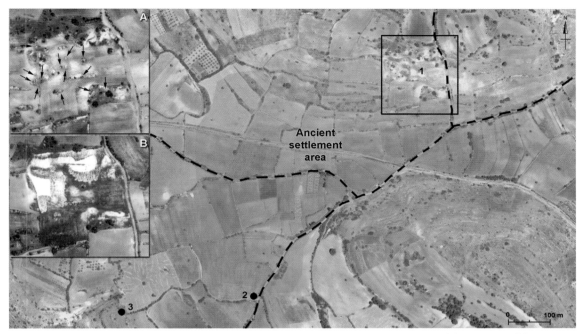

Fig. 2.14. Hierapolis territory. QuickBird-2 satellite image taken on July 2005 showing the location of the remains of the ancient settlement on the Gavurdamıarkası Tepe; the hypothetic ancient roads are added. In the two squares, details of the main necropolis in 2005 (A): the arrows indicate the tombs; and in a Pleiades satellite image taken on November 2013 (B): the last one shows the site quite completely destroyed by earthworks.

hill's flat summit and had different orientations, but for the most part had the entrance on the south, i.e. facing in the direction of the ancient settlement. Nowadays they are completely destroyed by earthworks made using bulldozers (Fig. 2.14.B). The chambers (Fig. 2.15) were rectangular in plan (about 1.70×2.80 m) and built using large limestone slabs, quite well finished on the inner face and only roughly hewn on the outer; it is not possible to verify the presence of funerary benches. The roofing consists of a barrel vault, constructed by means of two large slabs placed on the long sides, curving upwards and supported by two lunettes on the short sides; one or two other slabs are set on their long sides (0.30–0.35 m thick) close to the chamber's ceiling, functioning as the keystone of a vault. There are no clear elements to indicate how these chamber tombs were finished on the outside: the fact that the slabs were roughly worked on the outside presumably means they were not visible, but there is no stonework, which can be identified as belonging to some kind of superstructure. Therefore, it should not be ruled out that the chambers could have been covered with earth cones, even if not delimited by a *crepidoma*, of which no traces were found; in this case, they could be funerary tumuli. This interpretation could be also confirmed by the distance between the tombs and their irregular distribution. Near these tombs there are also some fragments of marble sarcophagi (Fig. 2.16.A), maybe originally

superimposed on simple basements in limestone blocks. Moreover, probably from this site, comes the fragment of a marble sarcophagus with columns and the inscription *Menandros* (Fig. 2.16.B), now in the Museum of Hierapolis (Ritti, Miranda, and Guizzi 2008, 296 no. 198). In fact, it was registered as coming from Uzunpınar, located 3 km to the south-east where many archaeological materials from the Gavurdamıarkası Tepe have been carried. The sarcophagus dates to the 2nd–3rd century AD and was re-used in the Byzantine age, when it was decorated on the back with a crux enclosed in a circle.

At the south-western edge of the area occupied by the ancient settlement (Fig. 2.14.2), on flat terrain near an ancient road heading south-westwards (nowadays a country path), two more funerary chambers were found; they have building characteristics and dimensions similar to the chambers of the Hellenistic tumuli in the necropoleis of Hierapolis. The first (Fig. 2.17.A), which is located north-eastwards and opened south-westwards, has a square funerary chamber (2.40×2.40 m), the walls of which (0.45 m thick) were built using travertine blocks dry-stone set and are preserved to a maximum height of 1.20 m. The other tomb, located about 10 m south-west of the first, has a rectangular chamber (2.30×2.45 m) with a well-preserved barrel vault, built using well-hewn and carefully set travertine blocks (Fig. 2.17.B). The chamber opened to the south-west (the door is 0.50 m

Fig. 2.15. Hierapolis territory. Four examples of chamber tombs on the top of Gavurdamıarkası Tepe

Fig. 2.16. Hierapolis territory. Fragments of marble sarcophagi from the top of Gavurdamıarkası Tepe: A) one documented in situ on 2006 and another (B) in the Museum of Hierapolis (from Ritti, Miranda, and Guizzi 2008, no. 198).

wide) and its maximum height is 1.20 m above the interment, which prevents verifying the existence of funerary benches. These two tombs, that are separated from the main necropolis, have different characteristics in comparison with the others tombs at the top of the hill and are probably two tumuli; in both cases the *crepidoma* is missing.

Fig. 2.17. Hierapolis territory. The two tumuli south of the settlement on Gavurdamıarkası Tepe and the isolated funerary chamber sited south-west of the ancient village.

Moreover, one isolated chamber tomb was found about 300 m west, it exhibits building features identical to the tombs at the top of the Gavurdamıarkası Tepe (Fig. 2.14.3). It is partially buried, has a rectangular plan (1.70×2.80 m) and a well-preserved barrel vault, built using large roughly limestone slabs (Figs. 2.17.C–D); two of the latter (0.35 m thick), with curved inner faces, are used as side walls and as the vault's rise, while a long slab closes the vault functioning as keystone. The door is located on the south-western side, while on the back wall two overlapping slabs (the topmost of which is curved at the end) support the vault. Also in this case, the slabs are rough on the outside and are finished only on the inside, and no remains of an eventual outer structure or a *crepidoma* are identified.

The main necropolis of the ancient settlement located at the site of Boyallı, about 2.5 km north-west of the modern village of Akçapınar (about 16 km north of Hierapolis and 4 km north-west of Gavurdamıarkası Tepe: Fig. 2.1), is on the flat area sited on its north-eastern edge (Fig. 2.18.A.1), along the ancient route to Thiounta (see below), nowadays a country path. The remains of the settlement (structures

in blocks, installations for olive oil and wine production, architectonical materials in marble and travertine, pottery, bricks, and tiles) cover a surface of about 19 hectares; it was in use from the late Hellenistic period to the Byzantine age (Scardozzi 2011, 122–3). The necropolis is characterized by the presence of some fragments of marble sarcophagi (Figs. 2.18.B–C), three of which are 'strigilati'. From this site comes also a marble sarcophagus with a relief representing *Marcus Aurelius Diodoros Zeuxion* and the masculine members of his family (Fig. 2.18.D). The sarcophagus, which in 1990 was moved to the Museum of Hierapolis, has also an inscription dated to the end of the 2nd century or first half of the 3rd century AD (Ritti 2002, 44–7; Ritti, Miranda, and Guizzi 2008, 290 no. 193). Moreover, in the necropolis some graves are also present, dug and encased using tiles, now partially destroyed by illegal excavations and agricultural works. Other tombs of the same typology are said by locals to be located about 150 m to the north of the northern limit of the settlement (Fig. 2.18.A.2), in a small valley not far from an ancient road nowadays preserved as a country path.

Fig. 2.18. Hierapolis territory. The ancient settlement of Boyallı: A) the location of the necropoleis (nos. 1–2) related to the settlement area, visualized on a QuickBird-2 satellite image taken on April 2007; two fragments of sarcophagi documented in situ in 2006 (B–C), and a sarcophagus in the Museum of Hierapolis (D).

The ancient settlement of Thiounta was identified by W. M. Ramsay during his travels of 1883 and 1888 in the area of the village of Eski Gözler (Ramsay 1883, 376; 1895, 124–5 and 142–6, nos. 30–31; 1928, 196–211), at the northern edge of the Uzunpınar plateau, about 20 km north of Hierapolis (Fig. 2.1). Thiounta is mentioned by some inscriptions of the Roman Imperial age, both from Hierapolis itself and the areas of the modern villages of Eski Gözler and Yeni Gözler (Ritti 2002, 41–54; Ritti, Miranda, and Guizzi 2008, 19, 98–100, 124, 140 nos. 32–3, 45, 54). (Scardozzi 2011, 123–4; Castrianni and Scardozzi 2012, 87–95). The remains of the ancient settlement, which were partially covered by landslides, nowadays extend across an area of about 6 hectares immediately to the north of the village of Eski Gözler, which was destroyed by an earthquake in 1976. The archaeological features (mainly travertine and marble blocks, bricks, tiles, and pottery) are on the terraces along the slope descending to the Maeander river (Fig. 2.19.A.1–2); they date from the late Hellenistic age to Byzantine times.

The presence of necropoleis on the hill of Bozburun, which delimits the settlement area to the west, are reported by locals. Here, chamber tombs nowadays collapsed, were visible some decades ago, but at present it is possible to identify only a few graves (damaged by illegal digs) in the southern and northern sectors of the site (Fig. 2.19.A.3–4); they were dug into the earth and coated by irregular shaped limestone slabs (Fig. 2.19.B). Moreover, the caves located in the cliff south of Eski Gözler (Fig. 2.19.A.5) were also identified along with other chamber tombs, but there are no elements that confirm this hypothesis.

Another two chamber tombs were identified during the archaeological surveys at the north-western limit of the Uzunpınar plateau, about 20 km north-west of Hierapolis (Fig. 2.1.5–6), in the area near the modern village of Çindere, which is not built on an ancient settlement (Ritti and Scardozzi, 2016, 841-2). In the territory surrounding the village, only the remains of a few farms were discovered; their archaeological materials dated from the late Hellenistic age to the Roman Imperial

Fig. 2.19. Hierapolis territory. The area of the ancient settlement of Thiounta: A) location of the main remains on a QuickBird-2 satellite image taken on April 2007 (nos. 1–2 settlement areas; nos. 3–5 funerary areas); B) an example of a grave.

times. Therefore, it is possible that the tombs belonged to the owners of the farms. The chambers are in a bad state of preservation and there are no elements that allow a more precise dating within the Roman Imperial period. The first one is on a terrace along the slope descending to the Maeander river, about 350 m west of the village (Fig. 2.20.A.1). It was built using large limestone blocks and is completely collapsed (Fig. 2.20.B); also a triangular lunette is preserved together with the blocks scattered on the ground surface. The second tomb is about 450 m north of Çindere (Fig. 2.20.A.2), along an ancient road descending from the north-western edge of the Uzunpınar plateau into the Maeander valley, with a south/north direction, nowadays preserved as a path. It crossed the river with an ancient bridge that in 2013 was submerged by a lake formed by a dam built downstream. It was built using large blocks (width of the walls 0.80 m), chiselled on the inner surfaces and coarse on the outer ones (Figs. 2.20.C–D); the chamber has a rectangular plan (2.40×1.60 m) and a height of 1.50 m. The door opens to the south-west and is 0.60 m wide; the roof is collapsed and was built using slabs rested on two triangular lunettes on the short sides of the chamber (the western one is still in situ).

Finally, another possible chamber tomb, built with blocks and characterized by a vaulted roof, is located at the northern periphery of a large settlement at the site of Fadılöreni, about 20 km north-east of Hierapolis (Fig. 2.1). This settlement is on the plateau immediately to the east of the valley of Thiounta (about 2 km away) and can be reliably connected to that village thanks to an ancient route partially cut in the bedrock. Its remains (structures in marble and travertine blocks, bricks, tiles, pottery) cover a surface of about 18 hectares and it was in use between the late Hellenistic age and the Early Byzantine period (Scardozzi 2011, 123–4; Castrianni and Scardozzi 2012, 93). The chamber is damaged by agricultural works and is mainly interred; its state of preservation does not allow an exhaustive examination of the remains.

Concluding remarks

The archaeological surveys performed in the territory of Hierapolis between 2005 and 2008 allowed us to collect rich funerary evidence, which provides new and important documentation about the necropoleis in ancient rural villages of southern Phrygia during the Hellenistic age and the Roman Imperial period. Unfortunately, it was not possible to acquire enough elements for a more precise dating of the archaeological evidence, due to the absence of archaeological excavations and the presence of illegal

Fig. 2.20. Hierapolis territory. The area of Çindere: A) location of two chamber tombs to the north-west (B–C) and west (D) of the modern village on a QuickBird-2 satellite image taken on April 2007.

digs and damages. However, the systematic exploration of the territory permitted the acquisition of many previously unknown data and even the documentation of archaeological remains before their destruction in the last few years.

Considering the topographical location of the necropoleis in relation to the settlements, they generally lie immediately outside the inhabited areas, along the roads; moreover, they often occupied peripheral areas (in many cases characterized by rocky outcrops) in relation to the fields used for agricultural activities and sheep farming. In some cases, for example in the hills close to Küçükdereköy or on the top of Gavurdamıarkası Tepe, the necropoleis lie in prominent locations dominating the settlement areas and the surrounding territories.

Different types of tombs were documented: tumuli, chambers built with blocks, pit graves excavated in the bedrock or in the soil and coated with limestone slabs or tiles, *chamosoria*, sarcophagi (even one *hyposoria*). Surely, one of the most important results of the research is the discovery of the tumuli in the territory of Hierapolis, isolated or within necropoleis; both small cemeteries belonged to a

small rural settlement and the large necropoleis belonged to villages which were spread over areas of almost 20 ha. These tumuli generally have the same building characteristics and dimensions of the Hellenistic ones located in the necropoleis surrounding Hierapolis, and take a prominent position in the cemeteries, sometimes becoming the main element influencing the orientation and distribution of the minor nearby tombs. It was easy to identify the examples from the Lykos valley (located in the Çukurbag area and near Küçükdereköy) due to the presence of the *crepidoma* built using large blocks. The hypothetical examples from the Uzunpınar plateau (near Eymir, south of Kurtluca, north of Belenardiç and at the Gavurdamıarkası Tepe) are nowadays remains of chamber tombs partially covered by earth mounds without *crepidoma*; the latter have not been preserved and it is impossible to know if they originally existed. Therefore, their identification as tumuli is based only on the characteristics of the funerary chambers; these are built with blocks, which are less refined compared to the tumuli in the necropoleis surrounding Hierapolis, but carefully constructed in comparison with the other chamber tombs of the Uzunpınar plateau, which are generally built using a rougher technique.

The discovery of these tumuli poses some questions regarding their owners and the meaning of their presence. Considering the distribution of the tumuli in the territory, it is possible to observe that they are within the probable boundaries of the chora of Hierapolis, at least inside its hypothetical extension in the Hellenistic and early Roman Imperial times. They seem to be absent only at the northern edge of the Uzunpınar plateau, but it is impossible to know if the chamber tombs that lie near the modern villages of Çindere and in Fadılöreni were originally covered by earth mounds; in fact, in both cases they are damaged by illegal excavations and agricultural works. In general, these tumuli could be considered as 'land-markers', showing the control of the territory by the aristocracy of Hierapolis. Unfortunately, we do not have finds and inscriptions from the tumuli, but it is evident that these tombs must have been associated with aristocratic families that in the Hellenistic age and at the beginning of the Roman Imperial times lived in the rural settlements, near their agrarian properties, and used the same funerary architectural typologies of the urban aristocracy. Finally, it is important to highlight that the presence of these tumuli have to be considered together with other aspects of the Greek colonization in the Hellenistic age, which were partially documented during the archaeological surveys and will be studied in more detail: in particular, the distribution of the rural settlements, the extensive occupation and exploitation of the territory (which is also characterized by a regular land division of the Hellenistic period identified in the northern sector of the Uzunpınar plateau, between the ancient villages of Boyallı and Thiounta: Scardozzi 2011, 120–2), or the integration between local communities and the colonists who founded Hierapolis.

Note

1   A previous presentation of these results, but taking only the Hellenistic tumuli into account, was in *Tumulus as sema: International Conference on space, politics, culture and religion in the first Millennium BC, Istanbul, 1–3 June 2009* (Scardozzi 2016). For the archaeological surveys in the territory of Hierapolis, see Scardozzi 2011 and 2012.

Acknowledgements

For his constant support and the valuable suggestions during the archaeological surveys in the territory of Hierapolis, I thank Prof. Francesco D'Andria. Special thanks also to Drs Laura Castrianni, Giacomo Di Giacomo, Imma Ditaranto, and Ilaria Miccoli, for their support during the field work activities and the data processing in the Laboratory of Ancient Topography, Archaeology and Remote Sensing of the Institute for Archaeological and Monumental Heritage in Lecce.

All photos (except Fig. 2.16.B, from Ritti, Miranda, and Guizzi 2008, no. 198; archive of the Italian Archaeological Mission at Hierapolis) were taken by the author, who also processed DEM in Fig. 2.1, map in Fig. 2.6, and satellite images (from Google Earth) in Figs. 2.2, 2.5, 2.11, 2.14, 2.18, 2.19, and 2.20.

Bibliography

Castrianni, L. and Scardozzi, G. (2010) Mossyna: The rediscovery of a 'lost city' in the territory of Hierapolis in Phrygia (Turkey). In *Proceedings of the 15th International Conference on 'Cultural Heritage and New Technologies' (Vienna, 2010)*, 616–33. Vienna, Stadt Wien.

Castrianni, L. and Scardozzi, G. (2012) I resoconti dei viaggiatori alla luce delle recenti ricognizioni archeologiche: la localizzazione di Thiounta e Mossyna nel territorio di Hierapolis. In D'Andria, Caggia, and Ismaelli (eds.), 81–107.

Castrianni, L., Di Giacomo, G., and Ditaranto, I. (2010–2011) Cartografia finalizzata alla ricerca archeologica da immagini satellitari ad alta risoluzione: un esempio dal territorio di Hierapolis di Frigia (Turchia). *Archeologia Aerea* 4–5, 387–90.

D'Andria, F. (2003) *Hierapolis di Frigia (Pamukkale). Guida archeologica*. Istanbul, Ege Yayınları.

D'Andria, F., Caggia, M. P., and Ismaelli T. (eds.) (2012) *Hierapolis di Frigia V. Le attività delle campagne di scavo e restauro 2004–2006*. Istanbul, Ege Yayınları.

D'Andria, F., Scardozzi, G., and Spanò, A. (eds.) (2008) *Hierapolis di Frigia II. Atlante di Hierapolis di Frigia*. Istanbul, Ege Yayınları.

Di Giacomo, G., Ditaranto, I., and Scardozzi, G. (2011) Cartography of the archaeological surveys taken from an Ikonos stereo-pair: A case study of the territory of Hierapolis in Phrygia (Turkey). *Journal of Archaeological Science* 38, 2051–60.

Philippson, A. (1914) *Reisen und Forschungen in westlichen Kleinasien. IV, Das östliche Lydien und südwestliche Phrygien*. Gotha, J. Perthes.

Ramsay, W. M. (1883) The cities and bishoprics of Phrygia. *Journal of Hellenic Studies* 4, 370–436.

Ramsay, W. M. (1887) Antiquities of southern Phrygia and the border lands. *American Journal of Archaeology* 3.3–4, 344–68.

Ramsay, W. M. (1895) *The cities and bishoprics of Phrygia, I, 1*. Oxford, The Clarendon Press.

Ramsay, W. M. (1928) *Asianic elements in Greek civilization*. London, John Murray.

Ritti, T. (2002) Documenti epigrafici dalla regione di Hierapolis. *Epigraphica Anatolica* 34, 41–70.

Ritti, T. and Scardozzi, G. (2016) Tra epigrafia e topografia antica: nuovi documenti epigrafici e 'iscrizioni ritrovate' dai villaggi del territorio di Hierapolis. In D'Andria, Caggia, and Ismaelli (eds.) *Hierapolis di Frigia VIII. Le attività delle campagne di scavo e restauro 2007–2011*, 807–48. Istanbul, Ege Yayınları.

Ritti, T., Miranda, E., and Guizzi, F. (2008) *Museo archeologico di Denizli-Hierapolis. Catalogo delle iscrizioni greche e latine. Distretto di Denizli*, Napoli, Liguori.

Ritti, T., Miranda, E., and Scardozzi, G. (2012) L'area sacra dei Motaleis e il santuario di Apollo Karios nel territorio di Hierapolis. In D'Andria, Caggia, and Ismaelli (eds.), 495–512.

Ronchetta, D. (2012) Necropoli Nord. Indagini nell'area della Porta di Frontino. In D'Andria, Caggia, and Ismaelli (eds.), 495–512.

Ronchetta, D. and Mighetto, P. (2007) La Necropoli Nord. Verso il progetto di conoscenza: nuovi dati dalle campagne 2000–2003. In D'Andria, F. and Caggia, M. P. (eds.) *Hierapolis di Frigia I. Le attività delle campagne di scavo e restauro 2000–2003*, 433–54. Istanbul, Ege Yayınları.

Scardozzi, G. (2011) Contributo alla ricostruzione della topografia antica della Frigia meridionale tra l'età ellenistica e l'epoca proto-bizantina: ricognizioni archeologiche nel territorio di Hierapolis. *Atlante Tematico di Topografia Antica* 21, 111–46.

Scardozzi, G. (2012) Ricognizioni archeologiche nel territorio di Hierapolis: gli acquedotti, le cave di materiali lapidei, gli insediamenti rurali, i tumuli funerari. In D'Andria, Caggia, and Ismaelli (eds.), 109–43.

Scardozzi, G. (2016) Tumuli in the ancient territory of Hierapolis in Phrygia. In O. Henry and U. Kelp (eds.) *Tumulus as sema: Space, politics, culture and religion in the First Millennium BC*, 589–99, pls. 274–85. Berlin and Boston, De Gruyter.

Scardozzi, G. (ed.) (2015) *Hierapolis di Frigia VII. Nuovo atlante di Hierapolis di Frigia. Cartografia archeologica della città e delle necropoli*. Istanbul, Ege Yayınları.

Vanhaverbeke, H. and Waelkens, M. (2002) The northwestern necropolis of Hierapolis (Phrygia). The chronological and topographical distribution of the travertine sarcophagi and their way of production. In De Bernardi Ferrero D. (ed.), *Saggi in onore di Paolo Verzone*, 119–45. Rome, Giorgio Bretschneider.

The South-East Necropolis of Hierapolis in Phrygia: Planning, typologies, and construction techniques

Donatella Ronchetta

Abstract

In the south-eastern area of Hierapolis in Phrygia, a city in Asia Minor founded in the 3th century BC, is a small, somewhat isolated necropolis, which appears to have been in use for three centuries. The necropolis is located on a small hill and is characterized by a uniform system, both with regard to its peculiar planimetric organization and to its periods of use. The funerary architecture, which is distinguished by rock-cut tombs with an architectural pediment façade and sarcophagi carved in the rock, follows local traditional Anatolian models. These tombs, devised as a novel project of funerary organization in the early period of the town's development, represent a significant document of funerary architecture as well as a unique testimony of funerary traditions and symbols.

Keywords: architectural pediment façade, Hierapolis in Phrygia, necropolis, rock-cut tombs

Hierapolis lies on a calcareous terrace extending in a north – south direction (D'Andria 2003). It is surrounded by burial grounds containing monumental burial buildings on three sides, conforming to the territory's orographic structure (Fig. 3.1). To the north, the largest and better-preserved burial ground flanks the road leading to Tripolis on the Meander. To the south, another burial ground, for the most part absorbed by the growing calcareous cliffs, extends along the road towards Laodicea and Colossae. The third part of the Hierapolitan necropoleis climbs the hill to the north-east, joining the North Necropolis. In addition, at the south-eastern edge of the city and immediately beyond the southern Frontinus Gate, we find, on a small terraced limestone rise, a complex of tombs and sarcophagi identified as the South-East Necropolis (Figs. 3.2–4).

The present presentation aims to provide a general reading of the planned layout of the South-East Necropolis, the choice of typological forms, as well as some construction techniques employed in most of the funerary buildings. To better illustrate the data, the catalogue at the end of the article includes some more detailed descriptions of a selected cluster of tombs (D1, D2, D3, D5, D8, D9-D4,

D10, D16). The selection was made according to typology, chronology, topographical position, and collected remains. These tombs were investigated archaeologically by the Italian Archaeological Mission, under supervision of the author.[1]

The necropolis hill has a conical form created by layers of travertine originating from the sedimentation of calcareous water which is typical of this territory, and rests on a bedrock of alluvial material. The tombs climb the hill in a spiral-like fashion (Fig. 3.5). Large boulders located at the foot of the northern and western slopes (at tombs D11 and D12) demonstrate that the specific and various nature of the travertine deposits, triggered by seismic and meteorological events, at times caused parts of the calcareous strata to break away.

Chronological phases of use

The south-eastern hill was used for burial purposes already during the city's early periods (late 2nd – early 1st century BC), to which tumulus D10 can be dated (Figs. 3.6–8),[2] and it was in continuous use as a burial ground until the 3rd

Fig. 3.1. Hierapolis, city plan. In the box the South-East Necropolis area (from D'Andria, Scardozzi, and Spanò (eds.) 2008, fig. on p. 53; by courtesy of the editors).

Fig. 3.2. Hierapolis. The South-East Necropolis, aerial view.

century AD, as indicated by the persistence of chambers featuring an architectural façade with pediment, carved out of the hillside,[3] by the presence of materials of the Augustan period found in tombs D2 and D9, and by an inscription (1st–2nd century AD) on the pediment of tomb D7 referring to the building as a '*heroon* of the ancestor'.[4] A tile-built tomb (*cappuccina* grave; 2nd century AD), located to the east of tomb D2, together with further funerary objects dating to the 3rd century AD, including

a Caracalla coin found in tumulus D10 (Yılmaz 1995), represent additional data confirming the persistent burial use of this small hill (Ronchetta 2007, 151). Below the hill, tomb D16, considering its typological and construction characteristics, can be identified as a burial of the *kamara* type;[5] furthermore, its unique decorative and technical apparatus, dates it to the late 2nd – early 3rd century AD, thus another testimony of the long life of this necropolis (Figs. 3.9–11).

Fig. 3.3. Hierapolis, South-East Necropolis. Topographical plan (E. Rulli 2014 del.).

Planning and division of the burial sites

The use of the hill as a burial ground appears to have been the result of a single design developed in a succession of phases according to the needs of changing tomb owners. Various building areas exploited the conformation of the terrain, partially carving the chambers one after the other from the rock face. A large number of sarcophagi were thus prepared by carving them directly from the bedrock or from separate boulders, especially along the eastern slope where the steep rock face was not suitable for building burial chambers (Figs. 3.12–13).[6] Two main burial sites, located on the same level, are bounded by a pathway leading from

Fig. 3.4. Hierapolis, South-East Necropolis. View from north.

Fig. 3.5. Hierapolis, South-East Necropolis. Morphological structure of the hill.

Fig. 3.6. Hierapolis, South-East Necropolis. Tumulus D10.

Fig. 3.7. Hierapolis, South-East Necropolis. Tumulus D10, dromos.

the city up to the top of the hill, skirting a ridge; along the southern face of this ridge, close to the pathway, there are signs of quarrying.[7]

The first burial complex, on the left of the pathway, is located on the north-western corner of the hill. It originally also contained the large boulders now resting at the foot of the

Fig. 3.8. Hierapolis, South-East Necropolis. Tumulus D10, first chamber and on site door of second chamber.

hill, boulders which show signs of slab quarrying (Fig. 3.14). A series of tombs and sarcophagi, cut into the bedrock, are placed in a semicircular arrangement on the narrow edge of the hill, well in sight of the city gates. They include a building with a barrel-vaulted ceiling (tomb D33) (Fig. 3.15).[8] To the north of tomb D33, and separated by a wide and deep crevice in the rock face,[9] is a small chamber (1.35×2.00 m) completely carved out of the rock (tomb D14). The tomb is badly damaged and from the preserved ashlars it is difficult to identify its covering system. Its off-axis entrance is preceded on the outside by a 'reception' area with low walls carved out of the rock (Fig. 3.16). The semicircular arrangement of the area allowed the sarcophagi to be carved out of the calcareous rock both here and at the summit of its southern ridge. Many sarcophagi have perfectly carved saddle roof covers, featuring rough angular *acroteria* and bosses used for hoisting (Fig. 3.17).

The second group of funerary buildings, on the right of the pathway, exploits the west-facing natural escarpment and contains a series of eleven burial chambers cut and regularized in travertine (D32, D31, D3, D1, D2, D27, D26, D25, D24, D23, D6). In order to make the tombs higher, blocks were added to the structure above the level of the rock surface. Each chamber was covered by a saddle roof, supported by monolithic, triangular pediment blocks, and composed of four or six large slabs, placed respectively length – and crosswise (Fig. 3.18). A sarcophagus (D15), positioned in front of the first tombs of the complex, used a peaked rock for carving both the coffin and an underlying base composed of three high steps (Fig. 3.19).[10] A second large sarcophagus (D28) was cut into the same spur used to create tomb D27, thus regularizing its front (Figs. 3.20–21).[11]

Fig. 3.9. Hierapolis, South-East Necropolis. Tomb D16.

Fig. 3.10. Hierapolis, South-East Necropolis. Tomb D16, entrance front and full view of barrel vault.

Fig. 3.11. Hierapolis, South-East Necropolis. Tomb D16, interior with sepulchral beds.

Fig. 3.12. Hierapolis, South-East Necropolis. Eastern slope with sarcophagi.

Fig. 3.14. Hierapolis, South-East Necropolis. Boulder (D11) resting at the foot of the hill, with signs of slab quarrying.

Fig. 3.13. Hierapolis, South-East Necropolis. Sarcophagi on the eastern slope.

Fig. 3.15. Hierapolis, South-East Necropolis. Tomb D33.

Fig. 3.16. Hierapolis, South-East Necropolis. Tomb D14.

Fig. 3.17. Hierapolis, South-East Necropolis. Sarcophagus in the semicircular arrangement of tombs.

Fig. 3.19. Hierapolis, South-East Necropolis. Sarcophagus D15.

Fig. 3.18. Hierapolis, South-East Necropolis. View of the first level of the hill, with rock-cut tombs.

Fig. 3.20. Hierapolis, South-East Necropolis. Sarcophagus D28.

Fig. 3.21. Hierapolis, South-East Necropolis. Tomb D27.

As previously mentioned, the hill's natural formation was exploited in different ways according to the required use: for quarrying of building material and for use as burial chambers. In addition, the presence of natural cavities in the south-east slope of the hill, was employed for the construction of two small underground structures (*chamosoria*)[12] closed by a sarcophagus. The first one (D5) is not connected to other burial structures; the internal space is rectangular in plan (1.20×2.55 m), but trapezoidal in cross section. The calcareous surface around the *chamosorion* was regularized and roughly cut blocks of stone (creating a step) added at the south-east end to receive a sarcophagus (Figs. 3.22–23); the structure was completed by the presence of a carved niche for lamps on the front. The second *chamosorion* (D4) was created closer to the eastern edge of the hill. It was placed next to tomb D9 with which it constitutes a single burial complex, carved from a large rock outcrop (Fig. 3.24). For a further description of this burial complex, see the catalogue below.

The ridge forming the eastern edge of the hill, at its south-east end, juts out towards the valley to the south and appears to have been completely exploited for the carving of sarcophagi of different sizes, mostly placed along the ridge, and covered by slabs placed flatly or with a low pitch (Fig. 3.25).

An outcrop of rock at the north-eastern tip of the hill produced a trench which was used both to quarry slabs and to carve some sarcophagi (as D21 and D22).[13] The porous rock surface of the sarcophagi was subsequently covered by a layer of plaster (Fig. 3.26).

Fig. 3.22. Hierapolis, South-East Necropolis. Chamosorion D5.

The large, detached boulder (D12) at the north-east corner of the hill certainly belonged to its highest ridge. The floor held three sarcophagi, two parallel ones and one placed at right angles; their finish is almost intact, thanks to the vertical position in which it landed (Fig. 3.27). A very thick layer of *opus signinum* shards hides the imperfections and the cracks in the stone, and subsequently a very thin layer of pure mortar plaster was laid to refine the surface and make it regular (Fig. 3.28).

From what is conserved of signs in the boulders and in the bed-rock surface of the hill (see, for example, the two sarcophagi D9 and D10) it appears that the cutting operations were well planned by first marking the perimeter of the slabs and then with the use of scalpels, saws and wedges, carefully dislodging them from the rock. The success of the work depended most likely very much on the experience of the stone cutter. The presence of plastered sarcophagi in the boulders to the north and east of the hill, demonstrates well that the detachment of the boulders happened after the burial ground had been formed.

Typology and construction techniques

The common design and construction principles of tombs are partly determined by ritual, economic possibilities, and local morphological features. A closer study of the architecturally and typologically better preserved tomb buildings in the western part of the South-East Necropolis may give us some insights into the importance of these factors.

For this group of tombs, the careful planning and design of façades hold a special importance since these buildings were considered as rock façades. The front part of the tomb which, from the bottom up was carved out of the rock, was completed in ashlar masonry which precisely defined the volume of the grave chamber. It also established the entrance with the components which typologically defined it: jambs and lintel with a simple trim, a well-finished door slab with fake wooden double shutter with crossbeams defining the panes, and tapered to ease swinging on metal hinges (Fig. 3.29).[14] The façade was topped by a horizontal cornice, generally less elaborate than the door lintel, carrying on the inside an L-shaped indentation to accommodate the overlying triangular pediment, often decorated with a central shield. The final element was the oblique *geison* which followed the sloping profile of the pediment or *tympanum*, placed on its top as a separate element or incorporated in the pediment block.[15]

In some buildings a further element completed the composition of the façade: an inscription which informed the reader of ownership and which was located within the pediment as a substitute for (tomb D2) or inserted in (tomb D7) the frequent shield decoration (Fig. 3.30), or was fitted as a marble *tabula* within the horizontal cornice (tomb D8) (Fig. 3.31).

D5

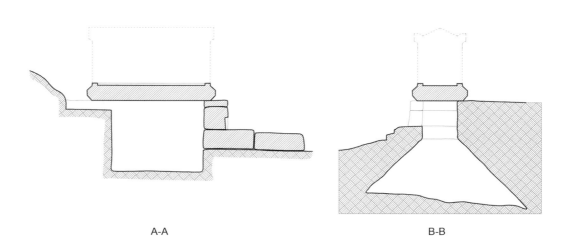

A-A   B-B

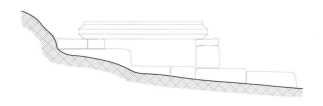

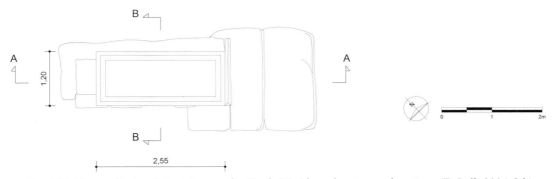

Fig. 3.23. Hierapolis, South-East Necropolis. Tomb D5. Plan, elevation and sections (E. Rulli 2014 del.).

Fig. 3.24. Hierapolis, South-East Necropolis. Tomb D9 and chamosorion *D4.*

Fig. 3.25. Hierapolis, South-East Necropolis. Sarcophagus D19 carved in a block of travertine.

Fig. 3.27. Hierapolis, South-East Necropolis. Collapsed boulder D12 with three carved sarcophagi

Fig. 3.26. Hierapolis, South-East Necropolis. Sarcophagus D22.

Fig. 3.28. Hierapolis, South-East Necropolis. Detail of a sarcophagus D12 in fallen boulder.

Fig. 3.29. Hierapolis, South-East Necropolis. Tomb D2.

Fig. 3.30. Hierapolis, South-East Necropolis. Tomb D7.

Fig. 3.31. Hierapolis, South-East Necropolis. Tomb D8.

Fig. 3.32. Hierapolis, South-East Necropolis. Area in front of tomb D2.

The layout of the external space, along with the definition of the areas of appurtenance, played a role in establishing the architectural features and manner of usage of the burial ground. This external space was characterized by sarcophagi placed in front of the tomb building, as was the case with tomb D2, which became structured as an area of respect thanks to the presence of sarcophagi, which defined the perimeter of the area. These sarcophagi were carved from the vestiges of waste rock deriving from previous quarrying and landscaping activities (Fig. 3.32).

The construction of the saddle roof of the tombs followed two basic techniques, both with reference to the way the roof slabs were placed. Both systems were used roughly equally; there is no particular preference shown for one technique over the other. We can, therefore, state that material costs and labour for laying the slabs did not differ. In the first technique a set of four slabs were placed lengthwise, two slabs for each side of the ridge of the roof. The slabs rested on the lateral walls and on off-sets in the pediment block. The joints between the slabs were carefully cut in order to optimize the point of contact and the slabs met at an acute angle at the ridge of the roof, touching each other at the lower edge, thus leaving above a small triangular space which could have been filled with soil to cover the top of the building (Fig. 33). The second system employed shorter slabs, usually six, three on each side of the roof, placed crosswise in parallel. The slabs rested securely on the top of the lateral walls at their lower end and leaned against each other at the top, each slab equipped with carefully cut contact surfaces. At the ridge of the roof the contact planes were cut vertically so that the tip adhered perfectly to each other. In order to increase the connectivity of the slabs, the contact surfaces of the two front slabs were further carved with a sawtooth-like motif (Fig. 3.34). The exterior surface of the slabs did not require any finishing as it was subsequently

Fig. 3.33. Hierapolis, South-East Necropolis. The two roof cover systems in tombs D1 (t.r.) and D2 (t.l.).

covered by soil. The oblique *geison* hid completely the roof slabs and highlighted the façade.

Inside the tomb the chambers were all equipped with deposition benches on three sides in a kind of 'triclinial' arrangement. However, within the framework of this typological archetypical solution of the grave chambers, natural restrictions and the personal inventiveness of the tomb owner resulted in some deviations from the norm. The deposition benches were made of slabs which were partly embedded in the walls, partly in each other, and/or

supported by indentations in the walls (tomb D1) (Fig. 3.35). Alternatively, they were carved directly out of the rock wall without any supports, as in the case of tomb D3 (Fig. 3.36). In still another case, in tomb D2, a sarcophagus was carved out of the solid bench and closed with a horizontally laid slab (Fig. 3.37).[16] A partial exception can be identified in tomb D9 which, possibly because of the incomplete nature of the building, has only two lateral benches fused together towards the back (Fig. 3.38);[17] Tomb D14 is a similar exception which, because of its small size, only had a single lateral bench.

Fig. 3.34. Hierapolis, South-East Necropolis. The second roof cover system in tomb D3.

Fig. 3.35. Hierapolis, South-East Necropolis. Sepulchral benches made of heavy slabs in tomb D1.

Fig. 3.37. Hierapolis, South-East Necropolis. A sepulchral bench and two sarcophagi carved into the rock in tomb D2.

Fig. 3.36. Hierapolis, South-East Necropolis. Sepulchral benches carved into the rock in tomb D3.

Fig. 3.38. Hierapolis, South-East Necropolis. Two klinai *partially carved into the rock in tomb D9.*

Concluding remarks

The architectural models employed in this necropolis – in particular the partly rock-cut chamber tombs and the sarcophagi carved directly out of the outcropping rock – are also present, as already outlined, in the North-East Necropolis, along the bottom of its steep slopes.[18] In fact, they belong to a traditional architectural theme in Anatolian funerary traditions and exemplify well the acquired skill of adapting plans to the site conformations and to the exploitation of morphological features. The goal

is attained by defining partly or totally the tomb volumes and by quarrying the respective building materials. The simplicity of architectural lines, volumes, and construction techniques stands in contrast to the great variety of typological, decorative, and constructive solutions in the North Necropolis,[19] the largest and the most representative burial ground at Hierapolis.

However, an overall assessment of the funerary architecture in the hilly Hierapolis areas gives a foretaste of the monumental architectural themes developed in the North Necropolis.

Comparing the contemporary tombs and burial plots of the North Necropolis with those of the surrounding hills it becomes evident that the typological forms and construction solutions, the personalized adaptions dependent on social needs and economic means are more strongly related to the area morphology in the hill necropoleis than in the North burial ground.

In conclusion, the small burial ground of the South-East Necropolis contributes to better understanding the availability of localization choices, of typological models and burial themes, as well as of symbolical representations and construction techniques, related to the funerary solutions in the initial stages of the life of the city between the 2nd/1st century BC and the 1st century AD.

Catalogue of some selected tombs

Tomb D1

The tomb buildings along the westerly facing terraces are characterized by rock-cut tombs with architectural pediment façades, among which is also the oldest in the series, tomb D1, still easily discernible (Figs. 3.39–41). Oriented south-west/north-east, with the entrance in the south-west, it was, like tomb D2, carved out of a large rock outcrop. Nonetheless, the latter was placed perpendicularly to the first tomb on a slightly higher level using a shared rock-cut wall.

The threshold and floor of D1 (2.90×3.32 m) are completely cut out of the rock, the threshold lying higher up compared to the present ground level. In its current state of preservation the façade cannot be reconstructed: nothing remains of the door jambs and lintel which would have defined it. The door itself, of which nothing remains, opened to the right as indicated by the offset of the right lateral *kline*.

The natural rock makes up the walls on the left and at the back, while the top of the right-hand wall and the façade were raised in ashlar masonry in order to bring them to the same roof level; the rock was not brought to a perfect horizontal level, instead the blocks of the walls were adapted to their irregularity. The saddle roof is composed of four very thick (0.40–0.45 m), but differently sized slabs placed crosswise, two on each side (Fig. 3.33). On the back wall, a triangular pediment slab of high-quality limestone completed the masonry while the pediment of the façade has been lost.

Inside, along the walls are three very thick (0.45 m, 0.41 m, 0.33 m) sepulchral benches in canonical layout (Fig. 3.35); the *kline* at the back rests on two rock overhangs, one at each end, while the lateral ones are embedded, as usual, in the back slab, but in addition rest on an overhang along each of the side walls.

The rock-cut walls and the ashlar masonry were wrought with a scalpel, but the result is rather rough due to the porous nature of the local stone. Nonetheless, the internal finishing work, emphasized by traces of rough mortar and small stones, would have smoothed the irregularities in the wall surfaces and most likely acted as a preparation for plaster. The sepulchral benches, the roof slabs, and the back wall pediment were more carefully worked, thanks to the use of better-quality materials (compact travertine).

Tomb D2

Tomb D2 (2.31×2.70 m), presenting an architectural pediment façade (Figs. 3.40–41), was carved out of the same rock outcrop as tomb D1, lying at a slightly higher level than its neighbour on the top of the hill, at the edge of the highest terrace. It was oriented north-west/south-east with the entrance in the south-east.

The limestone bedrock forms the floor, the walls, and the internal structures for depositions; worked blocks of compact travertine complete the façade and the roof, the latter composed of four slabs placed lengthwise, two on each side of the ridge (Fig. 3.33). On the façade (Fig. 3.29), the blocks formed jambs, lintel and cornice, a cornice which, judging from the thickness of the underlying lateral blocks, must have turned both corners. The façade is crowned by a triangular pediment carrying the inscription *tabula* inside a simple frame.[20] This pediment block, as well as the rear one, supports the roof slabs. The front block has an L-shaped indentation on the inside to receive the slabs, in this way strengthening the statics of the structure just like the oblique *geison* would have done, now lost, but which should have been positioned on top as a sort of coping. The floor of the tomb was lower than the bedrock outside and the threshold was made up of one slab.

The door, closed by a slab carved as a wooden double shutter with crossbeams bordering panes, was slightly tapered upwards; it opened to the right on metal hinges of which can still be seen the bottom seating and, in the lintel block, the hollow for the position of the higher one. In the door there are preserved holes for the lock, and the jambs carry indentations where the door would open.

In the grave chamber, the rock is worked leaving a shelf for burials: to the right it is used as a *kline*, while the shelves at the back and to the left were carved out as sarcophagi. Only the central sarcophagus still retains its closing slab (about 0.22 m thick) with rounded corners to facilitate its turning movement (Fig. 3.37).[21]

Inside, only the slabs of the roof, worked with a toothed chisel, were properly finished; the walls probably received a plaster coating although no traces of it have been found. Finishing touches were also carried out on the rear pediment, while the blocks above or next to the cut rock were only roughly carved, in the same way as the entrance wall. The jambs and the thresholds were

D1

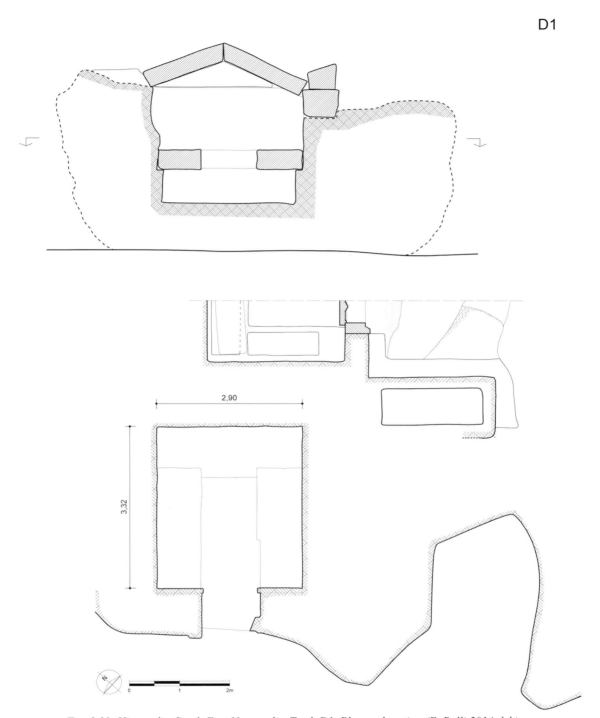

Fig. 3.39. Hierapolis, South-East Necropolis. Tomb D1. Plan and section (E. Rulli 2014 del.).

finished with a toothed chisel, as was the underside of the lintel. Externally, plaster was used on the façade (evidenced by many remaining traces) to hide the imperfections of the building material or where the rock and the worked blocks met. The plaster contained pottery fragments (*opus signinum*), which can still be seen between the wall and the burial bench at the front, in the joints between the front and the side juts of the rock, and between the jambs and the rock wall where they meet.

The reception area in front of the tomb is made up of a large rectangular space the sides of which are equipped with a continuous, low, rock-cut bench for sitting

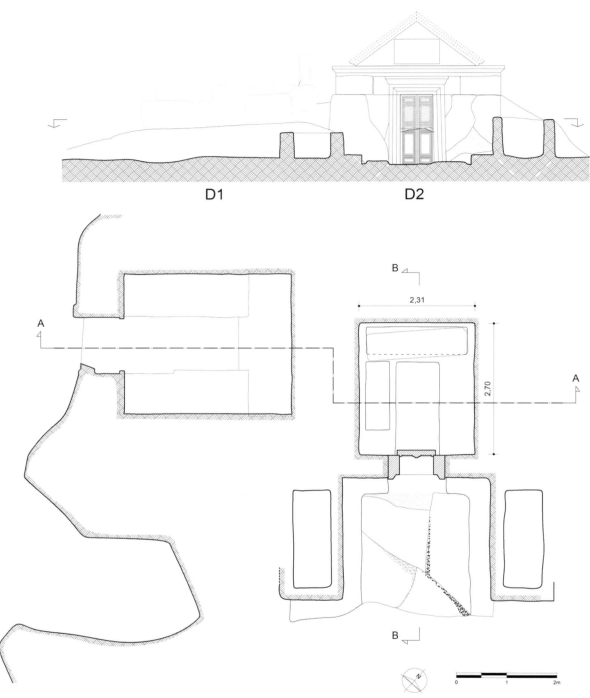

Fig. 3.40. Hierapolis, South-East Necropolis. Tombs D1 and D2. Plans and elevation of the reconstructed façade of tomb D2 (E. Rulli 2014 del.).

and behind, on each lateral wall, is hidden a rock-cut sarcophagus. In fact, the rock outcrop which constituted the tombs D1 and D2 follows the inclination of the hill, sloping down towards the south. This area was first regularized by cutting a groove to delineate the seating area. Presumably the intention was subsequently to cut down the central area to the level of the threshold of the tomb, but, as the work was never undertaken, we can only guess that it was filled with soil for some kind of garden cultivation (Fig. 3.32).

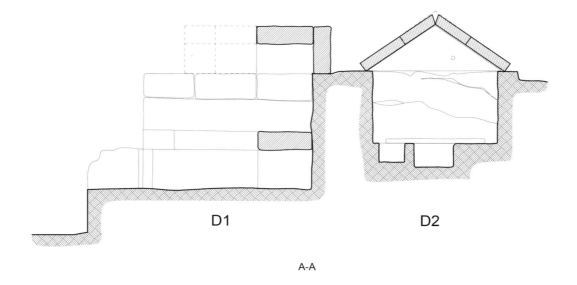

D1   D2

A-A

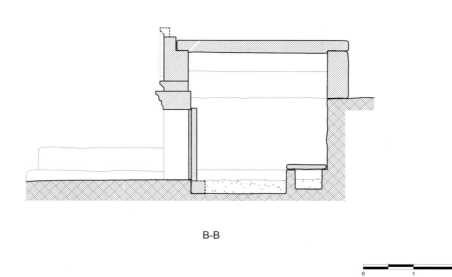

B-B

0   1   2m

Fig. 3.41. Hierapolis, South-East Necropolis. Above: Tombs D1 and D2; section A-A. Below: Tomb D2; section B-B (cf. Fig. 3.40) (E. Rulli 2014 del.).

The rock-cut sarcophagus (no. 21) on the western side was joined by another sarcophagus (no. 22), cut from a subsequent rock formation, and cleaved by a telluric movement, which must have affected the hill before it was used as a burial ground. On the eastern side, to the right of the entrance of the tomb, the rock also forms a thin wall, which was intended for yet another sarcophagus (no. 20) to be completed with the use of inserted slabs and the regularization of the rock face. In front of and aligned

with this sarcophagus, but on a level 0.60 m lower than the rock outcrop which ends here, a tomb *a cappuccina* (length 1.80 m) was found. This, made up of three '*bipedales*' tiles (*c.* 0.60×0.53 m) on each side and closed by a tile (0.34×0.31×0.05 m) at each end, was without a bottom and empty.

The presence of a tile-built tomb, datable to the 2nd century AD, and an extensive scatter of tile fragments bear witness to the continued use of this sepulchral area.

Tomb D3

The tomb is situated on the first terrace of the hill, to the west, and is part of the series of eleven tombs with architectural pediment façades for which this area is renowned. Probably it was not covered with earth and it stood out as a niche façade in the vertically rock-cut walls. The warmer colour of the cut rock, compared with the grey of the natural rock, made the tomb stand out among the other tombs as did also its saddle roof construction. Slightly rectangular in shape (2.52×2.68 m) (Fig. 3.42), completely carved out of the rock, the tomb was covered by six large and perfectly joined, contrasting slabs of different sizes placed crosswise, three on each side of the roof, all slabs with a worked extrados. Only the two slabs at the façade were tongued together, perhaps to prevent them from overturning (Fig. 3.34). The bottom slabs rested against the set-off top of the rock walls.

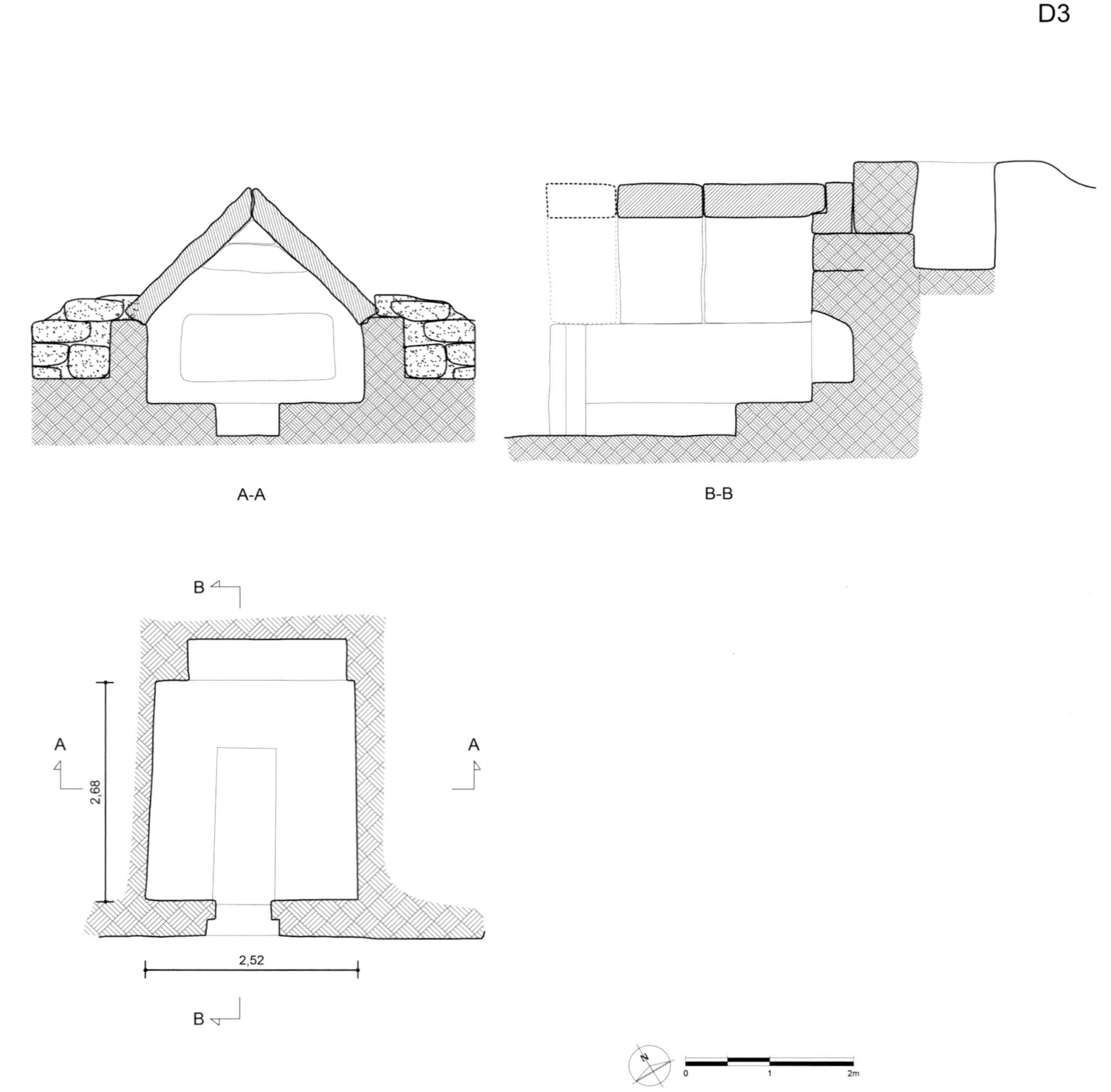

D3

A-A

B-B

B

A   A

2,68

2,52

B

Fig. 3.42. Hierapolis, South-East Necropolis. Tomb D3. Plan and sections (E. Rulli 2014 del.).

Inside, the three sepulchral benches were also carved from the rock to form a continuous shelf along the walls, obtained by lowering the central area of the chamber (Fig. 3.36). On the rear wall is situated a large niche (*c.* 1.83×0.80 m; depth 0.50 m); not suitable to hold a deposition it was perhaps used for sepulchral offerings.[22] Above it, the rock wall was levelled to form a triangular pediment including a rudimentary cornice. The façade was without architectural decoration, but an offset in the abutment face of the door opening suggests that originally jambs with belonging lintel and cornice were inserted. The tomb's triangular pediment block, decorated with a shield, lies now in ruins in front of the tomb. On the outside, the right rock wall – very porous and full of holes – has close to its bottom an overhang which may be traces of a rudimentary bench.

Tomb D8

This tomb (2.95×3.05 m), also with an architectural pediment façade, is found on the top of the hill facing south-east (Fig. 3.43). The chamber walls were carved from the rock and regularized with different size slabs. The rear wall was completed with two tiers of blocks on which the pediment rests, the lateral walls with a tier of blocks of different sizes and with dowel marks. The lateral blocks, trapezoid in section, were placed with the widest surface up, serving as support for the roof. The façade above the rock wall was made up of a single tier of blocks which included the moulded lintel; the jambs abutted the rock wall. The wall and lintel blocks were crowned by a simple moulded cornice which extended around the corners for the width of the corner blocks; the moulding was interrupted centrally above the door by a hole (0.40×0.32×0.09 m) left open to receive the inscription *tabula*, probably in marble and now lost (Fig. 3.31). According to tradition, the four roof slabs placed lengthwise, two on either side of the ridge, were supported by embeddings in the pediment blocks, which, on the front also accommodated the overhanging *geison*.

The concave moulding of the right-hand jamb allowed the door to open, as usual, to the right; the presence of hinges is testified in the lintel by preserved remains of sealing lead (removed).

Tombs D9–D4

Tomb D9, with an architectural pediment façade, oriented south-east, is located on the south-eastern border of the hill. It was built exploiting several outcrops of the rock and has a partly enclosed area in front, which holds *chamosorion* D4 (Figs. 3.24, 44–5).

Excavations carried out in 1986, examining the tomb interior, made it possible to recognize the morphology of the terrain on which the tomb rests. The façade with its threshold and the south-east wall of the enclosed area outside the tomb was formed from a single cut and regularized

boulder, in which the floor of the external area and part of the *chamosorion* was also hewn. A second boulder, detached from the first one by a rift of more than 0.10 m wide, was employed to form the south-east side wall, part of the back and south-west side wall, two large sepulchral benches, and the flooring of the tomb chamber. The back wall, in order to make a horizontal surface for the small triangular pediment block, was completed with a large block on top of the natural rock wall and supported by smaller stones. In addition, a tier of blocks of varying sizes were inserted on top of the rock-cut side walls to support the slabs of the roof. In the corners of the south-west wall, traces of mortar are visible, presumably used to smooth the highly irregular and uneven surface of the walls and subsequently finished with a layer of plaster.

The roof was made up of six large roof slabs, in porous travertine, quarried from the hill, placed crosswise, three on each side of the ridge. They rested on the irregular walls of the grave chamber and were kept in place by their own weights, without any particular joining technique except for a rough hewing of the slabs at the contact points with walls.

The façade wall was cut from the rock, including the slightly offset jambs. The lintel and the triangular pediment blocks (the last one including a bevelled cornice with fillet) were found badly damaged close to the tomb. In the lintel a 0.72 m wide cavity was cut to allow space for the door; in addition another small cavity was cut in order to insert the pivot of the upper hinge, and the upper face of the block was adjusted to receive the pediment-cum-cornice block. The pediment was worked in poor quality travertine. In the front wall, to the right of the entrance, appears a rather square cut (*c.* 0.17×0.17 m), which may have contained a decorative or commemorative grave marker. The door, also badly damaged, was worked in a more compact travertine stone than that of the bedrock. It presents the usual decoration which imitates a wooden door and the edges of its right side are rounded to relieve the rotation on the hinges, one of which is still functioning in the threshold; the jambs still contain the holes for the bolt. Among the materials found in the excavation was an iron key, *c.* 7 cm in length, perfectly preserved and with a ring 2.5 cm in diameter, from which runs a double lamina, with, on each side, three pins, of different lengths and thicknesses: the first pin is placed immediately after the ring and the other two together at the opposite end (Fig. 3.46). The key locked the door activating a mobile element within the lock.

The grave chamber (2.30×2.83 m) appears to have been planned to receive two lateral *klinai* carved directly out of the rock with an area of floor between them. In actual fact, the floor level of the tomb was deepened only just inside the door. Here the two *klinai* are clearly separated from each other and the floor is roughly hewn. Towards the back of

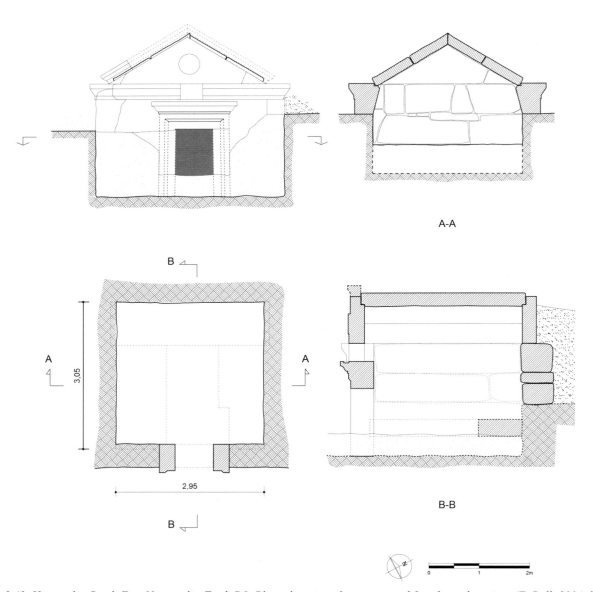

Fig. 3.43. Hierapolis, South-East Necropolis. Tomb D8. Plan, elevation of reconstructed façade, and sections (E. Rulli 2014 del.).

the chamber the separation of the funerary benches is only shown by a groove, cut by a chisel, without a continuation of the floor (Fig. 3.38). The fractures of the rock may possibly have been tamponed with mortar. The interior, due to the porosity of the rock and the irregularity of the walls themselves, seems definitely uncared for; a thick layer of plaster, of which some traces remain, may have enriched it somewhat.

On the left-hand side of tomb D9 is a small rectangular space (1.26×2.26 m) cut in the rock (*chamosorion* D4) and closed by a sarcophagus which, together with the tomb, forms a single burial complex carved out of a wide

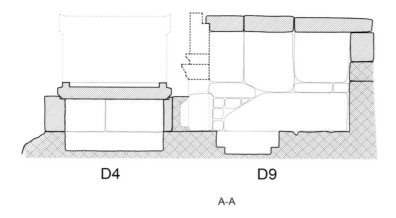

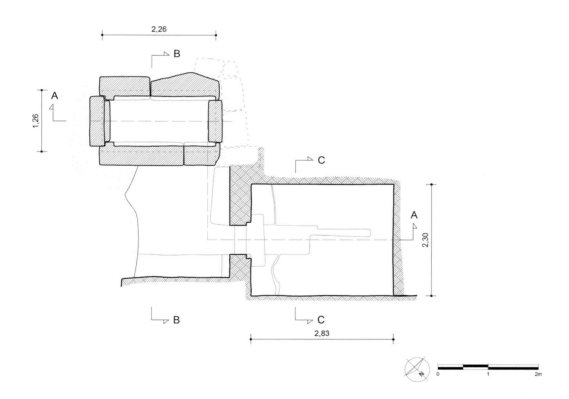

Fig. 3.44. Hierapolis, South-East Necropolis. Chamosorion *D4 and tomb D9. Plan and section (E. Rulli 2014 del.).*

rocky outcrop. The small internal space of the new tomb is entered from the south-east, where a specially worked slab acted as the door. The side walls of the *chamosorion* were each equipped with a squared, vertical stopper to perfectly accommodate the door slab.

The rock-cut chamber walls were regularized with overlying small blocks of different sizes, which carried two horizontally laid and well-squared stone slabs. They served as a base for the sarcophagus, which in this way was lifted up above the ground level. A fractured base is all that is left of the sarcophagus.

Tomb D10

The tumulus D10 rises on the lower slopes of the small hill; its orientation is north/south with entrance from the south (Fig. 3.6). The tomb is rather badly preserved, but from its remains still *in situ* it seems that it was originally equipped with two aligning chambers, one smaller and lower than the other, both vaulted, with plastered and painted walls and with a floor paved with slabs of travertine. The smaller, first chamber was preceded by an *anticella* covered with two flat slabs, a stepped *dromos*, and, at the front, a

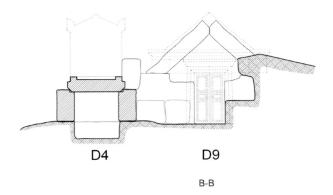

D4   D9

B-B

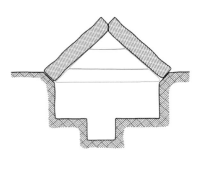

C-C

0   1   2m

Fig. 3.45. Hierapolis, South-East Necropolis. Chamosorion *D4 and tomb D9. Sections and elevation of the reconstructed façade of tomb D9 (E. Rulli 2014 del.).*

rectilinear *crepidoma*.[23] Subsequently the main entrance was restructured, probably affected by the new urban projects of the area in the late 1st century AD.[24] Some significant changes were made to the *crepidoma, dromos, anticella,* and the first chamber. The *dromos* was reduced and a new entrance opened towards the south-east, the barrel vault of the first chamber was replaced by a flat ceiling and the room was transformed into an *anticella* (Figs. 3.7–8). Some of the dismantled ashlars were reused in the restructured tomb building, both as wall blocks or as steps at the entrance, a slab from the flat ceiling of the old *anticella* and from the lunette block of the original barrel vault was used in the roof of the first chamber. Some ashlars also came to light during the digs at the edges of the building (Figs. 3.47–48).

In the first small chamber *(c.* 1.30×1.80 m) only the impost of the vault and its lunette blocks are at present

Fig. 3.46 Hierapolis, South-East Necropolis. Iron key found in tomb D9.

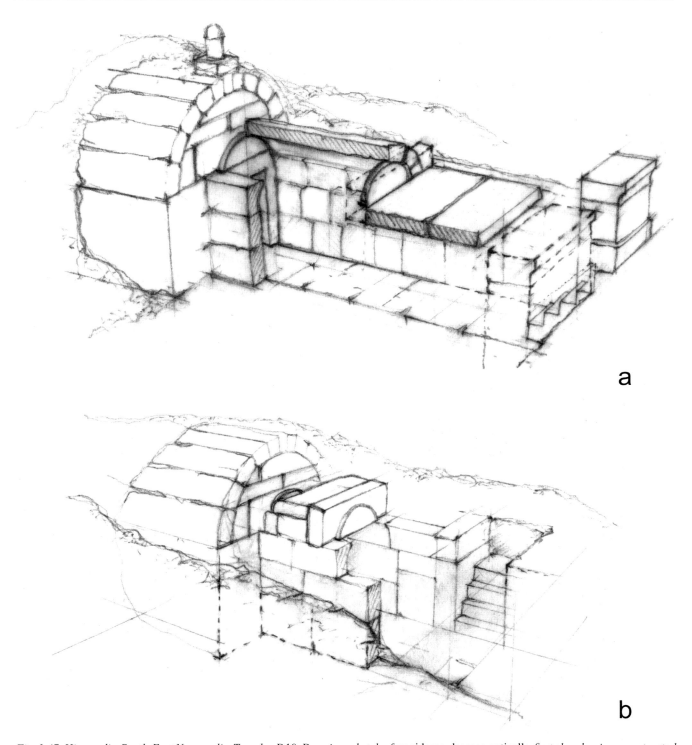

Fig. 3.47. Hierapolis, South-East Necropolis. Tumulus D10. Drawings sketch of crepidoma-dromos-anticella-*first chamber in reconstructed original project (a) and in present state (b) (E. Rulli 2014 del.).*

preserved. The original lunettes projected a few centimetres from the face of their respective blocks serving as a support for the barrel vault of the room. One of these blocks was turned round and serves now as lintel of the opening into the first room. The other lunette block remained *in situ* above the door into the larger second chamber.

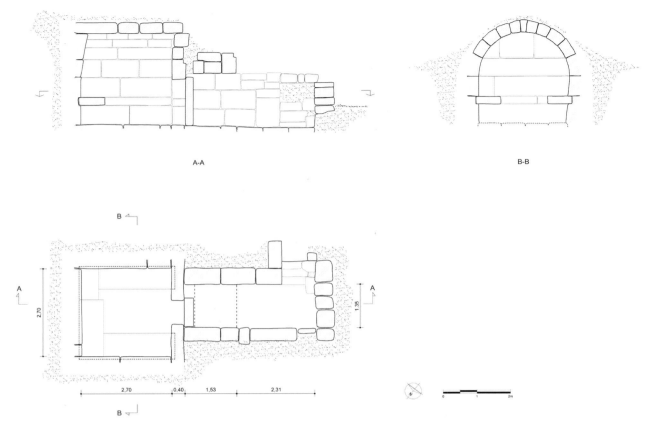

Fig. 3.48. Hierapolis, South-East Necropolis. Tumulus D10. Plan and reconstructed sections (E. Rulli 2014 del.).

The second chamber is square (2.70×2.70 m) and built of large slabs with perfect joints, but with a somewhat irregular face due to the rather porous building material. The vault is perfectly preserved and is made up of eleven tiers of ashlar masonry. All blocks of the interior were worked by a toothed chisel, perhaps creating a surface for plaster covering, as may be deduced from two fragments of painted plaster found in both chambers. The second chamber is furnished with deposition benches placed along three walls according to established custom.

As usual, a stone phallus erected on a marble base (about 0.83×0.70×0.35 m) represented the tombstone and, on either side of the façade, two marble stelae referred to the owners' names and events concerning their life and death (Guizzi 2007, 602; 2012, 669–72).

Tomb D16

This chamber tomb was covered with a barrel vault embedded in a flat roof structure supporting a sarcophagus, according to the *kamara*-type standards (Fig. 3.49).

The tomb, found at the foot of the hill on the western slope, is oriented north/south and, unlike the other rock-cut tombs which face the valley to the south-east or south-west, this one emerges with monumental characteristics and has its entrance to the north, towards the hill (Fig. 3.9). Most likely it was raised over stepped foundations crowned by a moulded socle, now underground. The corners of the façade were decorated with a pilaster each of full height, topped with simple capitals incorporated in the cornice; a block of the cornice, now in ruins, preserves the moulding.

The ashlar masonry was made of very compact blocks of travertine carrying ferrous veins, indicating that they were not quarried from the hill itself. The joints, both the horizontal and vertical ones, were neatly cut and the blocks were laid symmetrically in relation to the door. The jambs and lintel of the door were all moulded; the jambs, slightly bevelled, were framed by a rounded fillet, while the lintel had a more complicated moulding with corner *acroteria* and bevelling. All the blocks were probably linked by fasteners: the cavity of the vault (in the surmounting blocks) still shows two series of fasteners on the extremities of the blocks.[25]

The straight lateral walls hide the barrel vault that covers the chamber. Blocks set in place along the walls covered

D16

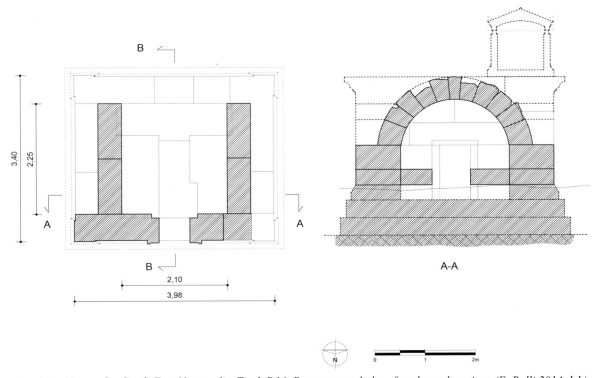

Fig. 3.49. Hierapolis, South-East Necropolis. Tomb D16. Reconstructed plan, façade, and sections (E. Rulli 2014 del.).

the extrados and created the flat roof. They were inserted in special cavities throughout the length of the perfectly laid ashlars of the vault, almost all of the same size, and all worked using a toothed chisel; only the keystone, in two pieces, is smaller and shows signs of the lewis used to put it in place (Fig. 3.10).

Inside, the chamber, almost square in shape (2.10×2.25 m), was filled with soil up to the level of the deposition benches, which run along under the impost blocks of the vault. The benches were perfectly worked, and as usual the back bench was anchored in the lateral walls, while the lateral benches were fastened at one end in the façade wall, the other resting on cavities cut in the back bench (Fig. 3.11). The right lateral bench shows a rectangular indentation for the door that would have turned on hinges: the right-hand jamb has, on its inside, a curved indentation to accommodate the movement of the door and the lintel still preserves the rectangular slot for the hinge pivot.

In front of the building lies a broken marble sarcophagus; most likely it was, according to custom, originally placed on the flat roof of the tomb. The coffin was decorated with roughly carved wreaths and circles while the fragmented lid was moulded and decorated with a shield on the pediment and with heads of a lion and a bull on the front bosses. The decorative theme of the sarcophagus follows a type from Aphrodisias (Işık 1984) datable to the mid-2nd – early 3rd centuries AD and thus is an important element in testifying to the longevity and continued use of the South-East Necropolis. Together with the diversity of planning and constructive technical set up of the building, it provides an important piece of information in defining that the tomb belonged to a later chronological period than other tombs with rock façades present in this necropolis.

Notes

1. In this burial ground the investigations started at the end of the 1970s and continued in many campaigns in the ensuing decades: Verzone 1980, 79-80; De Bernardi Ferrero 1985, 69–71; 1987; Ronchetta 2007; D'Andria, Scardozzi, and Spanò (eds.) 2008, 139, 143; Scardozzi (ed.) 2015, 191, 195, 196-198, 201.

2. A mound (tumulus) is a burial type used since prehistoric times not only in the Mediterranean region; it started to be employed in Hierapolis between the 2nd century BC and the 1st century AD featuring architectural features of archaic microasian patterns. The type is composed of a barrel vaulted quadrangular room with benches for the deposition of the deceased on three walls, hidden on the outside by a ground cone, surrounded by a *crepidoma* in which the entrance, structured in a *dromos* and a covered *anticella,* is located. Matz 1928, 271-80; Young 1981; Hanfmann 1983, 53 2; Fedak 1990, 16–20; Cormack 1997; Henry 2009, 81–102. About tumuli of Hierapolis, see Ronchetta; 2005, 171-2; 2016. The two *stelae* discovered in the mound area are dated between the late 2nd and early 1st centuries BC and support the dating of tomb D10; Guizzi 2007, 602; 2012, 669-72.

3. This type of tomb, derived from microasian tradition, appears and is further developed in Hierapolis from the 1st century BC; it is characterized by rock-cut chambers with an architectural pediment façade. This type of tomb is reported in the South-East Necropolis, in the North-East Necropolis and in some cases even in the North one. Akurgal 1955, 87–93; Kurtz and Boardmann 1971, 286, 288, 297; Roos 1985; 2006; Henry 2009, 55–79. About Hierapolis, see Schneider Equini 1972, 121; De Bernardi 1994, 348–50 figs. 4–6; Ronchetta 2007; 2008a, 93; 2008b, 139; Wenn, Ahrens, and Brandt, this volume.

4. Γάιου / Ἰούλιου Σεκούνδου / ἡρῷον τὸν προγόνων τοῦ Γάιου Ἰούλιου / Ἰουλιανοῦ καὶ παντῶν τῶν αὐτοῦ κ[λ]ή[ρ]ων 'Gaius Iulianus Secundus /The heroon, that of ancestors, belong to Gaius Iulius/Iulianus and to everybody [who are his heirs]' (reading of the author).

5. *Kamara* is a peculiar construction and design variation of a chamber tomb; it is composed of a vaulted ceiling, externally concealed and transformed into a flat roof for the placement of sarcophagi, see Catalogue below; Ronchetta 1999, 148–50. At Hierapolis the use of this word is confirmed by two grave inscriptions (Kubinska 1968, 94, 97). About this term, the corresponding architectural type and its Asia Minor diffusion, see Kubinska 1968, 94–9.

6. The rock outcrop was also exploited for building material for the tombs, as is well-evidenced in a large travertine boulder at the bottom of the hill along the northern front, in the entrance of tomb D2, and at many other points.

7. On the same rock face is a panel (0.52 m each side) which has in its centre a cavity used as a support for a *tabula*, perhaps a votive relief put up by the workmen labouring in this area, similar to the votive reliefs representing Heracles, Apollo, and a third deity, discovered in a primary placement in the quarries of the North Necropolis.

8. The vault, composed of very porous – and therefore very light – travertine blocks, rests on the walls of the chamber (*c.* 2.70×3.00 m) uncovered in the excavations, of which can still be seen the back wall and the door jambs, in travertine, abutting the rock walls.

9. The crevice reveals information of the interior nucleus of the geological hill formation, which seems to be composed of marl clay.

10. The travertine boulder, perhaps detached from the bed-rock, was smoothed off in steps on the front towards the valley and, on its top was cut the coffin (*c.* 2.00×0.64/0.69 m; depth 0.67 m), of which the front wall (raised 0.50 m), the interior faces and the top surface, on which the lid rested, are well defined, while the back and the long sides are irregular in shape. The lid of the sarcophagus, badly broken and found lying on the ground at the front, was probably cut from the same boulder as the sarcophagus. The lid has a slightly trapezoidal shape following the irregular shape of the coffin, its ridge out of axis. The two slopes of its lowly pitched roof run asymmetrically and are decorated with semicircular *acroteria* on the corners. The sarcophagus stands out, in its high position on top of the unusually high steps (0.40–0.50 m high).

11. The rock which descends towards the valley is cut in successive planes, obtaining thereby the definition of the façade of the chamber tomb D27 and the perimeter of the sarcophagus D28 (*c.* 2.60×1.45 m). Subsequently, after the rock surface had already been dug out, they detached the slab that made up the lid. The lid, a slab, 0.60 m in thickness, was found *in situ*.

12 The term *chamosorion* (χαμοσόριον) is used for a small chamber completely cut out of the rock/ground and closed by a cover of slabs, placed flatly or pitched; in this small burial ground sarcophagi act as covers. This funerary typology can be found spread around in western Asia Minor (Lycia, Caria, Cilicia) from the Hellenistic period until the Imperial age; Schneider Equini 1972, 105. For the use of the term see Heberdey and Wilhelm 1896, 37 n. 90.

13 Sarcophagus D22 (*c.* 2.10×0.70 m, depth 0.50–0.60 m) was cut using the rather unstable surface of the rock; in fact, one side of the sarcophagus was detached due to a fracture created following an earthquake or due to a subsidence of the rock caused by wind erosion of its clay layer underneath the travertine crust. On the rock surface, next to the sarcophagus D22, are traces of another sarcophagus.

14 For the doors, hinge and lock systems, and a preserved key, see below under the tombs D2 and D9.

15 The first group includes tombs D2, D3, D7, and D8, the second group all the other tombs with saddle roof.

16 For further descriptions, see the catalogue below.

17 See the catalogue below.

18 See *supra* note 4.

19 In the archaeological survey in the North Necropolis a large number of funerary types were identified: architectural pediment façade tomb, *aedicula* tomb, great barrel vaulted tomb, flat roof chamber tomb/*bomos*, *kamara,* podium to support sarcophagus, *hyposorion,* stepped podium to support sarcophagus/*bathrikon,* funerary exedra, which were further developed with different volumetric and decorative solutions, embedded in planned funerary areas. Verzone 1978; Ronchetta 1987; 1999; 2005; 2012; Ronchetta and Mighetto 2007.

20 Due to poor preservation the inscription could not be read.

21 The easy rotation may suggest that the closed sarcophagus could have been used as an *osteotheca.*

22 Tomb Tb16c in the North Necropolis, situated on the hill slope, had a similar niche carved in the rear wall of the grave chamber.

23 A similar solution with rectilinear *crepidoma* can be observed in tumuli in the North-East Necropolis, in a tumulus of the East Necropolis and in one surveyed in the North Necropolis; see De Bernardi Ferrero 1991, 135–6.

24 After the earthquake in AD 60 the present area was included in the new urban development of Hierapolis. The main street (*plateia maior*) was extended south and its terminal point marked by a gate put up by Iulius Sextus Frontinus, governor of the city in the years AD 84–86: D'Andria, Scardozzi, and Spanò (eds.) 2008, 143; Scardozzi (ed.) 2015, 197.25. This tomb is quite unique, since it is the only building, of the whole necropolis, fully made of masonry held together by fasteners.

Acknowledgements

All photographs are by the author except the aerial photograph (Fig. 3.2), published with the courtesy of the *Missione Archeologica Italiana a Hierapolis di Frigia.* All drawings are by E. Rulli, also these published with the courtesy of the *Missione Archeologica Italiana a Hierapolis di Frigia.*

Bibliography

Akurgal, E. (1955) *Phrygische Kunst.* Ankara, Archäologisches Institut der Universität Ankara.

Cormack, S. (1997) Funerary monuments and mortuary practice in Roman Asia Minor. In S. E. Alcock (ed.) *The early Roman Empire in the East* (Oxbow Monograph 95), 137–56. Oxford, Oxbow.

D'Andria, F. (2003) *Hierapolis di Frigia (Pamukkale). Guida Archeologica.* Istanbul, Ege Yayınları.

D'Andria, F. and Caggia, M. P. (eds.) (2007) *Hierapolis di Frigia I. Le attività delle campagne di scavo e restauro 2000–2003.* Istanbul, Ege Yayınları.

D'Andria, F., Caggia, M. P., and Ismaelli, T. (eds.) (2012) *Hierapolis di Frigia V. Le attività delle campagne di scavo e restauro 2004–2006,* 669–72. Istanbul, Ege Yayınları.

D'Andria, F., Scardozzi, G., and Spanò, A. (eds.) (2008) *Hierapolis di Frigia II. Atlante di Hierapolis di Frigia.* Istanbul, Ege Yayınları.

De Bernardi Ferrero, D. (1985) *I recenti lavori della Missione Archeologica Italiana a Hierapolis di Frigia, 1978–1980* (Quaderni de 'La ricerca scientifica', CNR, 112), 65–74. Rome, CNR.

De Bernardi Ferrero, D. (1987) Report on the activity of the Mission of Hierapolis in 1986. In *Kazı Sonuçları Toplantısı* 9.2, 228–9.

De Bernardi Ferrero, D. (1991) 1990 Yılı Hierapolis Kazısı. In *Kazı Sonuçları Toplantısı* 13.2, 131–40.

De Bernardi Ferrero, D. (1994) Frigya Hierapolis'i 1993 Kazı ve Restorasyanları. In *Kazı Sonuçları Toplantısı* 16.2, 345–60.

Fedak, J. (1990) *Monumental tombs of the Hellenistic age: A study of selected tombs from the pre-Classical to the early Imperial era* (Phoenix Supplementary 24). Toronto, University of Toronto Press.

Guizzi, F. (2007) La ricerca epigrafica: risultati dell'ultimo quadriennio e prospettive future. Iscrizioni sepolcrali e stele anepigrafi. In D'Andria and Caggia (eds.), 597–604.

Guizzi, F. (2012) Stele funeraria ellenistica di Apollonia. In D'Andria, Caggia, and Ismaelli (eds.), 669–72.

Hanfmann, G. M. A. (1983) *Sardis from Prehistoric to Roman times. Results of the archaeological explorations of Sardis, 1958–1975.* Cambridge and London, Harvard University Press.

Heberdey, R. and Wilhem, A. (1896) *Reisen in Kilikien ausgeführt 1891 und 1892 im Aufträge der Kaiserlichen Akademie der Wissenschaften* (Denkschriften der Akademie der Wissenschaften in Wien 44.6). Vienna, Alfred Hölder.

Henry, O. (2009) *Tombes de Carie. Architecture funéraire et culture carienne. VI –II s. av. J.C.*, 55–81. Rennes, Press Universitaire de Rennes.

Işık, F. (1984) Die Sarkophage von Aphrodisia. In B. Andreae (ed.) *Symposium über die antike Sarkophage* (Pisa 5–12 september 1982, Marburger Winkelmann-Programm), 243–81. Marburg, Lahn: Verlag des Kunstgeschichtlichen Seminars.

Kubinska, J. (1968) *Les monuments funéraires dans les inscriptions grecques de l'Asie Mineure.* Warsaw, PWN-Editions scientifiques de Pologne.

Kurtz, D. C. and Boardman, J. (1971) *Greek burial customs.* London, Thames and Hudson.

Matz, F. (1928) Hellenistische und Römische Grabbauten. *Die Antike. Zeitschrift für Kunst und Kultur des klassischen Altertums* 4, 266–99. Berlin, de Gruyter.

Ronchetta, D. (1987) Necropoli. In *Hierapolis di Frigia 1957–1987. Catalogo mostra*, 105–12. Milan, Fabbri Editori.

Ronchetta, D. (1999) Tecniche di cantiere nelle necropoli di Hierapolis di Frigia: alcuni appunti. In M. Barra Bagnasco and M. C. Conti (eds.) *Studi di archeologia classica dedicati a Giorgio Gullini per i quarant'anni d'insegnamento*, 131–67. Alessandria, Edizioni dell'Orso.

Ronchetta, D. (2005) L'architettura funeraria di Hierapolis. La continuità delle indagini dall'impostazione scientifica di Paolo Verzone alle attuali problematiche. In D. Ronchetta (ed.) *Paolo Verzone 1902–1986. Tra storia dell'architettura restauro archeologia*, 169–84. Turin, CELID.

Ronchetta, D. (2007) Lettura del progetto di pianificazione urbanistica di un particolare insediamento funerario: la necropoli Höyük a Hierapolis di Frigia. In C. Roggero, E. Della Piana, and G. Montanari (eds.) *Il patrimonio architettonico e ambientale. Scritti in onore di Micaela Viglino Davico*, 150–3. Turin, CELID.

Ronchetta, D. (2008a) Necropoli Nord-Est. In D'Andria, Scardozzi, and Spanò (eds.), 91–3.

Ronchetta, D. (2008b) Necropoli Sud-Est. In D'Andria, Scardozzi, and Spanò (eds.), 139, 143.

Ronchetta, D. (2012) Necropoli Nord. Indagini nell'area della Porta di Frontino. In D'Andria, Caggia, and Ismaelli (eds.), 495–512.

Ronchetta, D. (2016) Significance of the tumulus burial among the funeral buildings of Hierapolis of Phrygia. In O. Henry and U. Kelp (eds.) *Tumulus as sema: Space, politics, culture and religion in the first millennium BC*, 513–87, pls. 259–73. Berlin and Boston, De Gruyter.

Ronchetta, D. and Mighetto, P. (2007) La necropoli nord. Verso il progetto di conoscenza: nuovi dati 2000–2003. In D'Andria and Caggia (eds.), 433–55.

Roos, P. (1985) *Survey of rock-cut chamber-tombs in Caria. I. South-eastern Caria and the Lyco-Carian borderland* (Studies in Mediterranean Archaeology 72:1). Gothenburg, Paul Åströms Forlag.

Roos, P. (2006) *Survey of rock-cut chamber-tombs in Caria. II. Central Caria* (Studies in Mediterranean Archaeology 72:2). Gothenburg, Paul Åströms Forlag.

Scardozzi, G. (ed.) (2015) *Hierapolis di Frigia VII. Nuovo atlante di Hierapolis di Frigia. Cartografia archeologica della città e delle necropoli.* Istanbul, Ege Yayınları.

Schneider Equini, E. (1972) *La necropoli di Hierapolis di Frigia. Contributi allo studio dell'architettura funeraria di età romana in Asia Minore* (Monumenti Antichi 48), 95–142. Rome, Accademia nazionale dei Lincei.

Verzone, P. (1978) Hierapolis di Frigia nei lavori della Missione Archeologica Italiana. In *Un decennio di ricerche archeologiche* (Quaderni de 'La ricerca scientifica', CNR, 100), 391–475. Rome, CNR.

Verzone, P. (1980) Les travaux de restoration de Mission Italienne de Hierapolis. In *Kazı Sonucları Toplantısı* 2, 77–80.

Wenn, C. C., Ahrens, S., and Brandt, J. R. (this volume) Romans, Christians, and pilgrims at Hierapolis in Phrygia: Changes in funerary practices and mental processes, 196–216.

Yılmaz, S. (1995) Hierapolis (Pamukkale) kuzey ve güney giriş kapıları yakınlarında bulunan Roma mezarlarının restorasyonu ile güney Roma kapısı temizlik çalışması. In *Müze Kurtarma Kazıları Semineri* 5, 25–28 Nisan 1994 Didim, 130–1.

Young. R. S. (1981) *Gordion excavations (1950–1973). Final reports I: Three great early tumuli.* Philadelphia, The University Museum – University of Pennsylvania.

Tomb 163d in the North Necropolis of Hierapolis in Phrygia: An insight into the funerary gestures and practices of the Jewish diaspora in Asia Minor in late Antiquity and the proto-Byzantine period

Caroline Laforest, Dominique Castex, and Frédérique Blaizot

Abstract

The collective tomb 163d, recently excavated in the North Necropolis of Hierapolis, is characterized by the representation of a hanoukkiah *and an inscription mentioning Jewish owners. A general assessment of the knowledge of burial practices in Palestine and other places of Jewish diasporas is then proposed, with a particular focus on the treatment of the body. Although the history of the tomb is complex and shows at least two phases of occupation, the funerary dispositions observed in tomb 163d are examined by the mode of burial and the secondary depositions. The primary depositions, some of which were made in a wooden container, were in all parts of the grave. Some depositions preceding the Jewish phase were left* in situ, *but the majority, according to their volume, were moved about. These observations match well with elements known from Jewish communities settled in Hierapolis and in Asia Minor, showing a proximity between pagans and Jewish burials. In particular, the conclusions reject the idea that bones were collected in a separate container, as known in Palestinian tombs, and demonstrate that the movement and the skeleton's loss of individuality inside the tomb were admitted, making them inalienable, despite the change of users.*

Keywords: burial customs, collective grave, funeral management, Jewish communities, Hierapolis in Phrygia, late Antiquity, proto-Byzantine period.

Introduction

The Jewish diaspora from Asia Minor during Antiquity is above all known from historical and epigraphic sources but not from archaeological excavations. However, the exceptional discovery in the North Necropolis of Hierapolis in Phrygia of an untouched subterranean chamber associated with Jewish inscriptions contributes new knowledge about the funerary practices of the Jewish communities. As far as we know, this is the only Jewish tomb to have been excavated in Asia Minor. The fact that the chamber was not robbed gives the opportunity to analyze the funerary data in a different light by approaching the mortuary gestures on bodies, bones, and objects, which defined the use of the grave. The results of the excavation of this funeral chamber reveal that the burials spread from the first and at least until the 6th century AD, but as the Jewish inscription is dated to the 3rd century, the Jewish burials are related to the second phase of occupation occurring after the purchase of the tomb.

The purpose of our contribution is to document what could have been the funerary practices in the micro-Asian Jewish diaspora, by presenting the funerary gestures observed in tomb 163d. Firstly, a point will be made on existing studies on the funerary practices and gestures of the antique Jewish communities, followed by a methodical analysis of the grave 163d. The final aim is to put these results into a wider ancient Asia Minor context.

Investigating the funeral gestures and practices of the Jewish communities

During late Antiquity and the proto-Byzantine period the Jewish diasporas were present in almost all the Mediterranean countries (Rutgers and Bradbury 2006, 493; Hadas-Lebel 2011). However, our knowledge of their funerary practices is confronted with many problems, some of them valid also for non-Jewish graves, as, for example, past lootings and re-use(s) of the tomb, the lack of documentation and publications of old excavations, as well as funerary objects and stelae out of context (Stern 2013, 271). Nevertheless, the main problem is the identification of the Jewish graves in the diasporas. In general, the Jewish diasporas did not have a specific community area in which to bury their dead: they were settled among those of the 'Gentiles' and consequently were located outside of cities, in the urban outskirts (Rutgers 1992; Noy 1998). At the scale of the tomb, if the burial caves with one or several sub-rectangular chambers were the most common collective graves in Palestine (Hachlili 2005), the architecture of Jewish collective tombs is very close, if not similar, to pagan tombs in the diasporas. Few Palestinian influences in the tombs are visible, although there are some cases of *kokhim*, a kind of *loculi* cut perpendicularly into the walls, as at the site of Vigna Randini in Rome (Noy 1997, 86). Mural paintings or sarcophagus decorations did not always contain Jewish iconographic motifs nor did they contain Jewish symbols mixed with pagan motifs, which could have generated a debate on the Jewish nature of these graves (Rutgers 1998, 58–68). In the same way, in the diasporas, inscriptions on tombs are mostly in Latin or Greek; their contents, especially for onomastics and tomb protection measures, are similar to pagan inscriptions, although they could express a strong allegiance to Judaism (Rutgers 1995, 139–209; 1998, 57). In these inscriptions, the family ties are specified; in Palestine, the Jewish graves are above all family graves and lots of them were used for several generations (Hachlili 2005, 519, 524; 2007). The tomb is for the immediate family, but the extended family, such as the slaves and freedmen, could be buried inside, in particular in the regions where Roman influence was important (Rebillard 2003, 33). The few available osteological studies demonstrate indeed a well-balanced sex-ratio and access to the grave for young children (Hachlili 2005).

In short, our knowledge of Jewish burial customs depends on a limited number of archaeological discoveries, coming from the architecture of the tombs, the inscriptions and the symbolic motives engraved or painted on them (Rutgers 1998, 19; Künzl 1999, 43). If the diaspora of Rome, one of the largest diasporas, is one of the most known because of the long tradition of study of the Jewish catacombs in the *Urbs*, the archaeological material in other diasporas are much scattered and poor (Rutgers 1998, 139). As the diasporas are spread across the Mediterranean and connected to several waves of immigration, problems of geographical, but also chronological representativeness, can thus be evoked. The fact that only the monumental tombs are well identified as Jewish leads us to think that archaeologists do not know how to recognize simple graves belonging to the modest Jewish classes (Noy 1998; Stern 2013). This adds a bias: an over-representation of elites who could afford monumental tombs (Magness 2011). On the contrary, in the last decades a large number of tombs have been excavated in Palestine, and published, allowing some synthesis of the funerary practices during the Second Temple period and to a lesser extent, during late Antiquity (Hachlili 2005).

All these problems, as they shall be exposed below, will explain that few scholars really address the issue of the treatment of the body and the funerary gestures in Jewish diasporas. What we know about the primary deposits is quite limited. After death, the deceased were wrapped in rush mats or in shrouds (Hachlili 2005, 481). The deceased were always dressed; according to the wealth of the family, they could wear rich clothes or a simple cloth, but to be buried nude was considered a dishonour (Krauss 1934, 29). After the dead had been transported on a bier or a mattress to the cemetery, the corpse was placed in a wooden coffin or box (Hachlili 2005, 481); most young children were also buried in these containers (Krauss 1934, 29–31, 34). This is the most frequent mode of burial in the Palestinian sites of En Gedi (Hadas 1994) and in Beth Shea'rim (Mazar 1973, 222). More rarely these coffins could be of clay (Vitto 2011) or lead (Weiss 2010). The deceased were deposited on their back, with arms and legs extended (Hachlili 2005, 457). As known, cremation was forbidden by the Jewish law; when cremation deposits are discovered in graves or ossuaries, these are interpreted as a re-use of the grave by pagans, even if they leave the earlier deposits in situ (Avni *et al.* 1994, 216).

Contrary to the primary deposits, scholars have, for a long time, examined the question of secondary deposits for the Jews of Palestine, because of the rather frequent discovery of *ossilegium* in the burial caves. These containers, where the bones of the deceased were collected after decomposition (Hachlili 2005), were used in the late Second Temple period (1st century BC–1st century AD). Maintaining individuality became indeed fundamental in Jewish ritual, to permit the resurrection (Hachlili 2005, 524). Religious texts explained how to arrange the bones (Cohen-Matlofsky 1991). The bones in Palestine were grouped in different ways; they could be put in a stone, or more rarely, in a clay *ossilegium* (Kancel, 2009, 286), which sometimes carried the deceased's names and his family ties (Hachlili 2007, 263–7; Ilan 2007, 67). Wood containers are attested at the sites of En Gedi (Hadas 1994) and Beth Shea'rim (Mazar 1973, 223). However, the bones could also be grouped in the corner of a *loculus* or of the grave, or in small special *loculi*.

The Semahot (XII), the rabbinic mourning tractate (Zlotnick 1966), does not recommend the use of a shroud, because after its decomposition, the bones contained in it could mix with other individuals' remains (Krauss 1934, 10). Predominating in Jerusalem and its region (Kancel 2009), this practice seems to have been the standard in the 1st century AD, when the notion of individual place within the family took on more importance, subsequently, in the middle of the 3rd or in the 4th century, it became more sporadic and disappeared (Fine 2000; Hachlili 2005, 521–2; 2007, 277). Such secondary deposits were not maintained in the catacombs of Rome, but this is an argument *ex silentio* based solely on the lack of *ossilegia* (Krauss 1934, 2). The increasing dispersion of the Jews, however, led to the idea that the preservation of the deceased's bones was not relevant for the resurrection (Rahmani 1994, 205). Nevertheless, some containers for secondary deposits seem to have been found in the Gammarth catacombs, in Tunisia (Stern 2011, 333).

After the Second Temple period, the bones were placed in a central and common ossuary. In the grave of the family Eros, in the Beth She'arim necropolis, *ossilegia* dating from the Second Temple period rested on the ground and were covered by dislocated bones mixed with fragments of wooden coffins; this deposit, 0.50–0.60 m thick, was dated to the 5th century AD (Avni *et al.* 1994). However, some scholars, handling in a more or less detailed way the question of the varied practices around the secondary deposits, consider that the use of the collective ossuary and the *ossilegia* was contemporary, if only because not everybody could afford an *ossilegium* (Fine 2000, 74). Moreover, also a third stage in the mortuary treatment could have been practised: the bones placed in an *ossilegium* were subsequently shifted to a common ossuary, so that the *ossilegium* could be reused (Cohen-Matlofsky 1991). In fact, due to the long, continued use of the tombs, most of them contained both wooden coffins, *ossilegium*, and a common ossuary. This demonstrates the reluctance of Palestinian Jews to dispose of the bones of their ancestors from the family grave (Cohen-Matlofsky 1991), even when the grave was cramped or full.

Concerning the deposits of objects, it was in theory forbidden to feed the dead or to make offerings (Oberhänsli-Widmer 1998, 71). Furthermore, the burial should not be too expensive (Morris 1992, 118; Noy 1998, 83). Since the deceased was considered impure, objects used during the funeral ceremonies could not be reused in everyday life and were consequently deposited in the grave. Therefore, the artefacts discovered in the Palestinian graves are often limited to glass or ceramic containers, which were used to pour wine or perfumed oils on the bones during their gathering (Semahot, XII, in Zlotnick 1966), and to lamps used during the funeral ceremonies. A large number of them are often discovered in the graves, such as the

30 lamps found in Qiryat Tiv'on (Vitto 2011) or in burial cave 1 in Aceldama, dated to the 3rd century AD (Avni *et al.* 1994, 207). However, the archaeological reality shows that other kinds of objects were also deposited in the tombs. So, the presence of coins was not exceptional (Avni *et al.* 1994, 209; Greenhut 1994; Syon 2002). During the second half of the Second Temple period, the tombs could contain cosmetic utensils, spindles and whorls, modest jewels such as beads, bronze bells and box clasps (Hachlili 2005, 394–401). However, when a lot of artefacts, in particular jewels, are discovered, the tombs are not interpreted as Jewish graves (Vitto 2008a). The deposition of the goods in the tombs was as follows: the personal belongings were put in the coffin with the deceased, whereas containers, like bowls, craters, and *unguentaria*, were laid on the coffin or close to it (Hachlili 2005, 519). The personal objects of the deceased were later transferred with her/his bones to the common ossuary (Rahmani 1994, 193). The diachronic analysis of the goods' deposits in the necropolis of Beth She'arim revealed that oil lamps and objects were generally dated to before the middle of the 3rd century AD, although some artefacts were even dated to the Byzantine period (Weiss 2000, 225). From the 3rd century AD onwards, the Jewish graves contained fewer and fewer objects, reduced to the presence of lamps (Rahmani 1994, 205). On the funeral sites of diasporas, the data is very scarce. Many ceramic artefacts and lamps from the Jewish catacombs have not been recorded and analyzed (Rutgers 1998, 69; Stern 2011, 314). Studies on sarcophagi, gold glasses or marble chancel screens have shown that the Jewish objects were made in the same workshops as the objects from pagan graves and that Jews could use pagan objects and symbols, like amulets (Rutgers 1992).

Although no synthesis exists on the funeral practices of the Jewish diasporas and it is not always easy to make precise comparisons, the available data support a conclusion that the Jews of diasporas borrowed more from the local funeral practices than following the biblical rules (Rutgers 1992; Stern 2013). This phenomenon is maybe due to the fact that the funerals were a matter of family responsibility and not of the synagogue, the rituals being considered as a private choice (Davies 1999, 106). However, from the 3rd century AD, there is more evidence that the community was implicated in the funeral management, in particular to set up or organize the commemorations (Williams 1994, 171–3). Adopting the local burial customs is certainly not an isolated phenomenon but is also observed in other immigrant communities. Greeks settled in Alexandria continued to cremate their deceased ones, but some adopted the Egyptian practice of mummification (Empereur and Nenna 2001, 522–3). Even today, modern examples of Tunisian and Turkish immigrants show that the repatriation of bodies to their country of origin is the norm in the first few decades, before new funeral spaces are created in their country of settlement. This represents the symbol of a definitive establishment and integration (Chaib 1996).

Tomb 163d

General presentation of the tomb: some Jewish evidence

Description of the grave

Tomb 163d is situated in the North Necropolis of Hierapolis of Phrygia, above the road which led to Tripolis (Fig. 4.1). It is easily visible on the gentle hill slope and participates in

the monumental framework of the necropoleis skirting the city (Ronchetta and Mighetto 2007, 433; Laforest 2015). The tomb, oriented more or less north/south with the entrance in the south, belongs to a funeral complex consisting of one *bomos*, several sarcophagi resting on low platforms, and small chambers with unclear functions. The rectangular, saddle-roofed house-shaped tomb building itself, built in travertine ashlar masonry, is raised on a high podium and crowned by a saddle roof (Figs. 4.2–3). It belongs to a type developed in the 1st century AD (Ronchetta and Mighetto 2007). The tomb has chambers (3.15×2.75 m) on two levels, the lower one, in the podium of the tomb,

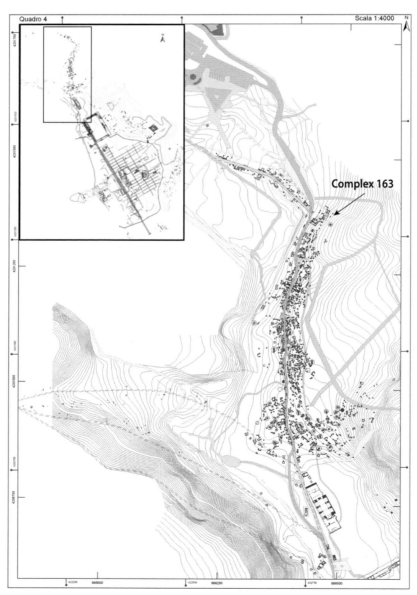

Fig. 4.1. Hierapolis. Map of the North Necropolis with position of tomb complex 163 signaled (modified after D'Andria, Scardozzi, and Spanò 2008, p. 55, quadro 4; courtesy of the Missione archeologica italiana a Hierapolis in Frigia).

Fig. 4.2. Hierapolis, North Necropolis. View of tomb 163d with surrounding sarcophagi, from west (Photo by J. R. Brandt).

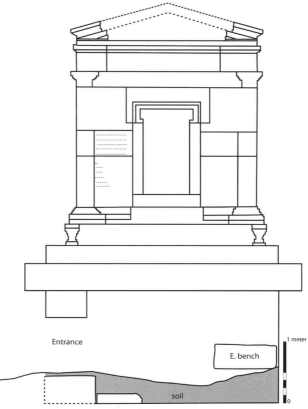

Fig. 4.3. Hierapolis, North Necropolis. The two superimposed chambers of tomb 163d from south (superior part integrated from Ronchetta and Mighetto 2007, 439 fig. 6; courtesy of the Missione archeologica italiana a Hierpolis in Frigia).

was partly subterranean dug in the bedrock. Both chambers have travertine benches along three walls in the normal Asia Minor fashion, the upper room having notches in the walls used as supports for a second row of elevated benches (not preserved). While the upper chamber was found empty, the subterranean was full of non-disturbed bones and soil. The chamber was indeed closed by a thick and sub-rectangular slab of travertine. The original access to this lower funerary chamber is unfortunately not documented, because the soil which covered the area in front of the opening, up to the current ground level, had been cleared without archaeological recording before our intervention.

Analysis of the inscriptions and discussion on the access to the grave

The façade of the tomb, decorated by finely carved lion pawns at its corners surmounted by simple corner pilasters and boasting a central, large door opening with finely profiled jambs and lintel. On the left, between the pilaster and the door jamb runs a first inscription:

Τὸ ἡρῷου σὺυ τῷ
ὑποκειμένῳ Θέμα-
τι κὲ τῷ περὶ αὐτὴν

τόπῳ Αὐρηλίᾳ Κο-
δρατιλλᾳ Αὐρ. Μαρ-
κέλλου καὶ Αὐρη-
λίας Πυρωνίδος καὶ
Αὐρηλίᾳ [—]
[—]’Ιουδαίων.

This text can be translated as: 'The heroon with the room situated below and the area all around belongs to Aurelia Kodratilla, to Aur. Markellos, and to Aurelia Pyronis, and to Aurelia [...], Jewish' (Ritti 2007, 606). It is thus specified that the owners possessed not only the tomb, that is to say both chambers, but also the space all around. As is often the case in Hierapolis, it is a common property (Ritti 2004, 486), because this tomb belongs to a group of four people: a woman (Kodratilla), two men (Markellos and Pyronis), and a person of unknown sex (due to bad preservation of the stone her/his name is illegible). The first owner to be named, Aurelia Kodratilla, is a woman; this suits Trebilco (1991, 231) well, who considers that the Jewish women had a certain financial independence and played a more active role in the public life of their family than was the case for non-Jewish women (Ritti 2004, 486). The persons named in the inscription do not reveal their family ties, but define themselves as Jews and thus share the same religion/ethnic group (Cohen 1999). The shape of the letters, but especially the onomastic Aurelia, allow us to date the inscription to the 3rd century AD, most probably to after 212 AD, when the *Constitutio Antoniana* gave Roman citizenship to all free inhabitants of the Roman Empire, and many adopted the family name of Emperor Caracalla, Marcus Aurelius Antoninus. In any case, the inscription cannot be earlier than the reign of Marcus Aurelius (161–180 AD) (Ritti, 2004, 464). Moreover, the custom to assert property of a tomb by listing the different owners becomes much more common in the 2nd and 3rd centuries AD (personal communication, T. Ritti, 2014). Two centuries after the construction of the tomb 163d, an inscription is thus engraved on the monument. The onomastics, as we have just seen, sweep away the possibility that the mentioned persons are the original owners, whose memory, in this case, should have survived for two centuries before being engraved on the stone. It is, therefore, highly likely that this inscription thus reflects a purchase of the tomb. A similar situation has recently been recorded in one of the tombs of the North-East Necropolis (Ahrens and Brandt 2016, 402). As for the first 'occupiers' of the tomb, no epigraphic information is preserved.

The second, less complete inscription, on the block immediately below the first one can be deciphered as follows: 'will be buried [... the son?] of Aureli [...] and of Aurelia Kodratilla and the sons of Ioustos; also me too, Doros, the father of Ioustos, but after my death that my sons have the right, and if somebody violates he will give to the very holy treasury [...] money' (Fig. 4.4). The writing is slightly different from the first inscription: it is less careful, less classic, but it is impossible to date it precisely. The mention of Aurelia Kodratilla's sons and the expanded access to the grave demonstrates that the inscription is later than the first one, certainly within a rather short lapse of time of not more than one or two generations. Two new males are named, Doros and Ioustos, while a number of individuals are mentioned under the term of their 'sons'. It seems logical to assume that Aurelia Kodratilla's sons also were Jewish, but it is not certain that Doros and Ioustos as well as their sons were. As is often the case in Asia Minor, the last part of the inscription threatens with fines to anyone who exceeds this access right or who would violate the grave (Ritti 2007, 606).

Finally, a third inscription is situated on the west wall, on the base of the *crepidoma* and thus above the entrance of the lower chamber. A candelabrum with nine branches is represented, below which are engraved three letters: the abbreviation of [*eu*] *log* [*ia*], which means 'blessing'

Fig. 4.4. Hierapolis, North Necropolis. The two inscriptions on the south façade of the upper chamber of Tomb 163d (Photo by J.R. Brandt)

(Fig. 4.5). Contrary to Ritti (2007, 606), we do not think that the candelabrum is a *menorah*, which has seven arms, but a *hanoukkia*, a ritual nine-armed candlestick used during the Jewish festival of the lights, Hanoucca (Ludwig 2004, 47). As for the 'blessing', it was until now unknown in Hierapolis, but is very frequent in Jewish inscriptions (Ritti 2007, 606). Neither the *hanoukkia* nor these three single letters can be dated, but it is known that the representation of the symbolic candelabrum is particularly common in the Greco-Roman world from the end of the 2nd century AD (Miranda 1999, 133).

The use of the lower chamber of tomb 163d

When the subterranean chamber was discovered in 2001, the benches, but also areas under and between them, were packed with bones, which were more or less covered with soil (Anderson 2007) (Fig. 4.6). The outer edge of the benches and the centre of the east bench were covered by just one layer of scattered bones, but up against the walls, the deposits were 0.30 m thick. In addition, the floor, both under the benches and in the space between them (the 'central space'), was covered with a 0.20–0.50 m thick layer of soil, also this was full of bones. The rocky floor itself did not carry any traces, in the shape of a pit or otherwise, of an ossuary, but a square slab of stone forming a step had been put by the entrance to make entry easier (Fig. 4.3). The Minimum Number of Individuals (MNI), calculated on the frequency of the preserved femurs (which gave the highest score), revealed that at least 293 people were represented in the lower chamber of the tomb.

Chronology

As it is often the case in collective graves, for several reasons it has not been an easy task to establish the chronology of the deposits. This is first of all due to the continuous reorganization of the deposits including datable objects and the mixture of the soils, but also the complex interweaving of skeletons, the absence of sediments, and thus stratigraphy in some areas. Radiocarbon analyses were taken of some selected articulated skeletons; because of poor bone conservation, only 13 samples gave a dating. They all came from skeletons on the south and north benches, although a skeleton (no. 31) overflows onto the east bench (Fig. 4.7).

According to these (2 sigma calibrated) the depositions range from 27 BC to 604 AD in date. As the purchase of the tomb dates to the 3rd century AD and as seven of the C14-dates reveal that the grave was still in use during the 6th century (skeletons nos. 4, 7, 6, 9, 18, 19, and 21), the Jewish ownership of the tomb could have been the longest one, lasting more than three centuries, subject to

the chamber not being reused by others who, at the present state of research, have not yet been identified. Indeed, the C14-datings and the stratigraphic analysis which resulted from it demonstrate that the majority of the deposits date to this late phase. However, two skeletons (nos. 11 and 31), resting directly on the benches, are clearly previous to the second phase of use. To this early phase could also belong skeletons nos. 13, 14, and 29. As this last skeleton directly overlaid the skeletons nos. 15, 16, 17, and 128, these skeletons may also date to the same early phase. A few layers may belong to the 2nd century, confirmed by a coin dated to the reign of Marcus Aurelius found on the bottom of the central area of the tomb. Even so, the majority of bodies and bones contained in this central space were all deposited after this date. The soil layers of this second phase are rather mixed; in fact, some artefacts from the Augustan period have been found in layers above those with artefacts from the 3rd century. The reassembly of many fragments, coming from different areas under the benches, into more or less complete objects may demonstrate that the filling of the floor areas happened within the same periods of time, from the early to the last phases.

Modes of burial: the primary deposits

Among the commingled bones, there were numerous articulated skeletons, in all 123. For our analysis it was first of all necessary to distinguish the primary deposits from the deposits in secondary position, which was done by observing the labile joints of the skeletons and their maintained anatomical logics. In some articulated segments, including only two or three bones, it was sometimes difficult or even impossible to determine with any certainty the nature of the deposit. In the end 67 skeletons were identified as primary

Fig. 4.5. Hierapolis, North Necropolis. Hanoukkia *engraved above the lower chamber of Tomb 163d (Photo by J.R. Brandt).*

Fig. 4.6. Hierapolis, North Necropolis. Interior view of the inferior chamber in 2003 (Photo by T. Anderson; courtesy by the Missione archeologica italiana a Hierpolis in Frigia).

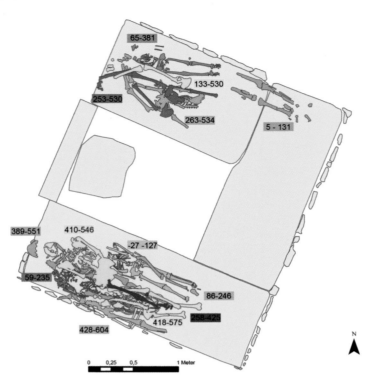

Fig. 4.7. Hierapolis, North Necropolis. Plan with the C14-dated skeletons (Drawing by CL).

deposits, 23 as hypothetical primary deposits, 7 deposits as in a secondary position and 24 as indeterminate (indet.). Their archaeothanatological analysis, which corresponds to the taphonomical study of the skeleton, allowed in the next instance to reconstitute the modes of burial and how the bodies and skeletons were managed (Duday 2009).

The large presence of primary deposits demonstrated that the chamber was not used as an ossuary, but that the deceased were brought into the subterranean chamber as newly dead corpses. Found in every zone of the grave, many skeletons, as already observed above, were fully articulated, others preserved as anatomical bits and pieces, a mixed find situation which demonstrates that the depositions were not simultaneous and that the reorganization of the skeletons had been numerous in each phase of interment (Fig. 4.8). The great majority of the most complete skeletons lay on the south and north benches. In general, the individuals were deposited on their back, although 12 examples show that a deposition on the stomach was possible (Fig. 4.9). The position of the forearms was, most of the time, along the body or placed on the abdomen whereas the lower limbs were always fully extended, not bent or crossed. The depositions followed the orientation of the bench on or under which the individual was placed, but there was no standard orientation and some bodies were even arranged so that the head on one skeleton lay next to the toes of another. A westerly orientation with the head looking east dominated,

best seen on the south bench, while on the north bench the orientation was more or less equally divided between west and east; on the east bench the orientations of the skeletons were north and south without a definite pattern. In the central space and under the south bench, the skeletons were more frequently oriented to the east, unlike under the north bench where a west orientation dominated; finally, under the east bench, all identified deceased were deposited with the head to the south.

Some skeletons showed some disturbances due to the decomposition of skeletons underlying new depositions, as, for example, the arm of skeleton no. 24 which fell into the rib cage of the underlying skeleton no. 25 when the thorax of this last skeleton collapsed under the weight of skeleton no. 24. These kinds of disturbances, as well as the shift of anatomical parts, from a taphonomical point of view, demonstrate well that the deposits of dead bodies could happen in a short period of time. Furthermore, the collapse of the body volumes (thoracic cage, gluteal masses) and the absence of unstable equilibria, as movement outside of the initial buried volume of the body, showed that all bodies decomposed into a void, meaning that even the bodies under the benches and in the central space were buried without a soil cover. The sediments in the tomb were thus not formed by human action, but by gradual infiltrations of fine soils over time after the moment of burial. As already mentioned, the disturbance of single bones or anatomical groups of

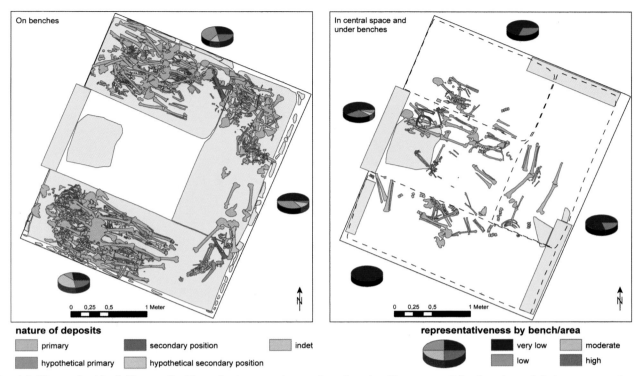

Fig. 4.8. Hierapolis, North Necropolis. Plan of primary and secondary deposits: The nature of the deposits and their representativeness (Drawing by CL).

bones was caused by the decomposition of the underlying skeletons. However, certain anomalies of some bone/joint positions and collapses as well as the observation of some lateral constraint effects indicate that the deceased, in many cases, were deposited in rigid containers. The impressive number of nails (5–9 cm long), but also of iron brackets carrying traces of oxidized wood, suggests that burials for some of the deceased were made in wooden coffins. At the same time, on the north bench, several skeletons presented taphonomical features which cannot be due to rigid containers, rather to soft ones, such as textiles or mats: the most convincing signs being legs straightened against the wall and/or strongly constricted shoulders.

Deposits in secondary positions and reorganization of the chamber

Deposits in secondary positions, due to physical interventions and reorganizations after the decomposition of the body, can take several forms. The body can remain individualized, i.e. more or less maintaining an anatomical logic, or, on the contrary, it can appear as an osseous heap constituted of disarticulated bones.

Only seven individualized depositions were positively identified as being in secondary positions. Some of them were indeed squeezed too close to a wall to be compatible with an *in situ* decomposition of the body. Other secondary deposits testify to an obvious attempt to group the bones of a single skeleton, as, for example, the gathering of ribs or the piling of tibiae and fibulas belonging to the same individual. These deposits have been found in every part of the funeral chamber (Fig. 4.8). In addition to these secondary deposits should also be mentioned the case of a young individual (6–18 months), whose bones were pushed up against the doorstep of the central space. Finally, the cremated remains of an adult were found slightly scattered under the east bench, but as cremation is not a Jewish practice (Avni *et al.* 1994, 217), it is reasonable to assume that these remains belonged to the first phase of occupation of the tomb.

Dislocated bones in the form of osseous heaps constitute the absolute majority of the human remains of the lower chamber of tomb 163d. In light of the density and the complexity of these deposits, correspondence factor analyses have been developed in order to compare each bench and the central space (Fig. 4.10). They demonstrated some characteristics according to areas in the tomb. The diagram shows that long bones and girdles were common on the south and north benches, where a heap actually was found in the north-west corner. The pattern is very different on the east bench with many skulls piled up in the north-east and south-east corners of the tomb, whereas the lower limbs occupied the centre of the bench, where leg bones were heaped. Small bones (hands, feet and in a lesser quantity,

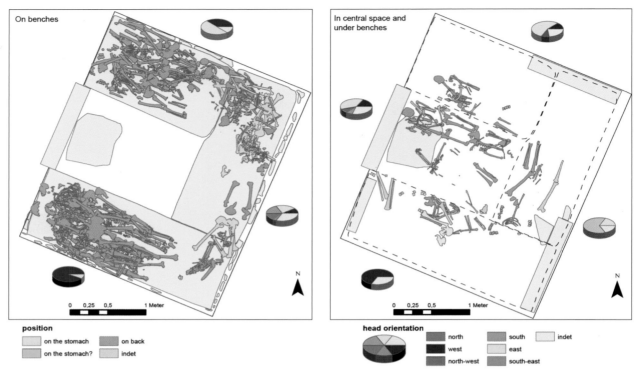

Fig. 4.9. Hierapolis, North Necropolis. Plan of primary and secondary deposits: Position of the skeletons and the orientation of the heads (Drawing by CL).

vertebrae) are over-represented in the central space and were probably thrown out of the benches. As for the zones under the benches, they presented an over-representation of voluminous bones, which seem to have been put away to save room (Fig. 4.10).

Grave goods

More than 1,260 artefacts or fragments of artefacts were discovered in all parts of the chamber, albeit only 20% of them were discovered on the benches. The artefacts included 22 ceramic *unguentaria*, 18 glass flasks, two small cusps, three jars, five lamps, a bronze bottle, six coins, a mirror, four intaglios, seven earrings, two rings, three *bullae* and five small bells, 34 game pieces and at least five pins, seven needles, and three spindles. They were lying among both disarticulated bones and articulated skeletons. As already noted, the artefacts were widely reshuffled and commingled, in particular in the central space and under the benches where early material overlaid late objects, demonstrating well how the depositions had been disturbed already in Antiquity. It was in no way possible, therefore, to attribute one or more personal artefacts to one particular individual.

None of these artefacts carry a Jewish symbol. One hundred and five artefacts were dated, but, with the exception of five artefacts (a lamp, two *unguentaria*, a coin, and a gemstone), no object could be attributed with absolute certainty to the Jewish phase of the tomb's lifetime, though

this does not exclude the possibility that more objects really belonged to the late phase of tomb. It is certainly possible that some prestigious objects (as, for example, intaglios, precious alabastra, and coins) had been in circulation for a long period of time before they ended up in the tomb. Furthermore, some objects, either very generically dated, or badly preserved and very fragmented, could belong to the Jewish phase. Consequently, it is impossible to conclude firmly that there were no objects deposited during the Jewish phase, but a serious decrease in the deposition of objects compared with the early phase is observed. For some reason the vessels and lamps used during the funerary ceremonies and thus rendering them impure were not left inside the burial chamber. This conclusion seems well consistent with what took place in other Jewish tombs of the same time, where the deposits of goods decrease especially after the 3rd century AD.

Discussion in the local context: Hierapolis and Asia Minor

The Jewish 'community' of Hierapolis

The ancient texts and the inscriptions depict a certain vitality and stability of the Jewish communities in Asia Minor. Except for some local episodes of tension, the authorities and the local population maintained rather good relationships with the Jewish diaspora, which kept a strong Jewish identity

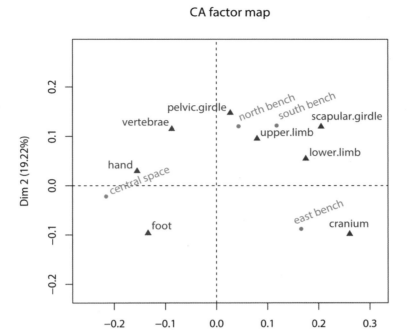

Fig. 4.10. Hierapolis, North Necropolis. Correspondence analysis of chamber areas and anatomical segments (Drawing by CL).

(Trebilco 1991). The tolerance of the Romans, who granted the Jews of Asia Minor some privileges (Cohen 1999, 58), indeed guaranteed them the maintenance of the most important aspects of their identity.

Most of the knowledge about funerary customs of the Jews, who lived at Hierapolis, comes from the study of the corpus of 23 Jewish inscriptions studied by E. Miranda (1999), and in the analysis of one particular case by P. A. Harland (2006). Few inscriptions from Hierapolis explicitly refer to a 'community' or an 'association', so this convention of language rather included all people who identified themselves as Jewish in Hierapolis (Harland 2006, 223). The Jewish community of Hierapolis was quite large and inscriptions portray a quite well-assimilated group in the Hierapolitan society, both culturally and organizationally. No archaeological evidence of a synagogue has been found, but it seems that the Jewish community must have possessed archives (Trebilco, 1991, 257; Harland 2006, 227). The Jews who settled in the Lykos valley played an important role in trade (Şimşek 2006) and some of them belonged to the wealthy association of dyers (Verzone 1987, 116).

From a funerary point of view, the Jewish graves were not concentrated in one sector of the necropolis, but instead were mixed with pagan tombs: the Jewish community of Hierapolis, therefore, had no reserved funerary space. Indeed, in Asia Minor in general, it has not been demonstrated that public authorities reserved particular areas for the Jews; however, at

Tlos in Lycia, a private benefactor, Ptolemaios son of Leukios, built a cemetery for the Jews of the city (Noy 1998, 81; Cohen 1999, 1–2).

In addition, the Jewish tombs in Hierapolis are similar to the pagan ones. The content of the inscriptions demonstrates that Jews often adopted local Greek names and they used the same customs in how to use and to protect the grave, as seen elsewhere in Asia Minor (Miranda 1999). With regard to the commemorative practices, it would seem that the Jewish inhabitants of Hierapolis could follow the Graeco-Roman rituals. At the turn of the 2nd century AD, for example, P. Aelius Glykon, both a Jew and a Roman citizen, with his wife Aurelia Amia, requested the guild of the carpet weavers that during two yearly Jewish festivals, their family sarcophagi should be crowned, a typical Graeco-Roman practice (Harland 2006). From Hierapolis are known four cases, a tomb and three sarcophagi, in which Jewish families purchased a burial from pagan Romans. The inscription on the tomb is damaged, but the inscriptions on the three sarcophagi are still legible; in none of the three cases was the original owner a Jew (Miranda 1999).

Burials and mortuary management in Ancient Asia Minor

In general, knowledge about burial customs and mortuary sequences are not analyzed in detail. Perishable containers are little known; sometimes, publications mention traces of wood and nails, suggesting the use of wood coffins, as in Attaleia, where the nails and wood residues were systematically discovered around the skeletons (Tosun 2009). But some scholars, referring to the small size of the burial chambers and the narrow and low openings of lower chambers, consider that it would have been difficult to insert a rigid container into them; consequently they judged the use of shrouds more likely (Tomasello 1991, 221; Spanu 2000, 172). Some rock-cut pillows on the benches, as, for example, in Elaiussa Sebaste, suggest that there were no rigid containers (Equini-Schneider 2003, 465), but their existence is, nevertheless, demonstrated in other chamber tombs. In tomb C128, in the North-East Necropolis of Hierapolis, on top of the skeletons, a lead blanket was discovered, which, combined with the discovery of numerous nails and wood fragments, certainly was the lid of a wooden coffin. With regard to soft containers, the only study, based on archaeothanatological methods, to confirm the use of shrouds in Asia Minor is the one on the late Roman necropolis of Porsuk (Blaizot 1999); the shrouds are highlighted by an extreme constriction of the bodies with, in particular, the internal rotation of the femorae. As for body position, the skeletons were generally buried on their back, with lower limbs extended, and, according to the site of Attaleia, in collective graves the bodies were deposited parallel to each other and to the walls of the grave (Tosun 2009, 196). Lastly, no particular orientation was favoured, as shown in the neighbouring site of Eudokias (Yalçınsoy and Atalay 2012, 226).

Most of the burial chambers have a limited size, which in one way or the other would have influenced the number of dead it was possible to bury in them. This observation raises the question on available space and on the deserved respect of the integrity of the first skeletons deposited in the tomb. Sometimes the management of the tomb remains faithful to the deceased's wishes. At Limyra, the excavation of a two-bench-chamber brought to light a single skeleton laid out on the right bench, while nine skeletons rested on the left one (Blakolmer 1989). This was interpreted as a sign of respect for the owner's wish not to see a body placed above him, a situation frequently signalled in tomb inscriptions (Henry 2003, 19). However, when from the end of the Hellenistic period the tombs were opened for wider access it was not possible to secure an undisturbed, perpetual rest for the original owner of the tomb. It became clear that the areas reserved for the dead could only be for a short period of time and that the reoccupation of the burial benches was perfectly acceptable. Therefore, more than the re-allocation of the bodies, it was, as shown in inscriptions, the non-authorized burials that people tried to avoid (Schweyer 2002, 40–1). Furthermore, the inalienability of the grave is limited to the conditions of use imposed by the founder (de Visscher 1963, 72), conditions which most often

ended after her/his line of heirs had died out or started a new family tomb, at which point the original tomb could be put up for sale. Thus, when the occupation of a burial place by the original families ends, monumental tombs and sarcophagi are frequently reused by others (Ritti 2006, 44). The time between two occupations of a tomb could be rather short (Schweyer 2002, 59) and the periods of re-use, as documented by grave goods, inscriptions, and C14-dates of the buried skeletons, could be many, over a long period of time (Cavallier 2003, 207; Zoroğlu 2012, 36; see also Korkut and Uygun, this volume). However, the question of the legality of these re-uses arose in some cases. If some new owners left the name of the original owner as a sign of respect for him/her, others erased it (Ritti 2004, 566–9). Besides, from the 3rd century AD onwards, the custom of giving precise indications on the use of the tombs and of the transmission of the tomb disappeared, most likely because lots of old tombs were used more or less legally (Ritti 2004, 456). After the 4th century AD at Hierapolis, perhaps after the devastating earthquake which hit the city (presumably) in the 360s, many family sarcophagi and tombs may have gone out of use and the control of the legal situation of tombs may no longer have been considered a priority (Ritti 2004, 567). So, if nobody maintained the tomb, one may question in the case of recorded later deposits, if someone had taken advantage of the less strict controls of burials and used the tomb illegally, i.e. contrary to the original legal restrictions.

In modern scholarly literature few, if any, examples can be cited on the management of the deceased after burial, but many situations can be observed in the tombs themselves. In a Hellenistic subterranean tomb in Kelenderis, for example, an adult and a child were heaped together on either side of the entrance to the tomb (Zoroğlu 2012). At Patara seven skulls were stored in a rock-cut niche (İskan and Çevik 1995). The skulls could be grouped in a corner of the tomb (Yalçınsoy and Atalay, 2012, 36), or, as again at Patara, the bones and objects were simply pushed to the back of the benches (İskan and Çevik 1995). In some cases, containers are used for collecting bones. At Iasos, a secondary deposit in an amphora has been attested (Tomasello 1991, 219), but most often, stone containers, *osteothekai*, are mentioned for this purpose. The bones, after decomposition, could also be stored under the benches or in a pit cut in the floor. In addition, the discovery of nails, brackets, and traces of wood indicates the existence of wooden boxes used to store the bones.

Concluding remarks
On funerary gestures

The archaeothanatological analysis of the articulated skeletons has highlighted two modes of burial in the tomb 163d: in rigid and wooden containers and in soft ones (shroud or mat). These are both used by the Romans in Asia

Minor and by the Jews in Palestine, the latter influenced by the Egyptians as a substitute for coffins (Hachlili 2005, 515). The position of the majority of the deceased in tomb 163d follows the standard pattern, on the back with legs and arms extended, or, in some cases, with the arms placed on the abdomen, a customary position known from other places both in Asia Minor and Palestine. However, in tomb 163d some were deposited on the stomach, in a position that is not mentioned, as far as we know, in publications on Jewish or micro-Asiatic graves.

No individualized and complete collection of bones was observed in tomb 163d, even if *osteothekai*, whose function could be the same as the *ossilegia* one, are widespread in Asia Minor (Thomas 1999; Thomas and Içten 2007). According to the taphonomic analysis no body with a preserved individuality was found in a secondary deposit in a wooden chest or in any other kind of perishable container. In the floor of the tomb there is no common pit either, this space being gradually filled during the Jewish phase with bodies as much as with dislocated bones. The south and north benches were the most favoured to receive primary depositions (which are later and more complete than the skeletons in other areas), while on the east bench, several primary deposits were removed before some voluminous bones were replaced there. A similar situation was also found at Tiberias, in Israel. A subterranean chamber of a Jewish mausoleum, built in the late 1st or 2nd century AD, contained several gradually piled bodies showing that they were pushed back to make space for new deposits, while another area contained both primary deposits and collected bones (Vitto 2008b).

The example of the grave 163d confirms the loss of importance of individuality. In the Jewish religion, resurrection is collective and the whole Jewish people will resurrect (Oberhänsli-Widmer 1998, 78), so it may, therefore, be that for the Jews keeping his/her individuality was not a necessary condition to resurrect, as the resurrection is anyhow assured. For instance, although the anthropological study of skeletal remains contained in *ossilegium* are not frequent (Kancel 2009, 291), the few anthropological studies published for Jerusalem and Jericho show that 48% of the ossuaries contain an individual, 37% contain two or three and two ossuaries will contain up to eight and 11 individuals (Kancel 2009, 291–2). In other cases, some bones were duplicated or were missing (Arensburg and Smith data in Hachlili and Killebrew 1999). Even when individuality seems to be a fundamental concept in the Second Temple period, half of the ossuaries include more than one individual and the care to collect the bones seems relative, since some bones are missing. To move bones has then as priority to respond to practical considerations concerning the management of the funeral chamber over a long time. The skeletons could remain 'abandoned' and incomplete, as in the tomb 163d.

The Jews from diasporas could here have been influenced in their belief by Graeco-Roman conceptions of death (Stern 2013, 278), where the integrity of the body is not debated.

However, in order to deepen our understanding of the modes of burial and the management of human remains after decomposition in Jewish communities, more precise comparisons with the local groups are necessary. Future excavations of graves shall have to go into more detail, for example, making systematic analyses of perishable containers of skeletons, establishing chronological types, as well as cross-checking information on containers, position, and orientation with the biological data. It is not enough to find non-looted tombs, the research also requires adequate methods of excavation and analysis.

On the purchase of the grave and the management of the first occupants' remains

The information yielded by grave 163d is in particular interesting with regard to the way the Jewish group chose to manage the bones of the first occupiers, for whom there is no evidence to indicate that they were Jewish. According to the stratigraphic data, when the Jewish group took over the tomb, in the 3rd century AD, they cleared the benches. Yet, radiocarbon dates of the skeletons combined with the date of tomb artefacts suggest that they left *in situ* some of the original occupiers of the tomb. In particular, some discreet remains of primary deposits (composed of a small set of articulated bones, as, for example, a few ribs) were found directly on the benches and could well have belonged to the first group. In some cases, the skeletons lying up against the walls (for example, skeleton no. 31) were simply less easily accessible. Likewise many *unguentaria* from the Augustan period were found in the corners of the chamber, on the benches, as if pushed to the side. In other cases, nevertheless, the deceased were close to the edge of the benches and were a priori already skeletonized (for example, skeleton no. 11; for the use of the term, see Duday and Guillon 2006). To sum up, the Jewish group made the choice not to clear the benches systematically. The north bench showed only few vestiges from the first phase: it was the most thoroughly cleaned bench. In the cleaning process many artefacts (datable from the Augustan period to the 3rd century AD) were placed, together with the bones, under the benches or in the central space. Consequently, the cleaning was certainly gradual, depending on the arrival of new corpses; some artefacts from the early phase were found above some burials belonging to the Jewish phase. Leaving in the tomb the bones of the previous owners together with their belongings, even if shuffled about, is coherent with the Roman burial system, which looks upon the grave as a *locus religiosus*. Throwing the bones outside would have been a criminal act according to the Imperial legislation (de Visscher 1963, 54; Ritti 2004, 527).

Identities and access to the grave

The management of the skeletons of the previous owners, as indicated by the second inscription, have been caused by an enlargement and change of funerary access to the tomb, whereby the depositions increased notably, bringing the total minimum number of individuals identified to 293. As expected for a family tomb, the depositions include men, women, and children. Even during the second phase of occupation of the grave it is impossible to know if the tomb only hosted Jews. Traditionally it is believed that a non-Jew could not be buried in a Jewish tomb (Davies 1999, 108; Rutgers 1995); yet excavations in Jerusalem of some tombs carrying non-Jewish names may challenge this idea (Avni *et al.* 1994, 214). However, in most cases, these names are suspected to belong to 'Gentiles' who were supportive of the Jewish religion.

In any case, neither the rabbinical laws nor Semahot contain rules separating Jews and non-Jews. In Aceldama, a grave presented an opposite situation to that of tomb 163d at Hierapolis. The tomb was owned by the 'Gentiles' who reused the grave in late Antiquity and did not disturb the vestiges of the earlier, late Second Temple period. They were identified as 'Gentiles' because on one hand, they practised cremation; the burnt and charred bone remains were placed in a Second Temple-period ossuary (which was the only disturbance of the grave), while on the other hand, jewellery and coins were deposited with the deceased. Since archaeologists balk at interpreting a burial as a Jewish one when it contains some jewels and coins, it is argued that these objects belonged to the 'Gentiles' (Avni *et al.* 1994).

The excavation and analysis of the lower chamber of tomb 163d at Hierapolis has, despite three major difficulties (firstly, a clear distinction between the tomb's two phases of use; secondly, the assertion that the tomb was not reused after the occupation by the Jewish family; thirdly, the lack of comparable, published material from Hierapolis), brought forward new evidence on funerary gestures, especially on modalities of tomb purchase and management of a collective tomb during late Antiquity and the proto-Byzantine period. This study highlights that a Jewish group could accept a great proximity between pagans and Jews in terms of mortuary management. In that sense, the data from the tomb 163d are in accordance with and complement the epigraphic context which testifies to the not insignificant participation of Jews in the organization and evolution of the funerary landscape of Hierapolis.

Acknowledgments

The authors thank the Maison des Sciences de l'Homme d'Aquitaine and the Région Aquitaine, which funded the present research works. They are also grateful to Prof. Francesco D'Andria, to have entrusted them the responsibility of the excavation of several funeral structures in Hierapolis. They express also their gratitude to the different specialists who graciously studied the objects discovered in tomb 163d. Finally, the authors thank sincerely the editors of this volume, whose *Thanatos* project has provided precious support, at many levels, to the study of tomb 163d.

Bibliography

Ahrens, S. and Brandt, J. R. (2016) Excavations in the North-East Necropolis of Hierapolis 2007–2011. In D'Andria, Caggia, and Ismaelli (eds.), 395–414.

Anderson, T. (2007) Preliminary osteo-archeological investigation in the North Necropolis. In D'Andria and Caggia (eds.), 473–93.

Avni, G., Greenhut, Z., and Ilan, T. (1994) Three new burial caves of the Second Temple period in Aceldama (Kidron Valley). In Geva (ed.), 206–18.

Blaizot, F. (1999) L'ensemble funéraire tardo-antique de Porsuk: approche archéo-anthropologique (Ulukişla, Cappadoce méridionale, Turquie). Résultats préliminaires. *Anatolia Antiqua* 7, 179–218.

Blakolmer, F. (1989) Die Grabung in Nekropole V. *Kazı Sonuçları Toplantısı* 11.2, 187–8.

Cavalier, L. (2003) Nouvelles tombes de Xanthos. *Anatolia Antiqua* 11, 201–14.

Chaib, Y. (1996) La mort sans frontière: sépulture et enterrement des musulmans en Europe. In J.-L. Bacqué-Grammont and A. Tibet (eds.) (1996) *Cimetières et traditions funéraires dans le monde islamique, Istanbul, 28–30 septembre 1991*, 135–48. Istanbul, IFEA – Société d'Histoire Turque.

Cohen, S. (1999) *The beginnings of Jewishness: Boundaries, varieties, uncertainties.* Los Angeles, University of California Press.

Cohen-Matlofsky, C. (1991) Controverse sur les coutumes funéraires des juifs en Palestine aux deux premiers siècles de l'Empire romain. *L'Information historique* 53, 21–6.

D'Andria, F. and Caggia, M.-P. (eds.) (2007) *Hierapolis di Frigia, I. Le attività delle campagne di scavo e restauro 2000–2003.* Istanbul, Ege Yayınları.

D'Andria, F., Caggia, M. P., and Ismaelli, T. (eds.) (2016) *Hierapolis di Frigia, VIII. Le attività delle campagne di scavo e restauro 2007–2011.* Istanbul, Ege Yayınları.

D'Andria, F., Scardozzi, G., and Spanò, A. (2008) *Hierapolis di Frigia II. Atlante di Hierapolis di Frigia.* Istanbul, Ege Yayınları.

Davies, J. (1999) *Death, burial and rebirth in the religions of antiquity.* New York, Routledge.

Duday, H. (2009) *The archaeology of the dead: Lectures in archaeothanatology.* Oxford, Oxbow Books.

Duday, H. and Guillon, M. (2006) Understanding the circumstances of decomposition when the body is skeletonized. In A. Schmitt, E. Cunha, and J. Pinheiro (dir.), *Forensic anthropology and medicine: Complementary sciences from recovery to cause of death*, 117–57. Totowa, Humana Press Inc.

Empereur, J.-Y. and Nenna, M.-D. (eds.) (2001) *Nécropolis 1* (Etudes Alexandrines 5). Cairo, Institut Français d'Archéologie Orientale.

Equini-Schneider, E. (2003) *Elaiussa Sebaste, un porto tra Oriente e Occidente II.* Rome, L''Erma' di Bretschneider.

Fine, S. (2000) A note on ossuary burial and the resurrection of the dead in 1st-century Jerusalem. *Journal of Jewish Studies* 51.1, 69–76.

Geva, H. (ed.) (1994) *Ancient Jerusalem revealed*. Jerusalem, Israel Exploration Society.

Greenhut, Z. (1994) The Caiaphas tomb in North Talpiyot, Jerusalem. In Geva (ed.), 219–30.

Hachlili, R. (2005) *Jewish funerary customs, practices and rites in the Second Temple period* (Supplements to the Journal for the Study of Judaism, 94). Leiden, Brill.

Hachlili, R. (2007) Funerary practices in Judaea during the times of the Herods: The Goliath family tomb at Jericho. In Kokkinos (ed.), 247–78.

Hachlili, R. and Killebrew, A. E. (1999) *Jericho: The Jewish cemetery of the Second Temple period*. Rome, L''Erma' di Bretschneider.

Hadas, G. (1994) Nine tombs of the second temple period at En Gedi. *Atiqot* 24, 1–75.

Hadas-Lebel, M. (2011) Les juifs en Europe dans l'Antiquité. In P. Salmona and L. Sigal (eds.) *L'archéologie du Judaïsme en France et en Europe*, 43–50. Paris, La Découverte.

Harland, P. A. (2006) Acculturation and identity in the Diaspora: A Jewish family and 'Pagan' guilds at Hierapolis. *Journal of Jewish Studies* 57, 222–4.

Henry, O. (2003) *Considérer la mort: de la protection des tombes dans l'antiquité à leur conservation aujourd'hui*. Istanbul, Institut français d'études anatoliennes Georges Dumézil.

Ilan, T. (2007) Ossuaries of the Herodian period. In Kokkinos (ed.), 61–70.

İşkan, H. and Çevik, N. (1995) Die Grüfte von Patara. *Lykia* 2, 187–216.

Kancel, D. (2009) Les ossuaires juifs au tournant de notre ère. Paris, École Pratique des Hautes Études, unpublished thesis.

Kokkinos, N. (ed.) (2007) *The world of the Herods. Vol. I of the International Conference 'The World of the Herods and the Nabataeans' held at the British Museum, 17–19 April 2001*. Stuttgart, Franz Steiner Verlag.

Korkut, T. and Uygun, C. (this volume) The sarcophagus of Alexandros, the son of Philippos: An important discovery in the Lycian city of Tlos, 109–120.

Krauss, S. (1934) La double inhumation chez les Juifs. *Revue des études juives* 97, 1–34.

Künzl, H. (1999) *Jüdische Grabkunst von der Antike bis heute*. Darmstadt, Wissenschaftliche Buchgesellschaft.

Laforest, C. (2015) Tomba 163d. In Scardozzi (ed.), 99.

Ludwig, Q. (2004) *Le Judaïsme*. Paris, Eyrolles.

Magness, J. (2011) Disposing of the dead: An illustration of the intersection of archaeology and text. In A. M. Maier, J. Magness, and L. H. Schiffman (eds.) *'Go out and study the land' (Judges 18.2). Archaeological, historical and textual studies in honor of Hanan Eshel*, 117–32. Leiden, Brill.

Mazar, B. (1973) *Beth She'arim Report on the excavation during 1936–1940*. New Brunswick, Rutgers University Press.

Miranda, E. (1999) La comunità giudaica di Hierapolis di Frigia. *Epigraphica Anatolica* 31, 109–56.

Morris, I. (1992) *Death-ritual and social structure in Classical Antiquity*. Cambridge, Cambridge University Press.

Noy, D. (1998) Where were the Jews of the Diaspora buried? In M. Goodman (ed.) *Jews in a Graeco-Roman world*, 75–89. Oxford, Clarendon Press.

Oberhänsli-Widmer, G. (1998) La mort et l'au-delà dans le judaïsme. In J.-C. Attias (ed.) *Enseigner le judaïsme à l'université*, 69–83. Geneva, Labor et fides.

Rahmani, L. Y. (1994) Ossuaries and ossilegium (bone-gathering) in the Late Second Temple period. In Geva (ed.), 191–205.

Rebillard, E. (2003) *Religion et sépulture. L'Église, les vivants et les morts dans l'Antiquité tardive*. Paris, Éditions de l'École des Hautes Études en Sciences Sociales.

Ritti, T. (2004) *Iura sepulcrorum* a Hierapolis di Frigia nel quadro dell'epigrafia sepolcrale microasiatica. In *Libitina e dintorni. Libitina e i luci sepolcrali. Le leges libitinariae campane. Iura sepulcrorum: vecchie e nuove iscrizioni. Atti dell'XI Rencontre franco-italienne sur l'épigraphie*, 455–633. Rome, Quasar.

Ritti, T. (2006) *An epigraphic guide to Hierapolis (Pamukkale)*. Istanbul, Ege Yayınları.

Ritti, T. (2007) La ricerca epigrafica: risultati dell'ultimo quadriennio e prospettive future. In D'Andria and Caggia (eds.), 583–618.

Ronchetta, D. and Mighetto, P. (2007) La Necropoli Nord. Verso il progetto di conoscenza: nuovi dati dalle campagne 2000–2003. In D'Andria and Caggia (eds.), 433–54.

Rutgers, L. V. (1992) Archaeological evidence for the interaction of Jews and non-Jews in Late Antiquity. *American Journal of Archaeology*. 96, 101–17.

Rutgers, L. V. (1995) *The Jews in late ancient Rome: Evidence of cultural interaction in the Roman diaspora*. Leiden, E. J. Brill.

Rutgers, L. V. (1998) *The hidden heritage of Diaspora Judaism*. Louvain, Peeters.

Rutgers, L. V. and Bradbury, S. (2006) The Diaspora, c. 235–638. In S. Katz (ed.) *The Cambridge history of Judaism, Vol. 4: The Late Roman Rabbinic period*, 492–518. Cambridge, Cambridge University Press.

Scardozzi, G. (ed.) (2015) *Hierapolis di Frigia VII. Nuovo atlante di Hierapolis di Frigia. Cartografia archeologica della città e delle necropoli*. Istanbul, Ege Yayınları.

Schweyer, A.-V. (2002) *Les Lyciens et la mort: une étude d'histoire sociale*. Paris, De Boccard.

Spanu, M. (2000) Burial in Asia Minor during the Imperial period, with a particular reference to Cilicia and Cappadocia. In J. Pearce, M. Millett, and M. Struck (eds.) *Burial, society and context in the Roman world*, 170–7. Oxford, Oxbow Books.

Stern, K. B. (2011) Keeping the dead in their place. Mortuary practices and Jewish cultural identity in Roman North Africa. In E. Gruen (ed.) *Cultural identity in the Ancient Mediterranean*, 307–34. Los Angeles, Getty Research Institute.

Stern, K. B. (2013) Death and burial in the Jewish Diaspora. In D. Master, A. Faust, B. A. Nakhai, L. M. White, and J. Zangenberg (eds.) *Oxford encyclopedia of the Bible and archaeology*, 270–80. Oxford, Oxford University Press.

Syon, D. (2002) The coins from burial caves D and E at Hurfeish. In Z. Gal (ed.) *Studies in Galilean archaeology*, 167–75. Jerusalem, Israel Antiquities Authority.

Şimşek, C. (2006) A menorah with a cross carved on a column of Nymphaeum A at Laodicea ad Lycum. *Journal of Roman Archaeology* 19.1, 343–6.

Thomas, C. M. (1999) The Ephesian ossuaries and Roman influence on the production of burial containers. In H. Friesinger and F. Krinzinger (eds.) *100 Jahre Österreichische Forschungen in Ephesos: Akten des Symposions*, 549–54. Vienna, Österreichische Akademie der Wissenschaften.

Thomas, C. M. and İçten, C. (2007) The ostothekai of Ephesos and the rise of sarcophagus inhumation: Death, conspicuous consumption, and Roman freedmen. In G. Koch (ed.) *Akten des Symposiums des Sarkophag-Korpus 2001* (Sarkophag-Studien 3), 335–44. Mainz, Zabern.

Tomasello, F. (1991) *L'acquedotto romano e la necropoli presso l'istmo*. Rome, Giorgio Bretschneider.

Tosun, A. (2009) Doğu Garajı-Halk Pazarı Mevkii Attaleia Kenti Nekropol Kurtarma Kazısı 2008 (Salvage excavations at the necropolis of Attaleia at Doğu Garajı-Halk Pazarı in 2008). *ANMED – Anadolu Akdenizi-Arkeoloji Haberleri* 7, 189–99.

Trebilco, P. R. (1991) *Jewish communities in Asia Minor*. Cambridge, Cambridge University Press.

Verzone, P. (1987) *Hierapolis di Frigia 1957–1987*. Milan, Fabbri.

Visscher, F. de (1963) *Le droit des tombeaux romains*. Milan, Giuffrè editore.

Vitto, F. (2008a) A late 3rd–4th century CE burial cave on Remez street, Qiryat Ata. *Atiqot* 60, 131–61.

Vitto, F. (2008b) A Jewish mausoleum of the Roman period at Qiryat Shemu'el, Tiberias. *Atiqot* 58, 7–29.

Vitto, F. (2011) A Roman period burial cave on Ha-Horesh Street, Qiryat Tiv'on. *Atiqot* 65, 27–61.

Weiss, Z. (2010) Burial practices in Beth She'arim and the question of dating the patriarchal necropolis. In Z. Weiss, O. Irshai, J. Magness, and S. Schwartz (eds.) *'Follow the wise'. Studies in Jewish history and culture in honor of Lee I. Levine*, 207–31. Winona Lake, Eisenbrauns.

Williams, M. (1994) The organisation of Jewish burials in Ancient Rome in the light of evidence from Palestine and the Diaspora. *Zeitschrift für Papyrologie und Epigraphik* 101, 165–82.

Yalçınsoy, H. and Atalay, S. (2012) Eudokias Antik Kenti Doğu Nekropolü: 267 Ada 18 Parseldeki Tonozlu Khamosorion Tipi Mezarları Kurtama Kazısı (East necropolis of Ancient Eudokias: Rescue excavations for vaulted chamosoria in Insula 267, lot 18). *ANMED – Anadolu Akdenizi Arkeoloji Haberleri* 10, 223–7.

Zlotnick, D. (1966) *The tractate 'Mourning' (Śĕmaḥotĕmaḥotctate 'Mourning' (Śĕma 18). amosoria in Insula 2*. New Haven and London, Yale University Press.

Zoroğlu, K. L. (2012) Kelenderis 2011 Yılı Kazıları (Excavations at Kelenderis in 2011). *ANMED – Anadolu Akdenizi Arkeoloji Haberleri* 8, 40–5.

Tomb ownership in Lycia:
Site selection and burial rights with selected rock tombs and epigraphic material from Tlos

Gül Işın and Ertan Yıldız

Abstract

This paper draws together archaeological, epigraphic, and historical evidence concerning burial customs in Lycia. The work concentrates in particular upon Tlos, one of the most important cities in Lycia. The rock-cut tombs chosen for study from the acropolis cover a period from the beginning of the 5th century BC to the 3rd century AD. These selected tombs at Tlos conform well with the formerly established archaeological facts regarding burial customs in Lycia. Tomb examples from Tlos are examined separately in terms of their periods and their architectural features.

The task involved gathering the published material. This material concerns foremost well-preserved tombs with inscriptions. They are classified according to their typological order and listed in chronological sequence separating tombs from the Classical period from those of the late Hellenistic and Roman periods. Consequently, it is understood that even the uninterrupted usage of family tombs throughout the years does not imply the existence of an unchanging burial tradition. Burial practice was usually transformed according to the necessities of the current conditions.

Keywords: burial customs, burial rights, Lycia, rock-cut tomb, Tlos.

There is a general consensus on the burial practices beginning by the 5th century BC in Lycia concerning some customs and institutions. The vital source for this consensus is the sepulchral epigraphic evidence dated from the late 5th century BC until the end of the 3rd century AD. Almost one thousand of these sepulchral inscriptions are in Greek. In addition, there are approximately 150 native, Lycian language inscriptions (Schuler 2007, 9; Colvin 2004, 45).

The concept of burial rights and customs has been discussed in Lycian archaeology by way of the evidence provided by the Lycian and Greek inscriptions mainly after the 1960s. Studies on Lycian graves and burial practices are quite rich, particularly the volume written by Bryce (1986), which provides the best methodological approach for the entire literary and epigraphic sources concerning Lycian burial traditions. A comprehensive study of social history based on epigraphic and archaeological material about mortality in Lycia was published by Schweyer (2002). The Kyaneai-Yavu studies by Hülden (2006; 2010) concerning the burial types and traditions in the region also contain very valuable additional remarks. Finally, the new monographies from Limyra edited by Borchhardt and Pekridou-Gorecki (2012), and Kuban (2012) will, doubtless be very much appreciated as well.[1]

On the other hand, despite new Lycian epigraphic discoveries, mostly of a formulaic character and with no richness of variety in their contents and the rareness of bilingual texts or more comprehensive inscriptions, the complete decipherment of the Lycian language still remains a problem. Maybe due to this, research over the last 25 years (since Bryce's book), particularly after the systematic survey conducted by the Austrian Academy of Science since

2002 under the title of 'The Lycian Inscribed Monuments Project',[2] has made no major changes to the claims put forward before.

Another difficulty is tracing the cultural continuity from the Classical to the Roman period. Any reflection of the political power of the Ptolemaic dynasty in the region cannot be recognized, seen or correlated to burial customs (Meadows 2006, 459–68). The scant quantity of securely dated sepulchral epigraphic material from Alexander's invasion to 167 BC and Rhodian control (Schuler 2007, 15) does not permit a clear understanding of any changes in burial institutions and organizations in the region. However, it is obvious that in understanding the cultural context, inscriptions or literary material are not the only evidence. Architectural details and artistic depictions on tombs sometimes display better information than others, while the best supportive evidence is, of course, full context intact tombs with well-documented untouched archaeological and anthropological evidence. The most important and unfortunately the weakest evidence however, is the archaeology in Lycia.[3] The Tlos excavation team in 2005 and 2007 was fortunate to excavate well-preserved, intact tombs with Lycian inscriptions. The detailed results of these finds will be published soon (Korkut 2013, 334–6; İşkan-Işık and Uygun 2009, 355–7 figs. 1–5). In this paper, due to the contradictory results of the find groups and the architecture of the newly unearthed tombs, the Tlos excavation team wished to share at least the general characteristics of these tombs, which provide a new

perspective on the known burial customs and rights in Lycia (Fig. 5.1).

Being more comprehensible and forming a better comparison between the burial rights and customs of the Classical and Roman periods, the Tlos examples were examined separately in terms of their periods, a distinction which was not made, to any extent, before. The selected rock tombs, mostly with inscriptions from the acropolis, were divided into three different categories according to their typology and phases of use.

1. Tombs that have particularly Lycian characteristics in their architectural features and also have Lycian inscriptions dated to the Classical period.
2. Tombs with Lycian architectural features having Greek inscriptions and dated to the Hellenistic or Roman Imperial period.
3. Tombs with architectural features of the Roman Imperial period and having Greek inscriptions dated to the Roman Imperial period.

Before focusing on the Tlos examples, a regional outlook with a summary of generally accepted burial customs and rights in Lycia is provided. As already mentioned, the most detailed analyses concerning Lycian burial customs to date have been made by Bryce (1986, 115–59) and Schweyer (2002, 45–89). According to Bryce, in the Classical period rock tomb ownership and the right of interment were privileges that were confined to a small proportion of the

Fig. 5.1. Tlos. Rock-cut tombs from the north-east of the acropolis.

total population of the country. The tomb owner made provision for multiple burials in his tomb; during the Roman period, however, the rules seem less rigid (Bryce 1979, 298).

An attempt is made to divide this accumulated knowledge into two chronological periods: first the Classical period, then the late Hellenistic and the Roman periods taken together. Recai Tekoğlu translated the Tlos Lycian language inscriptions[4] and Ertan Yıldız did the Greek ones.[5] The period between the early Hellenistic and mid-Hellenistic roughly up to 167 BC, when Rhodians took control of Lycia, could not be securely included due to the scarcity of dated sepulchral epigraphic material within this period in the region.

The Classical period

Right of use

- Tomb owners of the 5th and 4th centuries BC generally only provided for the burial of their wives and children, in only a few cases were other family connections provided for. These were restricted to a small group of lineal or collateral family connections such as the owners' mother (TL 86, 95, 127), grandchildren, siblings, and nephews/nieces. In some rare cases perhaps family servants or retainers appear to have been included as well.[6]
- According to the status or power of a family, it is possible to see the same person as an owner of two different tombs.[7]
- In some cases, the tomb owner left instructions indicating where the various occupants were to be placed in his tomb (Bryce 1986, 118). Although it is not a rule, in some examples the left side was reserved for the tomb owner and the men of the family, while the women were buried on the benches on the right side of the chamber. Accentuation of the left side of the tomb chamber with some architectural details, such as a bolster on a *kline* or a niche or carrying reliefs, for example from Myra and Xanthos, are known (Seyer 2003, 86).
- As a rule the inscriptions containing warnings and penalties mainly regard unauthorized interments and/or unauthorized use of reserved burial spaces within a tomb (Kloekhorst 2011; Bryce 1986, 118; Schweyer 2002, no. Myra 72; Avucu 2015).

Criminal offences and the protection of the tomb

- Twenty-four Lycian inscriptions mention an institution or council called *minti*, which seems to have been a secular authority connected with the organization of the tombs (TL 2, 3, 4, 11, 31, 36, 38, 39, 42, 46, 47, 50, 52, 57, 58, 75, 106, 114, 115, 118, 135, 139, 145, 149). Its chief function seems to have been to take care of the necropolis management and to monitor the instructions of the tomb owners (Bryce 1979, 298–9; Schweyer 2002, 47–50; Zimmermann 1992, 147–54; Hülden 2006, 339–40; Kuban 2012, 103–5).

- During the Classical period tombs were in most cases placed under the protection of deities; according to the tomb owner's choice, the local mother goddess, Trqqas, Maliya, Huwedri gods, or the confederate gods (Houwink ten Cate 1961, 94, TL 57) would punish the offender (Bryce 1981, 84–5).
- The Lycian inscriptions do not refer to specific criminal offences, such as damage to the tomb or stealing its contents. According to Bryce 'the criminal acts of this nature were no doubt subject to a different type of disciplinary process, which fell outside the scope of the sepulchral inscriptions' (Bryce 1986, 120).
- Specific monetary penalties for tomb violation were less known in Lycia during the period of Lycian inscriptions, but some inscriptions do indicate the amounts specified in the 'ada' formula (Neumann 2012, 410), which was probably a small amount of payment assumed by the *minti* on the tomb owners' behalf. According to Frei (1977, 66), 1 *ada* is equal to 1.25 Attic drachmae (Bryce 1976, 190).
- The carving of a new inscription or the erasure of an old one was not a criminal offence; there are no examples that refer to it as such. On the contrary, both the Classical and the later tombs with inscriptions present examples of this nature. Tomb no. 7 presents an example of an erased inscription from Tlos (TL 36, TAM II 1028) (Fig. 5.25).

Funerary rituals

- According to the sepulchral Lycian inscriptions, amongst the funerary rituals performed in honour of the tomb owner (Bryce 1986, 87, 127), annually organized animal sacrifices seem to have been a common practice (TL 84, 150, 74 b).
- Inscription TL 84 from Sura mentions a sacred place for offerings. According to the inscription, the tomb owner, Mizretiye, established an offering place, called *hrm̃mã* for himself (Hülden 2006, 323).
- Offering places in front of the rock tombs or open air rock compartments/spaces are common in the Lycian necropoleis; see, for example, Limyra (Kuban 2012, 106–9).
- The latest discovery at the acropolis of Tlos (Korkut 2015a, 89–94) presents us with a specific example of a rock-cut shrine with depictions of a bull and a horseman carved in relief. As the location of the cult area is on the northern slope where the acropolis and necropolis join, this rock shrine with its votive pits could easily be connected to the necropolis and its rituals. In this case it can be inferred that this rock shrine was a 'sacred offering place for common usage' within the necropolis.
- Cattle (*wawa*), sheep (*xava*), or cocks are the most common animals to be sacrificed (Hülden 2006, 311–5). A scene from the Harpy tomb monument at Xanthos exemplifies the sacrifice of a cock. The depiction on this

monument has been accepted as the earliest iconographic evidence of an animal sacrifice (Tritsch 1942, 49; Bryce 1980b, 42). Another well-known composition amongst the 4th century reliefs from Lycia is 'bull sacrificing', but on this disagreements prevail among the scholars. While İşkan (2004, 395–6) accepted the bull as a sacrificial animal to the storm god of Lycia Trqqas, Hülden (2006, 307–15) and Borchhardt (Borchhardt and Pekridou-Gorecki (eds.) 2012, 268–72, 304–8) explain the bull sacrifice through the cult of the dead. In addition, Borchhardt provides some examples of unearthed animal bones from the tombs at Limyra as evidence of the bull sacrificing ritual (Blakolmer 1993, 158; İşkan 2004, 395; Hülden 2006, 307–15; Borchhardt and Borchhardt-Birbaumer 1992, 99). In this case the location of the bull scene gains importance. If the location is related to the necropolis, the activity should be argued as having a funerary context.

- Although there are various examples of funeral banquet scenes in the Lycian tomb iconography (Işın 1995, 72–3), Lycian inscriptions do not mention banquets as part of the rituals in the necropolis (Hülden 2006, 318–20). If the archaeological context is not secure, finds like animal bones in the necropolis can easily be confused with animal sacrifice.

Hellenistic and Roman Imperial period

Right of use

- In exactly the same way as in the Classical period, tomb owners during the Hellenistic and Roman periods continued to be primarily concerned with the responsibility to provide for their immediate families (Bryce 1979, 298–9), but in the course of time, particularly during the Roman Imperial period, the family privacy gradually disappears and the tomb beneficiaries can easily derive from different family stems.[8] According to the degrees of kinship, these tombs were grouped into the following categories by Bryce: A. spouse, children, or other lineal descendants; B. parents; C. collaterals (including brothers/sisters, nephews/ nieces, uncles, aunts, cousins; D. in-laws (including parents-in-law, sons-/daughters-in-law, brothers-/ sisters-in-law), E. *threptoi* and related terms; F. slaves and freedman.
- As in the Classical period some tomb owners left instructions indicating where they preferred to be placed in the tomb; again accentuation of the left side is known (Adak and Şahin 2004, 99).

Criminal offences and the protection of the tomb

- By the end of the Classical period the Lycian term *minti* seems to have become practically defunct. It is referred to in Greek inscriptions, which are of comparatively early

dates, but the term *minti* is replaced by the Greek term *mindis* (TAM II 62). By the Roman Imperial period the institution of *mint* seems to have completely disappeared (Bryce 1986, 122; TAM II 62, 40).

- During the Roman Imperial period criminal acts such as the damaging of the tomb or the plundering of its contents were accepted as offences (Schweyer 2002, 259–60 (Myra 72)).
- In the Roman Imperial period the responsible institution or council related to the necropolis management seems to have been rather complicated. Penalties could be paid to religious bodies such as the Hierataton Tameion and the Temple or sometimes to secular bodies such as the Demos, Polis, Fiscus, Gerusia, Boule, or even Peripolion and Kome, but there was no specific institution with a particular name. In general, the fine varied from 500 denarii to 5,000 denarii, but there were several exceptions as well.
- It seems that using the same tomb for three generations was standard. On the other hand, in order to regulate the rights of the next generations some penalties had been imposed. Tombs were considered as entailed property. The authority regulated selling or alienating of the tombs during the Roman Imperial period.[9]
- Curse formulae are found in the Greek inscriptions dating from the Imperial period and in a number of cases offenders were threatened with divine retribution (Bryce 1981, 91–3). The curse statements can be in different formulations, for instance: A. the one who acts against (so and so), will be faithless; B. … will be impious; C. … will commit sin against all gods; D. … will be subject of the worst way to die; E. … will be cursed by all gods.
- An 'incentive payment', representing a fraction/part of the total penalty was frequently to be paid to the informant (Bryce 1981, 93); it was usually one third of the total amount.

Funerary rituals

- There must have been a funerary cult established, involving periodic sacrifices in honour of the dead. Several of the Greek inscriptions from the Roman Imperial period mention bird and animal sacrifices, such as cocks, fowls, goats,[10] or pigs (Schweyer 2002, 42), offered at a certain time of the year by the descendants or the heirs of the tomb owner – once before the harvest and once before the vintage (TAM II 245, 636, 637, 715).
- Although it is not known whether this ritual was common and exactly which period it can be dated back to, the Roman historian Valerius Maximus (2.16.3) mentions that it was customary for Lycian males to wear female garments as a sign of mourning for the departed. Plutarch

refers to a similar practice (*Consolatio ad Apollonium* 112F–113A). Probably related to this statement there is an interesting inscription from Tlos, which, according to Kolb (1976), mentions that a specific group of men, dressed as women, was permitted to participate in the women's cult for Dionysos.

These above-mentioned general remarks, mostly obtained from published epigraphic material from Lycia, should more or less also give some indication of the burial practices in Tlos. On the basis of the results of the survey conducted by Taner Korkut in Tlos and its periphery, there are a total of 87 rock-cut tombs on the acropolis and nine more around the city centre with a variety of forms, ranging in date from the 5th century BC to the 3rd century AD. There are only four tombs carrying relief carvings, two of them on the rock façade. Twenty-four of these rock-cut tombs have inscriptions, four of them in Lycian and the rest in Greek.

To work towards a better understanding, through a comparison of the literary and the epigraphic evidence with the archaeological evidence, the eleven most informative tombs have been selected for this paper, and are, as already mentioned above, examined below in three different categories according to their construction and phases of use.

Category 1: Tombs with particularly Lycian characteristics in their architectural features carrying Lycian inscriptions

Tomb no. 1 (Bellerophontes Tomb)

The tomb named Bellerophontes by Quintianus is located in a barely accessible location on the cliff (Figs. 5.2–3) (Korkut 2015b, 287–99). The tomb has an Ionic distyle in-antis temple façade with architrave and pediment. The anteroom gives access to two burial chambers with two doorways from the entrance, which was probably planned to avoid disturbance during later burials and to protect the tomb owner's privacy (Figs. 5.4–5) (Seyer 2006, 126). The room on the left measures approximately 2.10×2.05 m and contains three *klinai*. Following a typical pattern, the bolstered *kline* is placed on the right side of the tomb. The room on the right is almost the same in size and measures 2.06×2.10 m but the *klinai* are shorter and the workmanship is incomplete.

On the upper left side wall of the antechamber, in low relief, Bellerophontes rides his horse Pegasus, fighting against the Chimera.[11] The door on the left is decorated in reliefs with a lion at the top and a dog at the bottom. The right door, however, is simpler than the one on the left carrying only a single dog relief at the bottom. In between the two sliding real doors a pseudo-door in Doric order imitates the wooden construction with its metal-like decorations. A new discovery on the pediment is

Fig. 5.2. Tlos. Bellerophontes cliff: Rock-cut tombs from the north of the acropolis.

Fig. 5.3. Tlos. Tomb no. 1. Bellerophontes rock-cut tomb.

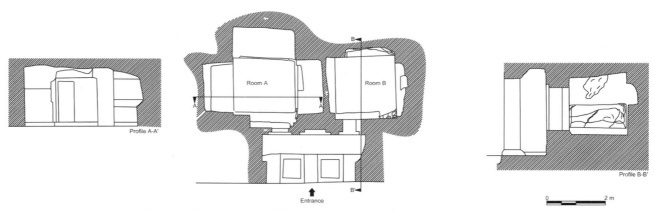

Fig. 5.4. Tlos. Tomb no. 1. Plan and sections of Bellerophontes rock-cut tomb.

a centred bust-relief ranked with the lions. The centred bust holds objects in both hands, in the left a double-axe and in the right what has been interpreted as a thunderbolt.[12]

The stylistic features of Bellerophontes as a horseman, the lions and dogs, and the centred bust in the pediment are all of an early 5th century BC date (480–450) (Korkut 215b, 289–93).

TAM I TL 22: Tomb of Hrikttibili and his wife

'Hrikttibili, the divine *uwehi*, and his wife (lies here)'

The Lycian inscription, which was carved on the upper left side, emphasizes not only the tomb but the whole cliff façade (Fig. 5.6). The placing of the inscription is very

Fig. 5.5. Tlos. Tomb no. 1. Interior view of the room A of the Bellerophontes rock-cut tomb.

Fig. 5.6. Tlos. Tomb no. 1. Lycian inscription on the façade of the Bellerophontes cliff (TL 22).

unusual, as is also the inscribed text itself. It is not written in the common formula well known from other 5th–4th century Lycian tombs. The most important difference from the other sepulchral inscriptions is the 'Godly' or 'Divine' adjunct, used after the name Hrikttibili. Hence we may think that this cliff was most likely reserved for this important man during the first building period, which

most probably includes the inscription and the upper left side of the Bellerophontes tomb; besides, accentuation of the left side of the tomb chamber with the depiction of Bellerophontes is noticeable.

The other chambers on the upper right and left end of the façade of the cliff are of late workmanship. These late tombs show that during the Roman Imperial period, when there was a shortage of space in the necropolis the cliff and the family of Hrikttibili seem to have lost their privileged position.

Tomb no. 2

As it has not as yet been excavated we do not know the architectural features of this tomb in detail, but it can be assumed that it has at least one burial chamber (Fig. 5.7). It is built in the very common manner of the classical simple house-type façade architecture and dated to the 4th century BC.

Tomb of the household of Ikuwe

'Ipresida son of Armanaza PN, father and member of household of Ikuwe (?) for his wife and children…'

The inscription was read by Recai Tekoglu (2002–2003, 104–14). According to the inscription the owner of the tomb was Ipresida, son of Armanaza (Fig. 5.8). He was a member of the household of Ikuwe; and he built this tomb for his wife and his children.

Fig. 5.7. Tlos. Tomb no. 2.

Fig. 5.8. Tlos. Tomb no. 2. Lycian inscription.

Tomb no. 3

Tomb 18.25, located in the east of the acropolis, was unearthed in 2005 (Fig. 5.9) (Korkut 2013, 335–6). The façade of the tomb shows simple Lycian house tomb architecture, as is the case with most of the others at

Tlos. The architrave has rounded beam-ends at the bottom and on top of it there are three fasciae. The traces of the wooden-like construction can also be seen on the main body of the chamber façade with the projecting ends. On the left there is a sliding door, which was closed when it was first discovered in 2005. The interior of the tomb is designed with U-shaped *triclinium* benches and there is a rectangular cavity right in the middle (Fig. 5.10). When it was entered, both the benches and the cavity were completely full of interments and their burial goods (Fig. 5.11). The burial gifts are dated from the early 3rd century BC to the early 1st century AD. So it is very clear that in the course of some 300 years the door of the chamber must have been opened numerous times.

Fig. 5.9. Tlos. Tomb no. 3.

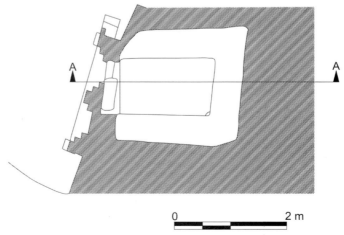

Fig. 5.11. Tlos. Tomb no. 3. Interior view.

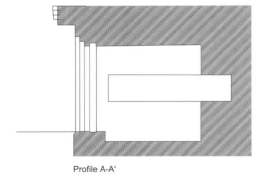

Profile A-A'

0   2 m

Fig. 5.10. Tlos. Tomb no. 3. Plan and section.

Fig. 5.12. Tlos. Tomb no. 4.

Tomb of Sikheriwale (New Lycian inscription, not published)

'Sikheriwale son of Ddew[ele]de, for himself and for the grandmother's descendant. *alahadali ada.*'

According to Tekoğlu's translation of the Lycian inscription the tomb was built by '*Sikheriwale the son of Ddweledes, for himself and for the offsprings of his grandmother*'. Tekoğlu adds that 'the grandmother' can also be interpreted as the 'mother in law'.[14]

The most confusing discovery of the tomb was the Lycian inscription. Before the tomb was opened, when the Lycian inscription and the Lycian house tomb façade architecture was first seen, it was expected that the tomb should be dated to a period covering the first three quarters of the 4th century BC, in other words, to a time before Alexander the Great. However, the find context of the tomb presented a Hellenistic dating. No furnishing prior to 300 BC was found.[15]

Tomb no. 4 (without inscription)

Another tomb was opened in 2007 (Fig. 5.12) (Korkut 2013, 334–6; İşkan-Işık and Uygun 2009, 357 fig. 5). It is almost of the same type but is somewhat smaller than tomb no. 3. The interior design was unusual. The tomb chamber was divided into two with a separation made from brick (Fig. 5.13). The find context is almost similar to tomb no. 3 and the burial gifts were dated from the beginning of the 3rd century BC to the early 1st century AD. (Fig. 5.14).

Fig. 5.13. Tlos. Tomb no. 4. Interior view; the separation made from brick.

Category 2: Tombs with Lycian architectural features having Greek inscriptions and dated to the Hellenistic or Roman Imperial period

Tomb no. 5

This tomb, with Lycian house-type façade architecture, is on the northeast corner of the acropolis cliff (Fig. 5.15). It differs from the others with its highly artistic reliefs (Fig. 5.16). The relief is placed on the east while the entrance gate of the tomb is on the south. It has a very simple interior design. It measures almost 2×2 m and has benches placed in an 'L' shape; one is on the left and

Fig. 5.14. Tlos. Tomb no. 4. Skeleton remains.

the other is on the rear wall (Fig. 5.17). The *kline* on the left is emphasized with a bolster.

The relief on the east is divided into two horizontal friezes. There are ten warriors fighting in pairs: three couples on the upper frieze and two on the lower. The victor is depicted as either displaying his rival's shield or trying to get hold of it; the vanquished, on the other hand, is rendered as lying down on the ground in all his nakedness. The style of the frieze presents the features of the early 4th century BC (Bruns-Özgan 1987, 232–5).

TAM II 600 Tomb of a woman (ignota)[16]

'The tomb (belongs to) [NN] from Kadyanda, who is the daughter of Alexandros (the son) of Alexandros, (who is the son) of Dionysios, following the legitimate concession given on the 13th of Audenaios under the archiereus (head-priest) Caesianus in the year ... and (this record was) put into the archives by Apollonios also known as Eirenaios. Into this (tomb), his father Alexandros, (the son) of Alexandros, (who is the son) of Dionysios; and Alexandros, (the son) of Eirenaios have already been (buried). She herself (NN) and her husband Eirenaios, the son of Zosibios will also be buried (here). Nobody else has permission to bury anyone (else here), otherwise the one (who will act against this) will give to the Tlosians' gerusia 1,000 denarii, one third of which the informant will receive.'

The inscription that was placed here dates from the Imperial period and has no relation to the earlier relief. It was inscribed while Caesianus was the archpriest of the Lycian *Koinon*. Although the years of his service as archpriest are unknown, he might be identified with Tiberius Claudius Caesianus Agrippa (Reitzenstein 2011, 237), who is known from an inscription in Sidyma which

Fig. 5.15. Tlos. Tomb no. 5.

Fig. 5.16. Tlos. Tomb no 5. Fighting warriors' relief.

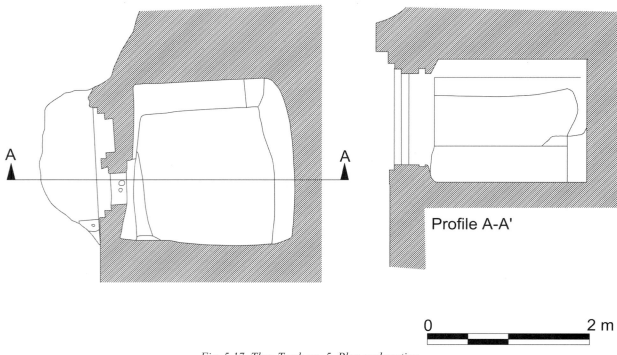

Profile A-A'

0   2 m

Fig. 5.17. Tlos. Tomb no. 5. Plan and section.

records that he repaired the roof of the *tetrastoa*, which was demolished probably by the earthquake of 141 AD (TAM II 179; Takmer 2010, 108), and from another inscription in Xanthos recording that he financed *themis* (TAM II 301–5).

The tomb is allocated to a woman (ignota) whose name is unknown (Fig. 5.18). Her father Alexandros and her son-in-law Alexandros were already buried in it; other than them, she, the tomb owner and her husband Eirenaios will be buried there as well. No one else will be allowed to be interred there; whoever acts against this will pay 1,000 denarii to the Tlosians' *Gerusia* and one third of the money will be paid to the informant.

This tomb is a very good example of the tombs that lose their meaning and privacy over the course of time.

Tomb no. 6

This is one of the most striking tombs in the acropolis in terms of size (Fig. 5.19). Its façade is about three times larger than the regular tombs, the tomb chamber, however, only measures 3.17×2.54 m (Fig. 5.20). On the façade there are two rows of quartet windows above the doorway. The benches are arranged in a U shape, in the *triclinium* design (Fig. 5.21). In contrast to the previous examples, the *kline* on the left is not accentuated. However, the benches on the left and on the rear wall both have bolsters.

Fig. 5.18. Tlos. Tomb no, 5. Greek inscription (TAM II, 2 600).

Fig. 5.19. Tlos. Tomb no. 6.

Fig. 5.20. Tlos. Tomb no. 6. Interior view.

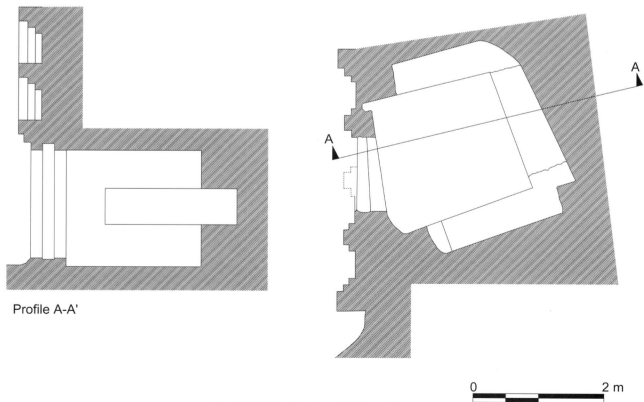

Profile A-A'

0   2 m

Fig. 5.21. Tlos. Tomb no. 6. Plan and section.

TAM II 599 Tomb of Eperastos[17]

> 'This tomb of (belongs to) Eperastos of Tlos, (the son) of Philokles, (and it is also) for his wife Nannis, for Soteris the heir of the same Eperastos and for the descendants of her (Soteris) in succession and for those to whom she may permit in writing; and also for her (Soteris's) husband Zosimos and her sister Syntrophia. To nobody else is permission given to bury (in this tomb), except if I or (my) heir Soteris permit, otherwise the one who authorizes or buries will pay to the Demos of Tlosians … drachms, half of which the informant will receive.'

Being one of the most pompous tombs in the acropolis, the smallness and the simplicity of the internal planning of tomb no. 6 is surprising. Nevertheless, the size of the interior is in accordance with the inscription. The tomb is allocated to a nuclear family of father, mother, and daughter.

A later addition to the inscription states that Eperastos is the tomb owner. It also adds that Eperastos not only allotted the tomb to his wife and his daughter, but to the descendants of his daughter as well. Besides, Eperastos also mentions that his son-in-law and his sister will be buried here. This shows that the girl's side of the family has taken the groom under their protection. If anyone other than an authorized person will dare to do anything against …, (…) amount of drachmae shall be paid to the Tlosians' Demos. Due to the use of drachmae instead of denarii as the unit of currency and the letter styles having late Hellenistic or early Roman features, the inscription can be dated to a period shortly before AD 43 when Lycia became a Roman province.

Tomb no. 7

This tomb is on the right side of the modern road going up to the city. The entrance to the tomb is on the south (Fig. 5.22). The façade is in the form of a classical Lycian house tomb whose roof with its round beams and three fasciae was cut from a separate block, applied afterwards and jointed into the rock. The burial chamber is rather small; it measures 2.63×2.01 m (Fig. 5.23). There is only one narrow *kline* measuring 1.76×1.10 m, carved out into the rock on the right side of the entrance (Fig. 5.24). On the doorpost of the façade there is an

Fig. 5.22. Tlos. Tomb no. 7.

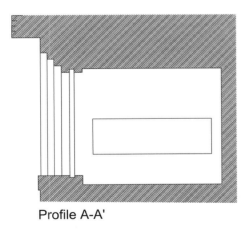

Profile A-A'

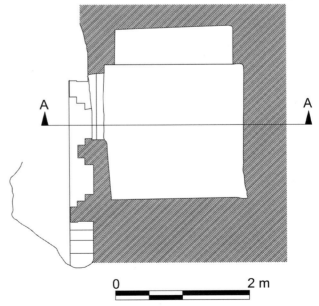

Fig. 5.23. Tlos. Tomb no. 7. Plan and section.

Fig. 5.24. Tlos. Tomb no. 7. Interior view.

Fig. 5.25. Tlos. Tomb no. 7. Erased inscription.

illegible, erased inscription and just next to the erased one is another inscription of eight lines which should be related to the change in the tomb's ownership (Fig. 5.25). The letter style of this later, added inscription may carry late Hellenistic features.

TAM II 639 Tomb of Meis[18]

> 'For Meis (the daughter) of Sarpedon, Semridarma (the daughter) of Androbios for her mother; and (with) Timarkhos and Sarpedon (the children) of Pherekles, for their grandmother (ordered built this tomb), because of her affection.'

The inscription mentions that Semridarma with her two children named Timarchos and Sarpedon built a tomb in honour of their beloved grandmother Meis. The inscription provides information about an exceptional example of a tomb built by a woman for her mother and no other name or family member is recorded as having the right of other interments. The inner arrangement of the tomb also supports this, as it has only a single *kline*.

Category 3: Tombs with architectural features of the Roman Imperial period and Greek inscriptions dated to the Roman Imperial period

Tomb no. 8

This tomb is on the cliff of Bellerophontes (Fig. 5.26). The façade architecture of the tomb presents Roman Imperial features with its very simple Doric doorframe, which was not cut from the monolithic rock block. Instead all the elements of the frame have been worked and placed separately. The tomb chamber measures 4.30×2.70 m (Fig. 5.27). On the back wall there is a *loculus*-like arrangement inside of it. The sidewalls are covered with plaster.

Fig. 5.26. Tlos. Tomb no. 8.

Fig. 5.28. Tlos. Tomb no. 8. Greek inscription.

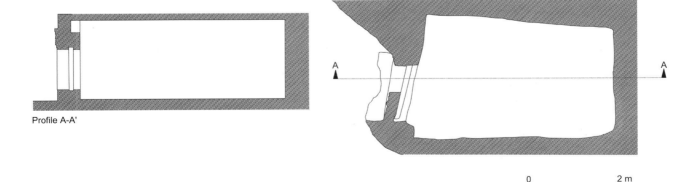

Profile A-A'

0   2 m

Fig. 5.27. Tlos. Tomb no. 8. Plan and section.

TAM II 605 Tomb of Iason, Menelaos and Aristippos (Fig. 5.28)[19]

'Lives,
Iason and Menelaos, the children of Menelaos the third (Menelaos son of Menelaos of Menelaos) and Aristippos son of Krateros (ordered) built this heroon, in which the klines are for themselves, (their) wives, (their) children who should be cognately from them, just as the will permitted. The first kline on the left side shall be for Iason; the first kline on the right side for Menelaos; The second kline on the left side for Aristippos; The second one in the right side (belongs) to Iason and Menelaos for the burial of their servants and for their descendants. No one else but us is authorized to approve, otherwise (the violator) shall pay 500 denarii to the Demos of the Tlosians. Menelaos shall approve the (burial) of Philoumenos (the son) of Arsasisinto the first kline on the right side that belongs to him (Menelaos).'

There is no certain evidence for dating, except that it is from before AD 212 as the names are without Aurelius/-a *nomen*. But the low fine and that it was paid to the Demos of Tlos may indicate an earlier date, probably before AD 141.

The tomb owners were two brothers, Iason and Menelaos, sons of Menelaos, and their friend Aristippos, son of Crateros. The tomb was intended for the use of the owners, their wives, their children, servants and their descendants, the arrangement being approved by the Demos.[20] The size of the chamber fits the tomb definition provided by the inscription. Particularly the arrangement of the two regular-sized double banks on the right and two on the left are obvious. As Bryce already formulated, *kline* A on the left was allotted to Iason, *kline* A on the right to Menelaos, *kline* B on the left to Aristippos, *kline* B on the right to Iason and Menelaos' *threptoi* (household slaves) and the children of their *threptoi*.

Fig. 5.29. Tlos. Tomb no. 9.

Tomb no. 9

This tomb is on the cliff of Bellerophontes (Fig. 5.29). It has very similar workmanship to the previous tomb no. 8. The inner arrangement is rather smaller measuring 3×2.80 m (Fig. 5.30). There is no trace of the *klinai*. It is obvious that the right of use of this tomb belonged to the privileged family members.

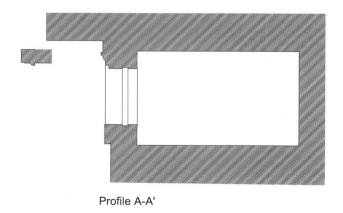

Profile A-A'

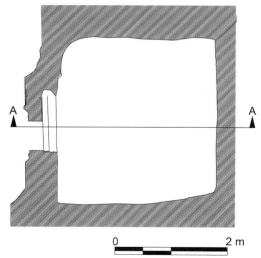

0   2 m

Fig. 5.30. Tlos. Tomb no. 9. Plan and section.

TAM II 602 Tomb of Zosimos[21]

'Lives,
Zosimos from Tlos (the son) of Neiketikos (who is the son) of Neiketikos of Lysanios (ordered) built this heroon for himself, for (his) children, wife, his descendants and for anyone he may authorize in writing. If anyone other than Zosimos, who built (the heroon), buries anybody (else), they will pay a fine of 1500 denarii to the most sacred treasury. But if Zosimos who built (this heroon) authorizes someone, the obtainer of the authorization shall have the authority for those the obtainer (of this authorization) might wish to bury.'

Though tomb no. 9 is plainer in terms of its lintel and doorpost craftsmanship compared to tomb no. 8, both tombs must belong more or less to the same time span within the Roman Imperial period.

According to the epigraphic information, the tomb was termed a *heroon* and was built by Tlosian Zosimos, for himself, for his children, for his wife and for their lineal descendants and for the people he allowed to be buried with a written document. If anyone buries someone without permission, a 1,500 denarii fine will be paid to the most sacred treasure.

It dates to before 212 AD due to the absence of Aurelius/-a *nomen,* but is probably close to TAM II 601 of AD 141, recording the same amount in the fine payable (see below, tomb no. 10).

Tomb no. 10

This is another Roman Imperial tomb building. The tomb chamber measures 3.61×3.98 m (Fig. 5.32). There is no trace of the *klinai* (Fig. 5.31). The features of the rock chamber and the multi-grooved architectural decoration of the doorpost and the lintel are very similar to tomb no. 8, but the added vaulted entrance just in front of the door differentiates it from the others and interest arises with its monumental effect.

TAM II 601 Tomb of ignota[22]

'(a) (This tomb belongs to) ...phoros (the son) of Pappos (who is the son) of Androbios; to Alexandros (the son) of K... also known as Sikouleinos; to Alexandros (the son) of an unknown father; to Hedia with her children; ...to Alexandros (the son) of Alexandros (who is the son) of Stephanos; to Daidalos also known as Eiphitos; to Kolakairos (the son of) Agrippinos also known as Stasithemis; to Arteimas (the son) of Arteimas; to Eutykhiane (the daughter) of Eutykhes, Claudia Vilia Procla's freedman; to Eutykhes, Claudia Veilia Procla's freedman and in the authorization given to Eutykhes for only six additional names; to those (six) he might consent, without permitting the subsequent generation of anyone (amongst the six) or (their) descendants.

Fig. 5.31. Tlos. Tomb no. 10.

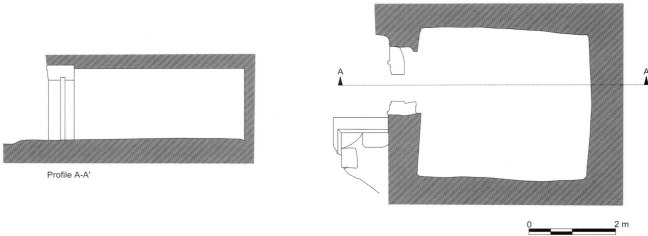

Fig. 5.32. Tlos. Tomb no. 10. Plan and section of Tomb no. 10.

(b) No one but us has the authority to authorize the joint burial right to anyone unmentioned or to bury anyone else, otherwise (the violator) should pay 500 denarii to the city of Tlosians. In the same way any unmentioned one has no authorization to bury anyone else, otherwise (the violator) shall pay to the city of Tlosians 1500 denarii, one third of which the informant will receive.

(c) The inscription itself and its security were recorded via the public registers, under Gaius Iulius Heliodoros also known as Diaphantos, archiereus of the Sebastoi (head-priest of the emperors).'

According to its inscription, tomb no. 10 was used by a minimum of 11 people; additionally one of these (Euthyces) had the right to approve six more external names. It is obvious that these beneficiaries belong to different family stems. The tomb presents no family privacy at all; this kind of tomb usage probably indicates economic difficulties.[23]

The inscription dates from the period when Gaius Iulius Heliodoros was archpriest of the emperors in AD 141 (Reitzenstein 2011, 198–200). Furthermore, the Vilia Procla mentioned in the inscription erected a statue of Hadrianus in AD 126 (Engelmann 2012, 188) and also renovated the proscenium of the theatre at Patara in AD 147 (TAM II, 408).

Tomb no. 11 Inscription belonging to an unknown 'monumental tomb' building

Although information concerning the origins of the inscription is lacking, it is very possible that this fragment belonged to a monumental tomb building, not to a rock-cut tomb. Nevertheless, as the text offers exceptional information and provides a wider perspective on the burial rights, the inscription has been included in this study.

TAM 604 (Private Collection, Fethiye)[24]

'The people who possessed each kline and who are the under-mentioned (ordered) constructed this heroon. The klines have been allotted (as follows): on the right, from amongst those who will (be put in) get in the (tomb) Asklepiades also known as Philoumenos, Glykon (the son) of Krateros of Krateros, Eleutheros also known as Apollonios, Kallinikos (the son) of Agathemeros, Polyneikos (the son) of Glykon; on to the left kline, Eirenaios (the son) of Damas (who is the son) of Damas, Zosimos (the son) of Zosimos, Apollonios also known as Symmachides (the son) of Apollonios (who is the son) of Apollonios, Philokyros (the son) of Apollonios (who is the son) of Apollonios (the son) of Symmaios, Eirenaios (the son) of Arteimas, Pompeius (the son) of Harpalos also known as Karpos; onto the middle kline, Apollonios and Zotikos, Epaphrodeitos (the son) of Epaphrodeitos (who is the son) of Epaphrodeitos, Stephanos (the son) of Epaphrodeitos, Alexandros (the son) of Alexandros (who is the son) of Eirenaios, Eirenaios (the son) of Eirenaios (who is the son) of Tilomas, Polyktetos (the son) of Apollonios (who is the son) of Apollonios (the son) of Symmasis. Only their wives, children and grandchildren (will have) the right of joint burial, but (it will) end for the next generation. In the hyposorion will be placed their (own) servants from their environment. No one else shall have the authority to give permission to anyone or to bury anyone, otherwise the violator shall pay to the city of Tlosians 1500 denarii, one third of which the informant will receive. No one else will place anyone onto the other (person)'s kline, which is unshared with him. Otherwise the one, who places, similarly shall pay to the city 300 denarii, one third of which the informant will receive'.

According to the inscription the tomb belongs to 17 or 18 men coming from different family stems.[25] Everyone has the right to inter their wives and children but no other descendants; besides, they all have the right to bury their households and their children as well. As it is seen with this extreme example, the total number of persons eligible for burial could be more than one hundred during the Roman Imperial period. The inscription itself was recorded via the public registers, under [NN] *archiereus* of the Sebastoi (head-priest of the emperors).

The amount of the fine is the same as for TAM II 601 and 602 (see above). That the payment is due to the city and the informant is given one third of the payment is also contained in TAM II 601 of AD 141. The increase in fines was perhaps associated with the earthquake of AD 141, so these three inscriptions might be from the end of the first half of the 2nd century AD.

General remarks on how the tombs of Tlos have contributed to the knowledge of Lycian funerary practice

The tombs of Tlos from the necropolis, situated in close proximity to the acropolis, have been carefully analyzed. The tombs' characteristics of period, type, and craftsmanship have provided interesting insights into Lycian burial customs.

Site selection and status

The Lycian inscription connected to the Bellerophontes tomb (tomb no. 1) honours the tomb owner Hriktibili as 'Godly', an uncommon practice in Lycia. It is assumed that in accordance with this honour, an *in-antis*, Ionic temple-like architectural choice was made deliberately. On the other hand, the privileged status of the tomb and its owner has continued throughout the Classical and Hellenistic periods. During the Roman Imperial period three new tombs were added to the same rocky area; consequently, the privileged status/position of the Hriktibili family was shared with other tomb owners.

Another example that shows how the status and privileges of a tomb owner can alter is tomb no. 5. This 4th-century BC Lycian tomb, whose owner was honoured with his warrior identity, evolved into a different tomb through the addition of an inscription in the Roman Imperial period. It became a tomb by which a woman honoured herself and her family.

On the other hand, the interior of tomb no. 6, one of the most pompous tombs in the acropolis in terms of its size, was planned on a small scale and the inscription allowed for a limited number of interments. This seems to have been a precaution taken against any future usurpation of the family's privileged status.

Inscriptions, tomb architecture, and dating

Tomb no. 1, which was dated to the 5th century BC (480–50), from the stylistic features of the depicted figures, is one of the earliest examples amongst the temple-like tombs in the region. This dating offers a new perspective for the much-debated typological dating of Lycian rock-cut tombs. In consequence the idea of dating the house-type tombs earlier than those of the temple-façade type should be questioned.

The newly found inscriptions of tomb no. 2 have contributed to the corpus of Lycian inscriptions.

The archaeological finds from the Sikheriwale tomb no. 3 proved that all the information regarding the dating of Lycian tomb architecture and Lycian inscriptions is in need of some revision. The new data showed that Lycian inscriptions were still being inscribed/carved on rock-cut tombs in the early 3rd century BC and that the tradition of house-type tombs could have continued for longer than had previously been understood.

Tomb no. 4, which was unearthed in 2007, is dated to the early 3rd century BC. It was documented that the tomb had been in use for 300 years.

The subsequently added inscription on the Lycian house-type tomb no. 6 is dated to a period shortly before AD 43, when Lycia became a Roman province.

Tomb no. 8 is dated to the 2nd century AD with its inscription and architecture. It is an important example in that its measurements and inner design match the instructions and advice that is provided in the inscription.

Rights of tomb use

The right of use of the remarkable tomb no. 6 belonged to a small family of father, mother, and daughter. But the father gave his daughter control over tomb's use. If she wished she could allot space for her husband and his sister. The inscription shows that the final rights concerning burial were completely surrendered to the daughter.

Tomb no. 7 was built for a woman named Meis by her daughter and grandsons. There is no inscribed record concerning the subsequent use of this tomb.

Tomb no. 10 is an example of Roman Imperial period practice, which does not limit its rights of use to one family, but allows people from different family lines to be buried together in the same tomb.

The otherwise unknown monumental tomb no. 11 is exceptional in its inscription detailing the tomb's use. It gives the names of 17 men with different lineages and their families the right of interment. This implies that around 100 people were given the right to be buried within this particular tomb.

Funerary rituals

In the Tlos excavations of 2013, on the northern slope where the acropolis and the necropolis merge, a very important cult site was located. There are 16 votive pits and depictions of a bull and a horseman carved on the rock. Taner Korkut (personal communication) has interpreted this site as an outdoor cult site dedicated to the native Lycian sky god Trqqas. On the other hand, in the light of the above-mentioned inscriptions it seems very probable that the site may also have been used for periodic offerings and sacrifices to the deceased. If this was the case, this recent archaeological discovery presents us with very important evidence as to the Lycian funerary practice outside the immediate area of the tomb.

Concluding remarks

The research on the selected tombs, prominent with their inscriptions, reliefs and architecture, revealed 700 years of continuing tradition in the usage of rock cut tombs at Tlos. The chronological analysis presents some social transformation particularly on the management and usage of these tombs. The right of use of the rock tombs, which was primarily for immediate families in the Classical period and some extended family members in the Hellenistic period, eventually changes during the Roman Imperial period. The family privacy disappears and the tomb beneficiaries can be from different family stems. The reason for the change in this common and unbroken custom during the Roman Imperial period is associated with the increase in the population and the adjustments in the management. Another important consequence of this study brings out some new aspects of the dating of the Lycian tombs. This new result is particularly related to the tombs that were examined under the first category in this study; tombs no. 1 and 3. Tomb no. 1 with its temple-like façade architecture is dated to the 5th century BC according to the stylistic feature of its relief. The Lycian-inscribed tomb no. 3 is dated to 300 BC by the archaeological finds from its excavation. This new suggestion for the dating argues for the possibility of a later date for the Lycian inscriptions on the rock tombs, for which the common dating is usually ascribed to the first half of the 4th century BC.

Acknowledgements

We would like to thank Professor Korkut for allowing us to work on this subject and to use the drawings and photos from the excavation archives, which were prepared by Çilem Uygun, Tijen Yücel, and Bayram Akdağ. We are also thankful to Mikhail Duggan for the English proof reading of this manuscript.

All illustrations are used with the kind permission of the 'Tlos Excavation' directorate.

Notes

1   An unpublished Master's thesis by Avcu 2014 at the University of Akdeniz in Antalya, concentrates on the penalties against tomb violations and the institutions related within Lycia.
2   Some preliminary results of this project are given; see Seyer 2009; 2007; 2005; 2004.
3   For the Patara necropolis excavations, see İşkan and Çevik 1997, 191–9; 1995; Işın 2007; for the Limyra necropolis excavations, see Kuban 2012.
4   R. Tekoğlu kindly permitted us to use the translation of an unpublished Lycian inscription on the tomb of *Sikheriwale*.
5   Up to now most of the Greek epigraphic material did not translate to any living languages. This study particularly prefers to give all the translations.
6   Schweyer 2002, 197–8; Lycian term *prnnezi* = members of the household = Greek *oixeioi*; see Bryce 1986, 116, 150–3; Neumann 2012, 409.
7   Borchhardt, in Borchhardt and Pekridou-Gorecki (eds.) 2012, 34 (Limyra TL 115 *Esedeplémi*); Bryce 1979, 279 (Karmylessos TL 7 and 8 *Triyétezi*).
8   Existence of different family stems in the same tomb can be exemplified with the inscriptions from Tlos see, TAM II 601, 604 etc.
9   Avcu 2014, 16; for the related inscriptions, see TAM II 752, 260, 124, 41, 624.
10   Schweyer 2002, 42–3; Hülden 2006, 312; for an example of goat sacrifice in Tlos, see Adak and Şahin 2004, 101–2.
11   For the meaning and the interpretation of Bellerophontes and Pegasus, see Borchhardt and Pekridou-Gorecki (eds.) 2012, 314–5.
12   According to Korkut 2015a, 98–102, 'double axe and thunderbolt are related to Trqqas and this belief gave way to Cronos during the Roman Imperial period in Tlos'.
13   Korkut will soon publish a detailed analysis of the Bellerophontes tomb.
14   I thank Tekoğlu for his comments.
15   There is some evidence for the Lycian inscriptions dated after Alexander the Great. TL 29 is a well-known example from Tlos; see Tekoğlu 2006, 1703–10. On the other hand, if we accept that the tomb was constructed in the 4th century BC, it is possible that the tomb could have been evacuated and sold to a new family by the end of the same century, in which case the tomb furnishings, not datable to before 300 BC, must belong to the new owners.
16   Sherk 1992, 225; Pembroke 1965, 225; Bruns-Özgan 1987, 156–7.
17   Pembroke 1965, 222; Naour 1977, 274.
18   Naour 1977, 277, 288; Adak and Şahin 2004, 99; Zgusta 1964, 308, 461; Robert 1978, 35; Balland 1981, 254.
19   Kubinska, 1968, 29, 110; Adak and Şahin 2004, 99–100. For the supervision of the Polis over the graves see Zimmermann 1992, 155; Kokkinia 2007, 169; Pembroke 1965, 221.
20   For the comments see Bryce 1980a, 172–3.
21   Zimmermann 1992, 194; Wörrle 1999, 359.
22   Pembroke 1965, 221; Naour 1977, 274; Balland 1981, 155; Sherk 1992, 225; Adak and Şahin 2004, 99.

23 The AD 142 and 144 earthquakes could have been the cause of dire economic conditions. For the dating of the earthquake see Erel and Adatepe 2007, 243–246.

24 Adak and Şahin 2004, 99; Rousset 2010, 2, 35; Pembroke 1965, 239.

25 For the comments, see Bryce 1980a, 173–4.

Abbreviations and bibliography

ANMED = News of Archaeology from Anatolia's Mediterranean Areas.

TAM I = Kalinka, E. (1901) *Tituli Asiae Minoris. Vol. I: Tituli Lyciae lingua Lycia conscripti*. Vienna, Hölder.

TAM II = Kalinka, E. (1930–1944) *Tituli Asiae Minoris. Vol. II: Tituli Lyciae, linguis Graeca et Latina conscripti*. Vienna, Hölder.

TL = Kalinka, E. (1901) *Tituli Asiae Minoris: Tituli Lyciae lingua Lycia conscripti*. Vienna, Hölder (revised and reedited by J. Friedrich, *Kleinasiatiche Sprachdenkmäler*, Berlin, 1932).

Adak, M. and Şahin, S. (2004) Neue Inschriften aus Tlos. *Gephyra* 1, 85–105.

Avcu, F. 2014 Lykia Bölgesi Mezar Yazıtlarında Mezar Cezaları ve Mezar Tahsilat Kurumları. Antalya, Akdeniz University Department of Ancient Languages and Cultures, Unpublished Master's thesis.

Balland, A. (1981) *Fouilles de Xanthos VII. Inscriptions d'époque impériale du Létôon*. Paris, Éditions Klincksieck.

Blakolmer, F. (1993) Die Grabungen in der Nekropole V von Limyra, Vorläufige Ergebnisse. In *Akten des II. Internationalen Lykien-Symposions*, Wien 6–12 Mai 1990, 2 Band (Denkschrift Wien 231 und 235), 149–62. Vienna, Verlag der Österreichischen Akademie der Wissenschaften.

Borchhardt, J. and Borchhardt-Birbaumer, B. (1992) Zum Kult der Heroon, Herrscher und Kaiser in Lykien. *Antike Welt* 23.2, 99–116.

Borchhardt, J. and Pekridou-Gorecki, A. (eds.) (2012) *Limyra. Studien zu Kunst und Epigraphik in den Nekropolen der Antike*. Vienna, Phoibos Verlag.

Bruns-Özgan, C. (1987) *Lykische Grabreliefs des 5. und 4. Jahrhunderts v. Chr.* (Istanbuler Mitteilungen, Beiheft 33).

Bryce, T. R. (1976) Burial fees in the Lycian sepulchral inscriptions. *Anatolian Studies* 26, 175–90.

Bryce, T. R. (1979) Lycian tomb families and their social implications. *Journal of Economic and Social History of the Orient* 22.3, 296–313.

Bryce, T. R. (1980a) Burial practices in Lycia. *Mankind Quarterly* 21, 165–78.

Bryce, T. R. (1980b) Sacrifice and dead in Lycia. *Kadmos* 19, 41–9.

Bryce, T. R. (1981) Disciplinary agents in sepulchral inscriptions of Lycia. *Anatolian Studies* 31, 81–93.

Bryce, T. R. (1986) *The Lycians in literary and epigraphic sources*. Copenhagen, Museum Tusculanum Press.

Colvin, S. (2004) Names in Hellenistic and Roman Lycia. In S. Colvin (ed.) *The Greco-Roman East, politics, culture, society*, (Yale Studies 31), 44–85. Cambridge, Cambridge University Press.

Engelmann, H. (2012) Inschriften von Patara. *Zeitschrift für Papyrologie und Epigraphik* 182, 179–201.

Erel, T. L. and Adatepe, F. (2007) Traces of historical earthquakes in the ancient city life at the Mediterranean region. *Journal of Black Sea/Mediterranean Environment* 13, 241–52.

Frei, P. (1977) *Schweizerische Numismatische Rundschau* 56.

Houwink ten Cate, P. H. J. (1961) *The Luwian population groups of Lycia and Cilicia Aspera during the Hellenistic period*. Leiden, E. J. Brill.

Hülden, O. (2006) *Gräber und Grabtypen in Bergland von Yavu (Zentrallykien). Studien zur Antiken Grabkultur in Lykien* (Antiquitas 45). Bonn, Dr Rudolf Habelt GMBH.

Hülden, O. (2010) Die Nekropolen von Kyaneai. Studien zur Antiken Grabkultur in Lykien II. In F. Kolb and O. Hülden, *Lykische Studien. 9. Die Siedlung Kyaneai in Zentrallykien* (Tübinger Althistorische Studien 9). Bonn, Dr Rudolf Habelt GMBH.

Işın, G. (1995) The easternmost rock tomb in Lycia: 'Topal Gavur' at Asartaş. *Lykia* 1, 68–78.

Işın, G. (2007) *Patara Terrakottaları. Hellenistik ve Erken Roma Dönemleri* (Patara V.1). İstanbul, Ege Yayınları.

İşkan, H. (2004) Zum Totenkult in Lykien II: Schlachtopfer und Libation an lykischen Gräbern. *Anadolu'da Doğdu. Fahri Işık Armağanı*, 379–417. İstanbul, Ege Yayınları.

İşkan, H. and Çevik, N. (1995) Die Grüfte von Patara. *Lykia* 2, 187–216.

İşkan, H. and Çevik, N. (1997) Patara 1995. *Kazı Sonuçları Toplantısı* 18.2, 119–99.

İşkan-Işık, H. and Uygun, Ç. (2009) Akropol. Tlos 2007 Yılı Kazı Etkinlikleri. *Kazı Sonuçları Toplantısı* 30.2, 355–7.

Kloekhorst, A. (2011) The opening formula of Lycian funerary inscriptions: mēti vs. mēne. *Journal of Near Eastern Studies* 70.1, 13–23.

Kokkinia, C. (2007) Junge Honoratioren Lykien und eine neue Ehreninschrift aus Bubon. In C. Schuler (ed.) *Griechische Epigraphik in Lykien. Akten des Internationalen Kolloquiums München, 24–26 Februar 2005* (Ergänzungsbände su den Tituli Asiae Minoris, 26), 165–75. Vienna, Verlag der Österreichischen Akademie der Wissenschaften.

Kolb, F. (1976) Zu einem heiligen Gesetz von Tlos. *Zeitschrift für Papyrologie und Epigraphie* 22, 228–30.

Korkut, T. (2013) Die Ausgrabungen in Tlos. In P. Brun, L. Cavalier, K. Konuk, and F. Prost (eds.) *Euploia. La Lycie et la Carie antiques: Dynamiques des territoires, échanges et identités* (Actes du Colloque de Bordeaux, 5, 6 et 7 Novembre 2009 Bordeaux), 333–44. Bordeaux, Ausonius.

Korkut, T. (2015a) *Akdağların yamacında bir Likya kenti: Tlos*. İstanbul, Ege Yayınları.

Korkut, T. (2015b) Tlos Antik Kenti Bellerophon Kaya Mezarı. In E. Okan and C. Atila (eds.) *Prof. Dr. Ömer Özyiğit'e Armağan*, 287–99. İstanbul, Ege Yayınları.

Kuban, Z. (2012) *Die Nekropolen von Limyra. Bauhistorische Studien zur klassichen Epoche*. Vienna, Phoibos Verlag.

Kubinska, J. (1968) *Les monuments funéraires dans les inscriptions grecques de l'Asie Mineure*. (Travaux du Centre d'Archéologie Méditerraéenne de l'Académie Polonaise, 5). Warsaw, Państwove Wydawnictwo Naukowe.

Meadows, A. (2006) The Ptolemaic annexation of Lycia: SEG 27.929. In K. Dörtlük, B. Varkıvanç, T. Kahya, and J. Des Courtils (eds), *The IIIrd symposium on Lycia. Symposium proceedings,* 459–70. Antalya, Suna-Inan Kıraç Akdeniz Medeniyetleri Araştırma Enstitüsü.

Naour, C. (1977) Inscriptions de Lycie. *Zeitschrift für Papyrologie und Epigraphik* 24, 265.

Neumann, G. (2012) Die Lykischen Grabinschriften von Limyra. In Borchhardt and Pekridou-Gorecki (eds.), 389–409.

Pembroke, S. (1965) The last of the matriarchs: A study in the inscriptions of Lycia. *Journal of the Economic and Social History of the Orient* 8, 217–47.

Reitzenstein, D. (2011) *Die lykischen Bundespriester, Repräsentation der kaiserzeitlichen Elite Lykiens* (Klio, Suppl. 17), 198–200. Berlin, Akademie Verlag Berlin.

Robert, L. (1978) Les conquétes du dynaste lycien Arbinas. *Journal des Savants* 165, 3–48.

Rousset, D. (2010) *De Lycie en Cabalide. La Convention entre les Lyciens Termessos prés d'Oinoanda* (Fouilles de Xanthos X). Geneve, Librairie Droz.

Schuler, C. (2007) Einführung: Zum Stand der griechischen Epigrafik in Lykien. Mit einer Bibliographie. In C. Schuler (ed.) *Griechische Epigraphik in Lykien. Akten des Kolloquiums München,* 24–26 Februar 2005, 9–27 (Ergänzungsbände zu den *Tituli Asiae Minoris,* 25) Vienna, Verlag der Österreichischen Akademie der Wissenschaften.

Schweyer, A.-V. (2002) *Les Lyciens et la mort. Une étude d'histoire sociale* (Varia Anatolica 14). Paris and Istanbul, De Boccard Edition-Diffusion.

Seyer, M. (2004) Likçe Yazıtlı Anıtlar Projesi: Ksanthos 2003 Yılı Çalışmaları Üzerine Bazı Düşünceler. *ANMED* 2, 85–9.

Seyer, M. (2005) Likçe Yazıtlı Anıtlar Projesi: Rhodiapolis, Karmilessos ve Pınara 2004 Yılı Araştırmalarının Bazı Sonuçları. *ANMED* 3, 153–7.

Seyer, M. (2007) Likçe Yazıtlı Anıtlar Projesi: 2006 Yılı Çalışmaları. *ANMED* 5, 123–6.

Seyer, M. (2009) Zur Ausstattung der Kammer lykischer Felsgraeber. *Istanbuler Mitteilungen* 59, 51–82.

Sherk, R. K. (1992) The eponymous officials of Greek cities IV. The register. Part III: Thrace, Black Sea area, Asia Minor (continued). *Zeitschrift für Papyrologie und Epigraphik* 93, 223–72.

Takmer, B. (2010) Stadiasmus Patarensis için Parerga (2): Sidyma I. Yeni Yazıtlarla Birlikte Yerleşim Tarihçesi. *Gephyra* 7, 95–136.

Tekoğlu, R. (2002–2003) *Die Sprache* 43.1, 104–14. Vienna, Wiener Sprachgesellschaft Harrassowitz Verlag.

Tekoğlu, R. (2006) TL 29: una nuova proposta di lettura. In R. Bombi *et al.* (eds), *Studi linguistici in onore di Roberto Gusmani III.* Alessandria, Edizioni dell'Orso, 1703–10.

Tritsch, F. J. (1942) The Harpy tomb at Xanthos. *Journal of Hellenic Studies* 62, 49.

Wörle, M. (1999) Epigraphische Forschungen zur Lykiens VII. Asarönü, ein Peribolion von Limyra. *Chiron* 29, 353–70.

Zimmermann, M. (1992) *Untersuchungen zur historischenLandeskunde Zentrallykiens.* Bonn, Dr Rudolf Habelt Gmbh.

Zgusta, L. (1964) *Kleinasiatische Personennamen* (Ceskoslovenská akademie ved. Orientalní ústav. Monografie 19). Prague, Verlag der Tschechoslowakischen Akademie der Wissenschaften.

The sarcophagus of Alexandros, son of Philippos: An important discovery in the Lycian city of Tlos

Taner Korkut and Çilem Uygun

Abstract

This paper focuses on an illegally excavated and looted Lycian-type sarcophagus of the Classical period that was recently investigated by a team of archaeologists, physical anthropologists, and epigraphers. Data produced from the archaeological excavation inside and around the sarcophagus demonstrated that it had been in use from the Classical to the early Byzantine period. The first phase of use dates to the Classical period. The second phase, characterized by an inscription recording 'Alexandros, son of Philippos' added to the north long face of the sarcophagus, dates to the Hellenistic period and it seems that the sarcophagus with its Hellenistic inscription was used until the 4th century AD. The third phase of use covered the 4th to the 7th centuries AD, during the late Roman and early Byzantine periods. The last period of use is defined by a nearly 1 m deep fill, in four layers, containing 34 burials. This newly discovered sarcophagus of 'Alexandros, son of Philippos' casts a fresh light on Lycian burial customs and the social status of the deceased.

Keywords: burial customs, Lycia, necropolis, sarcophagus, Tlos.

Introduction

The ancient city of Tlos (Fig. 6.1), in the foothills of Mount Akdağlar (Kragos) in the Xanthus Valley, in the modern village of Yakaköy, approximately 41 km east of Fethiye, was one of the major cities of ancient Lycia (Korkut 2013; 2015). The large territory of the city apparently extended as far as the cities of Xanthus and Pinara to the south, Kadyanda and Telmessos to the west, and Araxa and Oinoanda to the north. The city was also located on the crossroads of a number of routes that connected the coast with the Lycian hinterland.

According to a Greek myth, the name of the city was derived from Tloos, who along with Pinaros, Xanthos, and Kragos was one of the four sons of Tremilus and Praksidike. However, this myth about the origin of the name of the city of Tlos is not very clear. It is likely that the name Tlos was originally an equivalent of the Lycian word Tlawa, a word that has often been linked to the 'Land of Dalawa' found in the late Bronze Age Hittite sources (Carubba 1993, 13). For instance, a Luwian hieroglyph mentioning the land of Dlawa has been attested on a Hittite monument at Yalburt near Konya in central Anatolia (Poetto 1993, 70–4). Apart from references in written sources, there is no explicit archaeological evidence from the site to confirm that there was a settlement at Tlos in the late Bronze Age, although recent excavations in the stadium area near the acropolis of the city have begun to yield Iron Age finds and some remains possibly of late Bronze Age character. It should, however, be noted that settlements of pre-Bronze Age character are common within the territory of Tlos (Korkut 2014, 103–5). Archaeological soundings at the Tavabaşı Cave and the Girmeler Cave demonstrate human activity dating back to the early Neolithic period (Korkut 2014, 108–10; 2015, 22–5).

The ancient city of Tlos witnessed its golden age in the Classical period in the 5th and 4th centuries BC, when Tlos and other Lycian cities were a part of the Persian Achaemenid Empire (Bryce 1986). Following his arrival in 333 BC, Alexander the Great took control of Tlos and all of Lycia.

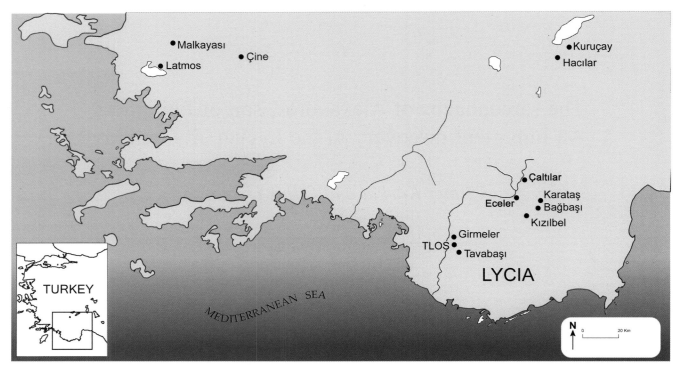

Fig. 6.1. Tlos. Map showing the location of the town.

After the death of Alexander, Lycia, from the beginning of the 3rd century BC, came under the control of the Ptolemaic dynasty, in 197 BC passing briefly to the Seleucids. In 168 BC all of the Lycian cities came together to form the Lycian League, in which Tlos was listed as one of the major cities with a right to three votes (Korkut and Grosche 2007, 79–81).

During the reign of Emperor Claudius, in AD 43 Lycia became part of the Roman Empire. Tlos managed to preserve its significance in the Lycian League under Roman dominion. According to the Roman milestone in the form of a monumental pillar (*Stadiasmus Provinciae Lyciae*), erected in the Lycian capital city of Patara as a dedication to the Roman Emperor Claudius, Tlos was accessible by a number of different routes during this period (Işık, İşkan-Işık, and Çevik 1998–1999). The city continued to maintain its importance in the early Byzantine period, at which time Tlos is listed among the important bishoprics of Lycia. The archaeological remains from the city confirm that Tlos maintained this importance until the 12th century (Korkut 2015, 42–7). The significance of Tlos within the borders of Lycia continued also during the Ottoman period (Hellenkemper and Hild 2004, 885–8).

Archaeological excavations, initiated in Tlos in 2005, placed special emphasis on the monumental structures located in the centre of the city and the areas close to it (Fig. 6.2). The excavations at present continue investigating structures such as the theatre, the stadium, the great baths, the temple of Kronos, the city basilica, and on the acropolis (Korkut 2015). In addition to these excavations, field surveys have

also been conducted in the city and its territory from 2005 (Korkut 2013; 2015). The focus of attention has been the pre-Classical and Classical past of the city and its territory.

The sarcophagus, the subject of this paper, was identified in an area to the east of the ancient city centre of Tlos (Fig. 6.3). The location in which it was found was evidently within the necropolis of the city, which had been used intensively from the Classical period into the Roman era. Although a series of sarcophagi, mausoleums, and a *chamosorion* grave had previously been documented in the necropolis, this sarcophagus had not been identified during our surveys due to a road employed by modern farmers passing over its top. In 2013, during the excavation season, an attempt was made to rob this sarcophagus by looters who succeeded in opening a hole in one of the narrow sides of the lid. This looting activity fortunately was noticed before any serious damage had been done to the sarcophagus. In consequence, its excavation was immediately included in the 2013 excavation program after the requested official permission had been obtained. The aim was to prevent any further destruction by grave robbers and obtain as much information as possible about the sarcophagus and its contents.

Carved from local limestone and oriented east/west this is a typical Lycian sarcophagus consisting of a rectangular coffer body (1.97×0.80×1.37 m) crowned by a gabled lid (2.40×1.22×1.40 m) (Fig. 6.4). The lid is preserved almost complete except for a small fracture in the ridge beam,

Fig. 6.2. Tlos. Aerial view.

while the rectangular coffer has several fractures on its north, west, and south sides. Both the interior and exterior of the rectangular body of the sarcophagus exhibit quite rough craftsmanship, excluding the smoothed northern side which carries a Greek inscription: ΑΛΕΞΑ[ΝΔΡ]ΟΥ ΤΟΥ ΦΙΛΙΠΠΟΥ carved over three lines (Fig. 6.5). On the lid there are in all six bosses for lifting, two on each of the long sides, one on each of the short ones. The long sides of the lid were also left smooth while the short sides have a 10 cm high border at the bottom end. On the short sides of the lid, a stylized pillar, in imitation of a wooden beam, rises above the bosses.

The excavation of the sarcophagus

In order to obtain as much information as possible about this sarcophagus, both the interior and its external surroundings were excavated. Four burial layers were identified (−60/−80, −80/−85, −85/−95, and −95/−110 cm) within the rectangular coffer of the sarcophagus, which contained an approximately 1 m thick fill (Fig. 6.6). The top, or first and latest surviving layer produced the skeletons of three individuals, the second layer contained only one skeleton, the third layer produced the skeletons of 11 individuals, and the bottom fourth layer contained the skeletons of further 19 individuals, in total 34 individuals. The skull (KF-3) found and drawn in the first layer originally belongs to the second layer. According to the bone fractures from the first layer the third individual (F-1) is

a foetus; the skull of it is shattered. The skulls of the foetus (F-2) found in the third layer and the two infants found in the fourth layer (I-1-2) are also shattered. The two adult individuals identified in the first layer had rubble stones placed under their heads to serve as pillows (Fig. 6.7). The two parallel lying skeletons, facing in opposite directions, have their arms bent at the elbows and placed on the abdominal cavity. Furthermore, skeleton 2 (KF-2) was intentionally placed in the space formed by the opening of the legs of skeleton 1 (KF-1), which could imply that these individuals, defined as middle adults, were buried simultaneously. The heads of the deceased in the first and the second layers were all placed in the west. In the next layer down one head appears in the east, the others in the west end, while in the bottom layer the heads were divided between both ends of the sarcophagus, though most of them in the west. The heads in the two lowest layers appear not to be in their original position but may have been repositioned in clusters as the result of some cleaning up of older burials to make space for new ones.

The osteological standards set by Buikstra and Ubelaker (1994) are adopted here in order to determine the age and the gender of the individuals buried in the sarcophagus. The age of the individual is divided into seven different age groups: foetal (prenatal), infant (0–3 years), child (3–12 years), adolescent (12–20 years), young adult (20–35 years), middle adult (35–50 years), and old adult (50+ years). In the first table below cases of uncertain allocation are signalled by a question mark (?).

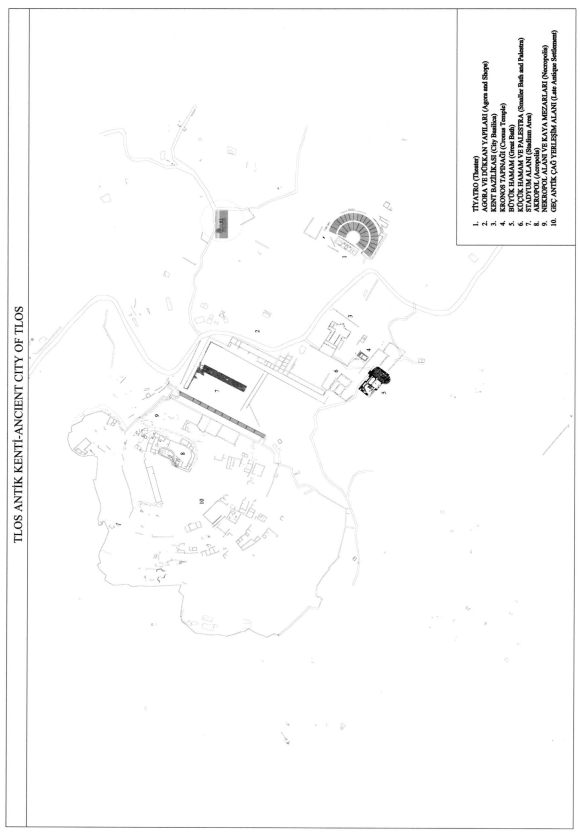

Fig. 6.3. Tlos. City plan of the ancient town and the location of the Alexandros sarcophagus.

Fig. 6.4. Tlos. The Alexandros sarcophagus (north side), carved from local limestone, at the end of excavations.

The data recording the gender, age group and level of the 34 individuals found in this sarcophagus are shown in Table 6.1, while the numbers in the age groups and gender of the individuals are shown in Table 6.2.

During the excavations conducted on the north side of the sarcophagus several bone fragments representing a skull, chin, and finger, together with several fragments of pottery were identified at the −50 cm level and below. These bone fragments fall into the category of infant, child, and middle adults, although they do not show any anatomical articulation. The find levels and the characteristics of the bones found outside the sarcophagus itself are listed below:

−50/−75: mandibles, some cranial bones, and some phalanges belonging to one child and a middle adult?

−60/−75: bones from a body belonging to one middle adult and one infant.

−65/−75: some cranial bones, phalanges and vertebral bones, as well as some animal bones.

−75/−115: some bone parts belonging to the infant found at a depth of −60/−75 cm, various bones and bone parts, in addition to animal bones.

−115/−140: body bones.

−140/−145: body bones and a tooth.

Evaluation of the archaeological finds

Most of the archaeological finds recovered from the inside of the sarcophagus consisted mainly of body shards of pots. The first of three diagnostic pottery fragments came to light

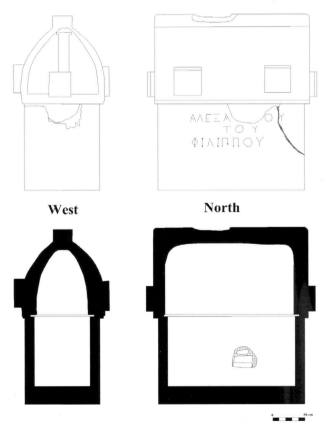

West   **North**

Fig. 6.5. Tlos. The Alexandros sarcophagus. Drawing of the sarcophagus with inscription (north and west sides).

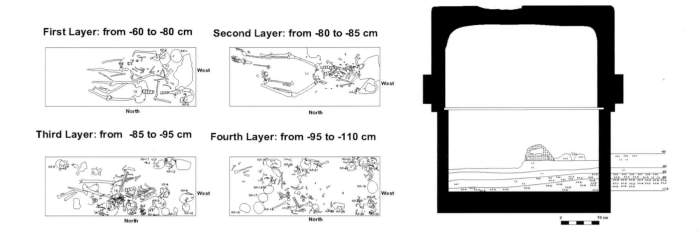

Fig. 6.6. Tlos. The Alexandros sarcophagus. Burial layers.

in the first layer, a black glazed *kylix* base datable to the Classical period (Fig. 6.8, no. 1). The second diagnostic fragment (Fig. 6.8, no. 2) came from the fourth, lowest layer, the body of a Hellenistic pot employed as a pillow for the child (KF-19). The third diagnostic example was a base of a jug (Fig. 6.8, no. 3), likewise from the first layer. In addition to the pottery, objects including glass, bronze, and iron were also found (Fig. 6.9). It is not possible to determine the original form of the glass fragments (Fig. 6.9, no. 1). Glass beads are of a wide variety of forms and colours; the round beads, embellished with glass frit in light or dark contrasting coloured layers, date from the 1st to the 3rd century AD (Fig. 6.9, no. 2). The same chronology is valid for the cylindrical and oval beads. Bracelets of iron and bronze with round sections were identified, as well as a spiral-shaped bronze ring (Fig. 6.9, no. 3). These metal finds also date to the period between the 1st and 3rd centuries AD.

The excavations conducted outside the sarcophagus provided more material than from its inside. The area in front of the inscribed northern flank of the sarcophagus, in particular, yielded diagnostic examples for dating. The pottery fragments recovered together with human bones near the sarcophagus range over a long period, from the 4th century BC to the 7th century AD. The earliest examples are black glazed body shards, followed by fragments of Hellenistic *unguentaria* (Fig. 6.8, no. 4), *lagynoi* (Fig. 6.8, no. 5), pots (Fig. 6.8, no. 6), cups (Fig. 6.10, no. 1), and *skyphoi* (Fig. 6.10, no. 2) datable from the 3rd to the 2nd century BC. A ring-stone depicting Heracles (Fig. 6.11), together with the Alexandros inscription, are both related directly to the use of the sarcophagus. On this ring, which features a black-and-white-veined onyx

stone, is a bust of Heracles wearing a wreath with his head slightly turned to his right. Both the position and physiognomy of the face are similar to the depiction

Fig. 6.7. Tlos. The Alexandros sarcophagus. The first, top layer with three skeletons.

Table 6.1. The gender, age, and excavation level of the skeletons found in the sarcophagus.

Individual	Sex	Age Group	Ground Level (cm)
1 (KF-1)	Male	Middle adult	−60/−80
2 (KF-2)	Female	Old adult	−60/−80
3 (F-1)	Unidentified	Foetal	−70/−80
4 (KF-3; İ-1)	Male	Middle adult	−80/−85
5 (KF-4)	Female	Old adult	−85/−95
6 (KF-5)	Female	Child	−85/−95
7 (KF-6)	Male	Young adult	−85/−95
8 (KF-7)	Male	Middle adult	−85/−95
9 (KF-8)	Female?	Adolescent	−85/−95
10 (KF-9)	Male	Middle adult	−85/−95
11 (KF-10)	Unidentified	Young adult	−85/−95
12 (KF-11)	Female	Middle adult	−85/−95
13 (KF-12)	Male?	Middle adult	−85/−95
14 (KF-13)	Unidentified	Child	−85/−95
15 (F-2)	Unidentified	Foetal	−85/−95
16 (KF-14)	Female?	Young adult	−95/−105
17 (KF-15)	Male	Middle adult	−95/−105
18 (KF-16)	Unidentified	Child	−95/−105
19 (KF-17)	Male	Young adult	−95/−105
20 (KF-18)	Male	Middle adult	−95/−105
21 (I-1)	Unidentified	Infant	−95/−105
22 (I-2)	Unidentified	Infant	−105/−110
23 (KF-19)	Unidentified	Child	−105/−110
24 (KF-20)	Male	Young adult	−105/−110
25 (KF-21)	Male	Adolescent	−105/−110
26 (KF-22)	Male	Young adult	−105/−110
27 (KF-23)	Male	Middle adult?	−105/−110
28 (KF-24)	Male	Middle adult?	−105/−110
29 (KF-25)	Male	Middle adult	−105/−110
30 (KF-26)	Unidentified	Child	−105/−110
31 (KF-27)	Unidentified	Child	−105/−110
32 (KF-28)	Female	Young adult	−105/−110
33 (KF-29)	Female	Old adult	−105/−110
34 (KF-30)	Female	Unidentified	−105/−110

Table 6.2. The age and gender distribution of skeletons found inside the sarcophagus.

Sex	Age Group (years)							
	Foetal (<birth)	Infant (–3)	Child (3–12)	Adolescent (12–20)	Young adult (20–35)	Middle adult (35–50)	Old adult (50+)	Unidentified
Male/Male?				1	4	10		
Female/Female?			1	1	2	1	3	1
Unidentified	2	2	5		1			
Total	**2**	**2**	**6**	**2**	**7**	**11**	**3**	**1**

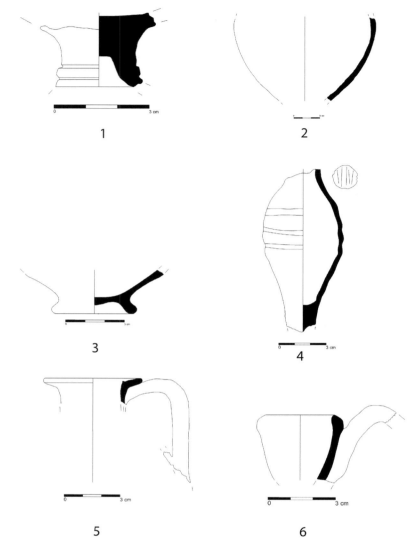

Fig. 6.8. Tlos. The Alexandros sarcophagus. Pottery from inside and outside the sarcophagus.

of Heracles by Lysippos. The deeply shaped eyes, the plasticity, the highlighted hair and beard, and other art/craft refinements, all suggest a date of the ring-stone to the beginning of the Hellenistic period.

The use of the sarcophagus during the Roman period is attested by the presence in the excavations of fragments of pottery and blown glass datable from the 1st to the 4th century AD. Open-mouthed bowls and plates with ring bases of a utilitarian character (Fig. 6.10, no. 3), a bowl imitating terra sigillata wares (Fig. 6.10, no. 4), *unguentaria* (Fig. 6.10, no. 5) and straight-rimmed glass cups (Fig. 6.10, no. 6; Fig. 6.12, no. 1) can all be dated to the 1st and the 3rd century AD, while the plates with roulette decoration, late Roman terra sigillata C and D groups (Fig. 6.12, nos. 2–3), are all from the 4th century AD.

The latest use of the sarcophagus is documented by late Roman and early Byzantine glass and pottery ranging in date from the 5th to the 7th century AD. The pottery includes *pithoi* with rims with relief decorations such as bead and incised and grooved wavy lines (Fig. 6.12, no. 4) and by cooking pots with flaring rims and bulbous bodies (Fig. 6.12, nos. 5–6). A coin from the reign of the Emperor Heraclius (622–623 AD) seems to confirm that the sarcophagus was still in use in the 7th century AD (Fig. 6.13).

The inscription

The inscription carved on the northern face of the sarcophagus reads: Ἀλεξά[νδρ]ου τοῦ Φιλίππου [(The tomb) of Alexandros, son of Philippos].

1

2

3

Fig. 6.9. Tlos. The Alexandros sarcophagus. Glass, bronze and iron objects from inside the sarcophagus.

The names Philippos and Alexandros are very common in the historical records. At first glance, the names recall Alexander the Great and his father. Alexander the Great is rarely called by his personal name in inscriptions or in the Greek sources, where he most often is recorded as 'Alexandros Basileus' or 'Alexander Magnus/Alexandros Makros'. For this reason, the inscription must be related to a family employing the names 'Alexander' and 'Philip'. For the interpretation and the dating of this inscription, some specific factors, such as palaeographic features and syntactic structure, have to be considered. Starting with the palaeographic features, the height of the letters in the inscription is 12 cm. This is quite unusual for an inscription, especially for an inscription on a sarcophagus. Such large letters often appear on architecture during the Roman period. However, writing in large letters on a stone monument is not peculiar to any period. The height of the letter may change according to the form and size of the monument and at the commissioner's request. Hence, there are some inscriptions containing large letters on the monuments of the Hellenistic period, especially in the

2nd century BC and later. Furthermore, some special letters used in the inscription are peculiar to the Hellenistic period, i.e. A, Ξ and Π. The letter 'Π', for example, that is 'pi', was written in two ways. The first 'pi' is the typical Classical and Hellenistic letter form in which the second leg of the letter is left short in mid-air. The other 'pi' has a shape which became standardized in the Roman Imperial period, when both legs of the letter were of equal length. The present type of 'pi' are transition letters found in 1st-century BC inscriptions.

In addition to these palaeographic features, the syntactic structure of the inscription forms a dating criterion. Indicating the patronymic, the father's name, with the masculine genitive article (ΤΟΥ), is a characteristic of Classical period standard inscriptions. For most of the Hellenistic period, the father's name was written with this article. During the Roman Imperial period, this article was used not for the father's name, but for the grandfather's. It is not possible to say that this inscription belongs to the Roman Imperial period due to this. However, there remains the possibility that this might be a clerical error and there should have been a KAI conjunctive after the TOY genitive article showing the father's name. However, there is no trace of writing in the second line of the inscription. If this conjunctive had been used, the translation would have been: 'Alexandros', also known as Philippos' (tomb)'. However, there is no evidence to show that it should be read in this way. All of the second line was left blank as would have been suitable for the writing of the TOY article. The inscription might have been added to the sarcophagus in the 3rd century BC based on this grammatical characteristic. There is the opportunity of evaluating the social status and personal name recorded in the inscription. Firstly, the names Alexandros and Philippos were very popular as Alexander the Great and his father Philip were historically important figures. During the Roman period there was a restriction on the use of 'tribu' and 'gentilicia' names. However, the use of 'gentilicia' names was allowed by the emperor and everyone had the right to use them, such as the name Aurelius. However, there was a restriction placed on the use of the name of Alexander. In contrast, he gave his own name to many cities. He was pleased that his name was given to monuments and buildings and the name Alexandros has been used from Alexander the Great until today by everyone.

The name Alexandros is not common in the Lycian region, so far only 23 individuals with this name have been registered, all from the Roman period. Until the discovery of this inscription at Tlos, no earlier use of the name Alexandros had been documented. Lycians spoke the Lycian language and gave themselves local names. Very few people gave Greek names to their sons, probably because the Greeks had little influence on their culture, but in the Hellenistic and later periods mixed families occur giving a much more varied picture in the use of personal names, especially in the Roman period when important changes happened to the Lycian culture. At that time, in addition to the local names, people

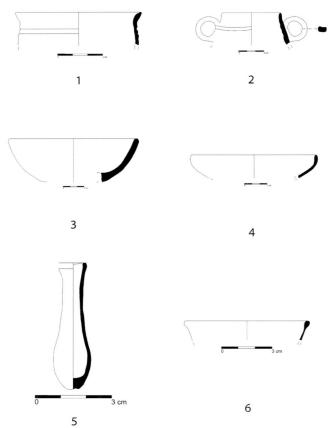

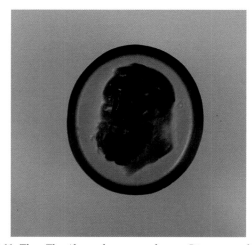

Fig. 6.10. Tlos. The Alexandros sarcophagus. Pottery and glass from outside the sarcophagus.

gave names such as Claudius, Aurelius, Marcius, Valerius, etc., which is why the name Alexandros is found frequently in the Roman period.

'Alexandros, son of Philippos' would fall within the normal range of names under normal conditions. However, this name

is an exception, not only in the inscriptions from Lycia and Anatolia, but for other Greek inscriptions elsewhere. This name serial is only present on the oracle inscriptions at Delphi. At CID (*Corpus des inscriptions de Delphes*) IV 117, there is a record of 'Alexandros, son of Philippos' in an inscription dated to 117 BC. Except for this example, no other matching name serials exist. The question remains, what was the reason for this exception. Did it result from admiration and special respect for Alexander the Great and his family? It is reasonable to expect that the grave owner's family may have expressed such admiration. A person with the name of Philippos might well have given the name of Alexandros to his son due to admiration for Alexander the Great. Such a family tree might not have been acceptable to the local population, but it was perhaps possible for a family which moved to Tlos from the West. As a result, it is reasonable to argue that the use of this name by the owner of the grave might not have been a casual incident. The use of the name seems to indicate that he was not only a person of importance in Tlos, but possibly in all Lycia in terms of social status.

Multiple uses of the sarcophagus

From the evaluation of the finds recovered from inside and outside the sarcophagus it can be concluded there were three phases of use of this sarcophagus. The first can be dated to the late Classical period with reference to the typology of the sarcophagus itself, with its gable-roofed lid typical for this period; the black glazed potshard found inside supports this date. The second phase of use dates to the Hellenistic period, when the inscription was added to its northern side: 'Alexandros, son of Philippos'. Both the ceramic finds and the ring-stone depicting Heracles date from the 3rd/2nd century BC. The grave probably continued to be used by the same family until the 4th century AD, when, as our excavations could testify, the sarcophagus was robbed and emptied.

According to the finds the third and final phase of use dates from the 4th to the 7th century AD, but it has not been possible to establish for how long the sarcophagus was abandoned between the second and third phases. The burials of this late phase were all unfurnished. While the skeletons of the lower layers in this last phase of the sarcophagus were much disturbed, the anatomical articulation of the skeletons in the latest, top burial layer was not destroyed; a Hellenistic pot and bricks from the hypocaust system of the Roman baths (abandoned in the 3rd century AD) were reused as supports for the heads of the deceased. This reuse of the bricks is another confirmation of the late date of the burials.

As a result, the excavation undertaken in and around this sarcophagus shows that the original Classical sarcophagus was reused first sometime in the Hellenistic period and then again in the early Byzantine period as a family burial (Fig. 6.14).

Fig. 6.11. Tlos. The Alexandros sarcophagus. Ring-stone depicting Heracles.

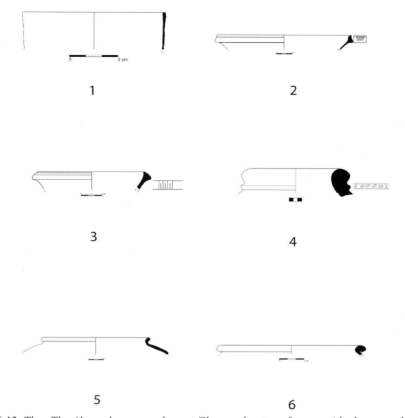

Fig. 6.12. Tlos. The Alexandros sarcophagus. Glass and pottery from outside the sarcophagus.

Fig. 6.13. Tlos. The Alexandros sarcophagus. Coin from the reign of Emperor Heraclius.

During the early Christian period the old city centre of Tlos was abandoned and a new city centre was established on the southern side of the acropolis. This new area was surrounded by a 'city wall' starting from the foot of the acropolis (see Fig. 6.3). However, the residential area extended beyond the city walls during the mid-Byzantine period. For the early Byzantine period in the city planning of Tlos therefore, the use of this area as a necropolis is unusual. It is clear that the sarcophagus found at the edge of the city was used as a necropolis throughout the Classical, Hellenistic, and Roman periods. Nevertheless, no archaeological remains representing the Byzantine period have so far been identified in this area. This is because the settlement was moved to the southern slope of the acropolis of Tlos in the early Byzantine period. As is understood from the name of the owner of the burial, the members of this family were ordinary people, even if once of some possible importance. Although it is not certain, the possibility that there was

Fig. 6.14. Tlos. The Alexandros sarcophagus. Recreated view of the position of the sarcophagus in the early Byzantine period.

another settlement area on the eastern side of the city centre in the early Byzantine period, should not be excluded. It is hoped that future research will illuminate this issue.

Acknowledgements

We thank the Turkish Academy of Sciences (TÜBİTAK) for supporting this work through the Project No. 111K227. We are grateful to Cenk Güner and Recai Tekoğlu who undertook the preliminary analysis of the bones and of the inscription. Our thanks are also due to Bilsen Özdemir, Tijen Yücel, Bayram Akdağ, and Kudret Sezgin for their contributions in the field and in the preparation of the illustrations.

Bibliography

Bryce, T. R. (1986) *The Lycians I. The Lycians in literary and epigraphic* sources. Copenhagen, Museum Tusculanum Press.

Buikstra, J. E. and Ubelaker, D. H. (1994) *Standards for data collection from human skeletal remains. Proceedings of a seminar at The Field Museum of Natural History* (Arkansas Archaeological Survey Research Series 44). Fayetteville, Arkansas Archaeological Survey.

Carubba, O. (1993) Dynasten und Städte. Sprachliche und sonstige Bemerkungen zu den Namen auf den lykischen Münzen. In Borchhardt and Dobesch (eds.) *Akten des II. Internationalen Lykien-Symposium, Wien 6.–12. Mai 1990*, 11–25. Vienna, Österreichische Akademie der Wissenschaften.

Hellenkemper, H. and Hild, F. (2004) *Tabula Imperii Byzantini 8: Lykien und Pamphylien.* Vienna, Österreichische Akademie der Wissenschaften.

Işık, F., İşkan-Işık, H., and Çevik, N. (1998–1999) *Miliarium Lyciae. Das Wegweisermonument von Patara* (Lykia, 4 [2001]). Antalya, Akdeniz Üniversitesi.

Korkut, T. (2013) Die Ausgrabungen in Tlos. In P. Brun, L. Cavalier, K. Konuk, and F. Prost (eds.) *Euploia. La Lycie et la Carie antiques: Dynamiques des territoires, échanges et identités* (Actes du Colloque de Bordeaux, 5, 6 et 7 Novembre 2009 Bordeaux), 333–44. Bordeaux, Ausonius.

Korkut, T. (2014) Tlos 2012 Kazı Etkinlikleri. *Kazı Sonuçları Toplantısı* 35.3, 103–18.

Korkut, T. (2015) *Tlos. Akdağlar'ın Yamacında Bir Likya Kenti*, Istanbul, Ege Yayınları.

Korkut, T. and Grosche, G. (2007) *Das Bouleuterion von Patara. Versammlungsgebäude des lykischen Bundes, Patara II 1.* Istanbul, Ege Yayınları.

Poetto, M. (1993) *L'iscrizione luvio-geroglifica di Yalburt* (Studia Mediterranea 8). Pavia, G. Iuculano Editore.

'Til death do them part': Reconstructing Graeco-Roman family life from funerary inscriptions of Aphrodisias

Esen Öğüş

Abstract

This article focuses on the funerary inscriptions at Aphrodisias in Caria, with the aim to illuminate social relationships in Graeco-Roman domestic life. These inscriptions are formulaic texts stating the type of the tomb, its owner, authorized individuals to be buried in the available spaces, and a fine that would be paid to the temple of Aphrodite if the tomb was violated. The preferences of the owners regarding the assignment of burial places to their relatives and acquaintances provide clues about domestic hierarchy. The main issues of focus are diversity of family sizes and types; gender roles and hierarchy; and the range of relatives and acquaintances. This line of inquiry showed that Aphrodisians were foremost concerned with providing a safe burial spot for their first-degree blood relatives, while also accommodating the other acquaintances, including ex-wives, foster children and slaves. Another important conclusion of the article is that the choice of burial spots was dependent on many factors other than who constituted as family, such as gender hierarchy, availability of space, socially acceptable ways of commemoration and the dominant funerary culture. In fact, the range of domestic relationships and family sizes reflected on funerary inscriptions make it difficult to define the typical ancient family.

Keywords: Aphrodisias, ancient family, domestic hierarchy, funerary inscriptions, gender roles.

The definition and constitution of the Roman family has been the topic of many studies, which have gathered epigraphic, literary, and archaeological evidence to understand how lives were led behind closed doors.[1] Funerary inscriptions provide one way of learning aspects of ancient *domus* life. While the format of these inscriptions vary in different regions of the Empire, in general, they include the name of the deceased, the individual(s) authorized to be inhumed in the tomb, and sometimes a description of the tomb itself. The inscriptions also sometimes disclose the nature of the relationship between the deceased and authorized individuals, and the assignment of the burial spaces to these individuals. However, there were several factors, other than the concern to secure an eternal resting place for oneself and acquaintances, which determined the assignment of places. These are regional customs, available finances, the desire to impress fellow citizens, care or neglect of the heirs, and the availability of grave space. Therefore, close scrutiny of funerary inscriptions should be approached with caution. First of all, the relationships expressed in funerary inscriptions do not necessarily provide a perfect mirror of daily social life, since one's familial obligations might be different than how one wishes to be commemorated after death. In other words, the proper mode of conduct after death might have been defined differently to that in life. Similarly, due to their selective and limited nature, it is inconvenient to derive from funerary inscriptions overarching conclusions about how the Roman family is characterized, whether it is 'nuclear' or extended, and how similar it is to our own.[2] In general, defining and categorizing the ancient family based on our own is not a useful effort.

However, there are still many convenient bits of information about ancient domestic life that may be derived from funerary inscriptions, especially when some of these inscriptions are accompanied by secure archaeological context. Above all, inscriptions shed light on:[3] (1) the diversity of ancient family types and sizes in the funerary realm (one-person or two-person households, extended families); (2) the range of relatives and acquaintances (parents, siblings, slaves, in-laws, etc.); and (3) gender roles and social hierarchy after death. With the aim of scrutinizing these three main aspects of domestic life, the present study focuses on the funerary inscriptions and archaeological evidence at Aphrodisias in Caria. Similar analyses were conducted in the west and other cities of Asia Minor (Saller and Shaw 1984; Martin 1996), but the advantage of Aphrodisias over the others is that some inscriptions were preserved in an established archaeological context. Therefore, the descriptive phrases denoting tombs in the inscriptions can sometimes be associated with physical remains, which allow for more concrete results on how household hierarchy was shaped based on the available burial space.

Aphrodisias and its funerary inscriptions

Aphrodisias is a Graeco-Roman city in the Meander valley, well-known in Antiquity for its sculpture 'school' attested by hundreds of locally produced marble statuary displayed in the city, and by the physical presence of sculpture workshops in the city centre.[4] The city, having been excavated since 1960, presents a well-documented archaeological and epigraphic record, the latter of which can generally be studied in relation to surviving monuments.[5] Funerary inscriptions, too, are preserved in large numbers. The majority of them date from the late 2nd and early 3rd centuries AD, a period that coincides with a surge of sarcophagus and tomb production (Smith 2008; Öğüş 2014). Some of the inscriptions are on marble blocks that were once the building blocks of the tombs, which were taken down, and built into the city walls in the mid-4th century.[6] Many other inscriptions are carved on sarcophagi, which are, to a great extent, preserved today, some even with their archaeological context.

Similar to many other cities in Asia Minor, funerary inscriptions ordinarily express no sentiments related to death, but are formulaic texts that announce the legal owners and the allocation of the burial places to the relatives and acquaintances of the tomb owner. The inscriptions also determine a fine to be paid to official institutions in the case of the violation of the tombs. In most cases, the fine was made payable to the goddess Aphrodite, a third of which was to be paid to the prosecutor. This latter obligation was to increase the incentive for prosecution of tomb raiders. The year and the month when the inscription was recorded in the property archive was mentioned, although it is almost not possible to convert these dates into modern terms.[7] For the dating, one has mostly to rely on the style of lettering and nomenclature, and if it survives, the style of the sarcophagus.

The inscriptions define the type of the burial place with special terms usually employed at other sites of Asia Minor as well. The main type of burial space associated with a tomb was the sarcophagus chest, which was called a *soros*. Sarcophagi at Aphrodisias are exclusively of marble, mostly with gabled lids, and they may be categorized into types according to their decoration. Garland and columnar sarcophagi are of the most common ones. Garland sarcophagi have a tabula in the centre of the chest specifically reserved for an inscription, and two hanging garlands on either side. Occasionally, busts of the owners were carved in the lunettes of the garlands. Columnar sarcophagi are decorated with columns that support an arcade, and standing human figures in the intercolumniations. While most of the columnar sarcophagi are not inscribed, some were designed for this purpose by leaving a blank space on the upper edge of the chest for the inscription. If the chest has an inscription at all, it starts on the lower edge of the lid, continues on the upper edge of the chest and inside the tabula (if it exists), and ends on the lower edge of the chest. The inscriptions could have been painted to be more visible.

Sarcophagi were family burials, usually used for the inhumation of more than one person. An inscription on a plain sarcophagus clarifies how several people could be inhumed in the same chest in practice, which was more sophisticated than simply tossing the bodies on top of each other (*IAph2007* 13.203, S-419; Smith and Ratté 1996, 27). According to the epitaph, the owner and his wife are to be inhumed at the bottom of the chest. When their sons die, they would place a slab called an *abakeion* (which is a term peculiar to Aphrodisias) over the bodies, and they would be inhumed in the second storey within the chest. In fact, small projections in the corners inside the chest of this sarcophagus and many others confirm the use of *abakeion*, which was probably of wood.

One sarcophagus or multiple sarcophagi were placed on various kinds of tomb buildings and platforms. The tomb buildings were not systematically excavated at Aphrodisias, and as mentioned, their ashlar blocks were taken down for the construction of the city walls. However, some *hypogeum*-type vaulted tombs and tomb platforms are preserved today. These lie outside the city walls in all directions, accessible through gates and major roads out of town. Similar funerary terms were used to describe the tomb structures in other cities such as Hierapolis, Termessos, and Elaiussa Sebaste, where the tombs are in a better state of preservation and allow for an understanding of the dismantled tombs at Aphrodisias (Kubińska 1968, 73–93, 101–8. See also Öğüş-Uzun 2010 and Turnbow 2011). Accordingly, the platform with burial niches underneath was called a *platos/platas/platē*; and the house-shaped building with a flat roof was called a *bōmos*. The sarcophagi on top of these structures were displayed openly to the public.

The tomb buildings and platforms provided subsidiary burial niches, masonry chests adjacent to each other, called *eisōstē*. Several patrons owned both the sarcophagus and the associated monument underneath with these burial niches. They often reserved the sarcophagus chest for themselves and their close family, such as a spouse and children, and the niches underneath to less close family, such as foster children (*threptoi*). This suggests that the sarcophagus chest itself was the most desirable and prestigious spot of burial.

After this brief information about the tombs and inscriptions at Aphrodisias, I would like to focus below on the information about family life that they can provide. There are about 390 funerary inscriptions from Aphrodisias that date from the 2nd century BC to the 7th century AD. Of these, 153 are carved on sarcophagi, and date from the 2nd to the 4th centuries AD. This study takes into consideration 53 that are adequately preserved to include information about the legal owners of the sarcophagi.

1. Diversity of family types and sizes

One can tell from funerary inscriptions the diversity of family sizes and the range of options for commemoration of individuals. The table below summarizes the range of family sizes on the 52 inscriptions on sarcophagi (Fig. 7.1). Accordingly, some sarcophagi were only for one person (Category 1). Others were for two, either husband and wife, parent and child, unrelated people, or people whose relationship is not disclosed (Category 2). Many were for the basic family unit consisting of parents and children (Category 3), and yet others authorized burial for extended families (Category 4).

As can be seen, few sarcophagi (*c.* 23%) were reserved for a single person (Category 1). In that case, it is possible that either the person lived alone, or that she/he wanted to be inhumed alone, because ample funds allowed additional sarcophagi for other family members. Similar reasons

		%
Category 1	12	22.6
Category 2	14	26.4
Category 3	24	45.3
Category 4	3	5.7
Total	53	

Fig. 7.1. Aphrodisias. Table showing the categories of family types and their numbers. Generated by the author.

probably apply to Category 2, when either the family consisted of two related or unrelated individuals, or when these individuals deliberately shared a sarcophagus chest for whatever reason. In fact, quite a lot of people (*c.* 26%) preferred to be buried this way.

Yet very few sarcophagi (*c.* 6%) were for an extended family including individuals beyond the first-degree blood-relatives (Category 4). An example is an inscription that reserved the sarcophagus chest for the husband, wife, son, grandsons, grandfathers and uncle (*IAph2007* 12.1109). The largest number of bodies that were physically unearthed from a sarcophagus chest at Aphrodisias (*IAph2007* 13.156, S-5) is 11, even though the inscription on this large garland sarcophagus grants the right of burial to a couple, their son and grandson (Fig. 7.2). Chests were commonly reused for subsequent burials, but since this sarcophagus was excavated in 1904, DNA tests and even systematic recording was out of question, so we will never find out whether the bodies belonged to the same family. On the other hand, it has been attested that sarcophagi standing on the same platform could belong to members of the extended family. For instance, a rescue excavation in 1989 discovered a *hypogeum*-type tomb and four sarcophagi lying on the platform above it (Tulay 1991). Two of these sarcophagi were inscribed, and placed next to each other. The original inscriptions were erased and new inscriptions were probably re-carved in the 4th century, when the sarcophagi were reused. Accordingly, the inscription on the unfinished double garland sarcophagus (*IAph2007* 13.602, S-8) (Fig. 7.3) and the plot underneath belonged to a 'Marcus Aurelius Zenon, son of Zenon, son of Ophellios Christos.' A fully finished double-garland sarcophagus (*IAph2007* 13.603, S-10) (Fig. 7.4) next to it belonged to 'Marcus Aurelius Attikos, son of Attikos, son of Ophellios Christos.' Therefore, the owners of these two sarcophagi were cousins, children of two brothers, and grandchildren of the same man, Ophellios Christos. Apparently, each of them purchased a 'second-hand' sarcophagus and placed these next to each other on the same tomb platform. While it is not clear who owned the two other uninscribed sarcophagi, it is tempting to think that they also belonged to relatives.

The majority, 24 out of 53 inscriptions (*c.* 45%), reserved the sarcophagus chest for the basic family unit composed of parents and children, by our standards the 'nuclear' family (Category 3). Regardless, it is difficult to say whether the widespread practice of inhuming close family within the same sarcophagus chest means that Aphrodisians led a 'nuclear' family life, or whether this practice had more to do with the dominant funerary culture. First of all, the statistics above only apply to the individuals inhumed in the sarcophagus, not those inhumed in the burial niches of the family tomb. The latter exhibit a wide range of kinship: from first-degree blood relatives

Fig. 7.2. Aphrodisias. Garland sarcophagus (S-5). Courtesy New York University Excavations at Aphrodisias.

Fig. 7.3. Aphrodisias. Unfinished garland sarcophagus (S-8). Courtesy New York University Excavations at Aphrodisias.

Fig. 7.4. Aphrodisias. Garland sarcophagus (S-10). Courtesy New York University Excavations at Aphrodisias.

to slaves. Moreover, not every couple had to have a child, nor did everyone have to be married into a conventionally defined family. For instance, should we not count a man, his son and his daughter-in-law authorized to be buried in a sarcophagus as a 'nuclear' family?[8] The inscription is silent on why he does not want his wife to be buried with them. Is it because she remarried someone else and was excluded from the family, or perhaps because she owned another sarcophagus of her own? Did the man have no other children, for instance daughters, or were the daughters buried with their own husbands? Therefore, it is not possible to conclude that the majority of Aphrodisians had a nuclear family, but it is possible to suggest that the majority was first and foremost concerned with securing a safe and prestigious burial place for their first-degree blood relatives.

2. Range of relatives and acquaintances

The inscriptions disclose that the household of an ancient family included several individuals, some of whom, like in-laws and ex-wives, are familiar to us, while others, like slaves and freed people, are culture-specific. Many of these acquaintances were given spaces in the family tombs, and their range, as explained below, show the diverse relationships surrounding the ancient family.

In-laws

Some in-laws were given authorization for burial in the family tomb (Chaniotis 2004, 408–9, S-466). There is no surviving evidence for mothers- and fathers-in-law, who presumably possessed their own tombs. However, an inscription on a marble block (*IAph2007* 12.322) is evidence of the preference accorded the son over his wife, the daughter-in-law. The block originally comes from a complex tomb structure with two sarcophagi that belonged to a distinguished family of a high priest. According to the inscription, the sarcophagus on the right was reserved for the high priest, his wife, his two sons and daughter. The wife and the children of one of the sons, Menodotus, were to be placed in the sarcophagus on the left. So not only were the family of Menodotus separated from each other after death, but the daughter-in-law and the grandchildren received the sarcophagus on the left. It is well-known that in Greek thinking, right was the preferred and superior direction over the left, which was often considered unlucky.[9] Therefore, the daughter-in-law and grandchildren were predestined to be inhumed in the less desirable sarcophagus in the family tomb.

Ex-wives and illegitimate children

Ex-wives were also provided a place in the family tomb. However, slight differentiation was made between the deceased and divorced ex-wives. The former were given an equally prestigious spot in the sarcophagus chest as current wives.[10] Divorced ex-wives, although not forgotten, were granted a burial niche underneath the sarcophagus, not in the sarcophagus itself. For example, an inscription (*CIG* 2824 = *IAph2007* 15.245) mentions two wives, both as *gunē*, one to be inhumed with the husband in the sarcophagus, and the other to be inhumed underneath the sarcophagus with her son. A second burial niche was reserved for the other children of the owner. Since the language is prospective (i.e. so-and-so *will* be buried), none of the 'wives' had died when the inscription was carved. One of the wives, therefore, may have been the divorced ex-wife (if not the mistress), and she is presumably the one that was inhumed in the niche underneath the sarcophagus, not inside the sarcophagus.

Illegitimate children and their mothers were not forgotten either. An inscription (*IAph2007* 12.19a) bestowed the right of burial to a man, his wife and his son. It reserved the burial niche, on the other hand, for the other son of the man. It is likely that this second son was illegitimate or was born by another woman. The text is sometimes more explicit about the illegitimate son; for instance when a sarcophagus chest itself was reserved for the male owner, his wife and his daughter, and the burial niche underneath was destined for a Theseus and 'the woman who gave birth to Theseus' (*IAph2007* 11.38). It is very likely that Theseus was born out of wedlock, and it is significant that the legitimate daughter comes ahead of the illegitimate son in social status. These examples not only show the diverse relationships in life reflected in the layout and allocation of burial spots, but also how families changed and expanded their structure over time, in other words how they diachronically evolved.

Foster children, slaves, and freed people

Further epigraphic evidence suggests that the burial places in the tombs were not always reserved for blood relatives, but sometimes for *threptoi*.[11] The *threptos* institution, for which there is no exact equivalent in Latin or English (foster or adopted children are common translations), was a well-established one that was common in many regions of Asia Minor.[12] Several inscriptions at Aphrodisias reserved burial niches for *threptoi*, sometimes together with blood relatives, and other times on their own. In the latter case, it is probably possible to assume that the tomb owner had no blood relatives.[13] However, many inscriptions allowed the foster children to be inhumed in the sarcophagi together with the owners, eliminating the distinction between fosters and blood relatives.[14]

Slaves were inevitable members of the ancient household, but as expected, few funerary inscriptions attest the burial of a slave. An inscription on a garland sarcophagus is one of the few such epitaphs at Aphrodisias, which announces that the sarcophagus belonged to a Philippos, who authorized the burial of himself and Zenas, possibly his son and a slave of Aurelius Chrysippos (Fig. 7.5).[15] The owner of the sarcophagus, in this case, may have been

Fig. 7.5. Aphrodisias. Garland sarcophagus (S-478). Courtesy New York University Excavations at Aphrodisias.

a freedman wealthy enough to afford a sarcophagus for himself and his natural son, Zenas, who could still have been a slave. The two male busts in the lunettes of the garlands, instead of the more common design with one male and one female bust, confirm that this was a specially commissioned sarcophagus. Moreover, the two males wear a chiton without the himation, a clothing choice which might allude to manual labour. A second inscription, on a garland sarcophagus (*IAph2007* 11.59, S-29), that belongs to a city councillor, grants the right of burial in the sarcophagus to himself, his son and his foster son. Drosis and Dionysios, presumably his slaves, were allowed to be buried in the burial niche facing the road.

There are also epitaphs that specifically mention freed people (*apeleutheros*).[16] One of them (*IAph2007* 11.34), on a fragmentary block, reserves the burial niches under a sarcophagus for the nephew and the freed slaves of the sarcophagus owner. Clearly, the nephew was part of the household, and interestingly, no hierarchical distinction was made between him and the ex-slaves. Another epitaph inscribed on a triple-garland sarcophagus (*IAph2007* 13.150, S-226) reserved the sarcophagus for the owner and his foster-brother (*suntrophos*), and the niches below for the fostered and freed slaves. Of course, there might have been other freed people who could have possessed their own funerary monuments, but may not have mentioned their social status on inscriptions.

3. Gender roles and hierarchy

Funerary inscriptions are also quite articulate on gender roles and hierarchy. Expectedly, most sarcophagi and burial places were owned by men. In one case (*IAph2007* 13.112, S-26), the owner conditionally allows his wife to be inhumed with him in the same chest, but only if she remains his wife throughout her life and gives birth to male children. However, 12 inscriptions attribute the ownership of a sarcophagus or a funerary monument to women.[17] In addition, in four other cases (*IAph2007* 13.147; 13.109; 13.206; 15.340), the monument was owned jointly by the husband and wife. An inscription on a fragmentary sarcophagus (*IAph2007* 11.413, S-773) specifically mentions that the woman owner was independent and free to conduct business (called *autexousia*), because she had *ius liberorum*, or the ability to act without a male guardian since she had three or more children. Otherwise, it was common practice both in the Greek East and Roman West for women to have a legal guardian, a *kyrios* (tutor), usually a husband or father, in order to own or dispose property (Gardner 1986, 20–1; van Bremen 1996, 205–36).

A female owner of a specially commissioned sarcophagus is not only unusual for the gender of its owner, but also the broader social class that she was associated with (Fig. 7.6). The sarcophagus (*IAph2007* 13.101, S-415) has an inscribed tabula in the centre, flanked by a standing man and a woman on either side, presumably representing the owners. According to the inscription, the sarcophagus belonged to Aurelia Tate. Her husband, Aurelius Aquilinos, had already been inhumed in the chest, and when they die, Aurelia Tate, her son and daughter will also be inhumed in the chest. The inscription, therefore, suggests that the wife commissioned the sarcophagus upon the death of her husband, who was later inhumed (possibly transported from an initial burial place) in the chest. What makes this chest additionally peculiar is the small professional relief underneath the tabula. The scene shows a workshop, possibly a glassblower's workshop, where two men are seated around a furnace, one of which blows into a long

Fig. 7.6. Aphrodisias. Sarcophagus with tabula and standing patrons (S-415). Courtesy New York University Excavations at Aphrodisias.

tube (Fig. 7.7). It is quite likely that Aurelia Tate's husband owned this workshop, therefore he was not necessarily of aristocratic pedigree, but earned a good living for his family to be able to afford a marble sarcophagus. The sarcophagus demonstrates that even in a family of sub-aristocrat standing, it was not unusual for women to possess property and secure burial places for their families.

Fig. 7.7. Aphrodisias. Detail of Fig. 6, professional relief. Courtesy New York University Excavations at Aphrodisias.

Despite such examples of female enterprise, other cases demonstrate a strongly specified gender hierarchy on the assignment of burial places. Several sarcophagus owners reserved the sarcophagus chest, the most preferable burial spot, for themselves and their children (both sons and daughters), but some were selective in favour of their son, as is exemplified by an inscription that authorizes three male heirs to share a chest (*IAph2007* 11.54, S-53). An even more discriminatory inscription is on a plain (undecorated) sarcophagus. According to it, the family had both sons and daughters, but only the sons were granted the right of burial in the chest (Fig. 7.8) (*IAph2007* 13.203, S-419; Smith and Ratté 1996, 27). This allocation is partly understandable if one assumes that the daughters would be married into another family, therefore inhumed with their own husbands. However, the inscription also allows the daughters, without any reference to their marital status, to be inhumed in the family tomb, only not in the sarcophagus chest, but in the round moulded base (?) (*speira*) of the sarcophagus, which is now lost. These examples of inscriptions, which prioritize sons over daughters, most explicitly reflect a sexist hierarchy in the funerary realm, which perhaps also extended into social life.

Fig. 7.8. Aphrodisias. Plain sarcophagus (S-419). Courtesy New York University Excavations at Aphrodisias.

Conclusions

This survey of funerary inscriptions at Aphrodisias aimed to provide a window into the complicated set of domestic relationships in the ancient city. As mentioned in the beginning, funerary inscriptions are not a perfect mirror of ancient family life, since they are silent on many issues, including why a burial place was deemed appropriate for a particular person; how the owner of the tomb made his/her decision on the assignment of places; and the various modes of social conduct that ruled these decisions. However, as this essay has shown, there are still many aspects of family life that funerary inscriptions shed light on. Accordingly, a sarcophagus chest was esteemed the highest in rank. Here, the patrons preferred to inhume themselves and their closest acquaintances, usually first-degree blood relatives. If they had to make a choice, their sons, rather than daughters, were inhumed with them. All the other members of the household, including second-degree blood relatives, in-laws, foster children and slaves, were also reserved spaces in the family tomb, although not in the sarcophagus chest itself. It is clear from this review that Aphrodisians provided burial spaces to their acquaintances, even their illegitimate children, as far as their budget allowed. This all-embracing and affectionate picture might be peculiar to the funerary realm, but is one substantial start to question the extent to which it also applied to domestic and social life.

Notes

1   'S' numbers are Aphrodisias excavation numbers given to sarcophagi: Saller 1984; 1996; Bradley 1991.
2   As has been attempted in Saller and Shaw 1984. An ensuing critical article is Martin 1996.
3   These points, considering new possibilities, were suggested in Martin 1996, 53–6.
4   'School' of Aphrodisias: Erim 1967; Squarciapino 1983. Sculpture workshops: van Voorhis 1998. These workshops provided sculptures for Italian patrons as well: Squarciapino 1983; Moltesen 1990.
5   For the latter, the most recent study is the corpus of inscriptions provided in *IAph2007*.
6   Two governors in the 350s and 360s were named for the project: Eros Monachius (Roueché 1989, 35–8, doc. 19); and Fl. Constantius (Roueché 1989, 42–5, doc. 23).
7   This is due to the fact that the year is named after a priest called *stephanephoros*, and the month is named after the month names in the Macedonian calendar, after an emperor name, or by numbering. However, some *stephanephorates* are held several times during the lifetime of a person, or are posthumous and continue many years after death. See the account of dating in Reynolds and Roueché 2007, 150–2.
8   Chaniotis 2004, 408–9, S-466. This inscription is counted as Category 3 in this study.
9   See the thorough discussion of the meaning of 'right' and 'left' in Lloyd 1962.

10   For instance S-715 (A. Chaniotis, unpublished field report, I 07.17).
11   For examples of inscriptions regarding the burial of *threptoi*: *CIG* 2818, 2825; Paton 1900, no. V; Reinach 1906, no. 163; S-730; S-13; *IAph2007* 11.59; 13.150, S-226. There are epitaphs elsewhere in Asia Minor that mention the burial of *threptoi*, see Kubińska 1968, 82–3: Tlos (*TAM* II, 604); Sidyma (*TAM* II, 213); Pinara (*TAM* II, 247); Patara (*TAM* II, 437); Antiphellos (*CIG* 4300g). In Phrygia, the term is discussed recently in Thonemann 2013, 140–1.
12   Cameron 1939. The institution is more like the Turkish tradition of 'besleme', which was commonly practiced in Anatolia up to very recent times.
13   S-730 (A. Chaniotis, unpublished field report I 07.14).
14   For example, *IAph2007* 11.59, 12.631, and 14.11.
15   Smith and Ratté 2008, 742–3, fig. 31, S-478. For other epitaphs mentioning burial of slaves, see Reinach 1906, no. 163; *IAph2007* 13.150, S-226; 11.59, S-29. Elsewhere in Asia Minor, see Kubińska 1968, 82–3: Tlos (*TAM* II, 611); Sidyma (*TAM* II, 213); Patara (*TAM* II, 438); Antiphellos (*CIG*, 4300i); Cyaneae (Petersen and von Luschan 1889, nos. 29 & 31).
16   For epitaphs with 'freed' people elsewhere in Asia Minor, see Kubińska 1968, 82–3: Tlos (*TAM* II, 213); Patara (*TAM* II, 438, 454); Cyaneae (Petersen and von Luschan 1889, nos. 29 & 31).
17   *IAph2007* 1.140; 13.618; 12.523; 12.631; 13.101, S-415; 13.104; 13.108; 13.153; 14.11; 15.5; 12.320; S-730 (A. Chaniotis, unpublished field report I 07.14).

Bibliography

Bradley, K. R. (1991) *Discovering the Roman family: Studies in Roman social history*. Oxford, Oxford University Press.
Bremen, R. van (1996) *The limits of participation. Women and civic life in the Greek East in the Hellenistic and Roman periods*. Amsterdam, J. C. Gieben.
Cameron, A. (1939) ΘΡΕΠΤΟS and related terms in the inscriptions of Asia Minor. In W. M. Calder and J. Keil (eds.) *Anatolian studies presented to William Hepburn Buckler*, 27–62. Manchester, Manchester University Press.
Chaniotis, A. (2004) New inscriptions from Aphrodisias (1995–2001). *American Journal of Archaeology* 108, 377–416.
CIG Corpus inscriptionum graecarum, 1828–1877. A. Boeckh *et al.* Berlin.
Erim, K. T. (1967) The school of Aphrodisias. *Archaeology* 20, 18–26.
Gardner, J. F. (1986) *Women in Roman law and society*. London, Routledge.
IAph2007 Inscriptions of Aphrodisias 2007 (By J. Reynolds, C. Roueché, and G. Bodard). Accessible http://insaph.kcl.ac.uk/iaph2007.
Kubińska, J. (1968) *Les monuments funéraires dans les inscriptions grecques de l'Asie Mineure*. Warsaw, Editions Scientifiques de Pologne.
Lloyd, G. E. R. (1962) Right and left in Greek philosophy. *Journal of Hellenic Studies* 82, 56–66.

Martin, D. (1996) The construction of the ancient family: Methodological considerations. *Journal of Roman Studies* 86, 40–60.

Moltesen, M. (1990) The Aphrodisian sculptures in the Ny Carlsberg Glyptotek. In C. Roueché and K. T. Erim (eds.) *Aphrodisias papers: Recent work on architecture and sculpture* (Journal of Roman Archaeology, Suppl. 1), 133–46. Ann Harbor, MI.

Öğüş, E. (2014) Rise and fall of sarcophagus production at Aphrodisias. *Phoenix* 68, 137–56.

Öğüş-Uzun, E. (2010) *Columnar sarcophagi from Aphrodisias: Construction of elite identity in the Greek East.* Unpublished thesis, Harvard University.

Paton, W. R. (1900) Sites in E. Caria and S. Lydia. *Journal of Hellenic Studies* 20, 57–80.

Petersen, E. A. H. and von Luschan, E. (1889) *Reisen im südwestlichen Kleinasien. Vol. II, Reisen in Lykien, Milyas und Kibyratis.* Vienna.

Reinach, T. (1906) Inscriptions d'Aphrodisias. *Revue des études grecques* 19, 79–150; 205–98.

Reynolds, J. and Roueché, C. (2007) The inscriptions. In F. Işık, *Girlanden-Sarkophage aus Aphrodisias. Mit einem Beitrag zu den Inschriften von J. M. Reynolds and C. Roueché,* 147–92. Mainz am Rhein, Verlag Philipp von Zabern.

Roueché, C. (1989) *Aphrodisias in Late Antiquity* (Journal of Roman Studies, Monographs 5). London, Society for the Promotion of Roman Studies.

Saller, R. P. (1984) *Familia, Domus,* and the Roman conception of family. *Phoenix* 38, 336–55.

Saller, R. P. (1996) *Patriarchy, property and death in the Roman family.* Cambridge, Cambridge University Press.

Saller, R. P. and Shaw, B. D. (1984) Tombstones and Roman family relations in the Principate: Civilians, soldiers and slaves. *Journal of Roman Studies* 74, 124–56.

Smith, R. R. R. 2008. Sarcophagi and citizenship. In R. R. R. Smith and C. Ratté (eds.) *Aphrodisias papers 4. New research on the city and its monuments* (Journal of Roman Archaeology, Suppl. 70), 347–95. Portsmouth, RI.

Smith, R. R. R. and Ratté, C. (1996) Archaeological research at Aphrodisias in Caria, 1994. *American Journal of Archaeology* 100, 5–33.

Smith, R. R. R. and Ratté, C. (2008) Archaeological research at Aphrodisias in Caria, 2002–2005. *American Journal of Archaeology* 112, 713–51.

Squarciapino, M. F. (1983) La Scuola di Aphrodisias (40 anni dopo). *Archeologia Classica* 35, 74–87.

TAM II Tituli Asiae Minoris Antiqua, Vol. II, 1944 (by E. Kalinka and K. F. Dörner). Vienna, Rohrer.

Thonemann, P. (2013) Households and families in Roman Phrygia. In P. Thonemann (ed.) *Roman Phrygia. Culture and society,* 124–42. Cambridge, Cambridge University Press.

Tulay, A. S. (1991) Kabalar Kurtarma Kazısı 1989. In *I. Müze Kurtarma Kazıları Semineri. 1990,* 25–39. Ankara, Ankara Üniversitesi Basımevi.

Turnbow, H. (2011) *Sarcophagi and funerary display in Roman Aphrodisias.* Unpublished thesis, New York University.

Voorhis, J. van (1998) Apprentices' pieces and the training of sculptors at Aphrodisias. *Journal of Roman Archaeology* 11, 175–92.

Social status and tomb monuments in Hierapolis and Roman Asia Minor

Sven Ahrens

Abstract

The article explores the social status of tomb owners in Roman Asia Minor following three angles of analysis. The first analysis is the demographic approach where available tomb spaces are compared to the number of deaths during the Roman Imperial period using Hierapolis of Phrygia as an example, resulting in the hypothesis that most Hierapolitans may have been buried in tomb buildings and sarcophagi. The second analysis calculates the working time needed to produce a limestone sarcophagus and possible prices compared to the income of lower social groups. The result leads to the assumption that lower social groups indeed could purchase a limestone sarcophagus if their need for self-representation or available burial space required it. The third analysis gathers about 500 inscriptions from Asia Minor, which contain a status or profession statement of the tomb occupants to differentiate which status group could afford or preferred which type of monument. The choice of a tomb monument not only depended on the wealth of the deceased or his family, but appears to have been specific to certain social or professional groups. Just as simple limestone sarcophagi could be purchased by groups with low social status, reliefs and inscriptions were commonly used in groups with limited financial means.

Keywords: Asia Minor, Greek inscriptions, Hierapolis of Phrygia, limestone sarcophagi, social status, tomb monuments, tomb prices.

Prices for tomb monuments are not known from Roman Asia Minor and are generally very rare in most parts of the Mediterranean. When inscriptions giving prices are found, often they cannot be connected with a specific monument or are published without detailed information on the archaeological evidence. Consequently, our estimates of which social stratum could afford which kind of monument are largely based on size, material, and elaboration of the monuments. Even though the economic situation of the deceased or his family must quite often have had a decisive impact on burial expenditure, the display of wealth of a tomb in the Imperial period through size, furnishings, and shape does not necessarily coincide with the occupant's position in the social hierarchy of a society (Flämig 2007, 88–9; Hope 2000, 178–9). In general, display of wealth in death could have been a means of social positioning of the dead and their descendants and it does not necessarily correlate with social status in life (Fahlander and Oestigaard 2008, 10). Consequently, burial grounds allow, to a certain degree, for inferences on economic status, but they do not mirror social hierarchy accurately (Kamp 1998). The question here at stake is which evidence can support suggestions on which social group of Roman Asia Minor would prefer or could afford which kind of monument. For this purpose, several approaches shall be presented.[1] The city of Hierapolis plays a dominant role due to the fact that probably no site in the Eastern Mediterranean preserves such a large number of graves compared to the size of the city, which makes demographic estimates much more accurate.

Demographic approach

Hierapolis provides a unique insight into the size and extent of ancient necropoleis. Hardly any site in the Mediterranean preserves as many tombs. Of course, many tombs have been destroyed and sarcophagi displaced, making it impossible to estimate the loss correctly. The survey in the North-East Necropolis and particularly a magnetometer prospection led to the assumption that the loss should be much less than 20% rather than more. This allows for the attempt to compare a population estimate with the number of tombs.

Several attempts have been proposed to estimate the size of ancient city populations. In Pompeii the population density has been estimated to be about 164 (Engels 1990, 82) or 166.15 (Storey 1997, 973) inhabitants per hectare. In Ostia the population density is supposed to have been about 317 (Storey 1997, 973) or 390 (Engels 1990, 82), in Alexandria it may have been over 326.[2]

The parameters for Ostia and Alexandria are only suitable for large, densely populated cities with multi-story housing blocks, they are far too high for a small-sized inland city like Hierapolis with an area *intra muros* of 72 hectares. A smaller city like Pompeii is much better suited for comparative calculations. Using the parameters of Pompeii, Hierapolis could have had a population size of almost 12,000 inhabitants. A density as low as 100–50 inhabitants/ha was used for an estimate of the population of Pergamon, where only those areas suitable for houses were considered,[3] and using this density would result in a population smaller than 7,200–10,800.

Another possibility for estimates is counting the housing units and *insulae*. Unfortunately very little is known of houses in Roman Hierapolis. With only one *insula* excavated, and remodelled in the early Byzantine period, the estimate can again only refer to calculations from Pompeii (Zaccaria Ruggiu 2007). Hierapolis had probably a little under 200 *insulae*, which could have been used for dwellings.[4] Most *insulae* had a surface area around 1,900 m², the *insulae* along the city limits may, however, also have been considerably smaller.[5] We do not know how high the houses were but it is realistic to assume that there was at least an upper floor and that shops would also have been used as dwellings. While a poor family certainly would have lived in a shop of 20 m², a rich family could have inhabited several thousand square metres with their slaves. Wallace-Hadrill (1994, 99) has calculated a hypothetical living space of 34–9 m² per inhabitant for Pompeii. For Hierapolis, this would mean 48–56 inhabitants per *insula* and for the whole city 9,600–11,200 inhabitants if all 200 *insulae* were of the same size and used for dwelling.

The maximum estimates from both calculations point to a population size of 11,000–12,000 inhabitants. Of course, it should be assumed that Hierapolis did not have a constant population and that the population grew together with the city extension during the late 1st and the 2nd century AD (D'Andria 2001, 101–8; Ismaelli 2009, 445–54). However, for the sake of simplicity and to get a maximum rather than a minimum number the following calculation will be based on a constant population of 12,000 inhabitants.

Based on the population size a crude death rate can be estimated. Scheidel (2001, 25) has recently concluded that it will never be possible to reach a conclusive result for the crude death rate in the Roman Empire because of the lack of satisfactory empirical data and possibly great regional variation. But using the frame of an estimated life expectancy of between 20 and 30 years, that should cover the probable average, the crude death rate was somewhere between 50/1,000 deaths/annum and 33/1,000.[6] That would mean between 400 and 600 deaths per annum in Hierapolis. According to this calculation between 120,000 and 180,000 Hierapolitans would have died during the 1st to 3rd centuries AD in a stable city population. There are roughly 800 tomb buildings[7] and up to 2,000 sarcophagi preserved in Hierapolis today. While almost all sarcophagi seem to be from the Imperial period, fewer than 100 of the tomb buildings are from the Hellenistic period. Excavations have shown that these tombs continued in use during the Imperial and even until the early Byzantine period (Schneider Equini 1972, 127–8). On the assumption that a small tomb chamber could house up to 200 bodies, and a sarcophagus could house five, there is space for about 170,000 burials in Hierapolis.[8] Additionally, many tombs in the North Necropolis are larger than the average or comprised two burial chambers. Space was also provided in the substructures of sarcophagi and in *chamosoria*. That calculation makes it not only possible but highly likely that most of the population of Hierapolis could have been buried in the mentioned burial monuments.

This can also explain why poorer burials in tile graves and cist graves, as excavated extensively in neighbouring Laodikeia, are almost unknown from Hierapolis before the late Roman period.[9] Of course, tomb buildings would be built by the few families with the financial means and then house the extended household including dependants, like freedmen and slaves. Tomb buildings of Lycia have, at least since the 5th/4th centuries BC and through the Imperial period, been used for the extended household, including foster children and dependants (Bryce 1979; see also Işin and Yıldız, this volume). Even though the inscriptions of Hierapolis do not give an extensive catalogue of those buried in a tomb building, other Roman Imperial inscriptions from Asia Minor do. They usually mention the family members, foster children, friends, freedmen, and slaves (*IK* 2, 527; 15, 1635; 16, 2217D, 2218C, 2291, 2313E, 2543, 2547; 23, 190, 196; 23, 232; 51, 20; 59, 9; 61.II, 378, 385; 61.II, 441. Family and slaves: *IK* 44.II, 231. For Aphrodisias, see Turnbow Awan 2008, 165). In some instances, a patron would build a tomb for dependants only (*IK* 23.228, 295; 61.II, 393). There are no indications for the existence of *collegia funeraticia* with collective tombs for their members; the tomb of a Hellenistic collegium of *metoikoi* in Iasos is a rare exception.[10]

Prices

With the hypothesis established that Hierapolitans were usually buried in tomb buildings, sarcophagus monuments and *chamosoria*, but rarely in poorer graves, it has to be discussed who in the society could afford to erect such monuments. For such a task, information on the price of tomb monuments and the individual income is crucial.

The *Freiburger Münsterbauhütte* has provided an estimate of work-time for a sarcophagus of soft limestone similar to that of C309 in the North-East Necropolis (Fig. 8.1) based on the assumption that the technique employed was manual drilling and sawing and a 40-hour work week.[11] The total work-time was calculated to 5–7 weeks (3–4 weeks for the chest, 2–3 weeks for the lid). That means 200–80 hours.[12] The quarrying of the sarcophagus in Hierapolis may have taken a skilled and two unskilled workers about 17.5 hours/cubic metre (Pegoretti 1863, 159). For a sarcophagus like C309 a stone block of roughly 2.2 m³ for the casket and a block of roughly 1.7 m³ for the lid would be needed,

amounting to about 70 hours in total. Since the estimates by Pegoretti used here are slightly higher than others, the proposed numbers can be considered a maximum estimate (See Barker and Russell 2012, 89 table 2). Of course, the average working hours/week in Antiquity is unknown, but it should certainly be assumed that it was much higher than today (Barker and Russell 2012, 85). For stonemasons the working day would be limited to daylight and be reduced to the mornings and evenings by the heat in summer. Assuming a 50–60-hour week for quarrying and shaping sarcophagus, C309 would have taken about 6–7 man-weeks, but certainly the production time was much shorter because several men would be working simultaneously during some of the production stages. A recent calculation for a simple marble sarcophagus assumes one month's work for one skilled and two unskilled workmen including the quarrying, obtaining a result slightly higher than our calculation (Russell 2010, 122). The cost would be drastically reduced by the help of unskilled day labourers because they usually earned about half of the wages of a skilled worker (Drexhage, Konen, and Ruffing 2002, 183; Frézouls 1977, 260–1; Rathbone 2009, 314) Some working time for the transport of the block, repairing and forging of tools, construction, maintenance, and moving of lifting devices, and the forging and inserting of metal clamps for the lid has to be added and there would

Fig. 8.1. Hierapolis, North-East Necropolis. Sarcophagus C309 (photo University of Oslo excavation project).

be the cost of approximately 2m² of land.[13] Additionally, inscribing the sarcophagus would have cost more working time. It is fair to assume three months of working time for an undecorated and uninscribed limestone sarcophagus as a maximum estimate. If the workmen worked as part of a private enterprise, the revenue for the patron or landowner must also be added to the price.

The size of sarcophagi in the North-East Necropolis varies greatly and must have had significant implications for the time used for quarrying, transport and shaping, and consequently for the cost of the final product. While a block of about 5.4 m³ of travertine was needed for one of the largest caskets in the necropolis, the block for the smallest casket was a little smaller than 1.4m³. With an average unit weight of 2,500 kg/m³, the travertine blocks quarried for these two sarcophagi must have weighed around 13.5 tons and around 3.5 tons. It seems natural that small sarcophagi would have been far cheaper than large sarcophagi and thus been within reach of those with a modest income.

According to that calculation, the cost for a sarcophagus, including a generous overhead, should not exceed 50% of the yearly income of a skilled worker and rather be much lower. This percentage seems to fall within the expected range of funerary expenditures in the Roman Empire.[14] A recent calculation of the living standard under the reign of Diocletian estimated that a family of an unskilled workman could live at a little over mere subsistence level, off an income from 250 working days per year, if the diet was very simple and mainly vegetarian (Allen 2009, 340). Living standards and average wages must have been higher earlier in the Empire and could even have been considerably higher than in most preindustrial societies (Jongman, 2007, 601–2, 608, 617). Even though this is a much more optimistic estimate than many others, it still does not consider that women and children would also make small contributions to the family's income if necessary, that small gardens were cultivated and that some inherited money may have existed (Kehoe 2012, 123). It should not be forgotten that one of the main benefits of the sarcophagus is that it could be used for multiple burials of entire families. A sarcophagus could save a small family from further expenditures on tomb monuments for two to three generations.

Calculating the cost in money, not in working time, is more difficult. The only prices for sarcophagi known are late Roman and early Medieval. Usually the costs from the 4th century for simple limestone sarcophagi from Salona are used to demonstrate that they were only affordable for wealthy people. The costs of the sarcophagi are respectively 10 and 15 solidi, but both were inscribed and the exceptionally long inscription of the 15 solidi sarcophagus must have cost some 5 or 6 solidi alone (Handley 2003, 39). These prices amount to about 45 and 67.5 g of gold (rounding the weight of a solidus to 4.5 g). This is equivalent to about 6.15 and 9.25 aurii (assuming a weight of 7.3 g for one aureus in the 2nd

century AD) or 154 and 231 denarii in the 2nd century AD. Wage lists for gold miners in Dacia and quarry workers in Egypt of the 2nd century AD point to a monthly salary of around 12 denarii (0.48 aurii) plus additional provisions, which in cases in Egypt consisted of one artaba of wheat and additional provisions for special groups of workers.[15] The assistants, usually members of the family of an Egyptian quarry worker, would earn 7 denarii (0.28 aurii). This means that a sarcophagus from 4th-century Salona would have cost an Egyptian quarry worker of the 2nd century around 13 to 19 months of wages without the provisions. A father and son together would have to save their wages (without provisions) for 8.6 and 13 months respectively, to buy such an inscribed sarcophagus. An average civilian in the Nile delta would hardly make more money than the assistant, about 6 denarii (0.24 aurii) per month, and would have to pay 25.7 and 38.5 months of salary for such a sarcophagus (Cuvigny 1996, 140).

According to a calculation using the Diocletian price edict as a base, a day labourer would have earned 1.16 grams of silver a day around 301 AD (Allen 2009, 331). At an exchange ratio of silver to gold of 1/12 to 1/14, his wages for 250 working days would be very close to those of the average Egyptian civilian of the 2nd century. However, the assistant of a goldsmith in 4th-century Egypt received a salary of 3 solidi per year and would have to save for 40–60 months to buy this sarcophagus. Obviously comparing the prices of late Roman sarcophagi with those of the 2nd century is an impossible task not only because of geographic, and chronological differences, but also because of a possible decrease of the wages in gold. Additionally the increase in the gold price in the late Roman and Byzantine period makes it more complicated. The average 4th-century price in wheat would be 80 artaba for a 10 solidi sarcophagus. In later centuries, the 10 solidi sarcophagus would probably already have cost 100 artaba or more. However, if the Salona sarcophagi were bought during the period of artificially low gold prices after Diocletian's price edict (Bagnall 1989, 70; Rathbone 2009, 319–20), the cost of a 10 solidus sarcophagus would have been equivalent to only 30 artaba. That would be 10 months' wages for a 2nd-century civilian in the Nile delta (with the artaba costing about 2 denarii at that time) and only six months for the Egyptian quarry worker.

Leaving the comparison with late Roman sarcophagus prices behind us, we could calculate that a limestone sarcophagus, produced by workers with the high labour costs as documented for Egyptian and Dacian workers in the 2nd century AD of about 18 denarii per month, would have cost between 54 denarii (assuming that the cost equalled three months' work) and 108 denarii (assuming that it is equivalent to three months of work plus 100% overhead).[16] That would mean a low-earning average civilian of the Nile delta would have to spend 9–18 months' salary to afford such a sarcophagus. How they would have scraped up this amount is difficult to say. But taking into account that a rural Egyptian household living under the poverty line was in 2007 able to spend an average of 15 times the yearly expenditure per head of the household on the wedding of one daughter,[17] the ancient cost of a sarcophagus seems more than reasonable for low-income households, even though it would place extreme financial strain on that family (Cf. Joshel 1992, 20). This argument is somewhat more optimistic than the prediction, that the limestone sarcophagi of Hierapolis 'were only affordable for the rather wealthy citizens' (Vanhaverbeke and Waelkens 2002, 142). Taking into account that the cost for the yearly commemorative rites, like the *stephanoticon*, could add significantly to the funerary expenses (Ritti 2004, 562–5; Judeich 1898, 129–30), it seems fair to assume that in a sarcophagus necropolis the distinction between monuments of poor and rich families would be made by inscribing foundations for post-funerary ceremonies on the sarcophagi and of course the ostentation of these ceremonies itself. Foundations of 150, 200, 300, 350, and 1,000 denarii can be found on sarcophagi of Hierapolis (Judeich 1898, 130; Amelung 2004, 419; Ritti 2004, 563) and might have reached, or indeed surpassed, the actual cost of the sarcophagus. Such a foundation would certainly have been out of reach for any sarcophagus owner with only a modest income.

The conclusion here is that a limestone sarcophagus would not be cheap, but when doing research on the owners of limestone sarcophagi we should always be aware that some of them could have been families with a relatively low income who had the desire to own a sarcophagus and spent some years' savings and some inherited money on a sarcophagus. This is even more probable in places like Hierapolis with an abundance of limestone resources in the immediate area of the city and a very active stone-working industry.

An estimate of the prices of tomb buildings, like tomb C92 in the North-East Necropolis (Fig. 8.2), would again need the working time calculation by a professional stonemason to get a rough estimate of its cost. With roughly 13 m³ of stone slabs, of which a part is finely dressed on one side only, while the back is roughly dressed, one would estimate at least four times the working time of a sarcophagus. Even though many tombs in Hierapolis are highly standardized in shape and size, some architects would still have been working together with the workshops and getting their share. Transport and lifting of the blocks would take considerably more time. The building site would also have to be prepared, bedrock had to be cut and terraces had to be built. Under the assumption that a sarcophagus cost the equivalent of three to six months' salary for a skilled worker, the tomb would cost at least one to two years' wages or rather more. It seems here that day labourers should be excluded as customers for tomb buildings while the family of a skilled worker would need

considerable savings or an inheritance to afford such a tomb. However, a tomb building would save a family any costs for a burial monument for hundreds of years. And it should also be kept in mind that used tombs were sold, and probably at a cheaper price.

There are some indications of the price of tombs in Africa (Duncan-Jones 1962, 90–1). Prices for stelae are between 96 HS and 5,000 HS. But since none has a date, it is not clear if the enormous price difference can be explained by the quality or size of the stelae or by inflation. Mausolea range between 1,000 and 80,000 HS. Only two of them are dated, both to the reign of Alexander Severus. They cost respectively 24,000 HS and 12,000 HS. Since nothing is known of the shape, type, size, and building material of these tombs the numbers are of no use for the question posed here. Inscriptions from Syria mentioning prices for tombs are either from a period of extreme inflation in the 4th century AD, out of its original context, or difficult to interpret (Sartre-Fouriat 2001, 194–7).

The only cost of a tomb in Asia Minor comes from the ruin of a small tomb building near Perge (Ormerod and Robinson 1910/1911, 220, 242–3 no. 23; *IK* 61, 76–7 no. 363). Besides a small pediment with a Triton relief, an Ionic architrave with two fasciae and a length of over 1.91 m was preserved. According to these elements, the monument could have been a quite elaborate temple tomb, far above the standard of the tomb buildings in the North-East Necropolis of Hierapolis. The deceased had contributed 1,500 denarii to the construction of the building while the rest of the costs was paid by his heirs. Consequently, the cost was higher than 1,500 denarii, but by how much is unknown. Tomb A from the necropolis under St Peter's in Rome can serve for comparison (Mielsch and v. Hesberg 1986, 9–10). It cost 6,000 HS, had probably a façade of 4.44 m width, a titulus and two small windows. The brick masonry, the door lintel and frame, and its size suggest that the mausoleum did not significantly differ from the other mausolea in the necropolis and probably provided burial space for a large household. It was certainly much bigger and more elaborate than the tomb buildings in the North-East Necropolis, but it is impossible to say if brick masonry in Rome would have been a cheaper building material than limestone in Hierapolis. A third tomb building, probably from the 2nd century, was a tomb tower in Umm el-Agerem in Tripolitania (Mattingly, Barker, and Jones 1996, 106–8, 111–2 fig. 7). The building was most probably a three-story so-called 'pillar tomb' with a ground area of 2.5×3 m and an elaborate decoration with pilasters and reliefs. The Punic inscription mentions that the donors paid 2,100 denarii and the builder, probably a member of the same family, 1,000 denarii. It should be stated here that 1,500 denarii could buy a monumental tomb building in 2nd-century Rome and probably also in Asia Minor. The mausolea in the North-East Necropolis of Hierapolis were considerably smaller than the tomb in Rome and certainly less elaborate than the tomb

Fig. 8.2. Hierapolis, North-East Necropolis. Tomb with saddle roof C92 (photo University of Oslo excavation project)

in Perge, which leads to the conclusion that there is a high probability that they were cheaper than 1,500 denarii.

Social status and occupations in inscriptions

Inscriptions mentioning social status or professions are widespread in Asia Minor but usually rare. There are two sites, Korakesion and Korykos, which offer a portfolio of profession and status inscriptions large enough for a demographic hypothesis (Hübner 2005, 91–6). These inscriptions, however, are Byzantine and detailed and extensive archaeological analyses on the monuments are still outstanding. In the case of Korykos, it seems that many Roman sarcophagi were reused and inscribed in the 4th century or later (Karaüzüm 2005, nos. 46, 49, 50, 54, 55, 57). Still inscriptions can be a good indicator for the customer groups of specific monuments, if collected in large numbers.

For the following Table 8.1, 31 funerary inscriptions from Hierapolis have been collected, which give the status of tomb owners and tomb founders.[18] Unfortunately, the older publications do not always specify the material of the sarcophagi, thus giving only limited information on this monument group.

According to the inscriptions in Table 8.1, it is mainly those with the highest offices in the city and merchants who stated their status on tomb buildings and marble sarcophagi and only one craftsman is identifiable on a tomb monument. Interestingly, leading figures of the gladiator schools are also represented. The 'white' sarcophagi in the inscriptions

designate with all probability marble sarcophagi, while the 'porous' sarcophagi are probably made of local limestone. Inscriptions on travertine sarcophagi also present members of lower social strata, however, none of these is of particularly low social status, but it should not be ignored that some of these sarcophagi may originally have been associated with tomb buildings.[19] Even the only freedman in this group is the freedman of an *asiarch*.[20] However, the fact that only one freedman appears in the funerary inscriptions and slaves are not present at all makes the result somewhat dubious. The example of Termessos, where freedmen and slaves have recorded their social status on – even ornamented – limestone sarcophagi by the dozens,[21] demonstrates clearly that this monument group was affordable to and popular in these social groups, at least at high-production sites like Termessos, where prices should be expected to be comparably low. This contributes to the suspicion that it was simply not common in Hierapolis for freedmen and slaves to inscribe their profession or status on their tombs,[22] while it was either obligatory or fashionable to do so in Termessos. A much larger, geographically widespread and less biased sample is necessary to assess which groups inscribe their status/occupation and on which kind of monuments.

Datasets 1, 2, and 3

Dataset 1 includes all inscriptions from *Inschriften griechischer Städte aus Kleinasien* (*IK*), which name the profession or a status statement of the buried person or of

Table 8.1. Funerary inscriptions from Hierapolis mentioning the profession or social status of the occupant.

Bomos, sarcophagus	Secunda rudis (second umpire of gladiators)	Ritti 2004, 587–8 no. 29; Ritti and Yilmaz 1998, 528–30 no. 22.
Bomos, sarcophagus	Secunda rudis	Ritti and Yilmaz 1998, 530–3 no. 23.
Bomos, 2 'white' sarcophagi	Councillor	Judeich 1898, 145 no. 234.
Bomos, 'white' sarcophagus	First secretary	Ritti 2004, 487
Tomb building	Councillor	Pennacchietti 1967, 300 no. 13
Tomb building, bomos, sarcophagi	Councillor	Pennacchietti 1967, 298–9 nos. 8–9.
Bomos, 3 'white' sarcophagi, 3 'porous' sarcophagi	High priests	Ritti 2004, 570 no. 2
Bomos, 'white' sarcophagus, 'old' sarcophagus	High priests	Judeich 1898, 151 no. 261; Ritti 2004, 570–1 no. 3.
Bomos	Overseas merchant	Judeich 1898, 92 no. 51; Ritti 2004, 573–4 no. 7.
Tomb building with saddle roof	Dyer	Unpublished, NE necropolis
Marble sarcophagus with relief	Porphyry merchant	Pennacchietti 1967, 313 no. 37
Marble sarcophagus with relief	Councillor	Pennacchietti 1967, 314–5 no. 39
Sarcophagus, travertine.	Councillor	Judeich 1898, 117 no. 145
Sarcophagus, travertine.	Councillor	Judeich 1898, 122–3 no. 163
Sarcophagus, travertine	Neokoros	Judeich 1898, 176 no. 347; Ritti 2004, 582–3, no. 22.
Sarcophagus, travertine, on platform	Freedman of an *asiarch*	Judeich 1898, 107 no. 110
Sarcophagus, travertine	Pedagogue	Ritti 2004, 594–5 no. 40;
Sarcophagus, travertine	Veteran	Judeich 1898, 152 no. 267; Ritti 2004, 583 no. 23.
Sarcophagus, travertine	Veteran	Unpublished, NE necropolis
Sarcophagus, travertine	Chandler	Judeich 1898, 98 no. 75; Ritti 2004, 487
Sarcophagus, travertine	Merchant (flour?)	Ritti 2004, 591–2 no. 36
Sarcophagus, travertine, on platform	Perfume merchant	Judeich 1898 262; Ritti 2004, 572 no. 5.
Sarcophagus, material unknown	Skutularios (gladiator?)	Ritti 2004, 487.
Sarcophagus, material unknown	Porphyry merchant and councillor	Judeich 1898, 121 no. 156; Ritti 2004, 487.
Sarcophagus, material unknown	Leader of the council	Pennacchietti 1967, 299 no. 10; Ritti 2004, 487.
Sarcophagus, material unknown	Painters	Ritti 2004, 487
Sarcophagus, material unknown	Herold, olympionike	Pennacchietti 1967, 308 no. 27
Sarcophagus, material unknown	Trainer	Judeich 1898, 244; Ritti 2004, 487
Sarcophagus, material unknown	Sculptor	Pennacchietti 1967, 313 no. 38
Sarcophagus, material unknown	Painter	Judeich 1898, 115 no. 135; Ritti 2004, 487
Sarcophagus, material unknown	Soldier	Judeich 1898, 98, no. 73

the donor. From about 6,000 funerary inscriptions published in the volumes, 484 inscriptions have been chosen, which most probably date roughly to the period from about the 2nd century BC, to the beginning of the 4th century AD (*IK* 3; 4; 5; 7.3; 8; 9; 10.I–II; 14; 15; 16; 17.I–II; 18.I; 19; 20; 22.I–II; 23.I; 24.II; 25; 26; 27; 28.II; 29; 31; 32; 33; 34.I; 36.I; 38; 39.I; 41.I; 44.II; 47; 48; 49.I; 53; 55; 56.I; 57; 58.I; 60.I; 61.II; 64.I; 66; 67; 68.III). The examples from the Hellenistic period and the 4th century are very few, and the overwhelming majority of the selected inscriptions should date to the Roman Imperial period. (Christian inscriptions have been completely excluded, if they date later than the 4th century AD.) Even though some volumes include rural areas, the series is mainly focused on cities, which makes it quite suited to compare to the situation of Hierapolis. Consequently, the statistics may give general information on the city populations, but not on the whole population of Asia Minor. This becomes obvious when comparing this dataset to dataset 3, which contains a considerable number of farmers, hunters, fishermen, etc.

Using inscriptions for statistical evaluations present numerous pitfalls to the researcher. Problems with the dataset used here are the following:

- Since the dataset is based on inscriptions, only those individuals who could afford an inscription are included. Why some groups are underrepresented may indicate their economic inability to purchase inscriptions, but it might also have other reasons, for example, that these groups preferred other forms of display.
- The dataset covers mainly the west and north-west of Asia Minor, while the east and south are not sufficiently represented (Fig. 8.3). The distribution of tomb types is not equal due to regional preferences with reliefs and stele more common in the north-west and sarcophagi in the south-west of Asia Minor.
- Tomb type and material are, in many publications, not further specified. Thus, altar, base, and column have been used as rough descriptions, but where a detailed description is lacking they may in cases refer to similar monuments. The repeated lack of the specification of material and ornamentation of sarcophagi is one of the main problems for archaeological work with the volumes of *IK*.
- Some dominant population groups of the sites used in the statistics may bias the result. In this dataset comparably large groups of military personnel in Alexandria Troas and Anazarbos, of gladiators in Smyrna and of Imperial freedmen in Ephesos have contributed to make the groups particularly dominant. The dominance of groups in the mentioned cities may in many cases, like Ephesos, be the result of a specific focus and particular discoveries of the archaeological excavations, rather than demographic realities. However, in none of the sites does this bias the final result significantly.

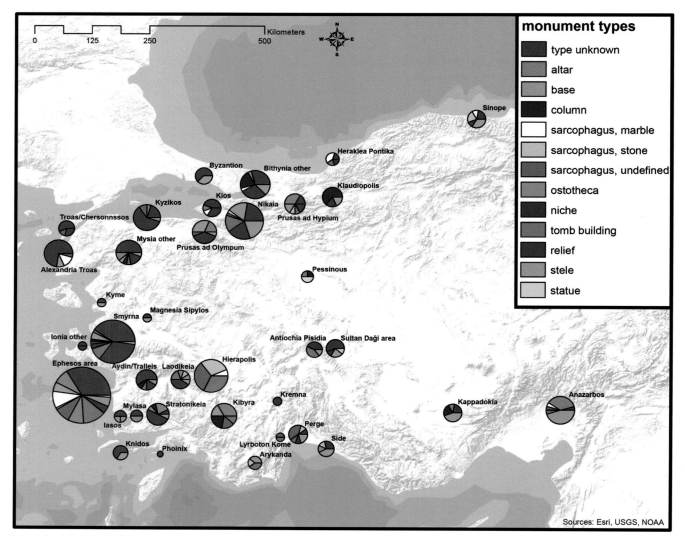

Fig. 8.3. Asia Minor. Geographic distribution of funerary monuments with inscriptions from IK with occupational titles and social status.

Only those inscriptions which name a status or profession have been included. Monuments with reliefs showing occupations or the mention of a guild or association have not been included, because reliefs can be difficult to interpret and those affiliated with, for example, the carpet weavers' guild, might well have been carpet traders, wool dyers and the like. In the sample, particular status statements or occupational titles were classified into groups. However, the individuals may have had very different financial and social status. While some craftsmen were leading a large enterprise, others may have lived on the margin of subsistence. It is also clear that individuals could belong to two groups. A craftsman could also be a member of the council, and a gymnasiarch may also have been a tradesman. The inscriptions of Rome with occupational titles represent, in the majority, slaves and freedmen, and less than half of these inscriptions can be linked to freeborn professionals (Joshel 1992, 46–7). The inscriptions with status statements in Asia Minor show a different distribution. Here slaves and freedmen, with occupational title or not, are far less dominant in the inscriptions. In the following samples I have chosen to classify all dependants in different groups, on the assumption that the leading dependant in a household or estate, the *oikonomos* or *pragmateutes*, or the Imperial freedman were socially and economically on a completely different level to a simple servant slave. Several Imperial freedmen, all from Ephesos, in the data set were also *dekadarchoi* and it can be assumed that the two *dekadarchoi*, without any other specification, may similarly have been freedmen, probably of high status. Where the *hieroi* should be placed in the hierarchy of dependants can be quite vague

and it is not clear which of them are slaves and which are freedmen.[23] The group 'elite' contains the political elite of the cities, particularly members of the council, and priests of the most important cults, but not leading figures of mystery cults. The group 'elite' certainly overlaps at times with the group 'administration', where mainly minor administrative offices of the cities and the administration of the Empire have been included, but also some rare clerks of professional or religious associations. The group administration and military again overlap and members of this group with military titles could have been employed in the civil administration of the cities. 'Entertainment' consists mainly of actors, dancers, theatre poets, and musicians, while intelligentsia consists of teachers, rethors, sophists, etc. Two categories are called private and royal household. They consist of household servants whose status is not clear.

The first chart (Fig. 8.4) shows all inscriptions with occupational title or status statements in the sample divided by monuments of known and unknown types. Interestingly the military, followed by the elite, gladiators and freedmen, were the groups who inscribed their status more than any other group. If the freedmen were grouped together with the *oikonomoi* and *pragmateutes*, Imperial freedmen and *dekadarchoi*, their group would even be the second largest group. This clearly shows that the indication of affiliation to one of these social groups was much more important than inscribing an occupational title. This also means that the epigraphic record does not represent a demographic reality, but rather social conventions and necessities. The following analysis should thus particularly focus on deviations from that model.

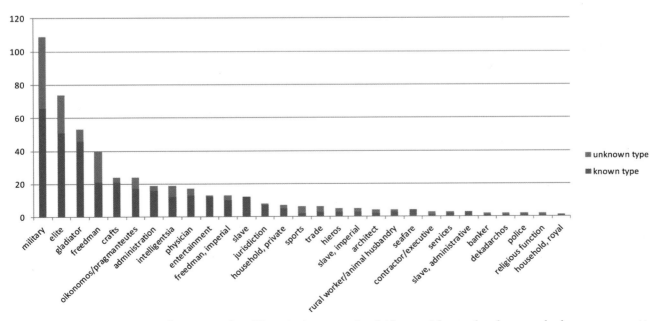

Fig. 8.4. Asia Minor. Comparison of inscriptions from IK mentioning occupational title or social status from known and unknown monument type.

Dataset 2 from a different source and on a different type of monument shall be analyzed to control whether specific groups would choose specific monument types. For this purpose 225 funerary epigrams with occupational title or social status recorded, were collected from Merkelbach and Stauber (1998–2004) (Fig. 8.5). Epigrams are a group of funerary monuments, which must have added significant costs to a burial. Firstly, composing the epigram and secondly, inscribing these often lengthy texts must further have increased the cost. Again, gladiators and the elite are in the top three groups. The next groups, however, deviate from the first chart (Fig. 8.4). Medical doctors, the entertainment, and the intelligentsia are also in the top five group. This indicates clearly that there were occupation groups, which preferred specific types of monuments. The dominant group of gladiators further demonstrates that the choice of a tomb monument was not mainly defined by the financial means of the customer, but also very much by social pressure or convention inside a group.

With this fact established, Dataset 1 can be further analyzed. Altars, bases, and columns are a group of monuments with a probably high overlap between the sub-groups. Only the total of all three sub-groups should be used to draw conclusions (Fig. 8.6). A part of these monuments is certainly or probably dislocated and may have been associated with other monuments, for example, as markers outside tomb buildings, next to sarcophagi or at other monument types. The distribution follows approximately the distribution of the total of all other monuments but there are few deviations from the general distribution. Gladiators did clearly prefer other monuments to columns, bases and altars. Craftsmen, who are absent, and freedmen/Imperial freedmen and intelligentsia are underrepresented, as shown in the comparison between this group and the total distribution of other tomb types in Asia Minor.

Inscriptions with occupational titles or status statements are underrepresented on tomb buildings, mostly tombs with chambers, weakening the result considerably (Fig. 8.7). As expected, the elite are on the top of the list as owners of tomb buildings. Interestingly the military follows next. Two particular deviations can be seen in this chart. The freedmen are not present among the strongest groups, only Imperial freedmen of which all but one are from Ephesos, which is valuable information for the situation in Ephesos, but must be excluded from a supra-regional survey in order to avoid bias. Gladiators have completely disappeared from the statistics. This strongly supports the assumption that freedmen and gladiators usually did not use tomb buildings. A list of all owners of tomb buildings defines them as an assumedly financially strong group.[24] Among the military there are no lowest ranks and the only two slaves are an Imperial slave and the slave of a consul. The list can more or less define what kind of social spectrum can be expected among the owners of tomb buildings.

Very different is the chart on stele and reliefs/relief stele, which represents the largest sample (138 inscriptions) in the dataset, increasing the reliability of the statistics (Fig. 8.8).[25] Here the gladiators are the largest group and the only group among the first ten which uses more reliefs and stele than other monuments. It is followed by the military and the freedmen, while the elite constitute a significant deviation to the general statistics by ranking much lower, using far fewer reliefs and stele compared to other monuments. Clearly stele were considered a less prestigious monument. At the same time the fact that gladiators score highest in the use of both reliefs and epigrams, even though many of the gladiator epigrams can be characterized as short and often standardized, demonstrates that relief and epigrams were commonly used by socially and financially deprived groups, when their occupation/status so required. Tomb monuments and inscriptions were an important means of providing an eternal remembrance (Häusle 1980, 64–91; Le Bris 2001, 176–7, 180–1). While members of the highest status groups would be remembered through honorary inscriptions, statues

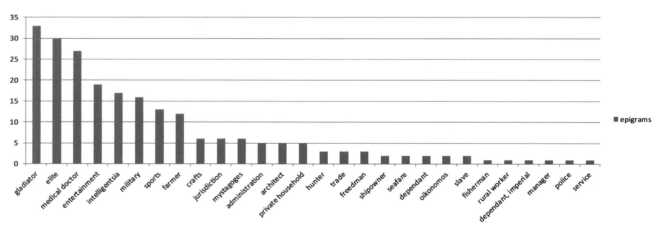

Fig. 8.5. Asia Minor. Epigrams from Merkelbach and Stauber (eds.) 1998–2004 with occupational title or social status.

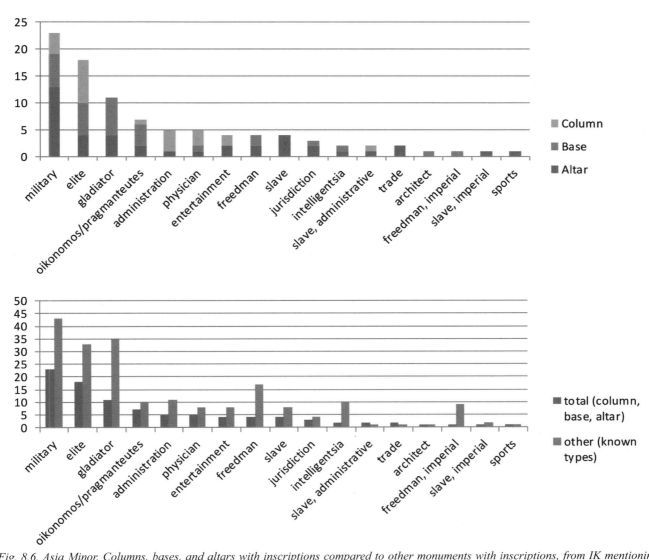

Fig. 8.6. Asia Minor. Columns, bases, and altars with inscriptions compared to other monuments with inscriptions, from IK mentioning occupational title and social status.

and festivals (Hopkins 1983, 247–55; Schmitt-Pantel 1982; Van Nijf 2001, 332–4), the inscribed grave marker would have constituted the only possibility for members of low status groups to eternalize their remembrance. The urge to erect a stele at any cost to gain that form of immortality must not be underestimated.

Dataset 3, a rough survey of reliefs published by Pfuhl and Möbius (1977–1979), supports the result concerning the gladiators (Fig. 8.9). It also adds the very strong groups of farmers (also landowners) and rural workers, hunters, fishers, etc. who are naturally underrepresented in the inscriptions from the cities. However, it needs to be emphasized that many of the reliefs do not have an inscription giving the precise profession and the interpretation of the tools or working scenes is quite often vague. There are some suggestive

deviances between the results from the *IK* and Pfuhl and Möbius (1977–1979), but a much more detailed survey would be needed to understand whether some professions did use relief rather than inscription on tombstones.[26]

Finally, the group of sarcophagi shall be examined (Fig. 8.10).[27] Unfortunately, the dataset suffers considerably from the lack of information on material and elaboration. Even though it seems probable that most of the undefined sarcophagi are made of limestone, the indications in the publications are not precise enough to define them here as stone sarcophagi. Clearly, the difference in costs between a marble sarcophagus with reliefs and a simple limestone sarcophagus must have been significant. Still the small sample gives some valuable information.[28] Firstly, the elite were by far the most common customers of marble sarcophagi.

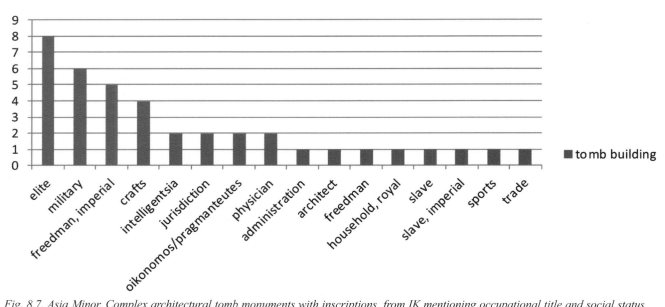

Fig. 8.7. *Asia Minor. Complex architectural tomb monuments with inscriptions, from IK mentioning occupational title and social status.*

There is one example of a marble sarcophagus owned by a cameleer, but because it is so exceptional, this case would certainly need more study. It might be that the sarcophagus was reused or that the cameleer was the owner of a larger caravan business. The most important divergence from the general distribution of monuments and inscriptions

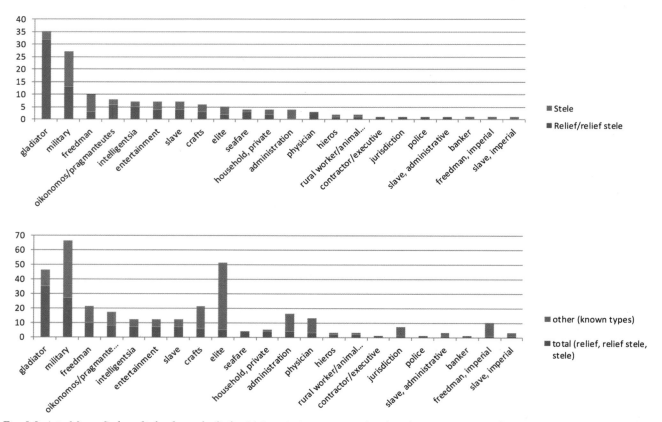

Fig. 8.8. *Asia Minor. Stele, relief stele, and reliefs with inscriptions compared to the other monuments with inscriptions, from IK mentioning occupational title and social status.*

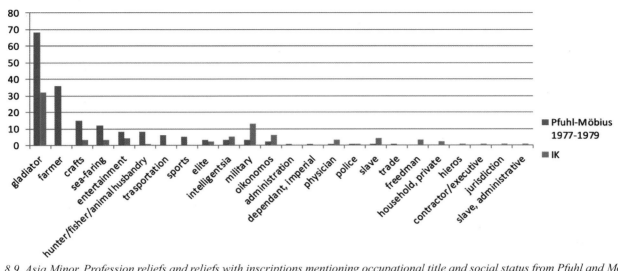

Fig. 8.9. Asia Minor. Profession reliefs and reliefs with inscriptions mentioning occupational title and social status from Pfuhl and Möbius (1977–1979) compared to relief stele and reliefs with inscriptions, from IK mentioning occupational title and social status.

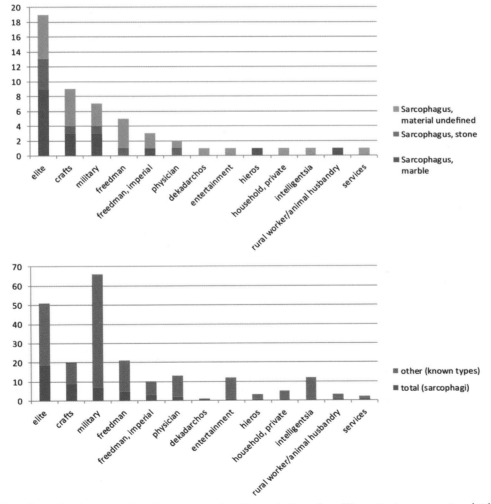

Fig. 8.10. Asia Minor. Sarcophagi compared to other monuments with inscriptions, from IK mentioning occupational title and social status.

is the strong group of craftsmen on the second place. Clearly, sarcophagi were a more common tomb monument for craftsmen than tomb buildings and craftsmen would also be able to afford marble sarcophagi.[29] Freedmen, who were very little represented among owners of tomb buildings, are the second largest group of owners of stone and undefined sarcophagi, on the same level as craftsmen. Together this indicates that sarcophagi – and particularly stone sarcophagi – were a monument type affordable also to lower social strata within the cities, as already shown in the case of Termessos.

Some afterthoughts

Three approaches have been used to assess the social status of tomb occupants. The results of the demographic calculation limited to the city of Hierapolis allow for the assumption that the entire population could theoretically be buried in tomb buildings, sarcophagi, *hyposoria*, and *chamosoria*. The second approach, the calculation of prices for limestone sarcophagi, was based on examples from several areas of the Mediterranean, and led to the assumption that these sarcophagi could have been a desirable investment for members of the lower social strata due to social pressure, planning for the next generations, or a particular individual urge for display. The third approach was the analysis of a comprehensive sample of status statements in funerary inscriptions from Asia Minor, used to further differentiate which status or occupation group could afford or preferred which type of monument. The result confirms that tomb buildings and marble sarcophagi are used mainly by those with the highest financial means, who are quite commonly part of the urban elites. Other complex monuments like stele, altars, reliefs, inscriptions, epigrams and limestone sarcophagi could also be purchased by groups with low social status and were commonly used in groups with limited financial means.

Besides these insights, interesting problems have emerged on the sidelines, which deserve a final summary. When comparing the situation in Hierapolis with the extensive finds of simple graves in Laodikeia it becomes clear that the appearance and composition of necropoleis could in certain cases rather be shaped by local or regional burial customs than by social stratification of the citizenry. The Hierapolitans were seemingly particularly fond of collective burials in standardized but impressive tomb monuments, while the Laodikeians used a much broader spectrum of tomb monuments and interment options. The example of Laodikeia invites us to interpret the diversity of burials as related to social and economic differences between the citizens. The uniformity in Hierapolis, however, makes a detailed interpretation of social aspects in the necropoleis impossible. In Hierapolis it might have been the abundance

of travertine and a highly productive stone industry that shaped the burial customs of the city and made the particular burial preferences possible.

Another interesting problem, which was briefly touched upon, is the diversity of the epigraphic habit of the same social group in two different cities. The example of the large group of slaves, who inscribed their limestone sarcophagi at Termessos, is in stark contrast to the lack of slaves in the funerary epigraphy of Hierapolis. The two epigraphic habits, be it a result of local traditions or of municipal laws, show that the same social group could have completely dissimilar burial rights or customs in different cities. Such divergences make the interpretation of social status in the funerary material difficult. While it is clear that slaves could own elaborate limestone sarcophagi in Termessos, it is very much unclear whether slaves owned tomb monuments in Hierapolis at all. The reasons for that can be manifold. Either they were not obliged or allowed to inscribe their status on tomb monuments, or they lacked the right or the financial means to purchase them. A fourth solution is that it was so deep-rooted a custom in Hierapolis that slaves were buried in the tombs of their masters, that the purchase of their own monuments was not considered an option.

Notes

1   I am very grateful to Tullia Ritti for many discussions and for freely sharing her views, knowledge, and unpublished information with me. This study has been made within the framework of the Missione archeologica italiana a Hierapolis in Frigia (MAIER) and the Norwegian excavation project in Hierapolis, and has been very kindly supported by the director of the MAIER, Francesco D'Andria and the Norwegian project leader J. Rasmus Brandt.

2   The number of slaves is unknown: Engels 1990, 82.

3   Rheidt 1991, 237. Rheidt considers a population of 40,000–60,000 inhabitants as probable for Pergamon in the 2nd and 3rd centuries. His calculation is based on an area of 435 hectares which could be used for houses and a population density of 100–150 people/ha.

4   This estimate is based on maps published in: D'Andria, Scardozzi, and Spanò (eds.) 2008.

5   D'Andria, Scardozzi, Spanò (eds.) 2008, 103 (G. Scardozzi): The width of the *insulae* is 25–28 metres and the length 70–76 metres, with most measuring around 26 and 72–74 metres.

6   A death rate of about 47/1,000 was calculated by Frier 1982, 247; for a life expectancy between 20 and 30 years, see also Duncan-Jones 1990, 103–4. Anthropological research in a family tomb in the North-East Necropolis of Hierapolis found the life expectancy close to 30 rather than 20, but, of course, the sample is too low to be extended to the whole population. This sample includes mostly Byzantine individuals and should consequently not be considered. I am grateful to Henrike Kiesewetter and Helene Russ for this information.

7   In the following, all architectural monuments that include chambers are termed tomb buildings. The term also includes architectural substructures for sarcophagi, which by far surpass the associated sarcophagus in size and/or elaboration, i.e. *exedrae*.

8   These numbers are based on unpublished results from recent excavations of several sarcophagi in Hierapolis and tomb 163d in the North Necropolis. The information was kindly provided by Henrike Kiesewetter, Caroline Laforest, and Helene Russ.

9   Şimşek (ed.) 2011. Comparing the necropoleis of Hierapolis with those of Laodikeia illustrates compellingly how impossible a comparison even between the necropoleis of neighbouring cities can be.

10 Sommer 2006, 193 (*REA* 65, 1963, 318–320 = *SEG* 18,450 = *IK* 28.2, 408). Against the existence of *collegia funeraticia* in general, see Bollmann 1998, 36.

11 I want to express my gratitude to Christian Leuschner, Freiburger Münsterbauverein e.V., who has made this estimate of work time.

12 Inscriptions, reliefs, a higher grade of elaboration of the lid and profiles and figural decoration of the lifting bosses would add to the working time. All these extras were not included in this calculation.

13 Taking the calculations of Barresi (2003, 175) as a base, the transport from the Hierapolitan quarries to the necropolis should not have cost more than 5 denarii.

14 Tomb expenditures of military personnel could exceed six months' to one year's wages: Duncan-Jones 1974, 79 table 2, 130 table 3.

15 Cuvigny 1996, 139–43. Incomes of 120–250 denarii per annum seem to have been quite common in the low income classes, see Jongman 2007, 601.

16 Some comparisons in Asia Minor: for 1,000 denarii, a statue (probably of marble) could be made and erected. In Hadrianic Smyrna, a sum of about 200 denarii seemed to buy a marble column of unknown size, while a sum of over 50 denarii/piece was paid for columns in Neronian Ephesos, see Barresi 2003, 161–3, 189. A marble basin (1.04×0.81×0.51 m) in Cirta (Algeria) cost 50 denarii and is about half the size of a sarcophagus casket (Barresi 2003, 158; Duncan Jones 1974, 111 no. 394). Of course, marble would be more expensive and, because of its hardness, require much more work than local limestone. Compared to these prices the sum of 100 denarii for a limestone sarcophagus in Hierapolis does not seem too low.

17 A 1999 survey on marriage expenditure in Egypt demonstrates that even much higher financial strain is accepted for the sake of tradition. Rural households living under the poverty line would spend on average 15 times of the annual per capita expenditure of the family on a marriage and urban households would spend nine times that amount, see Singerman 2007, 11 table 1-1. Compare further the situation in rural south India, where up to six times the annual income of a family is spent on dowries and an average of four months' family income on the wedding, see Bloch, Rao, and Desai 2004, 678.

18 I have received information on the material of some sarcophagi from Tullia Ritti.

19 The sarcophagus of a merchant from Aphrodisias has not been included, because the tomb was founded by a friend, see Judeich 1898, 153–54, no. 270.

20 As a sponsor and organizer of games and festivals, an *asiarch* must have disposed of outstanding financial means. On that aspect of the *asiarchate*: Friesen 1999, 286–8.

21 For a few examples among many others, see Heberdey (ed.) 1941, nos. 222, 224, 225, 228, 236, 258, 275, 284, 289, 293, 295, 310; İplikçioğlu, Çelgin, and Vedat Çelgin 1991, nos. 15, 20, 27, 31.

22 In general slaves are mentioned very rarely in epitaphs of the Lycos valley: Huttner 2013, 145–146.

23 On *hieroi* as part of the cult personnel in sanctuaries of Asia Minor, see Ricl 2003, 87–91.

24 Treasurer of the federal treasury, Smyrna: *IK* 23.I, 229; Asiarch, knight, Kibyra: *IK* 60.I, 149; *Bouleutes*, Ephesos: *IK* 16, 2222; *Euergetes*, politician, priest, Nikaia: *IK* 10.I, 89; Gymnastics teacher, *bouleutes*, Smyrna: *IK* 23.1, 246; Oil dealer*, prythane*, *bouleutes*, Smyrna: *IK* 23.I, 245; *Paidonomos*, Pessinous: *IK* 66, 130; Priest of Nemesis, Perge: *IK* 61.II, 366; *Prytane*, Priest, Arykanda: *IK* 48, 162; Architect, Arykanda: *IK* 48, 108; Butcher, Prusias ad Hypium: *IK* 27, 108; Silversmith, Inhisar/Söğüt: *IK* 10.II, 1257; Silversmith, Ephesos: *IK* 16, 2212; Smith, Laodikeia: *IK* 49.I, 98; Eunuch, *paidagogos* of a princess, Anazarbos; *IK* 56.I, 73; *Oikonomos*, Antiochia Pisidia: *IK* 67, 95; *Oikonomos* of a senator, Aydin?: *IK* 36.I, 194; Imperial freedman, Ephesos: *IK* 16, 2202A; Imperial freedman, Ephesos: *IK* 16, 2219; Imperial freedman, Ephesos: *IK* 16, 2261; Imperial freedman, Laodikeia: *IK* 49.I, 85; Imperial freedman, *dekadarchos*, Ephesos: *IK* 16, 2219; Imperial slaves of the Provinz Asia, Ephesos: *IK* 16, 2200A; Retor, sophist, Ahmetbeyler: *IK* 10.II, 1491; Sophist, Ephesos: *IK* 16, 2100; Jurist, Baciköy/Geyve: *IK* 10.II, 1231; Jurist, Kilciler/Göynük: *IK* 10.II, 1239; *Praefectus alimentorum, legatus augusti, praetor,* quaestor of the province Asia, and more, Ephesos: *IK* 17.II, 4355; *Cornicularius*, Lyrboton Kome: *IK* 61.II, 358; Courier (*Tabellarius*), Ephesos: *IK* 16, 2281A; *Decurio,* Antiochia Pisidia: *IK* 67, 187; Leader of a cohort, Pamukova/Geyve: *IK* 10.II, 1252; *Optio Tabellarius*, Ephesos: *IK* 16, 2222B; Medical doctor, Heraklea Pontica: *IK* 47, 33; Medical doctor, Antiochia Pisidia: *IK* 67, 71; Slave of a *consularis*, Kibyra: *IK* 60.I, 296; Olympionike, Ephesos: *IK* 15.V, 1626; Paper dealer (?), Smyrna: *IK* 2.I, 542.

25 Many tomb monuments are described in the epigraphic publications as reliefs, slabs with relief, etc., and are here treated together with relief stele, on the assumption that most of them would have had a similar function and value as relief stele.

26 Like occupations in transportation, seafaring, farming and other rural professions.

27 Known occupational titles and status statements in connection with *ostothecae* have been published in: Ahrens 2015, 192.

28 Marble sarcophagi: Prokurator, knight, Ephesos: *IK* 16, 2204A; *Bouleutes*, Ephesos: *IK* 16, 2227; *Bouleutes,* Ephesos: *IK* 16, 2253A; *Bouleutes,* Hypaipa?: *IK* 17.II, 3828; *Archiereus*, Ephesos: *IK* 17.II, 4354; *Proteuon*, Kios: *IK* 29, 39; *Bouleutes*,

Heraklea Pontica: *IK* 47, 10; *Bouleutes*, Alexandreia Troas: *IK* 53, 98; Imperial Priest, *euergetes*, Nikaia: *IK* 9, 116; Confectioner, *gerusiast*, Ephesos: *IK* 16, 2225; Dyer, Heraklea Pontica: *IK* 47, 15; (Son of a) linen-weaver or -dealer, Alexandreia Troas: *IK* 53, 129; *Hieroi*, Ephesos: *IK* 16, 2279; Imperial freedman, *dekadarchos*, Ephesos: *IK* 16, 2223A; Soldier, Ephesos: *IK* 16, 2232A; Protector, Kalchedon: *IK* 20, 71; Veteran, centurion, Sinope: *IK* 64.I, 121; Cameleer, Anazarbos: *IK* 5.I, 99.

Stone sarcophagi: *Gerusiast*, Ephesos: *IK* 16, 2446; Imperial priest, leader of festival, Prusias ad Hypium: *IK* 27, 72; Priestess (Diakonissa), Perge: *IK* 61.II, 456; (Wife of a) sculptor, Sinope: *IK* 64.I, 155; Freedman, Arykanda: *IK* 48, 120; (Son of a) centurion, Side: *IK* 44.II, 219; Medical doctor, Sinope: *IK* 64.I, 147.

Sarcophagi, material undefined: *Gerusiast*, Akilar?: *IK* 10.II, 1281 *Gerusiast*, Ephesos: *IK* 16, 2236C; *Bouleutes*, Assos: *IK* 4, 74; Knight, *scriba librarius*, Perge: *IK* 61.II, 409; Two priests, Lyrboton Kome: *IK* 61.II, 428; Baker, Ephesos: *IK* 16, 2226; Baker, Kyzikos: *IK* 18.I, 117; Saddler, Kyzikos: *IK* 18.I, 14; Wreath binder, Smyrna: *IK* 23.I, 504; Smith, Ilion/Kum Kale: *IK* 3, 171; Linen weaver, Prusa ad Olympum: *IK* 39.I, 104; *dekadarchos*, Ephesos: *IK* 16, 2404; Poet (made by a friend), Kibyra: *IK* 60.I, 211; Actor (*biologos*), Perge: *IK* 61.II, 449; (Wife of a) freedman, Ephesos: *IK* 16, 2244C; Freedman, Tire: *IK* 17.I, 3335; Liberta, Asagiyapici: *IK* 18.I, 407; (made for) freedmen, Perge: *IK* 61.II, 393; Imperial freedman, *dekadarchos*, Ephesos: *IK* 16, 2222C; Imperial freedman, Byzantion?: *IK* 58.I, 385; *Frumentarius*, Ephesos: *IK* 16, 2318; Soldier, Anazarbos, *IK* 56.I, 72; Veteran, Perge: *IK* 61.II, 454; Medical doctor, Ephesos: *IK* 16, 2304, Nurse (made by patron), Kibyra: *IK* 60.I, 317.

29   Four marble sarcophagi with status statements, occupational titles or reliefs from Aphrodisias represent a *pragmateutes*, a possible glassblower, a sculptor with a paint-dealer, and a freedman with a slave (Turnbow 2011, 170–2). This comparably high number of artisans and dependents could possibly be explained by a relatively rich 'middle class' and the large sculpture production of Aphrodisias, which might have led to lower prices for marble sarcophagi than elsewhere.

Abbreviations and bibliography

IK = Inschriften griechischer Städte aus Kleinasien
REA = Revue des études anciennes
SEG = Supplementum epigraphicum Graecum

Ahrens, S. (2015) 'Whether by decay or fire consumed …' Cremation in Hellenistic and Roman Asia Minor. In J. R. Brandt, M. Prusac, and H. Roland (eds.) *Death and changing rituals. Function and meaning in ancient funerary practice*, 185–222. Oxford, Oxbow Books.

Allen, R. C. (2009) How prosperous were the Romans? Evidence from Diocletian's price edict. In Bowman and Wilson (eds.), 327–45.

Amelung, W. (ed.) (2004) Inscriptiones judaicae orientis, Tübingen, Mohr Siebeck.

Bagnall, R. S. (1989) Fourth-century prices: New evidence and further thoughts. *Zeitschrift für Payrologie und Epigraphie* 76, 69–76.

Barker S. and Russell, B. (2012) Labor figures for Roman stone-working: Pitfalls and potential. In Camporeale, H. D. and Pizzo A. (eds.) (2012) *Arqueología de la Construcción III. Los procesos constructivos en el mundo romano: La economía de las obras* (Anejos de Archivo Español de Arqueología 64), 83–94. Merida, Instituto de Arqueología de Mérida.

Barresi, P. (2003) *Province dell'Asia minore: costo dei marmi, architettura publica e committenza*, Rome, L''Erma' di Bretschneider.

Bloch, F., Rao, V. and Desai, S. (2004) Wedding celebrations as conspicuous consumption. *The Journal of Human Resources* 39, 675–95.

Bollmann, B. (1998) *Römische Vereinshäuser*. Mainz, Verlag Phillip von Zabern.

Bowman, A. and Wilson, A. (eds.) 2009 *Quantifying the Roman economy*. Oxford, Oxford University Press.

Bryce, T. R. (1979) Lycian tomb families and their social implications. *Journal of the Economic and Social History of the Orient* 22.3, 296–313.

Cuvigny, H. (1996) The amount of wages paid to the quarry-workers at Mons Claudianus. *Journal of Roman Studies* 86, 139–45.

D'Andria, F. (2001) Hierapolis of Phrygia: Its evolution in Hellenistic and Roman times. In D. Parrish (ed.) *Urbanism in western Asia Minor* (Journal of Roman Archaeology, Suppl. 45), 97–115. Portsmouth RI, Journal of Roman Archaeology.

D'Andria F., Scardozzi G., and Spanò A. (eds.) (2008) *Hierapolis di Frigia II: Atlante di Hierapolis di Frigia*. Istanbul, Ege Yayınları.

Drexhage H. J., Konen H. and Ruffing K. (2002) *Die Wirtschaft des Römischen Reiches (1.–3. Jahrhundert)*. Berlin, Akademie Verlag.

Duncan-Jones, R. (1962) Costs, outlays and summae honorariae from Roman Africa. *Papers of the British School at Rome* 30, 47–115.

Duncan-Jones, R. (1974) *The economy of the Roman Empire. Quantitative studies*. Cambridge, Cambridge University Press.

Duncan-Jones, R. (1990) *Structure and scale in the Roman economy*. Cambridge, Cambridge University Press.

Engels, D. W. (1990) *Roman Corinth: An alternative model for the Classical city*. Chicago, University of Chicago Press.

Fahlander, F. and Oestigaard, T. (2008) The materiality of death: Bodies, burials, beliefs. In Fahlander, F. and Oestigaard, T. (eds.) *The materiality of death: Bodies, burials, beliefs* (BAR International Series 1768), 1–16. Oxford, Archeopress.

Flämig, C. (2007) *Grabarchitektur der römischen Kaiserzeit in Griechenland*. Rahden/Westfalen, Verlag Marie Leidorf.

Frézouls, E. (1977) Prix, salaires et niveaux de vie: Quelques enseignements de l'Edit du Maximum. *Ktèma* 2, 253–68.

Frier, B. (1982) Roman life expectancy: Ulpian's evidence. *Harvard Studies in Classical Philology* 86, 213–51.

Friesen, S. (1999) Asiarchs. *Zeitschrift für Papyrologie und Epigraphik* 126, 275–290.

Handley, M. A. (2003) *Death, society and culture: Inscriptions and epitaphs in Gaul and Spain, AD 300–750* (BAR International Series 1135). Oxford, Archeopress.

Häusle, H. (1980) *Das Denkmal als Garant des Nachruhms*. Munich, Beck.

Heberdey, R. (ed.) (1941) *Tituli Asiae Minoris, III. Tituli Pisidiae linguis Graeca et Latina conscripti, 1. Tituli Termessi et agri Termessensis*. Vienna, A. Hoelder.

Hope, V. (2000) Inscription and sculpture. In G. J. Oliver (ed.) *The epigraphy of death: Studies in the history and society of Greece and Rome*, 155–85. Liverpool, Liverpool University Press.

Hopkins, K. (1983) *Death and renewal 2: Sociological studies in Roman history*. Cambridge University Press, Cambridge.

Hübner, S. (2005) *Der Klerus in der Gesellschaft des spätantiken Kleinasiens*. Munich, Franz Steiner Verlag.

Huttner, U. (2013). *Christianity in the Lycus valley*. Leiden, Brill.

İplikçioğlu, B., Çelgin, G., and Vedat Çelgin, A. (1991) *Epigraphische Forschungen in Termessos und seinem Territorium I*. Vienna, Verlag der Österreichischen Akademie der Wissenschaften.

Işın, G. and Yıldız, E. (this volume) Tomb ownership in Lycia: Site selection and burial rights with selected rock tombs and epigraphic material from Tlos, 85–108.

Ismaelli, T. (2009) *Hierapolis di Frigia III: Architettura dorica a Hierapolis di Frigia*. Istanbul, Ege Yayınları.

Jongman, W. M. (2007) The Early Roman Empire: Consumption. In W. Scheidel, I. Morris, and R. Saller (eds.) *The Cambridge economic history of the Greco-Roman world*, 592–618. Cambridge, Cambridge University Press.

Joshel, S. R. (1992) *Work, identity and legal status at Rome: A study of the occupational inscriptions*. Oklahoma, University of Oklahoma Press.

Judeich, W. (1898) Inschriften. In C. Humann, C. Cichorius, W. Judeich, and F. Winter, (eds.) 1898, *Altertümer von Hierapolis* (Jahrbuch des deutschen archäologischen Instituts, Ergänzungsheft 4), 67–178. Berlin, Georg Reimer.

Kamp, K. A. (1998) Social hierarchy and burial treatments: A comparative assessment. *Cross-Cultural Research* 32.1, 79–115.

Karaüzüm, G. (2005) Doğu Dağlık Kilikia (Olba) bölgesi lahitleri (unpublished thesis, Mersin University).

Kehoe, D. (2012) Contract labor. In W. Scheidel (ed.) *The Cambridge companion to the Roman economy*, 114–30. Cambridge, Cambridge University Press.

Le Bris, A. (2001) *La mort et les conceptions de l'au-delà en Grèce ancienne à travers les épigrammes funéraires*. Paris, L'Harmattan.

Mattingly, D. J., Barker, G. W. W., and Jones, G. D. B. (1996) Architecture, technology and society: Romano-Libyan settlement in the Wadi Umm el-Agerem, Tripolitania. In L. Bacchielli and M. Bonanno Aravantinos (eds.) *Scritti di antichità in memoria di Sandro Stucchi* (Studi Miscellanei 29), 101–15. Rome, L''Erma' di Bretschneider.

Merkelbach, R. and Stauber, J. (eds.) (1998–2004) *Steinepigramme aus dem griechischen Osten I–V*. Munich and Leipzig, K. G. Saur Verlag.

Mielsch, H. and v. Hesberg, H. (1986) *Die heidnische Nekropole unter St. Peter in Rom. Die Mausoleen A-D*. Rome, L''Erma' di Bretschneider.

Nijf, O. van (2001) Local heroes: Athletics, festivals and elite self-fashioning in the Roman East. In S. Goldhill (ed.) *Being Greek under Rome*, 306–44. Cambridge and New York, Cambridge University Press.

Ormerod, H. A. and Robinson, E. S. G. (1910/1911) Notes and inscriptions from Pamphylia. *Annual of the British School at Athens* 17, 215–49.

Pegoretti, G. (1863) *Manuale pratico per l'estimazione dei lavori architettonici, stradali, idraulici e di forticazione per uso degli ingegneri ed architetti 1*. Milan, Domenico Salvi e Comp.

Pennachietti, F. A. (1967) Nuove iscrizioni di Hierapolis Frigia. *Atti della Accademia delle Scienze di Torino* 101, 287–328.

Pfuhl, E. and Möbius, H. (1977–1979) *Die ostgriechischen Grabreliefs*. Mainz, Philipp von Zabern.

Rathbone, D. (2009) Earnings and costs: Living standards and the Roman economy (first to third centuries AD). In Bowman and Wilson (eds.), 299–326.

Rheidt, K. (1991) *Die byzantinische Wohnstadt (Altertümer von Pergamon XV, 2)*. Berlin, De Gruyter.

Ricl, M. (2003) Society and economy of rural sanctuaries in Roman Lydia and Phrygia. *Epigraphica Anatolica* 35, 77–101.

Ritti, T. (2004) *Iura Sepulcrorum* a Hierapolis di Frigia nel quadro dell'epigrafia sepolcrale microasiatica. Iscrizione edite e inedite. In S. Panciera (ed.) *Libitina e dintorni (Atti dell'XI Rencontre franco-italienne sur l'épigraphie)*, 455–652. Rome, Edizioni Quasar.

Ritti, T. and Yilmaz, S. (1998) Gladiator e *venationes* a Hierapolis di Frigia. *Atti della Accademia Nazionale dei Lincei*, ser. 9, 10.4, 444–542.

Russell, B. (2010) The Roman sarcophagus 'industry': A reconsideration. In J. Elsner and J. Huskinson (eds.) *Life, death and representation: Some new work on Roman sarcophagi* 119–47. Berlin and New York, De Gruyter.

Sartre-Fauriat, A. (2001) *Des tombeaux et des morts: Monuments funéraires, société et culture en Syrie du Sud du Ier s. av. J.-C. au VIIe s. apr. J.-C.* Vol. 2. Beirut, Institut Français d'Archéologie du Proche-Orient.

Scheidel, W. (2001) Roman age structure: Evidence and models. *Journal of Roman Studies* 91, 1–26.

Schmitt-Pantel, P. (1982) Évergetisme et mémoire du mort. In G. Gnoli and J. P. Vernant (eds.), *La mort, les morts dans les sociétés anciennes* 177–88. Cambridge, Cambridge University Press.

Schneider Equini, E. (1972) La necropolis di Hierapolis di Frigia. *Monumenti Antichi, Serie Miscellanea* I.2, 95–142.

Şimşek, C. (ed.) (2011) *Laodikeia Nekropolü (2004–2010 yılları)*. Istanbul, Ege Yayınları.

Singerman, D. (2007) *The economic imperatives of marriage: Emerging practices and identities among youth in the Middle East* (Middle East Youth Initiative Working Paper 6, September) available at Social Science Research Network: http://dx.doi.org/10.2139/ssrn.1087433

Sommer, S. (2006) *Rom und die Vereinigungen im südwestlichen Kleinasien (133 v. Chr. – 284 n. Chr.)*. Hennef, Buchverlag M. Clauss.

Storey, G. R. (1997) The population of ancient Roma. *Antiquity* 71, 966–78.

Turbow Awan, H. (2008) The display of sarcophagi in the cemeteries of Aphrodisias. In R. R. R. Smith and J. L. Lenaghan (eds.), *Roman portraits from Aphrodisias*, 153–67. Istanbul, Yapı Kredi Yayınları.

Turnbow, H. N. (2011) Sarcophagi and funerary display in Roman Aphrodisias. Thesis, New York University.

Vanhaverbeke, H. and Waelkens, M. (2002) The northwestern necropolis of Hierapolis (Phrygia): The chronological and topographical distribution of the travertine sarcophagi and their way of production, 119–45. In D. De Bernardo Ferrero (ed.) *Saggi in onore di Paolo Verzone*, Roma, Giorgio Bretschneider.

Wallace-Hadrill, A. (1994) *Houses and society in Pompeii and Herculaneum*. Princeton NJ, Princeton University Press.

Zaccaria Ruggiu, A. (2007) Regio VIII, *insula* 104. Le strutture abitative: Fasi e trasformazioni. In F. D'Andria and M. P, Caggia (eds.) 2007 *Hierapolis di Frigia I: Le attività delle campagne di scavo e restauro 2000–2003*, 211–56. Istanbul, Ege Yayınları.

New evidence for non-elite burial patterns in central Turkey

Andrew L. Goldman

Abstract

Little is currently known about non-elite burial practice in central Anatolia. The funerary landscape of Roman Galatia has remained a comparatively elusive one; its population, largely composed of rural subject peoples with a mixed ethnic background, used few permanent markers to memorialize their dead. The key to identifying patterns of burial practice and commemoration among this non-elite populace clearly lies in a study of rural cemeteries, few of which have been investigated. Excavations at Gordion between 1950 and 1995 have provided a promising means to study the character of rural Galatian burial practice, in the form of 147 Roman-period burials belonging to three necropoleis which span from the mid-1st century to the early 5th century AD. Their examination has permitted an initial modelling of local, diachronic patterns of non-urban funerary activity, one that features the use of multiple construction types in a single necropolis, a preference for single interment, alignment along cardinal axes, and a substantial transformation of burial practice in late Antiquity. The publication of data from recent rescue excavations and established projects like Çatalhöyük has now provided an opportunity to test this model independently. Preliminary assessment of this new data appears to strengthen the case for certain widespread patterns of funerary activity within central Turkey which resemble those observed at Gordion. Examination of burial geography in rural Galatia has also revealed a widespread reuse of pre-Roman sites and cemeteries as Roman necropoleis, suggesting that rural non-elites chose to commemorate their own dead through proximity with monuments of Anatolia's pre-Roman past.

Keywords: cemeteries, Galatia, geography, Gordion, memory, Roman-period.

Introduction

In a paper delivered in 1997 on burial in Roman Asia Minor, M. Spanu candidly lamented the present state of mortuary archaeology in Anatolia. He pointed out that not only had Rome's Asiatic provinces garnered only cursory mention (when mentioned at all) in scholarly discussions of Roman funerary practice, but he also highlighted the 'almost total absence of extensive archaeological excavations of cemeteries' in Turkey, where conditions were 'only slightly offset by the state of preservation… of funerary monuments and by sporadic, restricted rescue excavations, mostly unpublished or locally published and therefore difficult to find' (Spanu 2000, 169). Nearly twenty years later, as the studies contained in this volume amply demonstrate, fundamental steps are being taken to remedy this state of research, including the expansion of excavation in Roman-period necropoleis, better recording practices, greater methodological and theoretical deliberation, and wider, more rapid publication of the results (for example, Devreker 2003; Şimşek 2011; Krsmanovic and Anderson 2012; Kelp 2013). Yet in spite of such promising developments, the general foci of fieldwork on Roman burial practice and ritual in Turkey have changed little since Spanu's much-deserved critique (Krsmanovic and Anderson 2012, 58–60). The preponderance of new fieldwork and scholarship on Roman cemeteries remains concentrated largely in the western and coastal regions of Turkey, as well as centred almost exclusively upon urban or elite burials and their associated monuments. No attempt has yet been made to construct the type of analytical models or even provisional studies like those produced for Imperial Roman provinces such as Britain and Pannonia, with the intent to examine extensive and diachronic variation in burial context and practice (Pearce 2013; Leleković 2012). Likewise, the types of associated issues outlined by J. Pearce (1992), such as burial and cultural identity, funerary ritual and the relationship of burial to the landscape, remained largely unaddressed within the provinces of Roman Anatolia.

This is particularly true for the inland provinces of the Anatolian plateau like Galatia, where little is yet known about Roman-period burial practice and commemoration. In comparison with its neighbouring regions, Galatia's funerary landscape is admittedly difficult to identify, investigate and characterize, largely lacking as it does the prominent material remains associated with Roman-period burials found elsewhere in Anatolia. This is due in at least part to the general pattern of regional settlement in Galatia, which contained few cities of any size. The vast majority of its population was of a non-elite and non-urban character, living in villages or estates scattered across the rural landscape (Mitchell 1993, 143–9). Once outside the immediate environs of Galatia's few cities (e.g. Ancyra and Pessinus), evidence for Roman-period necropoleis and associated monuments is relatively scarce, especially when compared to neighbouring funerary landscapes like that of western Phrygia, with its comparably plentiful Asiatic sarcophagi and rock-cut tomb façades (Kelp 2013, 79–86). While funerary sculpture and markers are certainly not absent from Galatia, the prevalent types which have been recorded – statues of crouching lions, Phrygian 'doorstone' stelae, tombstones with pediments and acroteria, and plain, inscribed *bomoi* – tend to be simple in form and found in proximity to urban areas (Temizer 1973, 82–5). This appears to be true even among the region's elite class, as indicated by the eventual cessation of Galatia's most elaborate pre-Roman form of funerary construction, the building or reuse of mounded tumuli. This burial tradition, with its origins in north-west Anatolia, remained popular with the region's Galatian elite through the end of the Hellenistic period, as indicated by the mounded tombs with elaborately constructed stone chambers at Karalar (ancient Blucium) and in Tumulus O at Gordion (Young 1955, 191–7; 1956, 251–2; Arık and Coupry 1935; Mitchell 1993, 57). Such construction and reuse does continue sporadically within or near the borders of Galatia into the early Imperial period: the Kocakızlar Tumulus near the village of Alpu (Midaeum) has been dated to the early or mid-1st century AD on the basis of coins and a sigillata bowl (Atasoy 1974), while two tumuli just to the north of Ankara's Esenboğa Airport, near the villages of Akkuzulu Köy and Kalecik, date on the basis of ceramic finds (*unguentaria*) to the second half of the 1st or early 2nd century AD (Anlağan 1968; Mermerci 1988). Yet while there is clear and growing evidence for the construction of later Imperial tumuli in Galatia

(Cinemre 2014a, 350), such evidence remains comparative scarce, suggesting strongly that the long-term popularity of this burial type had largely ceased by Roman times even among central Anatolia's elite (Goldman 2007, 317).

The relative absence of evidence for an overtly lavish tradition of funerary practice and commemoration across the landscape of Roman Galatia is probably owed less to the vagaries of preservation than to the underlying economic and demographic factors that prevailed among the population of this Roman province. As noted previously, Galatia held few urban centres, and the epigraphic record suggests that while agriculturally rich areas such as the Sangarius and Tembris valleys held numerous Imperial and private estates, wealthy landholders preferred to reside in Roman Galatia's cities or farther abroad (Mitchell 1993, 148–9). The general distribution of wealth and elite residency therefore seems the most likely factor for a clustering of funerary monuments around Galatia's few cities. The simultaneous scarcity of such monuments elsewhere in Galatia appears likely to be rooted in the character of its rural economy, which was overwhelmingly based upon agricultural and pastoral activities. S. Mitchell (1993, 154–7) has suggested that at the onset of the Imperial period, certain pre-Roman conditions – specifically the existence of a tribal social structure with greater transhumant behaviour – provided ambitious domainal landlords with an opportunity to carve out large rural estates and eventually to introduce the widespread cultivation of cereal agriculture. Mitchell has proposed, as was the case in neighbouring Asia and Bithynia, that local landowners were forced to sell off their holdings in order to meet the rising demands of Roman provincial taxation. Whatever the cause, village life came to predominate under this new system of rural land ownership, one which featured a tenant population who chiefly practiced large-scale dry cereal farming and animal husbandry. This populace, who possessed a far lower standard of living than Galatia's city dwellers, are likely to have had little capacity for constructing or purchasing funerary monuments, a circumstance that might well account for the scarcity of permanent memorials within the Galatian hinterlands (Mitchell 1993, 253–8).

Consequentially, if we are to ascertain the character of burial practice and its context in the Galatian landscape, we must turn to an investigation of rural cemeteries and their predominantly non-elite burials. Until relatively recently, such an investigation has not been possible owing to a general dearth of fieldwork across Galatia, where archaeological activity on Roman sites has concentrated almost entirely upon its few urban centres (i.e. Pessinus, Ancyra, Tavium). However, new excavations in the Roman cemeteries at Gordion during the 1990s, in combination with a reassessment of funerary contexts and materials unearthed and recorded there during the 1950s and 1960s, have created an initial opportunity to explore rural funerary practice and non-elite burial customs in Galatia. The resulting sample from the three cemeteries, composed of 147 graves dating between the first and fifth centuries AD, has provided the basis for constructing a much-needed diachronic model of local funerary activity in Galatia. Furthermore, the past decade has witnessed an increase in the publication of contemporary necropoleis at a variety of sites in central Anatolia, both the product of rescue excavations and fieldwork at established sites where study of Roman (or 'late') material previously languished. The fortuitous arrival of this newly published data, supplemented by online archives of the rescue projects and their finds, has presented the means for comparing local patterns observable at Gordion on a broader, regional scale. Following an overview of the Gordion data, mortuary patterns at four rural sites will be examined as a means to detect what (if any) regional patterns of funerary activity might be identified as widespread within Galatia. This paper will conclude by offering some tentative observations about burial patterns within central Turkey, what amounts to a necessary first step forward in creating a broader understanding of Roman mortuary activity in central Anatolia.

Before proceeding, however, a requisite note of caution must be interjected here, well-trodden territory for anyone familiar with synthetic publications about ancient mortuary patterns in the Roman world. Indeed, it has become de rigueur within articles on the subject of rural burial to preface discussion with a long list of necessary caveats, whose purpose is to concede limitations in the existing data, to address deficiencies in local conditions, and to pinpoint the needle's eye through which one's necessarily narrow threads of scholarly inquiry must emerge. This study is no more immune to the confines and vagaries that beset similar analyses of fieldwork elsewhere in Rome's provinces, with acknowledged impediments in the quality of recording and data of the type that Spanu and others have often cited. In his study of town cemeteries in Roman Britain, for example, S. E. Cleary (1992) provides a daunting list of empirical shortcomings that might well describe fieldwork conditions in Galatia, including an absence of statistically valid samples, cemeteries which have been excavated only in part, the slow adoption of palaeopathological investigation, and more. The data examined in this article suffers from similar weaknesses: publication of funerary data in Turkish periodicals is often only of preliminary character; field methodology and recording varies greatly in quality; valuable contextual information is often missing, as the cemetery and its associated settlement were not excavated in tandem; and so forth.

Likewise, one cannot proceed with a study of burial practice such as that proposed here without acknowledging the significant interpretive and methodological concerns that attend any investigation of burial and cultural identity. For Galatia, the subject of cultural identity is an exceedingly complex one in light of the province's high level of ethnic diversity. While such diversity is a characteristic of many regions of Anatolia and the Roman world in general, Galatia possessed an unusually heterogeneous mix that is well attested within Imperial epigraphic and literary sources. Among those living under Roman suzerainty in this vast province were people of Phrygian descent, tribal groups such as the Pisidians and Lycaonians to the south, and those descended from immigrants connected with successive waves of invaders, such as the Persians, Hellenistic Greeks, and Galatians of Celtic origin. The latter group was particularly influential following its arrival in the early 3rd century BC, and common speech across Roman Galatia, especially in the countryside, appears to have remained strongly Celtic through late Antiquity. Attested Celtic place names are fairly plentiful in Roman Galatia (e.g. Vindia, Acitorigiaco, Ecobrogis), and within funerary inscriptions family units often display an assortment of Celtic, Phrygian, Roman, and Greek names. Phrygian names in particular continue to be encountered commonly on the central plateau, and 'neo-Phrygian' curse formulae appear on tombstones with Greek inscriptions in western Galatia during the 2nd and 3rd centuries AD (Mitchell 1980, 1056–68; 1993, 50–1, 172–6).

As one might imagine, the discernment or isolation of ethnic characteristics from among the non-elite burials of Galatia represents a challenging task, since it has been demonstrated through post-processual critique that mortuary treatment by its very nature is not a direct reflection of a living people's social or ethnic identity. As J. Pearce (1992, 2–5) asserts, it is hard to achieve differentiation between burial practices of incomers and indigenous peoples in many situations, such as whenever one attempts to assign origin or cultural affiliation to burial practice, to identify local patterning in burial with a conscious ethnic identity or local group, and to distinguish local variation and the rate of persistence of local practice within a community. For the Roman period, one must also consider the relevance of Roman political geography and its partition of administrative units (here Galatia) with the grouping of burial evidence. Additionally, in the specific case of Galatia, any effort to discern changes in burial practice are impeded by a current lack of evidence for the region's pre-Roman, non-elite burial traditions. Furthermore, the general paucity of Roman-period funerary epitaphs and monuments in Galatia makes the study of migration during that period quite difficult (Pearce 2010, 81–2).

Consequentially, at present we cannot attempt to measure with any accuracy the rate or impact of multi-ethnic accultural practices, including those intrusive funerary forms which might be considered 'Roman'. Indeed, Imperial Rome is merely one of many non-indigenous, Iron Age cultures that must be considered when investigating the rural mortuary landscape of Galatia. By comparison and necessity, then, the analysis presented in this paper is a narrowly focused one, its objective to facilitate future discussion on themes of burial and cultural identity in Galatia, as a first step towards eventual engagement with and contribution to the broader theoretical and contextual discourse taking place elsewhere in the Roman world. Yet while the goal of this nascent analysis is admittedly a modest one, and while the excavation of non-urban cemeteries remains infrequent in Galatia, enough recent fieldwork has been completed and published to make such an initial inquiry a possible and beneficial exercise.

The Roman-period cemeteries at Gordion

Writing at the advent of the Roman Imperial period, Strabo informs us that the former Phrygian capital of Gordion (modern Yassıhöyük) had been reduced to a small village amid a landscape of ruins (*Geog.* 12.5.2: *kômê*). Recent fieldwork on the Gordion Citadel Mound has disproved Strabo's observation: stratified deposits confirm that the site in fact lay abandoned at the time of Galatia's annexation by Augustus (*c.* 25 BC), with resettlement taking place only during the late Claudian or early Neronian period, roughly

75 years later (YHSS 2) (Table 9.1). At that time a small auxiliary base was established atop the mound, providing a permanent military presence along the major highway which ran south-west from Ancyra to Colonia Germa and Pessinus (modern Babadat and Ballıhisar, respectively). The small base, never more than 3 ha in area, was occupied for roughly 75 years (YHSS 2:1–3) before its abrupt abandonment in the early Hadrianic period, possibly as a result of Hadrian's reorganization of the Empire's military forces (Goldman 2010, 136–8). An auxiliary presence remained based in the area, however, as indicated by a pair of early 3rd-century Roman military altars recovered in 2008 from the Sakarya river. Dedicated by members of the *cohors I Augusta Cyrenaica* to Caracalla and Geta (the latter's name erased), these altars have revealed the name of the unit stationed at the site, one known to have operated in Galatia in the 2nd and 3rd centuries (Darbyshire *et al.* 2009). Although the exact position of this second base in relation to the Sakarya river crossing has not yet been established, the site is most likely to be identified with the *statio* known in Roman road itineraries as Vindia or Vinda, which was located in the vicinity of Gordion (Goldman 2010, 136). Excavations atop the Citadel Mound indicate that the initial settlement was again occupied by the late 3rd century (YHSS 2:4–5),

Table 9.1. Gordion. The Yassıhöyük Stratigraphie Sequence (YHSS) (M. M. Voigt and A. Goldman).

YHSS Phase	Period name	Approximate dates
0	Modern	1920s AD
1	Medieval	10th–16th century AD
[Abandonment period		7th–10th century AD]
2	Roman	
2:5	Late Roman/Early Byzantine?	? 5th–7th century AD
2:4	Late Roman	Late 3rd–mid-4th century AD
2:3	Trajanic to Hadrianic/Antonine Era	*c.* 110/15–130/70 AD
2:2	Flavian to Trajanic Era	*c.* 75/80–110/15 AD
2:1	Late Claudian/Neronian to E. Flavian Era	*c.* 50–75/80 AD
[Abandonment period		*c.* 50 BC–*c.* 50 AD]
3	Hellenistic	
3A:3	Late Hellenistic	*c.* 100–*c.* 50 BC
3A:1–2	Middle Hellenistic	*c.* 260–*c.* 180 BC
3B	Early Hellenistic	333–*c.* 260 BC
4	Late Phrygian	*c.* 540s–333 BC
5	Middle Phrygian	*c.* 800–*c.* 540s BC
6A–B	Early Phrygian	*c.* 900–*c.* 800 BC
7	Early Iron Age	*c.* 1100–*c.* 900 BC
8	Late Bronze Age	*c.* 1500–*c.* 1200 BC
9	Middle Bronze Age	*c.* 1800–*c.* 1500 BC

perhaps in response to Gothic and Palmyrene raids into the heartland of Anatolia. Limited settlement activity, which included the creation of a necropolis amid derelict YHSS 2:4 buildings (the NWZ cemetery; see below), appears to have continued into late Antiquity until a final abandonment, in the 6th or 7th century AD.

In addition to producing a chronological framework for Gordion's Roman occupation, investigation of the YHSS 2 settlement has focused on the excavation and analysis of three contemporary necropoleis located on or around the Citadel Mound (Fig. 9.1): the South Lower Town (SLT) cemetery, containing 31 graves; the Common Cemetery (CC), with 51 graves; and the NW Zone (NWZ) cemetery, with 65 graves. Concentrated fieldwork in the necropoleis began in 1950, when Rodney S. Young and a team from the University of Pennsylvania initiated excavation at Gordion, and continued during the early 1990s under the direction of Mary M. Voigt. Each of the three burial areas associated with this small rural site has been excavated to a limited extent, and two of them – the SLT and CC cemeteries – contain plentiful evidence of both Roman and pre-Roman (YHSS 3–8) mortuary activity. The discovery in 1996 of a funerary stele dedicated to one Tritus, son of Bato, an auxiliary soldier from Pannonia, has raised the possibility that the settlement possessed a fourth cemetery, one which lay north-west of the Citadel Mound along the Roman road which ran north of the present course of the Sakarya river (Goldman 2010). Excavation has not yet been initiated in this area, however, so that its existence must remain hypothetical at present.

The results of the three investigated cemeteries and their contents are summarized below. While osteological and palaeodemographic data is available for the most recently excavated necropolis (i.e. the SLT cemetery), limitations in space precludes their discussion here (see Selinsky 2004; Goldman and Voigt 2016, forthcoming). Instead, the following sections will offer overviews of grave construction techniques, associated burial practice, orientation, demarcation, and range of offering types. In order to facilitate description and comparison between the Gordion necropoleis, a typology of grave construction form has been created (Table 9.2) using categories first defined for the Common Cemetery (in Goldman 2007) and later adapted to reflect practices in the Area A-B sample (in Goldman and Voigt 2016, forthcoming). In some cases the CC types were subdivided (e.g. Type 1 into 1a, 1b and 1c), and two new categories – Type 7: *busta*, and Type 8: cist graves – were added. While the sample size for the SLT cemetery is small, and the numbers for each variant type are even smaller, splitting (rather than lumping) is intended as a means of facilitating comparison with cemeteries elsewhere.

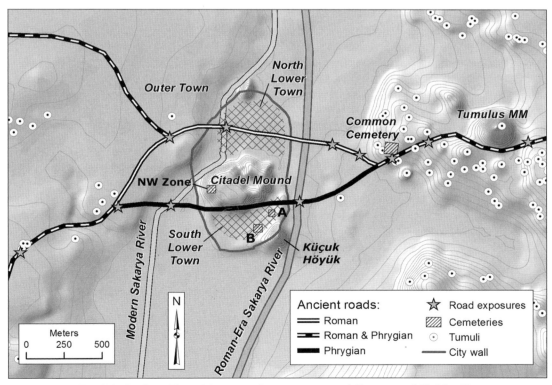

Fig. 9.1. Gordion. Plan of the site and its cemeteries with Roman road (B. Marsh).

Table 9.2. Gordion. Roman burial construction typology: Types 1–8 (A. Goldman and M. M. Voigt).

Type 1: Simple pits (3 variants)

1a   Simple pit, carefully cut in a rectangle or ovular shape

1b   Simple pit, with small slabs or small stones placed above body

1c   Simple pit, shallow, with small pile of rocks placed on stone slab or directly on body

Type 2: Stepped graves (2 variants)

2a   Stepped grave, with simple earth shelves, mudbrick cover

2b   Stepped grave, with paved stone shelves

Type 3: Mud-brick sarcophagi (2 variants)

3a   Mud-brick sarcophagus, roughly rectangular boxes

3b   Mud-brick sarcophagus, roughly rectangular boxes, with some stone lining

Type 4: Wooden coffin in pit or pit with wooden cover

Type 5: Tile and/or brick graves (2 variants)

5a   Grave constructed of tile and/or brick, used as lining

5b   Grave constructed of tile and/or brick, used as cover

Type 6: Chamber tomb

Type 7: Bustum, with cinerarium *for cremated ashes and bones*

Type 8: Cist grave, stone-lined with stone covert

The South Lower Town cemetery (1st–2nd century AD = YHSS 2:1–3)

The earliest of Gordion's three Roman cemeteries was laid out in the South Lower Town (SLT) area between the Citadel Mound and the Küçük Höyük (KH), a multi-storied middle Phrygian fortress and the siege mound built against it by the Persians in the mid-6th century BC (Fig. 9.1). Excavations on the top and slopes of the Küçük Höyük in the late 1950s and in Areas A and B during the mid-1990s exposed architectural remains dating to the middle and late Phrygian periods (YHSS 5 and 4), as well as a series of partial and comingled human and animal skeleton deposits (including complete human skeletons with evidence of trauma) dated to middle Hellenistic periods (YHSS 3A:1–2). The latter have been linked with ritual sacrifice carried out by the Galatian population who took control of Gordion around the mid-3rd century BC (Voigt 2012, 263–84).

Following the mid-1st century AD resettlement of the Citadel Mound (YHSS 2:1), burial activity in the SLT resumed and Roman burials were inserted alongside and occasionally cut into the pre-Roman deposits. A total of 31 Roman-period burials – 28 inhumations and three cremations, containing a minimum of 34 individuals – have been excavated in the SLT cemetery (Table 9.3). Ten of these were uncovered in Area A (LTG-1 to 10) (Fig. 9.2), 11 in Area B (LTG-11 to 21) (Fig. 9.3), and 10 along the top and slopes of the Küçük Höyük (LTG-22 to 31). An additional five graves were recorded in Areas A and B but could not be excavated owing to time and funding. As the graves from the KH area were casually unearthed and poorly recorded during the 1950s, discussion here will focus on the 21 graves from Areas A and B, those excavated with greater precision in the 1990s. Among this group, there are four categories of grave construction: Type 1 (pit graves), Type 2 (stepped graves), Type 3 (graves with mud-brick lining), and Type 7 (*busta*). It is in fact unclear whether or not even the 21 burials from SLT Areas A and B belong to one or several bounded necropoleis, since the orientation of grave shafts differs between the areas. However, similarities in grave construction types and in the dates of the funerary assemblage in Areas A and B as well as on the KH fortification walls suggest continuity in time and perhaps in space.

In regard to burial construction, Type 1 or pit graves were most common (10 of 21, or 47.6%). Three sub-types were defined based upon the way that stone was used within the grave pit, and the depth of the pit. In most cases, the shaft of a Type 1 grave was a carefully cut rectangle approximating the size of the body within it, but two graves were clearly oval. Shaft depth was *c.* 1.0–1.5 m. The deceased (mostly adults) were laid out in an extended dorsal position, with straightened legs, upper arms parallel to the body and lower arms crossed over the abdomen (as in LTG-1) (Fig. 9.4). The four Type 1b pit graves are differentiated from the unembellished Type 1a examples by the addition of stone slabs or smaller stones placed directly above the body, near the bottom of the grave shaft. Likewise, Type 1c consisted of infant graves – LTG-14 and 18 in Area B – where a small pile of rocks was placed upon a stone slab or directly atop the body which lay in a very shallow pit. These stones would have served both as a protective cover for the body and a marker for the grave.

Type 2 graves have a more complex construction: the body lay at the bottom of a deep shaft, placed in a secondary cut made below a narrow ledge or stepped shelf. Although numerous in CC necropolis (see below), the SLT cemetery contained only three stepped graves, all in Area A. The first two (Type 2a) were for children, and have simple earth shelves on which mud bricks were set to cover the body: (LTG-6 (Fig. 9.5), and LTG-9. The adult grave (Type 2b) is unique at Gordion in that the step to either side of the body was partially paved with stone (LTG-5).

Type 3 graves – seven of which were excavated in Areas A and B – consist of pits within which a sarcophagus of mud brick has been constructed around the body. In a majority of the examples, roughly rectangular boxes composed of irregular-sized, greenish mud bricks

Table 9.3. Gordion. SLT cemetery grave and burial types, with associated objects (A. Goldman).

Cat. #	Burial type[a]	Grave type†	Burial construction	Associated small finds (SF)
LTG-1	IN	1a	Simple pit	Ring with intaglio (SF 94-262)
LTG-2	IN	1a	Simple pit	—
LTG-3	IN	1a	Simple pit (2x)	Ring (SF 94-268), earring (SF 94-220), hobnails (SF 94-226)
LTG-4	IN	1b	Stone cover, wood coffin	Whorl (SF 95-7), hobnails (SF 95-37)
LTG-5	IN	2b	Stepped grave, w/paved stone shelves	Fibula (SF 95-22)
LTG-6	IN	2a	Stepped grave, w/mudbrick gable	Ring (SF 95-52), hoop earrings (SF 95-53), whorl (SF 95-26)
LTG-7	IN	1a	Simple pit	Arrowhead (SF 95-85), graffito (95-339)
LTG-8	IN	3a	Mudbrick coffin	Ceramic button (SF 95-182), hobnails (SF 95-199, 95-240), Textiles (no SF)
LTG-9	IN	2a	Stepped grave, with mudbrick gable	—
LTG-10	IN	1b	Pit with stone slabs (2x)	Hobnails (SF 95-266)
LTG-11	IN	3a	Mudbrick coffin	—
LTG-12	IN	1b	Pit with mudbrick cover	Anklet (SF 94-187), bead (SF 94-147), coin (SF 94-108: Galba or Flavian emperor), headband (stains only, no SF)
LTG-13	CR	7	Cremation	Whorl (SF 94-24), cremation pot (SF 94-59) and bowl (SF 94-58)
LTG-14	IN	1c	Pit covered with stones	Iron spatula (SF 94-68)
LTG-15	IN	3a	Mudbrick coffin	—
LTG-16	CR	7	Cremation	Cremation vessel (SF 95-190), iron strip with rivet (SF 95-253), textile impressions (SF 95-338)
LTG-17	IN	3a	Mudbrick coffin	Hobnails (94-334), nails (SF 94-164, 94-165), bead (SF 94-0)
LTG-18	IN	1c	Pit covered with stones	Glass (SF 94-131), iron bar (SF 94-159)
LTG-19	IN	3a	Mudbrick coffin	—
LTG-20	IN	1a	Simple pit	Coin (SF 95-50: Galba or Flavian emperor), fibula (fragmentary, no SF)
LTG-21	IN	3a	Mudbrick coffin	—

α IN = Inhumation; CR = Cremation. † Types 4 (wooden coffins), 5 (infant burials with reused ceramic or brick elements) or 6 (chamber tombs), which are present in the Common Cemetery, are not currently represented in the Southern Lower Town Cemetery.

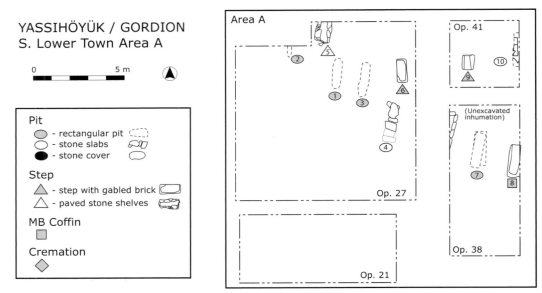

Fig. 9.2. Gordion. Plan of Area A, in the SLT cemetery (compiled by A. Goldman; digitized by T. Ross).

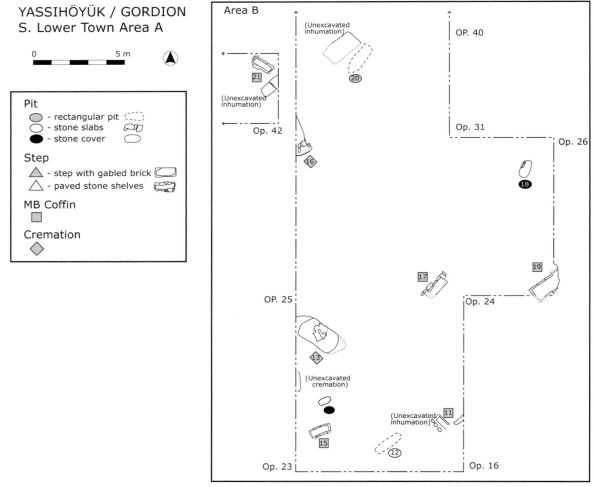

Fig. 9.3. Gordion. Plan of Area B, in the SLT cemetery (compiled by A. Goldman; digitized by T. Ross).

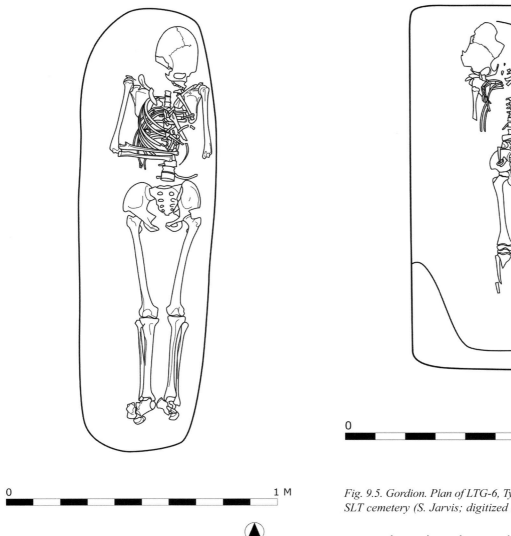

Fig. 9.5. Gordion. Plan of LTG-6, Type 2 stepped grave from Area A, SLT cemetery (S. Jarvis; digitized by T. Ross).

Fig. 9.4. Gordion. Plan of LTG-1, Type 1 pit grave from Area A, SLT cemetery (S. Jarvis; digitized by T. Ross).

(L. 0.25–55 m, W. 0.08–25 m) were assembled inside the burial trench (e.g. LTG-21) (Fig. 9.6). In all examples of graves with mud-brick coffins, skeletal position was identical to that found in the shaft graves, with bodies laid out on their backs, legs straight and arms bent over the abdomen. Type 3 burials were found in all of the SLT excavation units, but were more common in Area B. Adults, children and infants were interred in about equal numbers. Such crude mud-brick sarcophagi were likely constructed as a practical measure, as a means of better protecting bodies and their associated burial offerings from rodent disturbance.

Type 7 burials, which to date have been documented only within the SLT necropolis, are cremation pits or *busta*. At Gordion these consist of shallow rectangular trenches with rounded corners in which the dead were burnt and their

cremated remains subsequently gathered and secondarily interred in associated ceramic containers (*cineraria*). Two excavated *busta* from the SLT Area B (LTG-13, 16) and at least one more from the KH trenches have yielded urns and offerings that clearly identify them as Roman and belonging to the inhabitants of the YHSS 2:1–3 settlement. LTG-13, the best preserved of the two *busta*, is composed of a shallow pit placed on a north-west/south-east orientation, with a length of *c.* 2.25 m, a width of *c.* 1.15 m, and a depth of *c.* 0.5 m (Figs. 9.7–8). Remains of the *in situ* pyre, which left the roughly parallel sides of the trench fired bright orange in colour, lay in the approximate centre of the *bustum*. It consisted of an irregular patch of burnt, blackened soil *c.* 0.60 m in diameter which still containing small fragments of bone. Three patellas were identified in the total sample, suggesting that this is a multiple cremation, with a second individual present in this burial. Most of the cremated remains were gathered together once the pyre had cooled and placed in a common ware pitcher. Some sort of offering

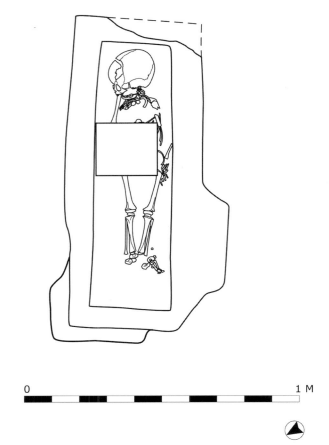

Fig. 9.7. Gordion. Photo of LTG-13, Type 7 busta from Area B, SLT cemetery (Gordion Archive).

Fig. 9.6. Plan of LTG-21, Type 3 mud-brick sarcophagus from Area B, SLT cemetery (S. Jarvis; digitized by T. Ross).

appears to have been placed in the cinerary vessel as well, since several of the bones displayed a greenish tinge, indicating contact with a copper alloy object, but no metal fragments were recovered from either the urn or the trough. Completing the burial practice, a small, roundish niche was cut into the north wall of the *bustum* and the pitcher, capped by a broken echinus bowl, was interred inside it to create a final resting place for the deceased. A small post-sized hole discovered *c.* 0.20 m to the east of niche seems to indicate that a grave marker of some type was installed there to identify the site of the collected remains (*cf.* Faber 1998, postholes at Cambodenum in southern Germany). Both the pitcher and the bowl have parallels with ceramics from contexts dating to YHSS 2:1–3 within the Roman settlement, indicating a probable date of the second half of the 1st century or early 2nd century AD (Voigt *et al.* 1997, 13–4; Sams and Voigt 1996, 437 and fig. 11). Slight variations can be observed between the two *busta*: the trough of LTG-16 maintains a different orientation (12° north-west, in comparison to 55° north-west), and its *cinerarium* was interred in the south-east corner of the trough, in a niche

that is encircled by rocks, cordoned off from the area of the funerary pyre. Yet the general ritual and furniture are quite similar, with the body burnt *in situ* – a 'hot' *bustum*, in contrast to a 'cold' *bustum* in which the ashes and bones were transferred from a nearby pyre (*ustrinum*) – and use of a wide-mouthed vessel to store the cremated remains in an immediately adjacent location (Matijasic 1991, 85–7).

The placement of small stones, wood, bricks and potsherds above the body is found in all grave types. A distinctive attribute of infant and child burials in the SLT cemetery is the placement of mud bricks or sherds above the body. In the case LTG-21 (Fig. 9.6), a child's body was placed in a crude mud-brick coffin and a square mud brick was then laid over the chest, while in LTG-6, another child burial was given a gabled mud-brick vault (Fig. 9.5). The latter can be considered an imitation of the 'tent' burial popular throughout the Roman Empire. Tent burials were normally constructed using fired roof tiles (*tegulae*) leaned against each other to form a triangular, tent-like cover (e.g. Pessinus Type B3, in Devreker 2003, 48–9). Unlike similar *cappuccina* burials in Italy, however, curved cover tiles (*imbrices*) are generally absent along the top and sides of the Galatian examples. At Gordion, where there is no evidence in the YHSS 2:1–3 settlement for either the manufacture or use of pan or cover tiles, the desire or need to fashion *tegula*-like structures appears to have driven residents to substitute available building materials (i.e. mud bricks) in order to fashion an acceptable likeness of this fashionable or familiar style.

Evidence for wood covers is also present in the SLT cemetery, in the burial of a child (LTG-17) and an adult (LTG-4). In the former, large iron nails (L. 0.087 m) with rectangular sections were found inside the Type 3 mud-brick coffin, while in the latter a line of nails was found beneath stone slabs, the nails extending down the centre of the long

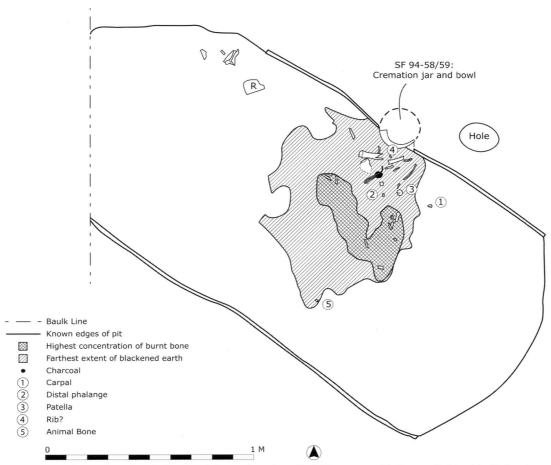

SF 94-58/59:
Cremation jar and bowl

Hole

- — – Baulk Line
——— Known edges of pit
▨ Highest concentration of burnt bone
▨ Farthest extent of blackened earth
• Charcoal
① Carpal
② Distal phalange
③ Patella
④ Rib?
⑤ Animal Bone

0   1 M

Fig. 9.8. Gordion. Plan of LTG-13, Type 7 busta from Area B, SLT cemetery (S. Jarvis; digitized by T. Ross).

axis of the body. Pseudomorphic evidence in the form of both vertical and horizontal wooden grain patterns preserved along the corroded nail shafts indicate that nailed wooden planks were clearly present in the grave, although it could not be determined whether they originally belonged to a complete coffin, a bier or a wooden cover. One possible interpretation of the line of nails is that that wooden planks were placed immediately above the body, held together by the nails.

While the limited area of investigation and non-contiguous character of the trenches render it difficult to detect any large-scale patterns of interment across the entire SLT cemetery, it is possible via the existing sample of burials and offerings to detect certain more localized characteristics relating to organization (orientation and distribution), content and dating. In SLT Area A (Fig. 9.2), the 11 recorded Roman burials (excavated and non-excavated) display a consciously arranged scheme of deposition, with the majority of burials maintaining a roughly consistent north/south orientation (between 0° north

and 17° north-east), similar equidistant placement (c. 1.25 m apart), and indications of elementary linear alignment (e.g. LTG-4, 7, and 8). Skeletal position within the graves is relatively uniform, with supine extended inhumations, heads normally placed to the north, arms crossed over the lower abdomen. In contrast, the 15 burials of SLT Area B (Fig. 9.3) are not only placed more distantly apart in most cases, but they also lie co-mingled along two opposing axes, with a dominant north-east/south-west group interspersed with a north-west/south-east group along the Area B's western edge. Notably, the dominant group consists largely of Type 3 inhumations (e.g. LTG-15, 17, 19), while the second, smaller group is the only one to contain cremations (LTG-13, 16). Among the inhumations, while skeletal posture (extended with arms crossed) is identical to that observed in Area A, the positioning of the crania varies greatly, observing no discernable pattern among the excavated sample. Exactly what this intriguing differentiation in alignment, construction type and body orientation might reflect is uncertain. Such variation could potentially be a sign of two distinct, unrelated phases of interment, datable evidence for which is currently lacking. Alternatively, the juxtaposition of a second axis might indicate the importation of ethnic or cultural practices new to the site or region, an effect of the influx of soldiers into Gordion. Although identification of long-distance migration through burial practice remains difficult, as J. Pearce (2010, 93) has recently asserted, the military function of the site does provide a viable context within which such a proposal can be offered.

Although on several occasions the Roman graves in the SLT cut into and disturb the pre-Roman burials, the fact that none in either Area A or B cut into or overlie one another indicates that some system of grave markers was employed throughout the necropolis for the purpose of organization and commemoration. This was certainly the case for two infant burials (LTG-14, 18) in Area B, where small piles of rocks were placed carefully over the body to demarcate the grave. Although difficult to recover, there is also evidence for rough stone markers for adult graves. No evidence for other permanent markers has been identified, however, suggesting that the inhabitants employed markers made of organic materials (i.e. wood) or that the site's post-Roman inhabitants subsequently scavenged any stone markers, removing all surface trace of the graves. Either case (or both) of such loss would help to explain the high degree of preservation among the graves of the SLT cemetery.

In terms of offerings, funeral gifts are rare in the SLT cemetery, especially for adults (Table 9.3). For example, with the exception of the cremation burials, no ceramic vessels (including lamps) were placed within or associated with the LTG burials, and only two coins were found, both from the early Flavian period. Glass finds are also infrequent at Gordion, although a well-preserved candlestick *unguentarium* was recovered from an infant's grave in Area B (LTG-18). J. Jones (personal communication, 2013) dates this find to the 2nd or early 3rd century. When datable, such grave goods belong to the 1st through 3rd centuries AD, indicating that burials began with or shortly after the establishment of the settlement in YHSS 2:1 (c. mid-1st century AD), and apparently continued after the population moved off of the Citadel Mound at the end of YHSS 2:3 during the second or third quarter of the 2nd century.

Items of personal adornment are more common in the SLT cemetery. There is good evidence that the deceased wore clothing at the time of interment, as attested by the calcified remnants of textiles, a button and the hobnails and eyelets of footwear recovered from several of the graves. The remains of such footgear are rarely encountered elsewhere in Roman Anatolia, but relatively frequent at Gordion. Hobnails were recovered from four of the 21 SLT burials, probably the remnants of *caligae* worn by soldiers and members of their families. These hobnails are even more frequently encountered in the CC burials, in just over one-third of the graves (18 of 51), and should not be understood merely as

the accoutrement of soldiers, since they are found in graves of men, women and children at Gordion (Goldman 2007, 313–4; 2014, 182). Other items of personal adornment (e.g. earrings, rings, a headband, an anklet or bracelet, fibulae) were associated with the graves of individuals identified as females by Selinsky (2004). Composed of silver, bronze and iron, most of these objects are quite modest in terms of their craftsmanship and materials, and likely represent everyday items used by the deceased. Such use is reflected, for example, in the worn silver alloy ring with an engraved gemstone found in the grave (LTG-1) of a woman aged 35–40 (Goldman 2014, 165). The ring bore a carved intaglio of yellow-orange carnelian with an image of a winged eros fishing, and around the lower band was a blob of organic adhesive (wax?), applied in order to shrink the band's width so that it would not slip off the smaller finger of its user. Much of the jewellery represents common types fashionable throughout the Roman Empire, such as the pair of silver earrings from LTG-6, with wire and granulation decoration in the shape of a flower (*cf.* Ergil 1983, 41 no. 99).

The Common Cemetery (2nd–4th century AD = YHSS 2:3–4)

The CC necropolis lies roughly 1 km south-east of the Citadel Mound, along the course of the Ancyra–Pessinus highway (Fig. 9.1). As the road travelled south-west from Ancyra and descended into the Sakarya basin, it neatly bisected a gentle, undulating ridge which forms the lower valley's eastern edge. Modern Yassıhöyük lies at the ridge's north-western end, while its spine is crowned by a concentration of over 30 Phrygian tumuli, the exploration of which was another primary target of R. S. Young's early fieldwork. Sixteen successive trenches laid across that slope between 1951 and 1962 eventually yielded c. 230 graves belonging to a span of over two and a half millennia, from the early Bronze Age through the late Roman period. The majority of the single- and multiple-use interments found there contained few offerings and remain difficult, if not impossible, to date. It is this profusion of non-elite burials which led excavators to name the necropolis 'the Common Cemetery'. At its heart, clustered just north of the highway, 51 Roman-period inhumations dating between the 2nd and 4th centuries AD were unearthed during the course of the fieldwork across an area c. 80 m². (Fig. 9.9). Since this cemetery has been published in detail elsewhere (Goldman 2007, 305–15) and many of its characteristics are analogous to those of the SLT necropolis, only a brief summary will be offered here.

Strong similarities between the two cemeteries and their contents can be observed among the grave construction types, orientation, spacing, demarcation (or lack thereof) and offering type and frequency. There is a comparable mélange of burial construction types, falling here into six separate categories (Fig. 9.10): pit graves (Type 1: 20

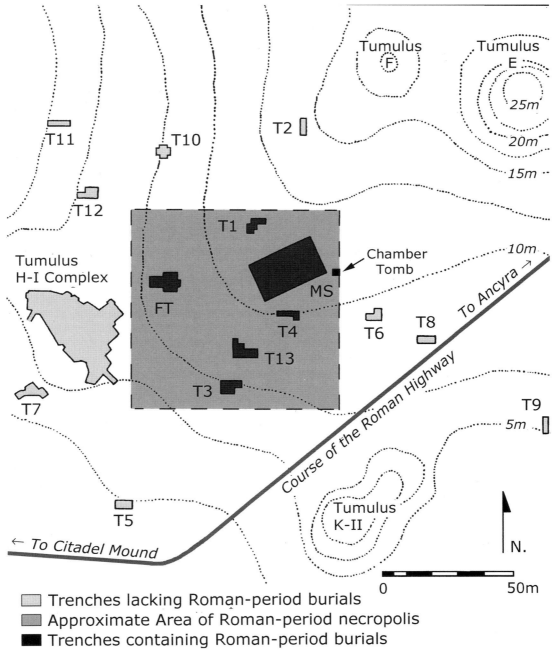

Fig. 9.9. Gordion. Map of the Common Cemetery with Roman-period burial area (original by G. Anderson; revised by A. Goldman).

examples), stepped graves (Type 2: 15 examples), graves lined with stone and/or mud brick (Type 3: six examples), wooden coffin burials (Type 4: six examples), graves lined with ceramic tile or brick (Type 5: three examples), and one example of a chamber tomb (Type 6). Although there is some clustering of construction types (e.g. Type 2 in the Museum Site Trench), all of the excavated areas containing

Roman burials had a mixed complement. Inhumations predominate here, as they do in the SLT necropolis, and there is a similar positioning of the skeletons, again supine and extended, with arms crossed over the abdomen and crania to the north. While cremation burials do occur in the Common Cemetery, they do not demonstrate the *busta* form with *cineraria* (Type 7), and there is no evidence that the

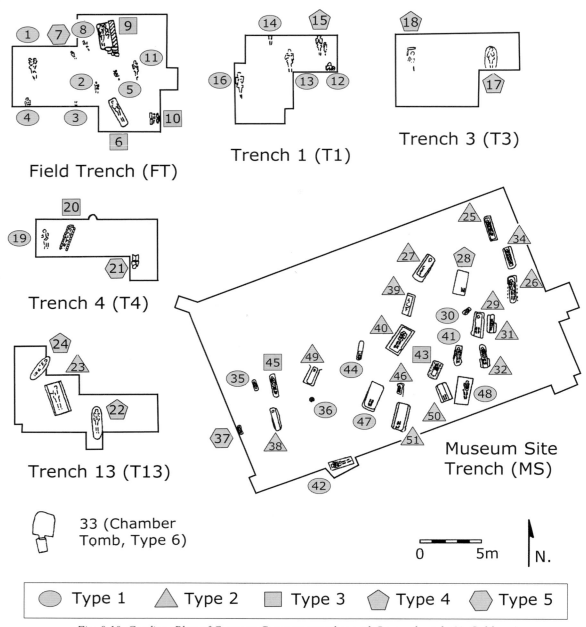

Fig. 9.10. Gordion. Plan of Common Cemetery trenches with Roman burials (A. Goldman).

rite was still practised there during the Roman period. The most common graves are single or multiple interments of the Type 1 pit category, followed by Type 2 step graves. Type 3 sarcophagi are present as well, three of which (CCG-9, 10, 43) incorporated small stone elements, while a fourth (CCG-20) was capped by an elaborate imitation of a *tegula* gable made of reused mud brick (Goldman 2007, 310, fig. 7).

The three categories of grave construction that are not found in the SLT but are present in small numbers here are

Types 4, 5, and 6. The three Type 5 graves (CCG-7, 21, 37) all belong to infants and children, and brick and tile are used in the lining and cover much like the use of stone in Type 1c. Although slightly more numerous, the use of wooden coffins (Type 4) appears to have been relatively infrequent as well, as were the mud-brick sarcophagi (Type 3). Perhaps not surprisingly, these more elaborate and presumably expensive burial types tend to be accompanied by offerings in higher numbers and of a higher quality, often in the form of jewellery and ceramic vessels (see below). In one Type 4 burial, CCG-24, the remnants of gold foil was found on the tops of the coffin nails. Interestingly, many of these coffins were placed more closely to the road, while simple pit burials (Type 1) tend to cluster more thickly further from the highway, a pattern which suggests that proximity to the roadway might have been reserved for those of higher status. Also close to the roadway is the single example of a chamber tomb, CCG-33. The tomb, which was found looted, consists of a rectangular access shaft (1.10×0.90 m) 0.75 m in depth, at the bottom of which, sealed by an upright limestone slab and cut deeply into the hardpan, was a nearly squared burial chamber (1.75×2.0 m) with a ceiling which sloped from the entranceway to the rear of the chamber (Goldman 2007, 308–12, fig. 8). Subterranean chambers with sloped ceilings and deep entrance shafts of this type are common at sites along the northern and eastern littoral of the Black Sea, such as Tanais, Zavetnoe, and Zolotoe, and perhaps we see here an outside influence from those regions (Arseneva 1977, 79–81; Firsov 1999, 3–4; Korpusova 1983, 104–8). Since only a single example of such a tomb has been identified, however, it is difficult to say anything beyond the fact that this type is an apparent rarity at Gordion.

In terms of spacing and orientation, alignment of burial in the Common Cemetery varies between north/north-west and north/north-east with only a single exception, CCG-42, an outlying pit grave oriented south-east/north-west. Orientation is thus fairly uniform here, laid out roughly north/south, with some slight variations, more akin to the orientation of SLT Area A rather than SLT Area B. There is no attempt to align the graves directly with the road, which runs north-east to south-west towards the Citadel Mound, as is sometimes the case in extra-mural cemeteries. Once again we find graves neatly separated from each other, suggesting that markers of some type (organic?) were in place, yet no trace of them remains.

In regard to offerings (Table 9.4), we see both a similar incidence of placement within the graves or on the bodies – nearly 75% of the burials had one object or more, as in the SLT cemetery – and a similar range of commonplace, daily objects with no explicit religious or ceremonial function. While a wider array of ceramic objects was recovered here (e.g. *unguentaria*, small amphorae, pitchers, cosmetic saucers), their numbers remain quite small and their presence infrequent. Glass vessels were also found in only small quantities, no lamps were recovered, and only a single coin was unearthed, a bronze half-*assarion* of early Severan date that was pierced and reused in a child's necklace. As in the SLT cemetery, the most common categories of object were those of personal adornment, including standard jewellery types such as gold, silver, bronze, and iron rings with engraved gemstones, gold and silver earrings, necklaces composed of beads, and a bronze bracelet. Also present in

small numbers were personal accessories, such as bronze bells and mirrors, carved bone hair-pins and bone spindles. While a small fraction of the objects do occur in precious metals (gold and silver), few are of outstanding quality or workmanship, and the general impression presented here is one of only moderate to low levels of wealth within this rural settlement (Goldman 2007, 312–4). In addition, this collection of offerings is useful in permitting us to distinguish the general span of burial activity in the CC necropolis, between the 2nd and 4th centuries AD, beginning at the period near or just following the abandonment of the Citadel Mound (YHSS 2:3) and continuing until or just after its reoccupation in YHSS 2:4. Thus while the earliest graves are perhaps contemporary with those from the SLT cemetery, this necropolis as a whole appears to be a largely successive burial ground, in use during the period when the Roman base had moved off of the mound.

The North-West Zone Cemetery (4th–5th century AD = YHSS 2:5)

A third necropolis has been explored at Gordion, a small but crowded late Roman cemetery lying on the north-western edge of the Citadel Mound itself, in the North-West Zone of excavation. Discovered and largely unearthed in 1950 during Young's preliminary season of excavation, the necropolis (*c.* 30 x 40 m) is composed of stone cist graves (Gordion Type 8), nearly all maintaining a strict east/west orientation (Fig. 9.11). Excavation in this area in 2004 and associated coin finds has confirmed that this cemetery, which is cut into the streets and structures of the YHSS 2:4 occupation phase, was created shortly after the area's abandonment in the mid-4th century AD and likely continued in use for two to three generations, until the mid-5th century AD (Sams and Goldman 2006, 45). As this cemetery, like the CC necropolis, has been discussed at length elsewhere (Goldman 2007, 301–5), a brief summary will be offered here.

The 65 graves unearthed to date in the North-West Zone cemetery represent the latest burials at Gordion, contained few gifts, and probably relate to the newly identified late Roman or early Byzantine (YHSS 2:5) occupation phase at Gordion. The graves are located in five separate clusters, which might indicate reserved family or clan areas within the necropolis. In contrast to the earlier cemeteries, grave construction and orientation is remarkably uniform in terms of shape, size and material, with cists lined with and covered by large limestone slabs set directly into the YHSS 2:4 surface. Construction materials were clearly scavenged from earlier and nearby structures, and lids of flat limestone slabs cover each grave (3–5 per adult cist, 1–3 for infants and children). Placement of the body is also consistent, laid in a dorsal posture with head to the west, legs extended, arms at the sides, and lower arms folded across the abdomen (with left arm typically placed over the stomach and right

Table 9.4. Gordion. CC necropolis grave and burial types, with associated objects (A. Goldman).

Cat. #	Burial type	Grave type	Burial construction	Associated finds
CCG-1	IN	1a	Simple pit	Hobnails, iron fragments (ILS 146)
CCG-2	IN	1a	Simple pit	—
CCG-3	IN	1b	Simple pit, stone over feet	—
CCG-4	IN	1a	Simple pit	—
CCG-5	IN	1a	Simple pit	Coin (C 341, early 3rd century), glass and stone beads (G 112–4), silver wire ring (ILS 148)
CCG–6	IN	3a	Mud-brick coffin	Hobnails
CCG-7	IN	5b	Tile grave	Bell (B 458), glass bead (G 119)
CCG-8	IN	1b	Simple pit with stone cover	Earring (J 73)
CCG-9	IN	3b	Mud-brick coffin, partially reusing pre-Roman wall	Silver ring with intaglio (J 76), glass beads and bronze clasp (G 115), hobnails
CCG-10	IN	3b	Mud-brick coffin, stone lining	Earring (J 74), hobnails
CCG-11	IN	1a	Simple pit	Hobnails
CCG-12	IN	1a	Simple pit	—
CCG-13	IN	1a	Simple pit	Hobnails
CCG-14	IN	1a	Simple pit	Hobnails
CCG-15	IN	4	Wooden coffin	—
CCG-16	IN	1a	Simple pit	—
CCG-17	IN	4	Wooden coffin or cover	Jug (P 740), saucer (P 741), spindle (BI 180), glass bottle (G 117), gold ring with intaglio (J 75), fibula, hobnails
CCG-18	IN	4	Wooden coffin with mudbrick cover	Unguentaria (2x: P 759–60), shoe eyelets (B 457)
CCG-19	IN	1a	Simple pit	Bronze ring (B 455), hobnails
CCG-20	IN	3a	Mud-brick coffin	Iron ring with intaglio (J 79), hobnails, shoe eyelets (B 547), iron nails (2x)
CCG-21	IN	5b	Tile grave (wood coffin?)	Tile
CCG-22	IN	4	Pit in bedrock, wooden cover	Unguentarium (P 1010)
CCG-23	IN	2a	Stepped grave	Unguentarium sherds, iron ring with intaglio (J 91), hobnails
CCG-24	IN	4	Pit with wooden cover	Unguentarium (P 1013), earrings (2x: J 92), saucer (P 1014), bone spindles (2x: BI 221), iron nails w/traces of gold foil

Table 9.4. Continued

Cat. #	Burial type	Grave type	Burial construction	Associated finds
CCG-25	IN	2a	Stepped grave	Amphora (P 2705), bronze ring with intaglio (P 1376), bone pin (BI 400)
CCG-26	IN	2a	Stepped grave	Iron ring with intaglio (ILS 355)
CCG-27	IN	2a	Stepped grave	—
CCG-28	IN	4	Coffin or wood-lined grave	Hobnails (ILS 360–1)
CCG-29	IN	2a	Stepped grave	Glass flask (G 282)
CCG-30	IN	1a	Simple pit	Amphora (P 2725)
CCG-31	IN	2a	Stepped grave	Hobnails
CCG-32	IN	2a	Stepped grave	Amphora (P 2733)
CCG-33	IN	6	Chamber Tomb	Alabastron (ST 485), iron knife (ILS 362), glass fragments
CCG-34	IN	2a	Stepped grave	—
CCG-35	IN	1a	Simple pit	Hobnails (ILS 370)
CCG-36	IN	1c	Simple pit, pitched rubble and sherd cover over body	Bronze ring (B 1402), bronze pin fragments
CCG-37	IN	5a	Stone- and tile-lined grave	Iron fragments, tile (A 216)
CCG-38	IN	2a	Stepped grave	—
CCG-39	IN	2a	Stepped grave	—
CCG-40	IN	2a	Stepped grave	—
CCG-41	IN	1a	Simple pit	Hobnails
CCG-42	IN	1a	Simple pit	Unguentarium (P 2755), jug (P 2767), hobnails, iron fragments
CCG-43	IN	3b	Mud-brick coffin	Glass flask fragments, glass bracelet (G 279), silver ring with intaglio (J 143), bronze mirror (P 1416)
CCG-44	IN	1a	Pit with mudbrick cover	Bone pin (BI 402)
CCG-45	IN	3a	Mud-brick coffin	Bronze ring with intaglio (B 1415), hobnails, textiles (?)
CCG-46	IN	2a	Stepped grave	Bronze necklace and ornament (B 1414), glass (G 281) and stone (BI 409a–b) beads
CCG-47	IN	1a	Simple pit	Silver ring with intaglio (J 144), glass bead (G 276)
CCG-48	IN	1a	Simple pit	Amphora (P 2747)
CCG-49	IN	2a	Stepped grave	Hobnails
CCG-50	IN	2a	Stepped grave	—
CCG-51	IN	2a	Stepped grave	Jug (P 2770)

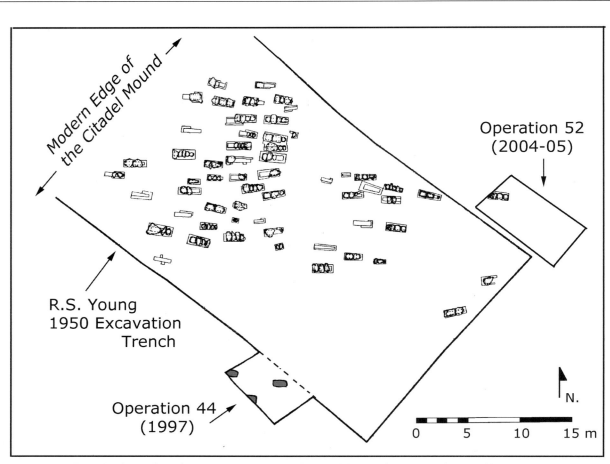

Fig. 9.11. Gordion. Plan of NW Zone cemetery on the Citadel Mound (E. B. Reed; revised by A. Goldman).

set parallel below it, usually over the pelvic area). The very few offerings found within these graves consisted of adornment objects such as glass bracelets, beads, a dove pendant, several flat-banded finger rings, and a twisted wire bracelet. Fragments of tightly woven, un-dyed wool textiles have been identified in one burial, the remnants of a shroud or the deceased's clothing. Overall, the necropolis is quite similar to other late Roman or early Byzantine cemeteries in central Anatolia, such as Dorylaion, Yalıncak (near Ankara) and Pessinus (Darga 1993, 484–5; Tezcan 1964, 18; Devreker 2003, 56). Such cemeteries with their comparative uniformity in construction and orientation appear to signify that that a major, widespread transition in burial practices had taken place across the region by late Antiquity, one that perhaps should be credited to the arrival and widespread practice of Christianity.

Patterns of burial in the necropoleis at Gordion

Among the 147 burials in the three Roman cemeteries discussed above, it is possible to identify certain distinct patterns of burial construction and funerary practice.

First, in terms of grave orientation, a stark difference can be noted between the earlier cemeteries (SLT, CC) and the latest necropolis (NWZ) at Gordion. Graves of the former adopt a predominantly north/south orientation – with slight shifting in alignment to the north-east and north-west (perhaps due to seasonal factors) – while the latter burials clearly and dramatically shift to an east/west orientation, with heads placed to the west. The proximity of the Roman highway does not appear to have been a factor, since that road runs on a north-east/south-west path through the Common Cemetery (Fig. 9.9). Within the SLT and CC necropoleis, it is worth noting not only that the north/south orientation is prevalent between these non-contiguous burial grounds, but that it was also maintained as an established local practice over the course of several centuries, until late Antiquity and the comprehensive adoption of a more uniform burial practice. As in many ancient cemeteries, one does encounter a minority of graves placed in an atypical fashion (e.g. CCG-42, LTG-21) or isolated geographically from the others, as 'outliers' (Morris 1992, 179–80). The pair of *busta* from SLT Area B appear to fall into these categories, and as such they might represent an importation of a mortuary practice from the Roman West, possibly by the troops brought to garrison the 1st-century auxiliary base. Although practice of cremation in Anatolia represents a poorly understood phenomenon, it is now recognized as widespread in Asia Minor (Ahrens 2014; Spanu 2000, 174) and has been explicitly documented at several sites, including 60 examples unearthed in the 'Acropolis' cemetery at Pessinus, where the ritual continued in use from the Hellenistic period until *c.* 300 AD. However, while a variety of cremation types are represented in that necropolis (e.g. simple deposition pits containing cremated remains, cremation graves lined with mud brick, *busta*), none appear to match those unearthed at Gordion, with the combination of *bustum* and intact *cinerarium* (Devreker 2003, 40–3). As a result, it is suggested here on the basis of variation in shape, contents, and orientation that these *busta* might represent non-indigenous burial practices and forms, perhaps an import from areas in the Roman West where cremation was widely practised, such as Moesia and south-eastern Pannonia in the Danubian region (Oţa 2007; Leleković 2012). In any case, the cemeteries at Gordion show a high level of internal uniformity in regard to orientation.

Second, if one excludes the late Roman graves of the NWZ cemetery, it appears to be the norm at Gordion to have multiple construction types mixed together in close proximity. While clusters of a single burial type can be identified in a few areas (e.g. brick sarcophagi in SLT Area A, stepped graves in the MS Trench), no single category is exclusively dominant in any section of the SLT and CC cemeteries. One constant among them, however, is the generous spacing between the graves, always placed at least a metre apart, occasionally disturbing pre-Roman but never Roman burials. Some type of marking system must have been employed, but aside from a few small piles of stones, all evidence for grave markers is largely lost. However, since only two Roman stelae (both with Latin inscriptions) have ever been recovered from Gordion or its vicinity, the use of more elaborate stone markers at the site does not appear to have been likely (Roller and Goldman 2002; Goldman 2010). This absence contrasts sharply with the numerous funerary inscriptions recorded at Pessinus, suggesting that that urban phenomenon did not translate to the rural landscape, even at a site of moderate prosperity located along one of the region's major transportation routes.

Third, in regard to burial practice, inhumation clearly predominates, although the SLT Area B *busta* indicate that cremation was also practised to a limited extent. Given the absence of cremation in the Common Cemetery, where burials on the whole appear to post-date the early Roman settlement phases, it seems reasonable to conclude that cremation rites at Gordion are contemporaneous with and limited to inhabitants of the 2:1–3 settlement. The decline

of cremation in favour of inhumation in the later Imperial period is well documented across the Empire, and the burials at Gordion appears to match pattern (Pearce 2010, 82). Among the inhumations, skeletal position is always extended and supine, with lower arms bent and placed across the stomach or abdomen. Single interment appears to have been the customary practice; examples of multiple burials are rare at Roman Gordion, and with one exception (CCG-33, the chamber tomb), they never exceed more than two individuals per grave. Only six of the graves contain more than one body. In nearly all of such cases we have an older female with an infant or neonate, and in the case of the two examples from the SLT cemetery – a young adult and neonate (LTG-3) and a young adult and infant (LTG-10) – the association between bodies would seem mostly likely to be that of mother and child. Secondary interments and multiple burials thus appear to have been quite rare at Gordion.

Finally, some patterns are observable in terms of burial offerings, placement of which normally occurred alongside the head or feet of the deceased on the body, with objects of personal adornment (e.g. rings, earrings) positioned on the body itself. Offerings tend to be utilitarian objects of relatively low quality, and many of the graves contained either no object or only the remnants of the hobnailed boots. Since relatively few offerings aside from personal jewellery or articles of adornment were recovered from any of the cemeteries, from inside or outside the graves, any prevalent customs or rituals associated with burial practice are not evident from our data. The low level of offerings is not surprising, as the deposition of grave goods across the Empire in general was on the decline by the 3rd century (Pearce 2010, 82). Even so, it is notable that the burial of infants and children was conducted with some care, and nearly all contained an object of some sort, such as a glass *unguentarium*, perhaps reflecting emotional attachment and feelings of loss on the part of the parent(s) or guardian(s).

In sum, it is possible to detect some general patterns among burial practices at rural Roman Gordion between the 1st and 4th centuries AD: burial objects and furniture are relatively simple and minimal, multiple interments are rare, the combination of multiple grave forms in a single necropolis appears to be common (with the exception of late Roman graves), and the graves are oriented and spaced with some care, although apparently lacking in permanent markers of any kind.

Recent fieldwork on Galatian necropoleis

While the analysis offered above is helpful in identifying local patterns of burial practice at Gordion, it must be recognized that various contextual factors could easily have affected the character of the sample in question, rendering it inappropriate or invalid for identifying and studying rural non-elite burial in the region. Among such factors are the site's function as a military base, the origin(s) of the local garrison, and the proximity of the highway. Differentiating between soldiers and civilians among the deceased is extremely difficult, since no weapons have been found in association with the graves and the 22 pairs of *caligae* were recovered from graves of men, women and children, indicating that they were worn across the population. In addition, attempts to trace ethnic identity among auxiliary forces on the basis of funerary finds are equally as problematic, as I. Haynes (2013, 135–42) has recently pointed out. Explicit identification of *origo* through funerary epitaphs like that of the Pannonian Tritus is quite unusual in Galatia, and the expense of ordering, conveying and erecting such a permanent monument must have been substantial, well beyond the means of the average Galatian townsperson or villager. Sadly, Tritus' tombstone was not found in situ, so that we cannot determine what type of burial was practised by the heir (Mersua) of this Pannonian soldier of the *cohors VII Breucorum*, a unit known to have been stationed in Germania Superior, Pannonia, and Moesia Superior prior to its arrival at Gordion, most likely during the later years of Trajan's Parthian War (Goldman 2010, 138–9). Whether such soldiers represent the agency behind the influx of fashionable 'Roman' construction types, such as the *tegulae* burials (or imitations thereof), remains uncertain and at present unknowable. As noted previously, pre-Roman burials have yet to be studied for Galatia, with the result that isolating the mechanisms by which 'Roman' types of burial spread across the rural landscape is presently a challenging, if not impossible task.

Given the site's military function and the periodic influx of garrison troops (at least during the settlement's early years), it is thus quite reasonable to ask whether we can even consider the Gordion cemeteries as practical candidates for identifying non-elite burial patterns in central Turkey. A case can be made that the primary unit at Gordion, the *cohors I Augusta Cyrenaica*, which was stationed in Galatia by the early 2nd century and into the 3rd century, might well have adapted to local funerary customs during their long-term assignment. They were certainly recruiting local Galatians into their ranks; funerary epitaphs from Ankara record men recruited from cities in southern Galatia, Iconium and Savatra (Bennett 2009, 113–17). Clearly the solution to assessing the representative character of Gordion's Roman burials depends upon engaging a larger body of material, an inquiry that has now become possible owing to a growing body of comparative data produced at sites in central Turkey. The primary focus of mortuary archaeology in the region remains urban necropoleis, like those at Pessinus and now the small city of Juliopolis, where a team from the Museum of Anatolian Civilizations in Ankara has excavated nearly 450 graves since 2009 (Krsmanovic and Anderson 2012; Arslan *et al.* 2012; Cinemre 2014b).

However, a series of recently published rescue excavations at a scattering of rural Galatian sites is slowly expanding our dataset, as are site descriptions and inventories of objects now available online at www.envanter.gov.tr, a development which greatly facilitates access to the material and preliminary dating of the finds and their contexts. While the quality of recording continues to be inconsistent, as is often the case with rescue work, it is possible now to assess some of the patterns observed at Gordion and to determine their legitimacy as a possible model on a broader, regional scale. A short summary of excavation background, burial construction types, grave orientation, skeletal position, offerings and topographic placement will be presented below for four recently published cemeteries which lie in or on Galatia's territorial boundaries: Boyalık necropolis, in the Gölbaşi district of Ankara province, with 32 burials excavated in 2008–9; Bahçeçik necropolis, in the Haymana district of Ankara province, with 16 burials excavated in 2009–10; the necropolis at Dadastana, which just west of the Ankara province (on the border of Bithynia), with 25 burials from 2009; and the Neolithic site of Çatalhöyük, with 79 'late' burials excavated between 2003 and 2008 in the 4040 Area.

Boyalık (Gölbaşi district)

In 2007 and 2008, a team from the Ankara Museum investigated a Roman-period necropolis at the village of Boyalık, in the Gölbaşi district due south of Ankara. Situated upon a small hill (Kartalkaya Tepe) to the immediate north-west of the village, the cemetery was originally brought to light in 2002 during construction work on a water tank and had subsequently suffered from looting. In a series of six contiguous trenches (A-1 to A-6), the excavators unearthed a total of 32 inhumation and cremation burials, 18 (M1-18) in 2007 and the remaining 14 (M19-32) in 2008. While a complete analysis of the cemetery and its funerary finds has yet to appear, two preliminary annual reports were issued which contain a descriptive summary of the burials and objects as well as an assortment of plans and photographs. The contents of those reports (Denizli, Kaya, and Çetin 2008; Çetin and Kaya 2010) are discussed below. In addition, an inventory of the recovered objects, including their measurements and photographs, has now been made available to scholars online. Even so, significant problems remain when attempting to analyze this sample of burials, including its relatively small size, the number of disturbed contexts (*c.* 20%), the widespread scattering of objects from ancient and modern looting activities, and the removal of all grave markers (many of which were observed in reused contexts in the village). From a broader perspective, we also lack evidence for the cemetery's associated settlement (now probably below the modern village) as well as an estimate of its original extent.

Despite these contextual problems and the incomplete state of the cemetery's publication, it is evident from the two preliminary reports that general patterns of burial practice, grave construction and offering types observable in the Boyalık necropolis parallel in many ways those observed within the SLT and CC necropoleis at Gordion. Both cremations and inhumations are present at Boyalık, with evidence for the two practices found in close proximity to each other (as they are in SLT Area B). While cremation represents a larger percentage of the excavated graves (nine of 32, 28%) at Boyalık than at Gordion, inhumation continues to predominate in both samples. Although the preliminary reports only contain scant information about the skeletons themselves, single interments appear to be the rule, as at Gordion. Among the inhumations, one observes a similar mixture of construction types, with a predominance of simple pit graves (= Gordion Type 1) and a seemingly random assortment from other familiar categories mixed in, including a stepped grave (= Gordion Type 2), five mud-brick-lined graves (= Gordion Type 3), and a tile grave (= Gordion Type 5). Notably, several of the cremations proved to be similar in form and function with the *bustum* (LTG-13) in SLT Area B at Gordion (= Gordion Type 7), with a combination of *cineraria* placed along the edges of pits with burnt orange sides. This discovery raises the possibility that the cremations in the SLT cemetery, previously posited as a military import to Gordion, might well represent a local (Galatian) rather than a migratory burial practice. Finally, although the sample is small in size, the largely modest burial offerings obtained from the Boyalık cemetery are virtually identical in terms of materials, quality and frequency to the finds from Gordion's Common Cemetery. These include glass and ceramic *unguentaria*, bronze rings with carved semi-precious stones, small gold loop earrings, glass and stone beads, and bone spindles and hairpins. Also present are ceramic jugs and small amphora of the type found at Gordion, along with nearly a dozen shallow bowls and dishes. Familiar categories of offerings that are missing at Gordion, specifically coins and lamps, are also absent at Boyalık. On the basis of such offerings, the Boyalık burials seem to be contemporary with the graves of the Common Cemetery and thus burial activity is likely to span from the 1st to 3rd centuries AD.

Yet some important differences between the graves at Boyalık and Gordion are apparent as well. Greater deviation is evident in the shape of the cremation pit, which varies from amorphous (M2) to round (M1) and rectangular (M28). In addition, there are several cremations clustered in a single trench (A6), within which only *cineraria* – reused urns or pithoi – are present (e.g. M21, M22, M24). In terms of grave orientation at Boyalık, adjacent areas of the cemetery display quite different patterns of orientation. For example, contiguous Trenches A3 and A4 contain mostly inhumations which follow a north/south orientation, with a single outlier (M10) on an east/west axis. Just north of these, however, in adjacent Trench A5, the six inhumations excavated between 2007 and 2008 are oriented either due east/west or north-east/south-west. In addition, many of the graves are placed in close proximity to each other, abutting or even cutting into each other to a minor degree. They display a far less ordered arrangement of interment than one finds at Gordion, resulting in a much more haphazard layout of burial at Boyalık. One might expect greater uniformity, as a brief mention in the 2007 preliminary report notes the presence and reuse of funerary stelae in various contexts in and around the village (Denizli, Kaya, and Çetin 2008, 139). Since these monuments appear in secondary contexts, it is impossible to say whether they originated from this necropolis or another in the vicinity. While no description of these stelae has been published, their mere presence does raise the issue as to whether Galatia's rural residents did in fact have the resources to purchase and erect permanent monuments, as opposed to the non-permanent or rudimentary markers that appear to have been in use at Gordion.

Bahçecik (Haymana district)

In the Haymana district to the south-east of Gölbaşi, on a high ridge near the small town of Bahçecik (*c.* 55 km south-west of Gordion), illegal excavations led to a 2009 investigation of a late Roman and early Byzantine extramural cemetery. According to the preliminary report on the rescue project (Arslan, Ateşoğulları, and Şahin 2011), a series of five non-contiguous trenches (A–E) were placed along the crest of the ridge over an area of *c.* 20×80 m, and a total of 16 burials were uncovered (two of which had been robbed). While the quantity of excavated burials here is much smaller here than at Boyalık, they represent an extremely consistent sample that parallels closely the signature of the cist graves (= Gordion Type 8) found in the North-West Zone Cemetery. As at Gordion, all of the graves were identical in construction – lined with stone blocks on the edges and topped with flat slabs – and placed on a strict east/west axis. Skeletons were placed in an identical posture as well, laid out in a dorsal position, with heads to the west and arms bent and placed across the abdomen. In addition, as at Gordion the graves were found to contain very few offerings, which consisted entirely of jewellery items such as glass beads, bronze pendent earrings and bracelets of glass and bronze. The excavators have tentatively dated the objects to the early Byzantine period (*c.* 5th–7th century AD), which would make this necropolis roughly contemporary with that of the North-West Zone on the Citadel Mound. Again, we have only a small sample here, but it is one which possesses a profile that provides more evidence for a substantial late Antique transformation in burial practice in the region of Gordion.

Islamalan, ancient Dadastana (Nallıhan district)

The village of Islamalan lies just east of the border between the Bolu and Ankara provinces and along the ancient provincial boundary of Bithynia and Galatia (*c.* 20 km west of Nallıhan and *c.* 95 km north-west of Gordion). The site was known in antiquity as Dadastana, a town which was chiefly famous for the visit of the Emperor Jovian in 364 AD and his subsequent death there, either of poisoned mushrooms or carbon monoxide poisoning from noxious charcoal fumes (Ammianus Marcellinus 25.10.12–3). A four-week rescue excavation in 2009 unearthed portions of a Roman and early Byzantine necropolis in a sloped, now forested area near the village. Eight trenches (T. I–VIII) were opened and a total of 25 inhumation burials, most in poor shape, were unearthed. No plan of the necropolis and its trenches were published in the brief preliminary report (Arslan, Ceminre and Erdoğan 2011), so that it is difficult to draw more than general conclusions about patterns of spacing. Published photographs of various trenches, however, do suggest ordered spacing in at least some areas that is not unlike that within the Gordion cemeteries. All of the graves discussed in the report were aligned on an east/west axis, and the skeletons (where preserved) were placed in a uniform posture, laid out in a supine and extended position with the head to the west and arms crossed over the abdomen. The majority of these burials consisted of simple pit graves (= Gordion Type 1), but once again we find a mix of construction types, with several examples of coffin burials (with wood fragments and nails), *tegula* burials (with actual tiles), and stone-lined cist graves (= Gordion Types 4, 5 and 8). Photographs and descriptions of the burial offerings indicate that the finds are of a similar type and quality as those excavated from the Gordion cemeteries and settlement (e.g. small amphorae, glass bracelets, bronze earrings, red-slipped *skyphoi*). Again, lamps and coins are not seemingly present, although here we find one class of object not at Gordion, specifically small bronze pendant crosses. Although the excavators offer no specific dates for the cemetery, the ceramics and other offerings suggest that burial activity took place between the 2nd and 6th centuries, with the result that at least a portion of these graves are contemporary with those of the Common Cemetery and North-West Zone cemetery.

Çatalhöyük, near Konya

The largest sample of non-elite burials in central Turkey to reach publication in recent years belongs to the Neolithic site of Çatalhöyük, where over 200 'late' burials have been excavated across the site. Data on the graves from the East and West Mounds has until recently been scattered throughout various publications, at a site where multiple teams are in operation and various outlets for publication have been used. Many of the graves have been discussed

upon their discovery within the project's annual reports, such as those unearthed from the Team Poznan (TP) Area since 2001 (Kwiatkowski 2009, for summary). Formal publication of these 'late' burials has proceeded more slowly, although one small group from the BACH Area – six graves excavated in 1997–1998 by the Berkeley Team in the upper layers of Building 3 – were recently published as a larger sample, with a more detailed analysis of the accompanying skeletal remains and grave goods (Cottica, Hager and Boz 2012). The complementary settlement(s) for these graves remains unidentified, but does not appear to have been located on either the East or West Mounds.

While a comprehensive study of the burials from the East and West Mounds has yet to appear, the recent publication by S. Moore and M. Jackson (2014) of 79 graves excavated within the 4040 Area of the East Mound between 2002 and 2008 now allows for a wider discussion of the necropolis, its organization and contents. While this large sample of 'post-Chalcolithic' burials is located at some distance from the Galatian heartland and Gordion itself (*c.* 225 km), nevertheless the site still lies within the former boundaries of Roman Galatia and the sample, from a non-urban settlement, is viable for comparison. As such, Çatalhöyük's 'late cemetery' represents a tremendous resource for investigating rural burial in central Anatolia. Significantly, although such 'late' burials are of peripheral interest to the Neolithic experts at the site, they have been meticulously excavated and methodically recorded by the various separate teams working on the East Mound. It is worth noting that while similar 'late' burials have been recorded at many Bronze Age and earlier sites in Anatolia (see below), few have been accorded the level of recognition or treatment that they have recently received at Çatalhöyük.

Moore and Jackson (2014, 606–13 and fig. 32.2) have currently grouped the 79 graves into four categories (Groups 1 to 4) on the basis of construction type, spatial location, body position and the presence of offerings. These four categories must be understood as provisional, since excavation of the site's 'post-Chalcolithic' burials has continued since 2008 and a larger sample will eventually become available for analysis. Group 1, clustered in the northern section of the necropolis, is composed of 28 rectilinear-cut graves lined with a variety of materials (wood, mud brick, tile). These graves, aligned in an east/west fashion, are considered the earliest of the burials. Fourteen contained burial offerings, personal effects that are almost identical in form and material to those found in the cemeteries at Gordion. These include ceramics (e.g. a cosmetic saucer, *unguentaria*), glass vessels (e.g. flasks, candlestick *unguentaria*), a bronze mirror, earrings of gold and bronze, bone spindles, stone beads, iron shoe hobnails, etc. Like the SLT and CC graves at Gordion, only a handful of graves held more than one or two objects. A single, badly corroded coin has been tentatively dated from the mid- to late 2nd century, while carbon dating from one grave (F. 1553) yielded a similar range, of the 1st century to the 3rd century. The six graves excavated in the BACH Area just to the north, which observe a similar orientation, display similar construction techniques and contain similar offerings, also appear to belong to this group of Roman Imperial burials. In comparison, Group 2 is composed of 33 sub-rectilinear graves clustered to the south that contained no offerings or personal objects but with some of the deceased placed in coffins and shrouds. These graves also observe an east/west alignment, although with some at a slightly more south-east/north-west angle. Like several of the Group 1 burials, several of those in Group 2 incorporated tiles or tile fragments in their construction, including one (F. 1476) which may have been used as a grave marker. These burials have been tentatively assigned to the early Byzantine period, on the basis of a single radiocarbon tested skeleton, dating *c.* 330–410 AD. Group 3 has an entirely different profile: clustered in the south-western part of the necropolis, these 10 graves – all single inhumations, lying on their right sides to face south, perhaps shrouded – are composed of very narrow cuts, either pit graves or lined with mud brick. On the basis of skeletal positioning, the excavators have suggested that these burials are Islamic, facing south towards Mecca. This group might well be linked to graves located to the south-west, in the TP Area, where radiocarbon dating has indicated burial activity spanning the mid-12th to 17th centuries. The 63 graves of the TP Area recorded by the end of 2014, however, differ somewhat in construction, using a 'niche grave' design which features an upper rectangular chamber and a second, lower niche chamber within which the body is placed (Kwiatkowski 2009; Filipowicz, Harabasz and Hordecki 2014). Finally, Moore and Jackson's Group 4 consists of eight outliers or burials in very poor condition, about which little can be said.

While Çatalhöyük's cemeteries still await comprehensive publication, several general, yet notable similarities and differences between the graves of the 4040 Area and those of the Gordion cemeteries can be observed from Moore and Jackson's substantial preliminary study. First, in regard to similarities, pit graves predominate once again, and among them one encounters a mixture of construction types during the earliest (Group 1) graves, with evidence for the use of wood-lining and mud-brick coffins. Second, there is again a strong predominance of single interments, with body posture in nearly all the Group 1 and 2 graves quite similar to that at Gordion (i.e. extended, supine, heads in one direction), with the exception that arms were placed extended along the sides. Third, the frequency of grave offerings is also similar, with a few richly endowed graves amid many more that contain very modest offerings or objects of personal adornment. One again finds that objects often accompany the infant and juvenile burials, the number of which appears curiously small in comparison with the adults. Notably, from the small sample in the BACH area, two lamps were recovered and four of the six graves held at least one object, suggesting that some variation in and frequency of offering type exists across the East Mound. The richest endowed grave, however, which contained a glass *unguentarium*, copper beads, a bone hair pin and a bone needle, is that of a 3–4-year-old child, a fact that fits well with the pattern of endowed infant/juvenile burials noted previously (Cottica, Hager and Boz 2012, 333).

The most significant difference which may be observed between the Gordion and Çatalhöyük cemeteries is grave orientation, with the latter having an east/west alignment that is predominant only in the later NWZ cemetery. In addition, 31 burials from the 4040 Area Groups 1 and 2 were either wood-lined or showed evidence of coffins, what is a far greater percentage than at Gordion. Moreover, the later graves at Gordion consist of stone cists, while there is an absence of stone-lined graves on the East Mound of Çatalhöyük. This difference might very well reflect the simple factor of availability; the Citadel Mound at Gordion possessed plentiful and easily accessible ashlars for reuse, while the Konya plain has fewer available sources of stone within easy distance of Çatalhöyük.

Burial geography in Roman Galatia

In addition to patterns noted in the cemetery profiles discussed above, one intriguing feature which stands out in this study of Gordion and its neighbouring cemeteries is a strong similarity in burial geography. All of the cemeteries – the three at Gordion and the four elsewhere – were placed on the top or slopes of prominent ridges, providing them with a high level of visibility in the landscape. There are a variety of practical reasons that might explain why this took place, such as a difficulty in cultivating such areas, leaving them free for alternative use, or a need to discourage attempts at robbery through situating the cemeteries in locations of greater visibility. Alternatively, such cemeteries could be laid out in imitation of Galatia's urban necropoleis. At Pessinus, for example, the burial grounds not only lay along the roads leading out of the city, which is typical of Roman graveyards, but also on prominent elevations overlooking the site (Krsmanovic and Anderson 2012, 62–3). As noted previously, the Common Cemetery and the South Lower Town Cemetery are placed on slopes – the former on the ridge south-east of the Citadel Mound and the latter on the Küçük Höyük fortifications – and close to or within sight of the Roman highway. The placement of Gordion's cemeteries might thus reflect a sensitivity towards the positioning of those in the nearest urban centre. Unfortunately, since the location of the complementary settlements has not been identified at the other four sites examined in this study, it is impossible to discern the relationship between cemetery and settlement in those cases.

What potentially differentiates the SLT and CC necropoleis at Gordion from the others, however, is the fact that they are incontestably fresh foundations, with burial resuming only after the mid-1st century AD, following the site's abandonment for a period of at least a century (Table 9.1). It can thus be argued that the locations of Gordion's necropoleis are the product of careful deliberation and selection by the site's new Roman-period residents, who had moved into an unfamiliar landscape and possessed a funerary *tabula rasa*, one seemingly unfettered by pre-existing local traditions or conditions. We must then seek to understand why the residents chose to return to pre-Roman burial areas and reactivate them for their own use. The fact that the areas were used for burial in earlier times was clearly understood by the Roman-period residents. In the SLT cemetery, several Roman-period graves do cut slightly into Hellenistic burials (e.g. the southern end of LTG-8). Perhaps more striking is the notable absence of Roman burials in those parts of SLT Areas A and B where earlier deposits of human remains cluster most densely (e.g. the southern half of SLT Area A). A similar pattern may be observed in the Common Cemetery, where areas of dense pre-Roman burial activity (e.g. Hittite pithos burials) were clearly encountered and generally avoided in the Roman period. Past activity in that necropolis was evident for all to see, the ridge towered over then as it is today by dozens of large and small Phrygian tumuli. Encircled by these early burial mounds (Fig. 9.9: Tumuli E, F, H–I, and K–II), the non-elite Roman burials were interred in such a specific location chosen quite carefully, in a manner that suggests a connection was being established with the cemetery's earlier, legendary residents – perhaps even to the Phrygian elite. Likewise, the decision by Gordion's new populace to reinitiate burial along the slopes and in the shadow of the towering KH fortifications is also conceivably linked to a desire to establish a connection with Gordion's former Phrygian inhabitants. This aspiration might equally explain the appearance of Roman, Byzantine, and Islamic burials at Çatalhöyük, where such graves cut directly into Chalcolithic and Neolithic structures across the entire site.

If one accepts the premise that Roman-period cemeteries in central Anatolia might have been established at specific locations for purposive reasons – that peoples of the Roman era placed their cemeteries with deliberation and care, in proximity to or directly upon features and monuments belonging to the pre-Roman landscape – then one might expect to find the phenomenon repeated elsewhere in Galatia. Indeed, even a cursory search reveals that there is an almost ubiquitous presence of Roman and successive Byzantine cemeteries on Neolithic, Chalcolithic and Bronze Age mounds and cemeteries in central Turkey. Such discoveries are hardly new; in the 1960s, excavations into a small mound adjacent to and north-west of the main early Bronze Age mound at Çavdarlı Hüyük exposed graves

of the 2nd and 3rd century AD (Akok 1965). Many such finds have gone unreported or under-reported, presumably of little interest to the pre-historians at work on their sites. Fortunately, Roman and early Byzantine burials at such sites are now being reported at an accelerated rate, although detailed publication of the later funerary finds continues to lag. Over the past decade, such sites include: the early and middle Bronze Age cemetery at Dede Mezarı, northeast of Afyon in the Bayat area, which also saw Roman interments (Üyümez *et al.* 2007); the early Neolithic mound of Boncuklu, south of Konya, where D. Baird encountered numerous late Roman and Byzantine burials (Baird 2008); Kültepe and its Karum, where 90 graves of Hellenistic and Roman burials were cut into the middle Bronze Age mound and town (Üstündağ 2009); the Neolithic and Chalcolithic mound at Tepecik-Çiftlik near Niğde, the top of which (Period 1) is cut into by Roman and Byzantine graves (Bıçakçı 2004); Porsuk Höyük, near Ulukışla in the Niğde area, where the French team has excavated over 130 graves dating from the 3rd to 7th century on the eastern end of the mound (Beyer *et al.* 2006); and others.

It is suggested here that such a pattern of burial on highly visible (often mounded) pre-Roman sites is not merely a geographical coincidence, but rather that it represents a more deliberate act on the part of Anatolia's Roman-period rural population. By seeking out and reusing ancient sites near their own habitations, the rural population of central Turkey appears to display a clear preference for embedding their own deceased among the ruins and dead of Anatolia's distant past. In their recent discussion about cemeteries at Pessinus, Krsmanovic and Anderson (2012, 82) propose that the placement of cemeteries should be understood as a control issue, in which burial placement is a means of 'making a material statement of genealogical descent and entitlement', and that various burial patterns at Pessinus (i.e. limited intercutting of graves, unified orientation and clustering) 'suggest some degree of maintenance and observation of such claims'. Such patterns are also apparent at Gordion, where there seems to be a conscious attempt to embrace the memory of the site's Phrygian past and to commemorate the dead through their physical interment within the Phrygian funerary landscape. Both the SLT and CC necropoleis possess a distinct boundary composed of pre-Roman monumental structures, and one might argue that the pattern is also repeated to some degree in the later NWZ cemetery, placed as it is among abandoned structures atop the Citadel Mound. If the Roman-period residents were seeking to bind themselves to their new home, to gain 'control' over the landscape by placing their dead among those of earlier times, they certainly chose the most appropriate venues at Gordion within which to do so.

Such a constructive funeral practice is perhaps reflected more concretely in the reappearance of Phrygian writing in the 2nd and 3rd centuries AD on over 70 rural tombstones from the borderlands between Asia and western Galatia. These 'neo-Phrygian' curse formulae, usually found to accompany traditional Greek epitaphs, represent a rather dramatic revival of a written language that does not appear to have been in use for several centuries (Mitchell 1993, 174). In a region like Galatia, where relatively few people could afford such tombstones for the commemoration of their dead and recognition of social identity, it is possible that such a capacity was served by the landscape itself, in areas which displayed overt connections to the past. If this is the case, we thus find our rural non-elites manipulating the landscape to construct a memorial relationship to the past, in ways not dissimilar to that observed in regions elsewhere in the Roman Empire, such as mainland Greece and Crete (Alcock 2002). For subjects of the Empire with little or no connection to their rulers, living in a conservative, rural region like central Anatolia, with its multi-ethnic population and rich mythical heritage, an attachment to the memorial power of the past – in Gordion's case, that of Midas and the Phrygians – might well have formed a significant and positive aspect of their constructed identity. Strabo's description of Roman Gordion (*Geog.* 12.5.2) is perhaps telling in this regard, as he specifically evokes a past landscape 'of the Phrygians, of Midas and of Gordios', a memorial connection with which the author, of Anatolian origin himself, is obviously familiar. If this tentative hypothesis is correct, then the rural cemeteries of Galatia will have much to add to the growing scholarly discussion about memory formation and manipulation in antiquity.

Conclusions

Clearly much work remains to be done on non-elite rural burial in Roman Galatia. Yet while the portrait which emerges from this multi-site analysis is hardly a comprehensive one, the data assembled here is helpful for distinguishing at least on a rudimentary level various patterns of burial practice in that region. It is encouraging to observe that certain common features or aspects can be detected among the rural non-elite burials at Gordion and elsewhere: a predominance of single interments; supine skeletal posture; grave orientation that demonstrates an axial preference for cardinal directions (i.e. north/south, east/ west); the mixing of multiple construction types during the Imperial era; a prevalence towards burial offerings that are personal possessions or objects of adornment in comparison to those which are commonly thought to reflect burial ritual (e.g. bowls, pitchers); a near total absence of burial objects that are ubiquitous elsewhere in the Roman world, in particular lamps and coins; the comparatively infrequent use of cremation during that period; and a seemingly abrupt, widespread move toward homogenization in burial practice

by the early Byzantine period. Admittedly, slight variations between and within cemeteries do exist in regard to these patterns, and we will need a far larger set of data from well-excavated, more fully investigated necropoleis before we can possibly assess whether the observations presented here are valid for widespread funerary activity in central Turkey. Nevertheless, while we remain at a very early stage in our investigation of rural Galatian burial practice, it is safe to say that through examination of the funerary remains at Gordion, Çatalhöyük and neighbouring sites, forward progress is at last being made – incremental though it might be – in understanding and appreciating the complex funerary landscape of central Anatolia.

Bibliography

Ahrens, S. (2014) 'Whether by decay or fire consumed...': Cremation in Hellenistic and Roman Asia Minor. In J. R. Brandt, M. Prusac, and H. Roland (eds.) *Death and changing rituals: Function and meaning in ancient funerary practices* (Studies in Funerary Archaeology 7), 185–222. Oxford, Oxbow Books.

Akok, M. (1965) Afyon – Çavdarlı Hüyük kazısı. *Türk Arkeoloji Dergisi* 14 (1/2), 5–34.

Alcock, S. (2002) *Archaeologies of the Greek past: Landscapes, monuments, and memories. The W. B. Stanford memorial lectures.* Cambridge, Cambridge University Press.

Anlağan, Ç. (1968) Akkuzulu Tümülüsü kazisi (Akkuzulu Tumulus). *Anadolu* 12, 1–7.

Arık, R. O. and Coupry, J. (1935) Les Tumuli de Karalar et la sépulture du roi Déiotaros II. *Revue archéologique* 6, 133–51.

Arseneva, T. M. (1977) *Nekropol' Tanaisa.* Moscow, Nauka.

Arslan, M., Ateşoğulları, S., and Şahin, Y. (2011) Haymana, Bahçecik Nekropolü kurtarma kazısı. *Müze Çalışmaları ve Kurtarma Kazıları Sempozyumu* 19, 235–54.

Arslan, M., Ceminre, O., and Erdoğan, T. (2011) Dadastana nekropolü 2009 kurtarma kazısı. *Müze Çalışmaları ve Kurtarma Kazıları Sempozyumu* 19, 327–40.

Arslan, M., Metin, M., and Ceminre, O. (2012) Juliopolis nekropolü 2010 Yılı Kurtarma Kazısı. *Müze Çalışmaları ve Kurtarma Kazıları Sempozyumu* 20, 177–214.

Atasoy, S. (1974) The Kocakızlar tumulus in Eskişehir, Turkey. *American Journal of Archaeology* 78, 255–64.

Baird, D. (2008) The Boncuklu project: The origins of sedentism, cultivation and herding in central Anatolia. *Anatolian Archaeology* 14, 11–2.

Bennett, J. (2009) The Cohortes Augustae Cyrenaicae. *Journal of African Archaeology* 7.1, 107–21.

Beyer, D., Chalier, I., Laroche-Traunecker, F., Lebreton, S., Patrier, J., and Tibet, A. (2006) Zeyve Höyük (Porsuk): Rapport sommaire sur la campagne de fouilles de 2005. *Anatolia Antiqua* 14, 205–44.

Bıçakçı, E. (2004) Tepecik-Çiflik: A new site in Central Anatolia (Turkey). *Architectura* 34, 21–6.

Çetin, N. and Kaya, V. (2010) Boyalık nekropolü 2008 Yılı kazı çalışmaları. *Müze Çalışmaları ve Kurtarma Kazıkarı Sempozyumu* 18, 79–90.

Cinemre, O. (2014a) 2012 yılı Akyurt – Kalaba Tümülüsü kurtarma kazısı. *Müze Çalışmaları ve Kurtarma Kazıkarı Sempozyumu* 22, 349–64.

Cinemre, O. (2014b) Juliopolis nekropolü 2012 yılı kazı çalışmaları. *Müze Çalışmaları ve Kurtarma Kazıkarı Sempozyumu* 22, 407–26.

Cleary, S. E. (1992) Town and country in Roman Britain? In S. Bassett (ed.) *Death in towns. Urban responses to the dying and the dead*, 29–42. Leicester, Leicester University Press.

Cottica, D., Hager, L., and Boz, B. (2012) Post-Neolithic use of Building 3 (Space 86) and Spaces 88, 89. In R. Tringham and M. Stevanović (eds.) *Last house on the hill: BACH area reports from Çatalhöyük, Turkey* (Monumenta Archaeologica 27), 331–43. Los Angeles, Cotsen Institute of Archaeology Press.

Darbyshire, G., Harl, K., and Goldman, A. (2009) 'To the Victory of Caracalla': New Roman altars at Gordion. *Expedition* 51.2, 31–8.

Darga, A. M. (1993) Sarhöyük – Dorylaion kazıları (1989–1992). *Kazı Sonuçları Toplantısı* 15.1, 481–501.

Denizli, H., Kaya, V., and Çetin, N. (2008) Boyalık nekropolu 2007 yılı kurtarma kazısı. *Türk Arkeoloji ve Etnografya Dergisi* 8, 133–40.

Devreker, J. (2003) *Excavations at Pessinus: The so-called Acropolis*. Ghent, Academia Press.

Ergil, T. (1983) *Küpeler: Istanbul Arkeoloji Müzerleri küpeler katalogu/Earrings: The earring catalogue of the Istanbul Archaeological Museum*. Istanbul: Ali Rıza Baskan Güzel Sanatlar Matbaası.

Faber, A. (1998) *Das römische Gräberfeld auf der Keckwiese in Kempten. II. Gräber der mittleren Kaiserzeit und Infrastruktur des Gräberfelds sowie Siedlungsbefunde im Ostteil der Keckwiese. Cambodunumforschungen VI* (Materialhefte zur Bayerischen Vorgeschichte A75). Kallmünz/Opf, Lassleben.

Filipowicz, P., Harabasz K, and Hordecki, J. (2014) Excavations in the TPC Area. *Çatalhöyük Archive Report 2001*, 72–9.

Firsov, K. (1999) The Roman period necropolis at Zavetnoe in southwest Crimea. Burial structures and mortuary ritual. In M. Rundkvist (ed.) *Grave matters. Eight studies of the first millennium AD burials in Crimea, England and southern Scandinavia* (British Archaeological Reports S781), 1–18. Oxford, Archaeopress.

Goldman, A. (2007) The Roman-period cemeteries at Gordion in Galatia. *Journal of Roman Archaeology* 20, 299–320.

Goldman, A. (2010) A Pannonian auxiliary's epitaph from Roman Gordion. *Anatolian Studies* 60, 129–46.

Goldman, A. (2014) The octagonal gemstones from Gordion: Observations and interpretations. *Anatolian Studies* 64, 163–97.

Goldman, A. and Voigt, M. M. (2016 forthcoming) Investigating Roman Gordion: Report on the 1993–2005 excavations. In *American Journal of Archaeology*.

Haynes, I. (2013) *Blood of the provinces. The Roman auxilia and the making of provincial society from Augustus to the Severans*. Oxford, Oxford University Press.

Kelp, U. (2013) Grave monuments and local identities in Roman Phrygia. In P. Thomemann (ed.) *Roman Phrygia. Culture and society*. Cambridge, Cambridge University Press.

Korposova, V. N. (1983) *Nekropol'Zolotoe*. Kiev, Naukova dumka.

Krsmanovic, D. and Anderson, W. (2012) Paths of the dead – interpreting funerary practice at Roman-period Pessinus, Central Anatolia. *Melbourne Historical Journal* 40.2, 58–87.

Kwiatkowska, M. (2009) Byzantine and Muslim cemeteries at Çatalhöyük – an outline. In T. Vorderstrasse and J. Roodenberg (eds.) *Archaeology of the countryside in Medieval Anatolia* (Papers from the Netherlands Institute for the Near East 113), Leiden, Netherlands Institute for the Near East.

Leleković, T. (2012) Cemeteries. In B. Migotti (ed.) *The archaeology of Roman southern Pannonia* (British Archaeological Reports, International Series 2393), 313–57. Oxford, Archaeopress.

Matijasić, R. (1991) *Campus Martius*. Pula, Archaeological Museum of Istria.

Mermerci, D. (1988) Kızıleşik Tümülüsü kazısı. *Anadolu Medeniyetleri Müzesi Yıllığı* 1987, 23–32.

Mitchell, S. (1980) Population and the land in Roman Galatia. *Aufstieg und Niedergang der römischen Welt* 2.7.2, 1053–81. Berlin and New York, Walter de Gruyter.

Mitchell, S. (1993) *Anatolia: Land, men and gods in Asia Minor. Vol. I: The Celts and the impact of Roman rule; Vol. II: The rise of the church*. Oxford, Clarendon Press.

Moore, S. and Jackson, M. (2014) Late burials from the 4040 area of the East Mound. In I. Hodder (ed.) *Humans and landscapes of Çatalhöyük: Reports from the 2000–2008 season* (Çatalhöyük research project 8; British Institute of Archaeology at Ankara monograph 47; Monumenta archaeologica 30), 603–20. Los Angeles, Cotsen Institute of Archaeology at UCLA.

Morris, I. (1992) *Death-ritual and social structure in classical antiquity*. Cambridge, Cambridge University Press.

Oţa, L. (2007) *Busta* in Moesia Inferior. *Acta Terrae Septemcastrensis* 6.1, 75–99.

Pearce, J. (1992) Burial, society and context in the provincial Roman World. In Pearce, Millett, and Stuck (eds.), 1–12.

Pearce, J. (2010) Burial, identity and migration in the Roman world. In H. Eckardt (ed.) *Roman diasporas: Archaeological approaches to mobility and diversity in the Roman Empire* (Journal of Roman Archaeology Supp. 78), 79–98. Portsmouth (RI), Journal of Roman Archaeology.

Pearce, J. (2013) *Contextual archaeology of burial practice. Case studies from Roman Britain* (British Archaeological Reports, British Series 588). Oxford, Archaeopress.

Pearce, J., Millett, M., and Stuck, M. (eds.) (1992) *Burial, society and context in the Roman world*. Oxford, Oxbow Books.

Roller, L. E. and Goldman, A. (2002) A Latin epitaph from Gordion. *Zeitschrift für Papyrologie und Epigraphik* 141, 215–20.

Sams, G. K. and Goldman, A. (2006) Gordion 2004. *Kazı Sonuçları Toplantısı* 27.2, 43–56.

Sams, G. K. and Voigt, M. M. (1996) Gordion archaeological activities, 1994. *Kazı Sonuçları Toplantısı* 17.1, 433–52.

Selinsky, P. (2004) An osteological analysis of human skeletal material from Gordion, Turkey. Master's thesis, University of Pennsylvania.

Şimşek, C. (ed.) (2011) *Laodikeia nekropolü (2004–2010 yılları)*. Istanbul, Ege Yayınları.

Spanu, M. (2000) Burial in Asia Minor during the Imperial period, with a particular reference to Cilicia and Cappadocia. In Pearce, Millett, and Stuck (eds.), 169–77.

Temizer, R. (1973) *Ankara 50*. Ankara turizmi, eskieserleri ve müzeleri sevenler derneği yayınları 5. Ankara, The Ankara Society for the Promotion of Tourism, Antiquities and Museums.

Tezcan, B. (1964) *Yalıncak Village Excavation in 1962–1963*. Ankara, Middle East Technical University Arkeoloji Yayınları.

Üstündağ, H. (2009) Kültepe/Kanesh (Turkey), season 2007. *Bioarchaeology of the Near East* 3, 31–5.

Üyümez, M., Koçak, Ö., İlaslı, A., Çay, T., and İşcan, F. (2007) Afyonkarahisar'ın doğusunda önemli bir Orta Tunç Çağı nekropolü: Dede Mezarı. *Belleten* 262, 811–41.

Voigt, M. M. (2012) Human and animal sacrifice at Galatian Gordion: The uses of ritual in a multiethnic community. In A. Porter and G. Schwartz (eds.) *Sacred killing. The archaeology of sacrifice in the ancient Near East*. Winona Lake (IN), Eisenbrauns.

Voigt, M. M., DeVries, K., Henrickson, R. C., Lawall, M., Marsh, B., Gürsan-Salzman, A., and Young, Jr., T.C. 1997 Fieldwork at Gordion: 1993–1995. *Anatolica* 23, 1–59.

Young, R. S. (1955) Grave robbers' leavings. *Archaeology* 8, 191–7.

Young, R. S. (1956) The campaign of 1955 at Gordion. *American Journal of Archaeology* 60, 249–66.

Reflections on the mortuary landscape of Ephesus:
The archaeology of death in a Roman metropolis

Martin Steskal

Abstract

Since 2008 the Austrian Archaeological Institute has been carrying out intense field-archaeological research in the necropoleis of Ephesus. This research has included a combination of methods, such as survey, excavation, architectural study, geophysics, geoarchaeology, and biological anthropology. In terms of Ephesus, we can consider our research approach to be holistic meaning that it covers a wide chronological, topographical, and contextual spectrum.

This paper addresses several aspects of our interdisciplinary research and focuses on the Roman period. It discusses topics such as the organization of the mortuary landscape, the transformation of areas into architectural landscapes of the dead, burial sites in their intra- and extra-urban contexts, means of self-expression, general tendencies in the social attitudes towards death and funeral practices as well as visual memorialization of status – outward-oriented representation versus a more pensive, family-focused memorial.

Keywords: Asia Minor, death, Ephesus, mortuary landscape, necropolis, society.

This paper deals with the necropoleis of Ephesus with a focus on our new research on the so-called Harbour Necropolis (also: West Necropolis).[1] However, this study goes beyond the borders of this densely structured mortuary area located on either side of the harbour channel: Based on the systematic research on this particular necropolis that started in the year 2008 we have significantly broadened out our research since the year 2011 – not so much through excavations, but mainly through surveys increasing our data considerably. In terms of Ephesus, we can now view our research approach to be holistic meaning that it covers a wide chronological and topographical spectrum (Fig. 10.1). Together with a team from different fields of the humanities and the natural sciences the following topics are currently under study: size, extent, and chronology of the Ephesian necropoleis; typology of the burial architecture; appearance and quantity of non-burial architecture in the necropoleis; development of burial house types; organization of the built environment in the Ephesian necropoleis and comparison with other necropoleis in Asia Minor; contrast between intra- and extra-urban burials; staging of death; performance of rituals; health, age expectancy; cause of death; family cohesion; and origin. The discussion of these research topics shall lead to a general understanding of the Ephesian mortuary landscape.

The largest Ephesian cemetery is the so-called Harbour Necropolis (Figs. 10.2–3). It is currently being studied through the lens of stratigraphic excavations as well as geophysical, geoarchaeological, palaeopathological, and archaeozoological methods, and extensive and intensive surveys. This cemetery of at least 45 hectares was in use from the 2nd century to the 6th century AD. The burials of the Harbour Necropolis do not follow the orthogonal street grid of the city, but instead are situated along a complex system of paths and are primarily oriented towards the harbour channel. In this context the channel functions as the burial street. The statistical analysis indicates that the average size of the burial houses is 8 m². The majority of the burial houses are made of dressed stone masonry (*opus vittatum*) and were plastered inside and out (Fig. 10.4).

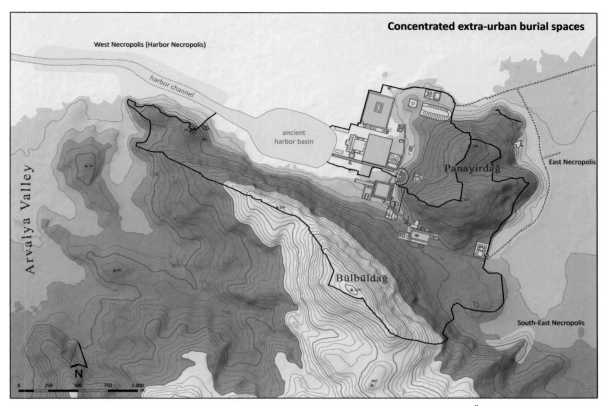

Fig. 10.1. Ephesus. Concentrated extra-urban burial spaces (Map: C. Kurtze, ÖAI).

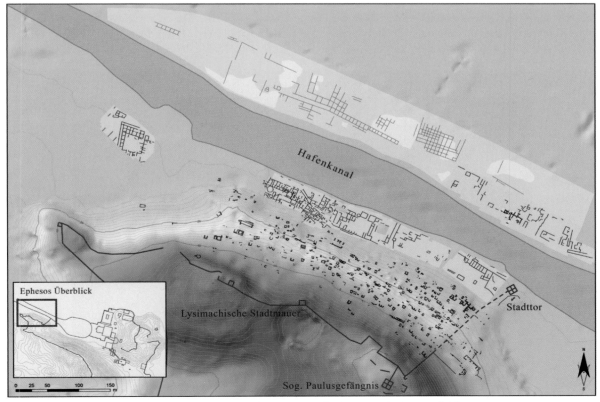

Fig. 10.2. Ephesus. Harbour Necropolis (Design: C. Kurtze, ÖAI).

Fig. 10.3. Ephesus. Harbour Necropolis from the north (Photo: L. Fliesser, ÖAI).

Numerous burial houses also had wall paintings and were shaped with niches. A burial house could usually be closed with a single wooden door and contained three graves. These were either richly decorated or semi-finished sarcophagi or imitations of sarcophagi. The imitation sarcophagi were made by constructing a small brick wall in the same width

of a sarcophagus with a marble revetment. The graves were sometimes also simply cut into the local bedrock as *chamosoria* and then covered with stone slabs.

The burial houses can be reconstructed as follows: They were either cubical structures with a flat roof that was used as a terrace (i.e. *solaria*) or also buildings with an arched

Fig. 10.4. Ephesus. Harbour Necropolis. Burial house 09/10 (Photo: L. Fliesser, ÖAI).

vault, such as those visible in Annemurium or Elaiussa Sebaste. The question of their origin is obsolete due to simplicity of the architectural types. They developed at the same time in multiple places in response to certain functional requirements, according to the zeitgeist and through general technological conventions. In addition to burial houses there were also detached sarcophagi that were not integrated into the burial houses. They make up about 6% of the entire mortuary architecture.

To summarize the results of the topographical analysis: within the Harbour Necropolis measuring at least 45 hectares' evidence of 355 burials was found on the surface. If we include the data of the geophysical prospection and the aerial photographs, we can identify more than 1,000 burials, i.e. 1,000 burial houses or freestanding sarcophagi.

As in any modern large-scale cemetery there are not only burials inside the cemetery, but also facilities for the infrastructure, such as stonemasons, stores, storage rooms, areas for celebrations, a complex system of paths and much more. As a result, only about 2/3 of the architectural evidence can be identified reliably as burial sites. In terms of the organization of the necropolis it has been established that the development first began along the slopes. Once space began to run out on the highly visible slopes in the course of the 2nd century, the adjoining plain was developed for burials in the early 3rd century. Thus, a burial on the slope does not necessarily represent a hierarchically better location simply due to its better visibility. The development of new burial areas consciously followed the internal organization of the cemetery according to the available space.

Our surveys have not only been focused on the Harbour Necropolis, but we have attempted to record the entire Ephesian mortuary landscape. In 2011 and 2012 the remaining cemetery areas of the Roman period were studied by means of surveys. In total we were able to identify two additional densely used cemetery areas: the South-East Necropolis along the Sarıkaya about 10 hectares in size (Figs. 10.5–7) and the large East Necropolis along the Panayırdağ measuring 35 hectares (Figs. 10.8–10).

In these two necropoleis we discovered 74 (South-East Necropolis) and 99 (East Necropolis) burial sites on the surface[2] and recorded them geodetically. Therefore, we can estimate that in the Roman period Ephesus had at least 90 hectares of densely used extra-urban burial land with more than 1,200 individual burial sites. Again, these are primarily burial houses and freestanding sarcophagi. The number of burials at each site is of course far higher.

Unfortunately, we do not know very much about the older, i.e. pre-Roman burial sites (Fig. 10.11). In particular, the Hellenistic necropoleis of Ephesus have completely disappeared from the landscape and only individual finds attest to their former existence. The reason for this can be sought in the humble size of the simple Hellenistic burials, but also the massive agricultural and geological changes of the landscape that have made it impossible to locate the Hellenistic cemeteries. The known pre-Hellenistic burial sites were primarily located in close proximity of the old processional way – at this point in time located outside of the settlement.

The burial sites of Antiquity are primarily located in extra-urban contexts. The legal framework for the prohibition of intra-urban burials is frequently mentioned in the literary sources (cf. Cicero, *de legibus* 2.58. Cf. Schrumpf 2006, 63–4; Berns 2003, 27; Burkert 1977, 295; Young 1951, 67–134; Schörner 2007, 11–19; see also the contributions in Henry 2013). In this context burial was viewed as a necessary hygiene measure that removed the corpse from the domestic space and it was cremated or interred outside of the settlement (cf. Kolb and Fugmann 2008, 14; Cormack 2004, 38). Thus, visitors to a city always had to first pass through a cemetery before they entered the city. The numerous burial houses and free-standing sarcophagi truly transformed the area into an architectural landscape of the dead. The location of cemeteries in close proximity to major traffic routes and their conspicuous presence permanently confronted the living with all the uncertainties, expectations, and individual ideas of the afterlife that humans have always connected with death. The burials did not only function as monuments of memory; through differences in location, appearance, size, and furnishings they also served as vehicles for status display and expression of social hierarchies. This extra-urban context was finally recalled by the emperor Leo VI around 900 AD (Leo VI *Novellae* 53; on the sources cf. Schrumpf 2006, 64 note 168) and a situation was acknowledged that appears to have already become common (Ivison 1996, 102). With the exception of the extra-mural burials, Ephesus appears to have always encouraged intra-urban burials (cf. Steskal 2013). Already in the late Hellenistic period the privilege of burying outstanding citizens within the city limits is attested for Ephesus (on this privilege cf. Varro, *De lingua Latina* 6.49; Kolb and Fugmann 2008, 15; Cormack 2004, 38). In this context burial monuments were understood in the same way as personal honorific monuments (cf. Hesberg and Zanker 1987, 9–20; Berns 2003, 20, 24, 27–30, 52). Their location within the city was intended to arouse the maximum amount of public attention and the burial within the city limits marked an exceptional honour. The functional differentiation between a burial monument and an honorific monument is not possible within this context.[3] This outward focused form of self-representation seems to have changed with the reign of Augustus (cf. Hesberg 1992, 37–42; Berns 2003, 25–6; 79–81). The process that led to the abandonment of the elaborate and outward focused self-representation had multiple reasons. The competition among the elites lost most of its political incentive owing to the overbearing power of the Imperial court. The rest of the population became more aware of its membership to its own class and social group as well as its upward boundaries.

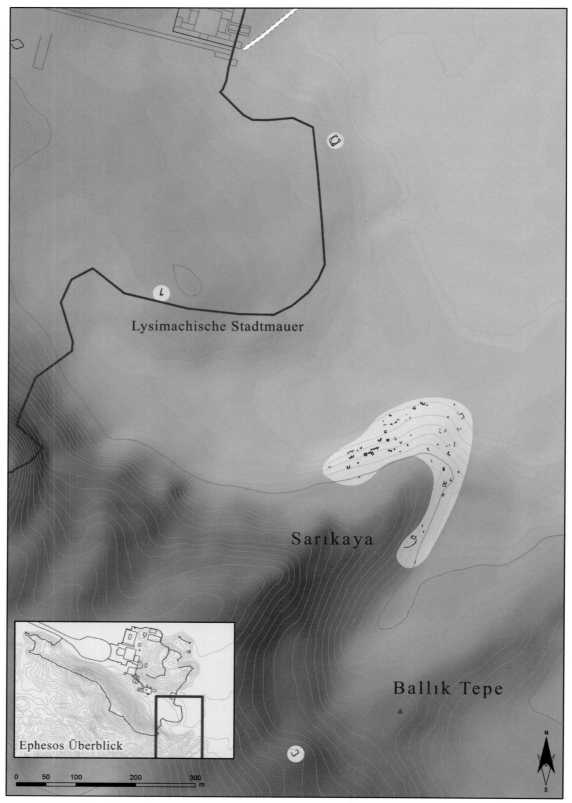

Fig. 10.5. Ephesus. South-East Necropolis on the Sarıkaya (Map: C. Kurtze, ÖAI).

Fig. 10.6. Ephesus. South-East Necropolis from the west (Photo: M. Steskal, ÖAI).

Luxurious self-representation went out of fashion in the early Imperial period influencing the norms for burial sites and monuments. A general tendency in the social attitudes towards death and funeral practices is noticeable: The aggrandizing outward-oriented representation was exchanged for a more pensive, family-focused memorial (Schrumpf 2006, 74; Berns 2003, 140–1, 147; for Ostia cf. Heinzelmann 2000, 118–22). As a result the outward status definition took place primarily through the extravagance of the funeral procession, the *pompa funebris*, and the

Fig. 10.7. Ephesus. South-East Necropolis. Burial house 129/12 (Photo: L. Fliesser, ÖAI).

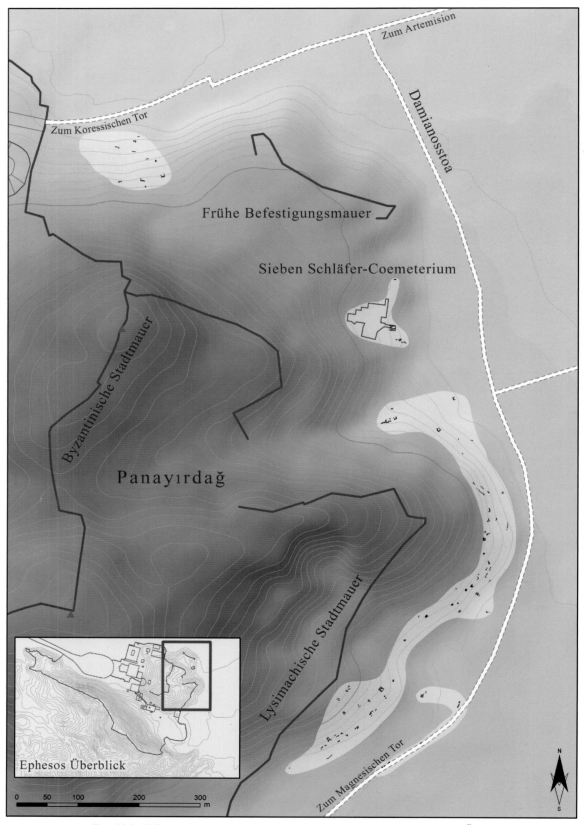

Fig. 10.8. Ephesus. East Necropolis along the Panayırdağ (Map: C. Kurtze, ÖAI).

Fig. 10.9. Ephesus. East Necropolis from the east (Photo: N. Gail, ÖAI).

funeral itself. The events after the death of an individual were suited to fulfilling the need for self-definition in a variety of ways. Henceforth, the funeral demonstrated the rank in society and to a lesser extent the architecture of a burial house. The staging of the death for those left behind still required a visible location. This necessity was not met by an individual burial but by the location, the ensemble. This process appears to have been more strongly

Fig. 10.10. Ephesus. East Necropolis. Burial house 167/12 (Photo: L. Fliesser, ÖAI).

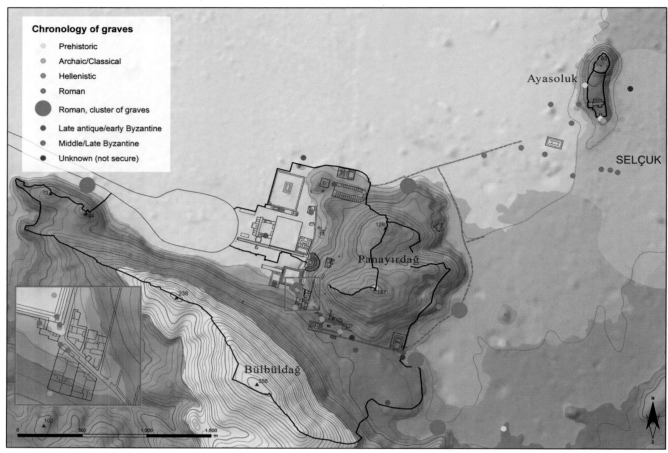

Fig. 10.11. Ephesus. Diachronic map of the town's burial spaces (Map: C. Kurtze, ÖAI)

articulated in Rome and the West and did not take place in the East with the same consequences. The interest of the once dominant social figures for a visual memorialization of their status within the urban cityscape decreased. While these early examples are limited to honorific burials and monuments and thus only pertained to an elite minority, the later examples can be classified as a real change in burial customs (cf. Stone and Stirling 2007, 17). In the mid-Byzantine period[4] intra-mural burial rituals were mainly maintained in areas surrounding churches. Although we must assume that even in the Byzantine period the majority of the burials took place in an extra-urban context (see also Lightfoot, and Wenn *et al.*, both this volume), the incorporation of cemeteries into the urban setting and their connection to Christian churches signals a clear break with Antiquity and marks a central element in the definition of the Medieval city (cf. Ivison 1996, 99; Dagron 1977, 1–25; Ariès 1981, 29–92).

In the course of our research on the Harbour Necropolis we addressed the question of whether the intra-urban spaces were systematically used as burial grounds following the abandonment of the largest cemetery in the 6th century.

The core of the issue was whether the inhabitants of the post-classical city perceived the former city walls which had lost their function and were partially destroyed as a sacred border or barrier – comparable with the *pomerium* of Rome. In the course of the survey we were not able to find any systematic burials in the surveyed areas located within the city walls. This is surprising because the fortified city was diminished in size at the beginning of the mid-Byzantine period and the formerly intra-mural city quarters now became extra-mural. And yet, the old border appears to have played a certain role (cf. also Wenn *et al.*, this volume).

But where are the cemeteries of the 7th, 8th, and 9th centuries? Even in the smaller Byzantine city there will have been enough space for burials. Unfortunately we must acknowledge that our current understanding of the location of the cemeteries at the beginning of the mid-Byzantine period is quite vague. On the one hand we are aware of the cemetery surrounding the Cemetery of the Seven Sleepers (Zimmermann 2011a, 365–407; 2011b, 160–6; Zimmermann and Ladstätter 2010, 149–58; 203–7; Miltner 1937; Pillinger 2001; 2005; Jobst 1972–1975, 171–80). On the other hand, we can assume that the cemeteries surrounding the intra-urban churches developed a lot earlier, but based on the few or complete lack of burial goods they cannot be properly dated. In addition, most of the churches and thus the possible cemeteries have not been sufficiently studied archaeologically. Until these sites have been studied more carefully, the phenomenon of early Byzantine intra-urban burials located around churches remains a working hypothesis for Ephesus.

But let us again return to the Ephesian necropoleis of the Roman period, the period for which we have the best evidence; if we take a look at the burial houses, their uniformity is striking (Figs. 10.4, 7, 10). They do not differ greatly in size, architectural construction, and appearance. Since the walls of many of the burial houses were constructed in a single bond, it is evident that building firms were operating in the cemeteries that built burial houses on a large scale and then sold them. The location and neighbourhood appears to have been the deciding factor for obtaining the highest possible price. It was no longer possible to advertise the social rank or provide a record of the life of a person with such a generic architecture. There was the option of course of setting oneself apart from other deceased by exquisitely decorating the interior of the burial house with elaborate wall paintings or a conspicuous sarcophagus (Fig. 10.12). Since the burial houses were usually closed these design elements were only visible during the funeral with the exception of later burials of other family members in the same grave. If we consider the incredible effort made from the time of death of a person up to his or her burial (cf. Rife 2012, 153–232; Schrumpf 2006) it is evident that the lying-in repose, the funeral procession, and the funeral celebration were the actual elements defining status (see also Rife *et al.* 2007, 175–6).

Historically phases reoccur when tendencies towards uniforming burial practices are noticeable. This does not mean that the actual ownership was equally divided. The relationship between a necropolis and society is not direct, but instead it passes through an ideological filter (cf. Graepler 2006, 139). If several burials in a necropolis are only sparsely decorated and marked, this does not signalize an egalitarian society because even societies appearing to be very simple follow strict hierarchies. In our case we can propose that the society is attempting to portray itself as egalitarian or that the social competition of its members was being channelled in a different direction. The uniformity is likely also an expression of a collective approval, that generally made social norms a lot more binding than legal regulations.

Of course there are also limits to the archaeological analysis: the archaeological record is fragmentary and not every human action is reflected by the objects of the material cultures. For example the study of burial goods is very difficult, in particular attempts to infer from them the social status of the deceased (Fig. 10.13). Due to depositional and post-depositional activities burial goods are some of the most difficult objects to understand. What is a burial good actually? Is it an object that the deceased owned and wore during his or her lifetime and that was then placed in the grave by a mourner? Or is it an object that was the property of a mourner and then was given to the deceased? The quantity and quality of the finds does not provide accurate information about the economic power of the deceased. A necropolis can reflect the structure of a society, its values, hierarchies, and the social personality of the deceased. Generally it conveys a structured picture of the community that it refers to. The burial system is a system of collective representation (cf. D'Agostino 2009).

Fig. 10.12. Ephesus. Harbour Necropolis. Burial house 01/05 (Photo: M. Steskal, ÖAI).

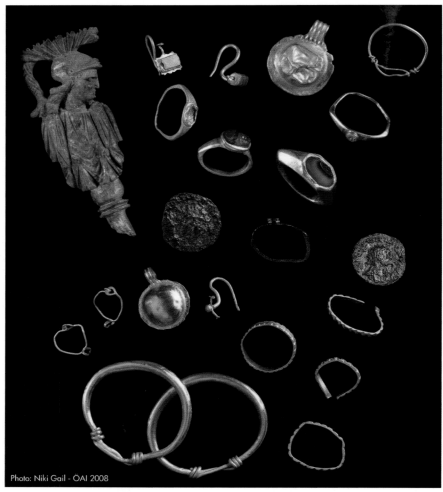

Fig. 10.13. Ephesus. Harbour Necropolis. Grave goods from burial house 01/08 (Photo: N. Gail, ÖAI).

Commemoration is a strategy to defy death. Every medium is only as good as it helps to remember. Greeks as well as Romans invested a lot of energy and intelligence in activities that were supposed to ban death through memoria. The success of this strategy far exceeds the imagination of those once involved. Even two millennia later they still demand our attention and are the subject of discussions.

Acknowledgements

The author would like to thank the director of the excavations in Ephesus, S. Ladstätter, for the publication permission. I am indebted to N. Zimmermann for numerous suggestions and discussions. I am grateful to Ch. Kurtze for the production of the maps (Figs. 1, 2, 5, and 8), to L. Fliesser and N. Gail for their photographs (Figs. 3, 4, 7, 10, respectively 9, 13). and to N. M. High-Steskal for the revision of the English manuscript. The study of the necropoleis of Ephesus is funded by the Austrian Science Fund (FWF-project P22083-G19).

All illustrations are published with the courtesy of the Österreichische Archäologische Institut Wien (ÖAI), of which the present author is a member.

Notes

1. Cf. Steskal 2013; in print; Steskal, Taeuber, and Zimmermann 2011, with an extensive bibliography to the Ephesian necropoleis. A general overview is provided by Pietsch 1999, 455–60; Trinkl 1997; Groh *et al.* 2006, 109–12.
2. A much greater number is of course still expected below ground. Again, geophysical exploration carried out in the area of these necropoleis proves this assumption.
3. Berns 2003, 24–5. Contra: Thür 2009, 13–14. Cf. Steskal 2013 on the ambiguous character of honorific monuments in Ephesus, like the Octagon, the Memmius Monument, the Pollio Monument, and the later Library of Celsus.
4. The periodization of the late Antique and Byzantine period varies. The following division is used in this paper: late Antiquity 284 to 7th century AD; early Byzantine period 395 to 6th century AD; mid-Byzantine period 7th century to 1204 AD; late Byzantine period 1204–1453 AD.

Bibliography

Ariès, P. (1981) *The hour of our death*. New York, Knopf.

Berns, C. (2003) *Untersuchungen zu den Grabbauten der frühen Kaiserzeit in Kleinasien* (Asia Minor Studien 51). Bonn, Habelt.

Burkert, W. (1977) *Griechische Religion der archaischen und klassischen Epoche*. Stuttgart, Kohlhammer.

Cormack, S. (2004) *The space of death in Roman Asia Minor* (Wiener Forschungen zur Archäologie 6). Vienna, Phoibos.

D'Agostino, B. (2009) Archäologie der Gräber: Tod und Grabritus. In A. H. Borbein, T. Hölscher, and P. Zanker (eds.) *Klassische Archäologie. Eine Einführung*, 313–31. Berlin, Reimer.

Dagron, G. (1977) Le Christianisme dans la ville byzantine. *Dumbarton Oaks Papers* 31, 1–25.

Graepler, D. (2006) Gräber. In T. Hölscher, *Klassische Archäologie*, 129–39. Darmstadt, Wissenschaftliche Buchgesellschaft.

Groh, S., Lindinger, V., Löcker, K., Neubauer, W., and Seren, S. S. (2006) Neue Forschungen zur Stadtplanung in Ephesos. *Jahreshefte des Österreichischen Archäologischen Institutes in Wien* 75, 47–116.

Heinzelmann, M. (2000) *Die Nekropolen von Ostia. Untersuchungen zu den Gräberstraßen vor der Porta Romana und an der Via Laurentina* (Studien zur antiken Stadt 6). Munich, Pfeil.

Henry, O. (ed.) 2013 *Le mort dans la ville. Pratiques, contextes et impacts des inhumations intra-muros en Anatolie, du début de l'Âge du Bronze à l'époque romaine* (2èmes Rencontres d'archéologie de l'IFÉA. Istanbul 14–15 Novembre 2011). Istanbul, Institut Français d'Études Anatoliennes Georges Dumézil.

Hesberg, H. von (1992) *Römische Grabbauten*. Darmstadt, Wissenschaftliche Buchgesellschaft.

Hesberg, H. von and Zanker, P. (1987) Einleitung. In H. von Hesberg and P. Zanker (eds.) *Römische Gräberstraßen. Selbstdarstellung – Status – Standard. Kolloquium in München vom 28. bis 30. Oktober 1985* (Bayerische Akademie der Wissenschaften. Philosophisch-Historische Klasse. Abhandlungen 96), 9–20. Munich, Verlag der Bayerischen Akademie der Wissenschaften.

Ivison, E. A. (1996) Burial and urbanism at Late Antique and Early Byzantine Corinth (c. AD 400–700). In N. Christie and S. T. Loseby (eds.) *Towns in transition. Urban evolution in Late Antiquity and the Early Middle Ages*, 99–125. Aldershot, Scolar Press.

Jobst, W. (1972–75) Zur Bestattungskirche der Sieben Schläfer in Ephesos. *Jahreshefte des Österreichischen Archäologischen Institutes in Wien* 50, Beiblatt 171–80.

Kolb, A. and Fugmann, J. (2008) *Tod in Rom* (Kulturgeschichte der antiken Welt 106). Mainz, Zabern.

Lightfoot, C. S. (this volume) Christian burials in a pagan context at Amorium, 188–95.

Miltner, F. (1937) *Das Coemeterium der Sieben Schläfer* (Forschungen in Ephesos 4.2). Baden, Rohrer.

Pietsch, W. (1999) Außerstädtische Grabanlagen von Ephesos. In H. Friesinger and F. Krinzinger (eds.) *100 Jahre österreichische Forschungen in Ephesos. Akten des Symposions Wien 1995* (Österreichische Akademie der Wissenschaften, Philosophisch-Historische Klasse. Denkschriften 260), 455–60. Vienna, Verlag der Österreichischen Akademie der Wissenschaften.

Pillinger, R. (2001) Kleiner Führer durch das Sieben Schläfer-Coemeterium in Ephesos. *Mitteilungen zur Christlichen Archäologie* 7, 26–34.

Pillinger, R. (2005) Martyrer und Reliquienkult in Ephesos. In B. Brandt, V. Gassner, and S. Ladstätter (eds.) *Synergia. Festschrift Friedrich Krinzinger I*, 235–41. Vienna, Phoibos.

Rife, J. (2012) *The Roman and Byzantine graves and human remains* (Isthmia 9). Princeton, The American School of Classical Studies at Athens.

Rife, J. L., Moore Morison, M., Barbet, A., Dunn, R. K., Ubelaker, D. H., and Monier, F. (2007) Life and death at a port in Roman Greece. The Kenchreai cemetery project, 2002–2006. *Hesperia* 76, 143–81.

Schörner, H. (2007) Sepulturae graecae intra urbem. *Untersuchungen zum Phänomen der intraurbanen Bestattungen bei den Griechen* (BOREAS – Münstersche Beiträge zur Archäologie, Beiheft 9). Möhnesee, Bibliopolis.

Schrumpf, S. (2006) *Bestattung und Bestattungswesen im Römischen Reich. Ablauf, soziale Dimension und ökonomische Bedeutung der Totenfürsorge im lateinischen Westen*. Bonn, V & R Unipress.

Steskal, M. (2013) Wandering cemeteries. Roman and Late Roman burials in the capital of the Province of Asia. In Henry (ed.), 243–57.

Steskal, M. (in print) Defying death in Ephesus. Strategies of commemoration in a Roman metropolis. In *Proceedings of the symposium 'Cityscapes and monuments of remembrance in western Asia Minor', University Aarhus/Denmark*. Oxford, Oxbow Books.

Steskal, M., Taeuber, H., and Zimmermann, N. (2011) Psalmenzitat, Paradieskreuze und Blütenmotive. Zu zwei neu entdeckten Grabhäusern mit spätantiker Malerei in der Hafennekropole von Ephesos. *Jahreshefte des Österreichischen Archäologischen Institutes in Wien* 80, 291–307.

Stone, D. L. and Stirling, L. M. (2007) Funerary monuments and mortuary practices in the landscapes of North Africa. In D. L. Stone and L. M. Stirling (eds.) *Mortuary landscapes of North Africa* (Phoenix Supplementary Volume 43), 3–31. Toronto, University of Toronto Press.

Thür, H. (2009) Zur Kuretenstraße von Ephesos – Eine Bestandsaufnahme der Ergebnisse aus der Bauforschung. In S. Ladstätter (ed.) *Neue Forschungen zur Kuretenstraße von Ephesos. Akten des Symposiums für Hilke Thür vom 13. Dezember 2006 an der Österreichischen Akademie der Wissenschaften* (Österreichische Akademie der Wissenschaften, Philosophisch-Historische Klasse. Denkschriften 382). 9–28. Vienna, Verlag der Österreichischen Akademie der Wissenschaften.

Trinkl, E. (1997) Die Nekropolen von Ephesos. Ein Überblick. *Forum Archaeologiae* 4/VIII/1997 (http://homepage.univie.ac.at/elisabeth.trinkl/forum/forum0897/04necro.htm).

Wenn, C. C., Ahrens, S., and Brandt J. R. (this volume) Romans, Christians, and pilgrims at Hierapolis in Phrygia: Changes in funerary practices and mental processes, 196–216.

Young, R. S. (1951) Sepulturae Intra Urbem. *Hesperia* 20, 67–134.

Zimmermann, N. (2011a) Das Sieben-Schläfer-Zömeterium in Ephesos. Neue Forschungen zu Baugeschichte und Ausstattung eines ungewöhnlichen Bestattungskomplexes. *Jahreshefte des Österreichischen Archäologischen Institutes in Wien* 80, 365–407.

Zimmermann, N. (2011b) Die spätantike und byzantinische Malerei in Ephesos. In F. Daim and S. Ladstätter (eds.) *Ephesos in byzantinischer Zeit*, 125–72. Mainz, Verlag des Römisch-Germanischen Zentralmuseums.

Zimmermann, N. and Ladstätter, S. (2010) *Wandmalerei in Ephesos von hellenistischer bis in byzantinische Zeit*. Vienna, Phoibos.

Christian burials in a pagan context at Amorium

Christopher S. Lightfoot

Abstract

During the course of twenty-five years of fieldwork at Amorium numerous tombs, tomb markers, and skeletons have come to light. Despite the site's long occupational history most of these remains belong to the Roman and Byzantine periods and can be dated to the first millennium AD. They provide evidence for a range of different burial practices, some of which are outlined here and others in the paper presented by Prof. Arzu Demirel. Because the city survived the end of Antiquity and continued as a thriving centre of population until the late 11th century AD, Amorium affords the opportunity to study the transition from pagan to Christian burials. Of particular interest is the way in which the Byzantines exploited the Roman cemeteries both for building materials and as space for continued use for burials.

Keywords: Amorium, burials, cemeteries, Christian, pagan, tombs.

Archaeological evidence suggests that the site of Amorium has been occupied, almost without interruption, since at least the early Bronze Age, *c*. 3,000 BC (Lightfoot *et al.* 1999, 347–8). The prehistoric mound is certainly one of the largest in the whole of eastern Phrygia, and the site (Fig. 11.1) later developed into one of the more important Roman cities in the region and had the right to mint its own coins from the late Republic onwards (Katsari 2012, 30–1). Excavations, however, have failed to reveal much evidence for the early periods of occupation, including any likely Hittite settlement in the second millennium BC, the Iron Age Phrygian town, or even the Hellenistic and Roman city. For the period stretching from late Hellenistic to early Byzantine times (1st century BC–6th century AD) the most common and conspicuous surviving material comes in the form of funerary markers and tombs. Relatively little else of the pre-Byzantine city has survived.

The earliest tomb is probably the tumulus located in a prominent position to the south-west of the centre of the site, which may date to the 2nd–1st century BC (Lightfoot *et al.* 2009, 208–10, pl. 7). It comprised a single rectangular chamber built of large ashlar blocks, with a stone barrel roof and a broad *dromos* on the east side (Fig. 11.2). For the Roman period there are the extensive cemeteries that encircle the ancient city (Lightfoot and Lightfoot 2007, 154–61; Yaman 2012, 332). They contain a variety of different tomb types, including monumental tombs, chamber tombs, and simple rock-cut cist graves. The most common variety of grave marker is the limestone stele or *bomos* of the Phrygian doorstone type (Yaman 2008).

Study of the necropoleis and funerary archaeology of Amorium was not a major priority of the excavation team that worked at the site between 1987 and 2009. However, either from necessity, when a rescue dig was required (Pugsley 1995), or from a desire to provide as complete a picture of the site as possible and so to understand something of the dynamics of the place and its population, a number of limited excavations have been undertaken (Yaman 2012), and a comprehensive surface survey of visible tombs outside the city walls also was carried out in 2004–2006 (Lightfoot, Koçyiğit, and Yaman 2006, 80; 2008, 456–7). The cemeteries, which extend far beyond the walls that encircle the Lower City, have also attracted the misguided nocturnal attention of treasure hunters. Four tombs were thus discovered and looted in fields to the south-west of Amorium in November 2011. They were subsequently excavated by a

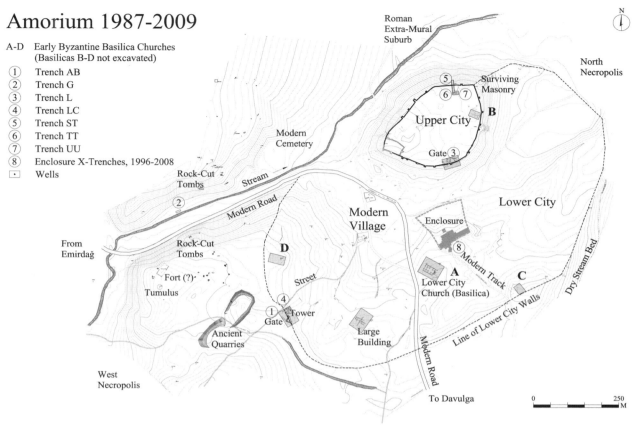

Fig. 11.1. Amorium. Topographic plan showing excavated areas (courtesy of the Amorium Excavations Project).

Fig. 11.2. Amorium. Tomb chamber inside tumulus in west necropolis (courtesy of the Amorium Excavations Project).

team from the Afyonkarahisar Museum, and two doorstones (T3364 and T3365; Fig. 11.3) that had been reused as roofing slabs in one, possibly early Christian, tomb were deposited at the Amorium Dig House. It may be doubted whether more than a few, if any, tombs now remain intact. Indeed, the Roman cemeteries contain many tombs that were used and reused over a number of centuries; others were plundered in post-Roman times for the offerings they contained or the materials from which they were built.

Four principal church sites have been identified, only one of which has been excavated so far.[1] Numerous undisturbed tombs, usually containing multiple burials, were found in and around the so-called Lower City Church, all of which

Fig. 11.3. Amorium. Two doorstones from tomb excavated in December 2012 (courtesy of the Amorium Excavations Project).

belong to the mid-Byzantine period; that is, the 10th–11th centuries AD (Lightfoot *et al.* 2005, 249–52; Ivison 2010, 326–38). Other remains are not formal burials and the most dramatic examples of this category are the two skeletons found in destruction layers in the so-called Enclosure that can be dated to an historical event, namely, the siege and sack of the Byzantine city of Amorium by the Arabs in August AD 838 (Ivison 2012, 63). Both skeletons showed clear signs of traumatic injuries to the head (Demirel 2012). Other dismembered human remains, found at different locations in the Lower City, probably belong to the same catastrophe (Brayne and Roberts 2003, 169, 173–4; Ivison 2012, 63). Three other bodies have also been found in the former latrine of the early Byzantine bathhouse. The exact circumstances of their death and deposition remains unexplained, but evidence was found suggesting that the bodies had lain exposed for some considerable time before they were buried (Brayne and Roberts 2003, 171–3, 176–9). It has been assumed that they may represent some of the last Byzantine inhabitants of Amorium in the late 11th century AD.

The site was thereafter reoccupied by the Seljuks and eventually came under the control of the Ottomans. The modern village was established in 1890 by Islamicized Slavic refugees from the Balkans, who created a cemetery, still in use today, to the north of the centre of the site. Some Roman and Byzantine *spolia* are to be found in the modern cemetery, but much more funerary material from Amorium is to be found in the neighbouring villages (Waelkens 1986, 212, no. 534, pl. 78; Fig. 11.4). For the past century Amorium has been a convenient and plentiful source of

Fig. 11.4. Amorium. Triple doorstone reused as a public fountain in Hamzahacılı (courtesy of the Amorium Excavations Project).

stone, and the villagers exploited this resource by selling it for reuse elsewhere. From the pre-modern Turkish period come only three burials, two on the Upper City mound (Brayne and Roberts 2003, 159–60, pl. XI/1–7), and one from the south-east courtyard of the Lower City Church. The last was orientated with his head to the south, clearly in the direction of Mecca (Fig. 11.5). So, despite the site's long occupational history, little remains of its deceased inhabitants, except for the Roman and Byzantine periods.

It is clear that some Roman tombs survived and continued to be used at least until the 6th or 7th century AD. Such is the case with two rock-cut tombs, MZ 1 (Pugsley 1995; Brayne and Roberts 2003, 161–8) and MZ 90 (Lightfoot *et al.* 2009, 210, pl. 8), that have been excavated. There is also a monumental tomb, MZ 2 (Kelp 2013, 76, figs. 4.6–4.7, wrongly identifying the tomb as T1073), that employed as part of its façade two doorstones, both of which were found, one broken, the other overturned (T1074, Fig. 11.6), still in situ. The tomb is of a type that is widespread in eastern Phrygia. Another monumental tomb was even more elaborate, including a long decorative frieze (Harrison 1989, 169, 171, pl. XLV *b*). Two fragments of the frieze (T933 and T934) had been taken to the neighbouring

village of Suvervez, perhaps in the 1920s; they were retrieved and brought to the Dig House for safekeeping in 1996. It should be noted that the decoration includes a fish (T933; Fig. 11.7), and several doorstones at Amorium also are carved with fish (for example, fish hanging from sticks in the two lower panels on the left-hand doorstone in Fig. 11.4). It has been argued that these merely represent a pastime and welcome addition to the diet for the local population and in most cases probably have no Christian connotations (Yaman 2008, 61–2, fig. 6). Fragments of marble (for example, T1549; Lightfoot and Arbel 2003, 526, pl. 10) and limestone sarcophagi indicate that some of the monumental tombs probably contained freestanding funerary receptacles, while others clearly comprised a number of rock-cut graves, as in the case of Tomb MZ 3 (Lightfoot *et al.* 2009, 210–11, pl. 9).

Many tombs were, however, destroyed during Byzantine times, and in this respect Amorium suffered a different fate from many other Roman cities in Asia Minor. Because it remained a large and dynamic centre of population between the 7th and 11th centuries AD, the Roman cemeteries were both reused and exploited for building materials by the later inhabitants of Amorium. Many of the typical Phrygian

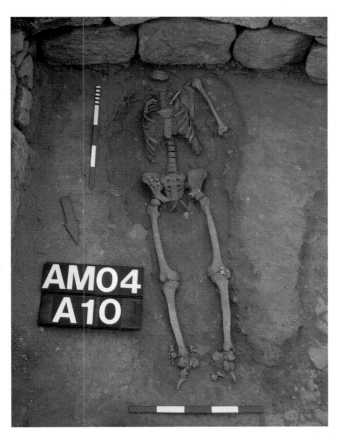

Fig. 11.5. Amorium. Post-Byzantine burial in southeast courtyard of the Lower City Church (courtesy of the Amorium Excavations Project).

Fig. 11.6. Amorium. Doorstone from tomb MZ 2 after removal to the Dig House (courtesy of the Amorium Excavations Project).

Fig. 11.7. Amorium. Detail of decorative frieze from a monumental tomb (courtesy of the Amorium Excavations Project).

doorstones were moved into the city in Byzantine times. The Upper City fortification wall in its early Byzantine phase (probably 7th century AD) made extensive use of these large blocks in its foundations and lower courses, some of which are still *in situ* (Fig. 11.8). Remarkably, other Roman tombstones found use as building material in the Lower City. There is, for example, an epitaph on a large plain slab, T1951 in structure 9 (Lightfoot, Koçyiğit, and Yaman 2006, 78, pl. 4), that was reused in the construction of a mid-Byzantine room adjacent to the south side of the former early Byzantine baths. Similarly, one tomb (Tomb 62) on the south side of the church was constructed using doorstones (Lightfoot *et al.* 2010, 134, pl. 1, wrongly numbered as Tomb 57; Lightfoot 2012, 180, fig. 7.3, wrongly numbered as Tomb 65). Unusually, it contained a single interment, an adult male, who may have had the name of Leo (Lightfoot *et al.* 2010, 134, fig. 1; Schoolman 2010, 378, 380, fig. 4). In neither case can it be proven when the Roman material was brought in from the cemeteries. It may have been during the reconstruction of Amorium after AD 838, but it is equally possible that they were taken from earlier Byzantine structures within the city that were either already ruined and abandoned or actively demolished after the mid-9th century AD.

Several of the tombs at the Lower City Church incorporate fragments of early Byzantine church furnishings (for example, Narthex Tomb 4; Lightfoot *et al.* 2005, 245–6, fig. 15; Fig. 11.9) and even some early Byzantine tombstones. The most notable of the latter is the slab (T2114) commemorating the *heroon* of Etherios, apparently a native of Amorium, reused in the west end of Tomb 19 in the narthex (Lightfoot, Koçyiğit, and Yaman 2008, 445, pls. 4–5; Ivison 2010, 321, fig. 13). A relatively small number of early Christian tombstones and sarcophagi are known at Amorium (Lightfoot, Koçyiğit, and Yaman 2008, 453, pl. 11; Lightfoot *et al.* 2009, 214, pl. 5). It is likely that these Christian tombs were also located in or around the church (or, perhaps, one of the other churches in the city) during its early Byzantine phase, before it suffered extensive damage at the hands of the Arabs in AD 838. However, it is not clear where these tombs were located, nor where the tombstones came from for reuse in the mid-Byzantine phase of the Lower City Church. It does raise the question of whether early Christian graves were deliberately desecrated by the later inhabitants, presumably with the cognisance of the Church, or whether material from abandoned and demolished tombs was regarded as appropriate for reuse in a new sacred setting.

Even if some of the Byzantine inhabitants of Amorium were buried at churches located within the city walls from the 6th to the early 9th century AD, it would seem that many others found their final resting place in the Roman

Fig. 11.8. Amorium. Part of the Early Byzantine fortification wall around the Upper City, showing reuse of Roman doorstones (courtesy of the Amorium Excavations Project).

Fig. 11.9. Amorium. Middle Byzantine tomb 4 in the narthex of the Lower City Church, showing reuse of Early Byzantine architectural elements including two Ionic impost capitals (courtesy of the Amorium Excavations Project).

necropoleis outside the walls. The continued practice of extra-mural burial implies that they were content to share the same ground as their pagan predecessors (see also Steskal and Wenn *et al.*, both this volume), some of whom may even have been their own ancestors. But it is also clear that they had no compunction about disturbing earlier burials, demolishing their tombs, and reusing the materials for their own graves. Amorium thus provides good evidence for the ways in which Phrygian doorstones were reused for later tombs in the same cemeteries, and doorstones proved to be ideal as building material for constructing new tombs, where they were used either as side and end panels or as lids. It is not, however, an isolated phenomenon, and parallel cases can be found at a number of other sites in western central Anatolia (Yaman 2012, 335), notably at Pessinus (tombs 2.63 [5th century AD], 3.81 [undated], and 6.87 [possibly second half of 4th century AD]) and at Aezani in, as yet, unpublished tombs.[2] One important factor in the restructuring of the cemetery was the desire to provide tombs with an east/west orientation, with the head at the west end of a grave and feet to the east. This became common, although not exclusive, in Anatolia, being associated by modern scholars (Rush 1941, 1–22, 236–53) with a desire to face eastwards at the general resurrection. Likewise, Phrygian doorstones, as well as being in abundant supply, may have been chosen because of their apparent Christian symbolism. The door is represented as a panel, divided vertically into two leaves, each of which has two recessed sections usually of unequal height, giving the central frame the shape of a cross. The doorstone could thus have been taken as the sign of the cross and a means of warding off evil, just as some slabs were inserted into Byzantine fortifications to protect the defenders against attack. Certainly, the unique tomb, MZ 94, was constructed reusing doorstones to create four compartments with dividers arranged in the shape of a cross (Yaman 2012, 332–5, figs. 1–3).

Unfortunately, the physical remains provide little evidence to shed light on the social and economic status of the deceased or the religious divisions within the city. Amorium was known to have contained Christian groups such as the Athinganoi, a Judaising sect that was numerous in Phrygia in the 9th century AD (Starr 1936, 95). It is, however, unknown whether such people were buried alongside Orthodox Christians, but such factors may have played an important role in the choice of a burial site. Nevertheless, the size and importance of Amorium as a Byzantine city offers a great opportunity to learn about mortuary archaeology at the site during the long centuries of its existence. Between the 6th and late 11th centuries funerary practices gradually evolved, and these changes can be traced in the archaeological record. They include the reuse of early Byzantine tombstones and the allocation of special areas for the burial of foetuses and infants no older than 18 months (Lightfoot *et al.* 2009, 203, fig. 2;

Ivison 2010, 338, fig. 32). Both practices have significant implications for Orthodox liturgy, eschatology, and canon law (see Talbot 2009, 283–308). Scientific analyses of the physical remains may also shed light not only on the general health but also the ethnic origins of some of its inhabitants. Such work remains to be done in the future, but already new excavations in 2014 under the direction of Assoc. Prof. Zeliha Demirel Gökalp of the University of Anatolia, Eskişehir have begun to reveal more graves around the large church (Basilica B) on the Upper City mound.

Notes

1 Recent excavations (2014–2015) have been carried out at the east end of Basilica B on the Upper City, revealing several tombs, presumably also belonging to the mid-Byzantine period (Demirel Gökalp *et al.* 2016, 201–2, fig. 2, illus. 2).

2 The use of spolia in tomb construction at Aizanoi is now documented (Özer 2014, 325).

Bibliography

Brayne, K. and Roberts, J. A. (2003) Human skeletal remains, 1993–2001. In C. S. Lightfoot (ed.), *Amorium reports II: Research papers and technical studies*. (British Archaeological Reports, International Series 1170), 159–84. Oxford, Archaeopress.

Daim F. and Drauschke J. (eds.) (2010) *Byzanz – das Römerreich im Mittelalter, Teil 2,1 Schauplätze* (Monographien des Römisch-Germanischen Zentralmuseums 84). Mainz, Römisch-Germanisches Zentralmuseum.

Demirel, A. (2012) Two weapon-related skull traumas from the Enclosure, 2008. In Lightfoot and Ivison (eds.), 387–94.

Demirel Gökalp, Z,. Erel. A. C., Tsvikis, N., and Yaşar, H. Y. (2016) 2014 yılı Amorium kazısı. *37. Kazı Sonuçları Toplantısı, Erzurum, 11–15 Mayıs 2015*, vol. 3, 199–214. Ankara, T .C. Kültür ve Turizm Bakanlığı.

Harrison, R. M. (1989) Amorium 1988, the first preliminary excavation. *Anatolian Studies* 39, 167–74.

Ivison, E. A. (2010) Kirche und religiöses Leben im byzantinischen Amorium. In Daim and Drauschke (eds.), 309–43.

Ivison, E. A. (2012) Excavations at the Lower City Enclosure, 1996–2008. In Lightfoot and Ivison (eds.), 5–151.

Katsari, C. (2012) The Roman provincial coins of the city mint. In C. Katsari, C. S. Lightfoot and A. Özme, *The Amorium mint and the coin finds: Amorium reports 4*, 26–56. Berlin, Akademie.

Kelp, U. (2013) Grave monuments and local identities in Roman Phrygia. In P. Thonemann (ed.), *Roman Phrygia. Culture and Society*, 70–94. Cambridge, Cambridge University Press.

Lightfoot, C. S. (2012) Business as usual? Archaeological evidence for Byzantine commercial enterprise at Amorium in the seventh to eleventh centuries. In C. Morrisson (ed.), *Trade and markets in Byzantium*, 177–91. Washington D.C., Dumbarton Oaks.

Lightfoot C. and Arbel, Y. 2003 Amorium kazısı (2001). *24. Kazı Sonuçlar Toplantısı, Ankara, 27–31 Mayıs 2002*, vol. 1, 521–32. Ankara, T. C. Kültür ve Turizm Bakanlığı.

Lightfoot C. S. and Ivison E. A. (eds.) (2012) *Amorium reports 3: Finds reports and technical studies*. Istanbul, Ege Yayınları.

Lightfoot, C., and Lightfoot, M. (2007) *A Byzantine* city in *Anatolia: Amorium, an archaeological guide.* Istanbul, Homer.

Lightfoot, C., Koçyiğit, O., and Yaman, H. (2006) Amorium kazıları 2004. *27. Kazı Sonuçları Toplantısı,* cilt 1, *Antalya, 30 Mayıs – 03 Haziran 2005,* vol. 1, 77–88. Ankara, T. C. Kültür ve Turizm Bakanlığı.

Lightfoot, C., Koçyiğit, O., and Yaman, H. (2008) Amorium kazısı 2006. *29. Kazı Sonuçları Toplantısı, Kocaeli, 28 Mayıs – 01 Haziran 2007,* vol. 1, 443–66. Ankara, T. C. Kültür ve Turizm Bakanlığı.

Lightfoot, C. S., Arbel, Y., Ivison, E. A., Roberts, J. A., and Ioannidou, E. (2005) The Amorium project: Excavation and research in 2002. *Dumbarton Oaks Papers* 59, 231–65.

Lightfoot, C. S., Drahor, M. G., Gill, M. A. V., Kaya, M. A., Roberts, J. A., Ülker, F., Yazıcı, İ., and Young. S. (1999) The Amorium project: The 1997 study season. *Dumbarton Oaks Papers* 53, 333–49.

Lightfoot, C., Ivison, E., Koçyiğit, O., and Şen, M. (2010) Amorium kazıları, 2008. *31. Kazı Sonuçları Toplantısı, Denizli, 25–9 Mayıs 2009,* vol. 1, 133–57. Ankara, T. C. Kültür ve Turizm Bakanlığı.

Lightfoot, C., Ivison, E., Şen, M., and Yaman, H. (2009) Amorium kazısı 2007. *30. Kazı Sonuçları Toplantısı, Ankara, 26–30 Mayıs 2008,* vol. 1, 201–26. Ankara, T. C. Kültür ve Turizm Bakanlığı.

Özer, C., (2014) Aizanoi 2012 yılı çalışmaları. *35. Kazı Sonuçları Toplantısı, Muğla, 27–31 Mayıs 2013,* vol. 2, 324–42. Muğla, Sıtkı Koçman Üniversitesi.

Pugsley, P. (1995) Rock-cut tomb. In C. S. Lightfoot and E. A. Ivison (eds.), Amorium excavations 1994, the seventh preliminary report. *Anatolian Studies* 45, 97–102.

Rush, A. C. (1941) *Death and burial in Christian Antiquity.* Washington D.C., Catholic University of America Press.

Schoolman, E. M. (2010) Kreuze und kreuzförmige Darstellungen in der Alltagskultur von Amorium. In Daim and Drauschke (eds.), 373–86.

Starr, J. (1936) An eastern Christian sect: The Athinganoi. *Harvard Theological Review* 29, 93–106.

Steskal, M. (this volume) Reflections on the mortuary landscape of Ephesus: The archaeology of death in a Roman metropolis, 176–87.

Talbot, A.-M. (2009) The death and commemoration of Byzantine children. In A. Papaconstantinou and A.-M. Talbot (eds.), *Becoming Byzantine. Children and childhood in Byzantium,* 283–308 Washington D.C., Dumbarton Oaks Research Library and Collection.

Waelkens, M. (1986) *Die kleinasiatischen Türsteine. Typologie und epigraphische Untersuchungen der kleinasiatischen Grabreliefs mit Scheintür.* Mainz, von Zabern.

Wenn, C. C., Ahrens, S., and Brandt J. R. (this volume) Romans, Christians, and pilgrims at Hierapolis in Phrygia: Changes in funerary practices and mental processes, 196–216.

Yaman, H. (2008) Door to the other world: Phrygian doorstones at Amorium. In O. Özbek (ed.), *Funeral rites, rituals and ceremonies from prehistory to Antiquity,* 59–68. Istanbul, Institut Français d'Études Anatoliennes.

Yaman, H. (2012) Small finds for the dating of a tomb at Amorium. In B. Böhlendorf-Arslan and A. Ricci (eds.), Byzantine small finds in archaeological contexts. *Byzas* 15, 331–42. Istanbul, Deutschen Archäologischen Instituts.

Romans, Christians, and pilgrims at Hierapolis in Phrygia: Changes in funerary practices and mental processes

Camilla Cecilie Wenn, Sven Ahrens, and J. Rasmus Brandt

Abstract

There are many ways a deceased person can be brought to rest, and the shape and position of the funerary structures are likewise many. In the North-East Necropolis of Hierapolis can be followed a funerary history spanning at least 14 centuries from the 3rd/2nd century BC to c. AD 1300. The aim of the present paper is not to describe the physical evidence of funerary activities and changes, but to investigate if, in the periods of funerary changes, the changes can also be related to mental processes or cognitive concerns in the population who made the burials.

Keywords: *ad sanctos*, anxiety, coins, cremation, identity, legitimization, memory, pollution, reuse, views

The archaeological unit called a burial is normally composed of the funerary structure itself, the skeleton of the deceased and private belongings and offerings following the deceased. These can tell us something of death disposal and organization, of burial types and typology, of object typology and function, of chronology, of historical and cultural contexts, of characteristics of death rituals, and of the life and death of the buried person as extracted from osteological, DNA, and isotope analyses. The present volume can give examples of most of these funerary elements. But in a diachronic perspective can the funerary elements and changes of them also reveal some changes in mental processes among the inhabitants in the ancient Roman and Medieval society at large?

The point of departure for the present investigation will be the North-East Necropolis at Hierapolis (Figs. 12.1–2), where the three authors have worked in annual campaigns since 2007. The North-East Necropolis was mainly laid out in the Roman Imperial period, and stretches along the hill and hillsides above the town, including the area of the later sanctuary of St Philip, the apostle. It developed along the foot of the hill from scattered Hellenistic tombs, which were raised along a secondary road leading from the north towards the town (Ahrens 2015b; Hill, Lieng Andreadakis, and Ahrens 2016; Ronchetta 2015). It is one of several necropoleis in Hierapolis, which were laid out in a large semicircle to the north, east, and south of the town from the Hellenistic period onwards. Some of these necropoleis are oriented towards main access roads, like the North, the North-West, the South-East, the South-West, and the lowest part of the North-East Necropolis. The East and the largest part of the North-East necropolis are instead located on the hills and slopes above the town without connection to traffic arteries (Schneider Equini 1972; Ronchetta 2008; 2012; 2015; this volume; Ronchetta and Mighetto 2007).

Tombs with a view in a time of stability

The first major funerary change in Hierapolis involves the layout and location of the necropoleis in general. In the Hellenistic and early Imperial periods the tombs clustered along the main roads into the town, as we find them in many a Roman settlement, the exception being the South-East Necropolis, on a detached small hill at the southern end of the calcareous terrace on which the town was founded, facing and overlooking the road (Ronchetta, this volume). The tombs

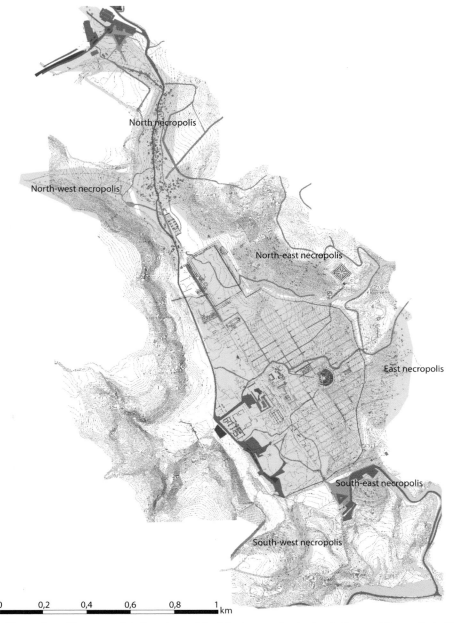

Fig. 12.1. Hierapolis. Town plan with necropoleis (courtesy of the Missione archeologica italiana a Hierapolis in Frigia; with adaptions).

communicated directly with the passers-by through their architecture, decorative elements, and inscriptions (Ahrens 2011). The road, so to speak, was the arena for the population of the Hellenistic and Roman Imperial periods to manifest their wealth and social status. Even though the Hellenistic tumuli and most of the later tomb monuments were highly standardized in size and shape and thus did hardly stand out from the other monuments in the necropoleis, they would still demonstrate the affiliation of the tomb owner and his family with the part of the citizenry of Hierapolis that actually could afford such monuments. The building boom along the

streets of tombs, particularly the North Necropolis, continued unbroken until the 4th century AD.

From the later 1st and early 2nd century AD, however, the hilltops with slopes north-east and east of the town became the new development areas for family tombs. Here, overlooking the town, the wide Lykos river valley below, and the high mountains beyond, the dead citizens could rest in silent peace, withdrawn from the busy life along the town's main road approaches.

The marked topographical change in the location of the tombs from along the main access streets to the

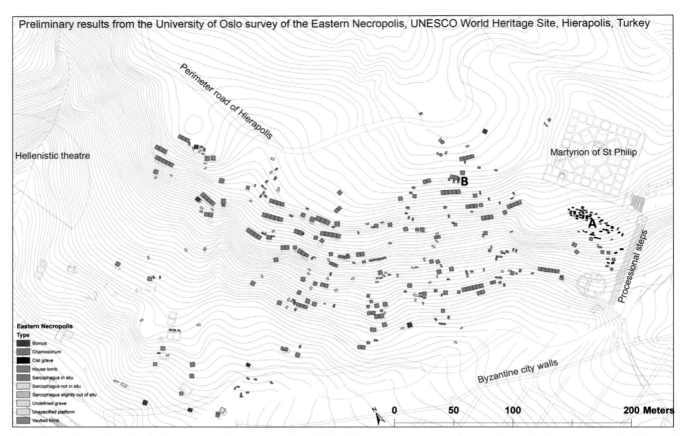

Fig. 12.2. Hierapolis, North-East Necropolis. Survey plan 2010 of the necropolis including to the right the Martyrion building of St Philip. A: Early to Mid-Byzantine cemetery; B: Area of the Oslo University excavations (plan: David Hill; archive of the Oslo University Excavations at Hierapolis).

town in the north and south of the city and to the hills of the North-East and East Necropoleis may reflect the Greco-Romans' new attitudes towards the experience and use of the surrounding landscape, as expressed, for example, through their pastoral poetry (Moore 2007, 90). In addition, the view enjoyed from the necropolis corresponded perfectly with accounts, from the Roman Imperial period, of the ideal landscape view (Ahrens 2011). and it reflected the contemporary Roman villa ideology which praised the relaxed *otium* life of the countryside in contrast to its negation, the town life's hectic *negotium* (see, for example, Mielsch 1987, 94–7; Ackerman 1990, 37–42). It must also be assumed that the view from the tombs onto the hometown played a significant part in the choice of burial places in the North-East Necropolis. The connection between tomb and hometown is very pronounced in the written sources. The desire for a burial in the hometown is a common topic in funerary inscriptions from Asia Minor and the place of the tombs of the ancestors was frequently mentioned in Greek literature and funerary inscriptions as a definition for the hometown. Some sources even point out, in particular, the view of the hometown

from a tomb (Ahrens 2011). At the same time, the tombs, being not too distant, could easily be seen from the town and its immediate surroundings, each day reminding the citizenry of the inextricable connection between their ancestry and their hometown.

By withdrawing to more peripheral areas, like the hills and slopes of the North-East and East Necropoleis, the tombs no longer became objects of personal manifestations. In early Imperial times self-representation had moved into the arena of the town, where social and political status was displayed through public burials and by statues and inscriptions. These public honours were granted as a tribute to the most notable citizens in gratitude of their private financing of public and religious buildings, of festivals and games, and of general charity. According to preserved inscriptions from Asia Minor, this kind of euergetism reached its peak in the 1st and 2nd centuries AD (Zuiderhoek 2009, 17–22).

The Roman tomb complex chosen for the excavations in the North-East Necropolis lies near the top of the hill, less than 70 m west of the sanctuary of St Philip, as the crow flies (Figs. 12.2:B; 12.3–7). It included three saddle-roofed

Fig. 12.3. Hierapolis, North-East Necropolis. Aerial view of the tombs in the North-East Necropolis selected for excavation, seen from the south (2009). Above the tombs C91 (still not excavated) C92 and C311 with the sarcophagi C308-10, and C212 (badly preserved). The row of five tombs below the excavation area carry from left to right the numbers C93–C97 (courtesy of the Missione archeologica italiana a Hierapolis in Frigia).

house tombs, two combined (C92 and C311) and one alone (C91) (Verzone 1961, 35; 1965, 373 fig. 7; Laforest *et al.* 2013, 147–8; Ahrens and Brandt 2016, 401–6). The tombs are typical examples of saddle-roofed house tombs as found throughout the necropolis. In general, they measured in plan roughly 3 × 3 m, and in height around 3.80 m (measured from bedrock). They were built in a modular system with large interlocking blocks of local limestone (Schneider Equini 1972, 121, pls. 22a–b; D'Andria 2003, 191–2), and they were placed alone or in terraced units of 2–6 tombs. The type became standardized in a system found in many tombs in Asia Minor: they had a saddle roof, a panelled door made of a single stone block, and inside they were equipped with stone benches on three walls and with an extra bench on a higher level on the back wall. In connection with the double tomb (C92 and C311) were placed four sarcophagi (C308, C309, C310, and C212, all carved from local stone), three along the front and one on the east side. According to the finds inside the tombs, correlated with radiocarbon dates from a selection of the skeletons, the monumental tombs appear to have been raised some time in the late 1st/2nd century AD, but in most, if not all tombs the depositions

were not limited to this period, they continued long after this date.

The saddle-roofed house tombs occasionally carry inscriptions with the names of their owners, as in the case of the three investigated tomb buildings: tomb C92 belonged to Eutyches, son of Apollonios, from Lagina (pediment inscription: Gardner 1885, 345 no. 69; Judeich 1898, 157 no. 281), tomb C311 to the dyer Patrokles junior (Fig. 12.8) and tomb C91 to Attalos *laparos* (the lanky) (Pennacchietti 1966–1967, 295 no. 5), a tomb which later was sold to Aurelios Artemonidos (Fig. 12.9). Apollonios, the son, and Ariste, the daughter of Eutyches, were buried in separate sarcophagi (C309, respectively C308) outside their father's tomb, both sarcophagi carrying the standard Roman inscription formula stating that whoever violates the tomb shall pay fines in one case to the *gerousia*, in the other to the *boule*, the *gerousia* and the tax office (Pennacchietti 1966–1967, 294–5 no. 2; 296 no. 4). The inscriptions testify firstly to the importance of family ancestry in a strict sense; secondly, they show concern for the legitimate grave use, and the anxiety of disturbance of the grave as an ancestral nucleus preserving the memory

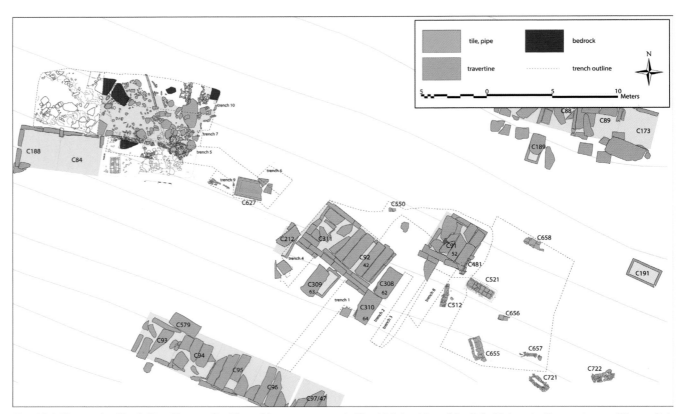

Fig. 12.4. Hierapolis, North-East Necropolis. Plan of the same area as in Fig. 12.3 (archive of the Oslo University Excavations at Hierapolis).

of its past members. This was an expression of cognitive concerns or social anxiety which lasted all through the Roman period. In addition, the continued use of the same inscription formula for centuries, reflects a socially and politically stable society.

Fig. 12.5. Hierapolis, North-East Necropolis. Tomb C92 with the sarcophagi C309 (t.l.) and C310 (t.r), seen from south (archive of the Oslo University Excavations at Hierapolis).

Tombs in a time of conflict

Tombs and sarcophagi

The building of new monumental tombs and the carving of new stone sarcophagi seem to stop in Hierapolis in the course of the 4th century AD, if not before (on the sarcophagi, see, for example, Vanhaverbeke and Waelkens 2002). However, a large portfolio of coins in the tombs investigated in the North-East Necropolis, supported by unpublished radiocarbon dates of skeletons, and by tombs from the North Necropolis, demonstrates that burials continued in the old monumental tombs until the late 4th or 5th centuries AD, and in a few cases into the 6th century AD (Travaglini and Camilleri 2010, 227: tombs 33 and 55). One sarcophagus (C310) outside the tomb of Eutyches (C92) received a Christian inscription (the two Greek letters Alpha and Omega finished the standard violation precautions) in the 4th century, when it was reused for burials (Pennacchietti 1963, 131–3 no. 1; 1966–1967, 295 no. 3; Robert and Robert 1967, 545) (Fig. 12.10). The names in this late Roman inscription were erased when the sarcophagus was reused a third time, at an unknown date. In the sarcophagus of Eutyches' son (C309) a coin of Justin II (565–574) may derive from a secondary burial of the late 6th century or later. Also in the North Necropolis there is evidence that

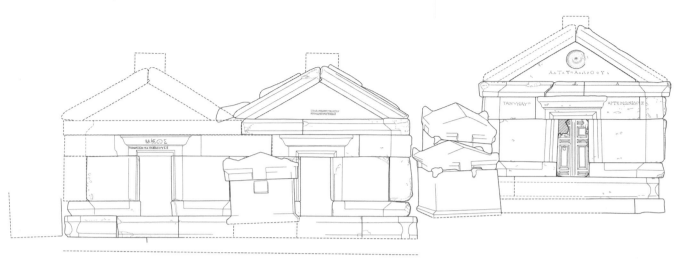

Fig. 12.6. Hierapolis, North-East Necropolis. Drawing of the facades of the tombs (from left): C311 (reconstructed), C92 and C91, and the sarcophagi (from left), C309, C310, and C308 (behind) (drawing: Sven Ahrens; archive of the Oslo University Excavations at Hierapolis).

the monumental tombs were still used for burials during the early Byzantine period (Schneider Equini 1972, 127; Laforest *et al.*, this volume).

In Hierapolis there is hardly any evidence for the employment of simple tile or cist graves in the Roman Imperial period. In the South-East Necropolis one grave, possibly(?) from the 2nd century AD, has been found (see Ronchetta, this volume), to which should be added the undated, possibly late Roman grave 156a in the North Necropolis (Anderson 2007, 473–5; Moore 2013, 80). This changes in late Antiquity

Fig. 12.8. Hierapolis, North-East Necropolis. Tomb C311: inscription on the lintel block of the tomb: BAPHEWS/ TOMNHMEIONTOPATROKLEOYSB (archive of the Oslo University Excavations at Hierapolis).

Fig. 12.7. Hierapolis, North-East Necropolis. Inside view of tomb C92, seen from south-west (archive of the Oslo University Excavations at Hierapolis).

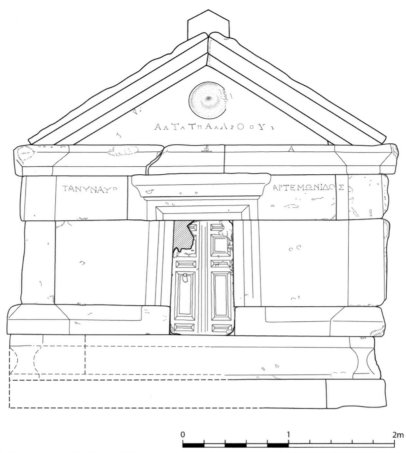

Fig. 12.9. Hierapolis, North-East Necropolis. Tomb C91 with the inscriptions: ΑΛΤΑΤΠΑΑΛΡΟΟΥΥ (pediment); ΤΑ ΝΥΝ ΑΥΡ / ΑΡΤΕΜΩΝΙΔΟΣ (wall); [Η] ΡΩ Α (top edge of the architrave) (drawing by Sven Ahrens; archive of the Oslo University Excavations at Hierapolis).

Fig. 12.10. Hierapolis, North-East Necropolis. Sarcophagus C310 with an inscription finishing in the last line with two Greek letters: Alpha (left end) and Omega (right end), seen from west (archive of the Oslo University Excavations at Hierapolis).

when in the North-East Necropolis, such simple graves come gradually into use in the course of the 4th century, as can be seen in an area taken into use immediately east of Attalos' tomb (C91) (Fig. 12.11). By shifting large amounts of soil the sloping area was changed into an approximately horizontal terrace in which the new, simple tile graves were dug (Figs. 12.4; 12.12). Before being filled Attalos' tomb was emptied of its many burials and the bones presumably redeposited in some unknown place.

When parts of the necropoleis of Aphrodisias were dismantled to build the 4th-century town walls, the bones seem to have been treated with comparable respect (Staebler 2008, 193–4), and we may imagine that the same happened to the tombs disturbed when the more or less contemporary Byzantine walls were built in Hierapolis. A written source supports the assumption that the bones were treated respectfully during construction works in the 4th century: when writing against grave-robbing Gregory of Nyssa (Letter 31 to Letoius bishop of Melitene, 7a) explains that the dismantling of tombs

Fig. 12.12. Hierapolis, North-East Necropolis. Section of the large heap of soil dumped around Attalos' tomb (C91) (t.l.), in which can still be seen cist grave C521, seen from west. Notice the curving dark bands in the soil, demonstrating the filling actions of the build-up (archive of the Oslo University Excavations at Hierapolis).

Fig. 12.11. Hierapolis, North-East Necropolis. Cist grave C721; seen from west (archive of the Oslo University Excavations at Hierapolis).

and the reuse of the building material in publicly important building projects was considered pardonable by his coevals, 'if someone spares what deserves respect and leaves the interred body intact, so that the shame of our nature is not exposed to the sun, and only makes use of the stones from the facing of the tomb in order to build something else.'[1]

After the tomb had been emptied the door of Attalos' tomb was subsequently closed and sealed with mortar (as found in situ during the excavations) (Fig. 12.9) and the tomb covered by soil up to its pediments. It is not known who was responsible for this move or for what reason it was done, but it is worth noting that the depositions in Eutyches' tomb (C92) also stopped at about the same time, or soon afterwards, while those in Patrokles' tomb still continued, perhaps for some decades more. The intervention in Attalos' tomb and the discontinued use first of Eutyches' tomb and later of Patrokles' tomb stand as a clear emblem of a change in funerary customs at Hierapolis and corresponds chronologically to drastic changes in the monumental topography of Hierapolis around the year AD 400.

At that moment, in the centre of the town, started the destruction of the temple of Apollo, the oracular, patron god of Hierapolis (D'Andria 2007, 19–25; this volume; Semeraro 2007, 172, 191–6; 2008). It was gradually torn apart stone by stone, some reused for other buildings, some sent to the lime kilns, some hacked to pieces, and the site of the temple was eventually turned into a dump for building materials. In a similar way the round temple, the *tholos*, of the Pluto sanctuary nearby was demolished. The cave with the entrance to the Underworld was screened off and hidden behind a tall *spolia* wall; at the same time started also the gradual demolition of the sanctuary *theatron* above the cave (D'Andria 2013, 175–7). Instead, outside the town in the area of the North-East Necropolis, a new sanctuary grew up in connection with the tomb of the apostle Philip (D'Andria 2011–2012; 2014; this volume; Caggia 2012; Caggia and Caldarola 2012). The sanctuary, giving a new Christian topography to the town, was composed of a bath complex at the foot of the hill, a ceremonial stairway leading up to the tomb of the apostle, in the 4th century already isolated and incorporated into a building structure, in the 6th century turned into a three-aisled basilica built around the tomb (which was a standard saddle-roofed house tomb like the others found in the North-East Necropolis), and from there another stairway led up to the 60 × 60 m square monumental early 5th-century Martyrion of St Philip enclosing an octagonal church space, presumably raised on the spot where Philip was martyrized (for a plan of the sanctuary, see D'Andria 2011–2012, 8, fig. 3; this volume, Fig. 1.10). The sanctuary became the new *locus sancti* of the town, and, in opposition to the pagan sacred areas, it also became a place of gravitation for burials, presumably attracted by the *ad sanctos* concept (on this see below). One such area of burials was laid out on the partly terraced slope just to the south-west of the Martyrion and contained some 80 documented more or less preserved cist graves, built of roughly cut stones and spoils from destroyed monumental tombs (Ahrens and Brandt 2016, 395–400) (Figs. 12.2:A; 12.13). The badly preserved graves were in general oriented north-west/south-east following the orientation of the Martyrion and the cemetery terrace, and in general they measured around 0.50/0.85×1.45/1.90 m. All skeletal remains, found 0.10–0.20 m below surface level, were poorly preserved and dislocated. Unfortunately, the soil conditions in Hierapolis have usually a very negative impact on the preservation of bone material (Kars and Kars 2002, 174–7). In addition, natural collapse or man-made destruction of the grave covers, as well as multiple burials inside the same grave, and possibly reuse of the graves, have certainly contributed to the dislocation of the skeletal remains. Dislocated skeletal remains from the surrounding cemetery may also have ended up in the graves, but on this definite evidence is lacking. Indeed, none of the skeletal remains are articulated or allow for further inferences on the number, position, sequence, and orientation of the skeletons in the burials.

Radiocarbon dates of selected skeletons from the Martyrion cemetery may be attributed to a long period of use (Ahrens and Brandt 2016, 399). Four samples (*c*. AD 400–7th century) coincided with the lifetime of the Martyrion before it collapsed in the mid-7th century earthquake, the same earthquake which also razed the town; two other samples point to a possibly limited continuity: one sample of the 7th–8th centuries coincides with the collapse of the building and later demolition activities (phase V), while the date of the other sample to the 9th–11th centuries agrees with the reactivation of the old building on a much smaller scale in the same centuries (phase VI) (D'Andria and Gümgüm 2010, 97; cf. also Verzone 1960, 13; D'Andria 2008; R. D'Andria 2012, 635, fig. 1).

In the early Byzantine period the new peri-urban sanctuary of St Philip became the new pole in a changed burial topography and in which monumental tombs were subdued to the monumentality of the temples of the new religion. The burials became individualized, detached from the strong family bonds of the past. The tomb was no longer a solid, eternal container of family members, but a temporary resting place for the individual awaiting resurrection at the end of time. The pagan grave inscription bids a permanent 'Farewell', the Christian a temporary 'May you live (in God)', as death was not considered the end of life, but as the beginning of true life (Rush 1941, 254–6). In addition, the terms *heroon* (as used, for example, on the cornice above the door of Attalos' tomb) and *mnemeion* (as used in the inscriptions of the tombs of both Eutyches and Patrokles), alone or in combination, were gradually displaced by the term *koimeterion* (place of sleep) (Trombley 2013, 353).

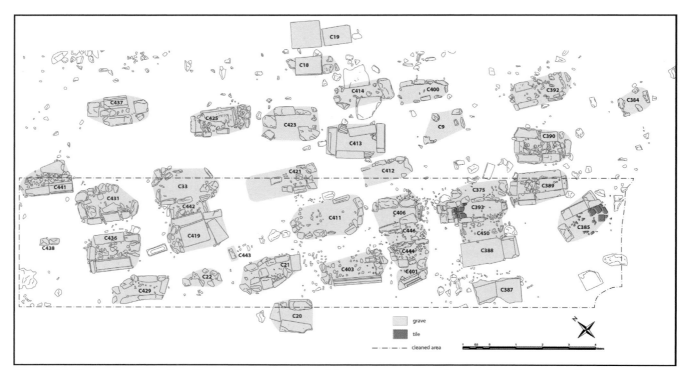

Fig. 12.13. Hierapolis, North-East Necropolis. Plan of part of the cemetery area south west of the Martyrion of St Philip, marked A on Fig. 12.2 (archive of the Oslo University Excavations at Hierapolis).

Significantly, the old formulas stressing ownership and punishment of tomb violation gradually disappear; the importance of memory and ancestry seems to fade, creating a new attitude towards death in which family bonds are replaced by a new religious bondage (see, for example, also Yasin 2009, 61; Moore 2013, 80; D'Andria, this volume); in fact, on Christian gravestones (as witnessed at Corinth) the family name disappeared, the stones were, as it seems, 'purposely designed as short-term memorials for the deceased' (Ivison 1996, 107). Furthermore, with the Christians the position of their graves was no longer determined by the areas of the old extra mural necropoleis and the place of the family tombs, but by their location in cemeteries, which gradually became administered by churches (Ivison 1996, 105). A burial in proximity to or even inside a church offered the possibility to be included in church ceremonies, like the general celebration of the liturgy, processions, or even masses for the dead (Ivison 1993, 26–7; Marinis 2009, 165–6; Yasin 2009, 64). Such a location quite often meant a burial close to the relics or even the body of a saint – *ad sanctos* – offering the advocacy, protection, and blessing provided by the saint, like the following inscription from Tyana exemplifies: 'I, a woman of many sins, (…)suoa, rest close to you. You strike me with awe, you, who is in the mind of (…) the good one: Pray for me!' (Berges and Nollé 2000, no. 144).

In Rome, according to the Law of the Twelve Tables from the mid-5th century BC interments had to be located outside the ritual borders of the city (Cicero, *de legibus* 2.58). This custom was generally observed until the 5th–6th centuries AD when the tombs appeared more frequently inside the borders of cities of the West Mediterranean (Meneghini and Santangeli Valenzani 1995; Gonzalez Villaescusa and Lerma 1996, 40; Haug 2003, 273–88; Leone 2007a, 198–200; 2007b; Piepoli 2008, 579–80). This new situation is normally explained pragmatically as the result of war and sieges, or ideologically connected to the moving of relics of dead saints and martyrs from outside the city to their name-churches inside. Both explanations may carry some truth, but also a third explanation shall be considered, one which operates on the mental or cognitive level (Brandt 2012, 151–2).

Burials were regarded as polluters of the sacred space of the town, and burial practices contained elements of purification through the execution of animal sacrifices. When such acts were banned by emperor Theodosius in AD 391 and 392 (*Codex Theodosianus* 16.10.10–12), it must have created a cognitive crisis among the pagans as this eliminated their possibility to be purified. The Christians, however, considered their dead bodies as sacred and holy, permanently purified at the moment of baptism. For them the pagan pollution dogma accordingly made no sense. We have no method to discover if the Christians were the first to bury their beloved ones inside the ritual borders of

the Roman cities, but the Theodosian decree and Christian theology were certainly important elements in the process, which gradually transformed the ancient classical cityscapes from a sacred space into an area of holy places, defined by the churches, which gradually became the new grounds for interment of the dead (Brandt 2012).

An important technique used in this transformation process was for the Christians to appropriate pagan symbols and give them a new meaning. We shall see one example of this below discussing the presence of coins in the tombs. Since pollution was a central concept in Roman burial ideology, the word had no place in a Christian funerary context. By redefining who polluted who, in what way and when, the Christians created a positive attitude towards death and the word pollute changed its meaning. In *Codex Theodosianus* 16.10.10 of AD 392, one of the decrees regulating the use of temples and animal sacrifices, it is stipulated that *nemo se hostiis polluat* – 'no person shall pollute himself with sacrificial animals.'[2] Seen with pagan eyes this decree was senseless – how could the killing of a sacrificial animal be considered an act of pollution, when animals were sacrificed just in order to purify polluted humans and things? By using the word pollute, a strong word in Roman juridical language, Theodosius gave a reason as to why animals could not be sacrificed as part of religious practices.

Such was the situation in the Roman West, but we should be careful in applying Roman traditions to the Greek East. Plutarch (*Lykurgos* 27) tells that the Spartans were the only people in Greece to consciously bury their dead inside the town walls, but no law text or inscription has been preserved which clearly states that burials could not be made inside the defined town area. However, it was normal practice also in the East to bury the dead outside (Schörner 2007, 11), but perhaps for other reasons than those flagged in the West. If a law existed it may have been transferred orally (Schörner 2007, 19), alternatively it was observed with some exceptions due to traditions and/or taboos (for one kind of death taboos in Greek funerary practices, see, for example, Brandt 2015, 111–2). One exception was apparently that important men and benefactors of the town could receive the honour of being buried inside the town borders. It should, however, be noted that Sikyon once asked approval to do so from the oracle at Delphi (Schörner 2007, 15–16); some restrictions of sacral character must, therefore, have been felt even among the Greeks, whether connected to pollution or not.

On two occasions Roman emperors, first Hadrian (*Corpus Iuris Civilis. Digesta* 47.12.3.5) and then Antoninus Pius (*Scriptores Historiae Augustae: Pius* 12.3), issued decrees which made the burying of people *intra muros* illegal (Schörner 2007, 18), underlining that from a Roman point of view this was a custom which could not be accepted. In the Greek East the latest burials recorded archaeologically or by written sources inside a town's territory, happened in the late 2nd/early 3rd century AD (see Schröner 2007, 209–88 catalogue). Accordingly, there is reason to believe that in late Roman times the Roman burial dogma was observed also in the East, perhaps in the 4th/early 5th centuries AD even reinforced by the new walls built around many towns in Asia Minor, as, for example, at Aphrodisias, Blaundos, Laodikeia, Perge, Sagalassos, Sardis, Selge, and Hierapolis (Jacobs 2012, 118, table 1; Jacobs and Waelkens 2014, 94; on Hierapolis, see Scardozzi 2006, 119–22; 2008; 2015). Late Roman and Byzantine law codes, like the *Codex Iustinianus* from the 6th century AD, consistently repeat the Roman prohibitions against intramural burial (*Codex Theodosianus* 9.17.36; *Corpus Iuris Civilis. Digesta* 47.12.3.5). It is, of course, difficult to assess whether these laws still were obeyed or if there are other reasons why in the early Byzantine times burials *intra muros* are a rare occurrence in the East. It seems, though, that in Asia Minor people must have abided by the law more faithfully than in other comparable regions within the Roman Empire (for other areas, see, for example, Achim 2015, 288, and her list of regions with literary references; add Fox 2012). This law-abiding character may also be the reason why no churches at Hierapolis, not even the cathedral built in the late 5th/early 6th centuries AD, have burials *intra ecclesiam* before the mid-Byzantine period. According to *Codex Iustinianus* (1.2.2) not only burials *intra muros*, but also burials in *apostolorum vel martyrium sedem* were not allowed (Achim 2015, 290). It can be doubted that a concept like the Western burial church existed in the East (on Western burial churches, see, for example, Yasin 2009, 61–100). A comparable situation can be observed in Greece (Laskaris 2000, 147). In fact, as far as can be said at the current state of research, burials within the town area did not occur till the mid-Byzantine period at Hierapolis.

The change from built monumental tombs and stone sarcophagi in the Imperial period to individual cist graves and reuse of Roman tombs in late Antiquity appears quite decisive. There were, however, also pagan funerary rites which seemingly continued into the late Roman period and even later. Two of these rites were the deposition of coins in tombs and cremation.

Cremations

Let us start with cremations, which was a common burial rite in Asia Minor, particularly in the Hellenistic and Roman Imperial periods (Ahrens 2015a). Hierapolis followed suit, as documented in the North Necropolis (Ahrens 2015a, 210.53; Anderson 2007, 477–8; Okunak 2005, 57; De Bernardi Ferrero 1996, 91; Verzone 1961–1962, 637). Finds of cremated bones in tomb C92, and of an *ostotheca* from the Imperial period in the sanctuary of St Philip demonstrate that cremation was also practised in the North-East

Necropolis. Recent discoveries in Patrokles' tomb (C311) (Figs. 12.4; 12.6) amplify notably the evidence for cremation burials, as in this tomb about one third of the depositions registered (out of a MNI of 103) are cremations (see Kiesewetter, this volume). Four charred or burnt bone samples were tested for radiocarbon dates, giving the following results (2 sigma calibrated):

Sample no. 403073: AD 90–100, 125–250

Sample no. 403072: AD 255–300, 315–405

Sample no. 405530: AD 265–275, 330–420

Sample no. 403071: AD 420–570[3]

On the basis of these we can state with some certainty that cremation was practised from the 2nd/3rd until the 5th/6th centuries. Though it is important to add that cremation in these periods, and especially in the latter part, was not exclusive, it alternated with inhumations, as the overlapping radiocarbon dates of the unburnt skeletons demonstrate:

Sample no. 405527: AD 5–125

Sample no. 403270: AD 180–190, 215–340

Sample no. 405526: AD 235–385

Sample no. 405529: AD 235–385

Sample no. 405525: 255–300, 315–405

An equally late date for cremations in Asia Minor has till now only been established for burials at Pessinus, also there alternating with inhumations: 3rd/4th centuries; even one late 5th century, but explained as belonging to deceased with a foreign cultural background (Lambrechts 1969, 127–31, 135).[4] While cremation was still practised in some regions of Asia Minor in the 3rd century AD (Ahrens 2015a, 203), it is exceptional after that century. An increase of inhumation during the 2nd century and the decrease of cremation in the 3rd century in Rome and in the Empire in general can currently not be linked to new religious ideas (Scheid 2007, 19, 25).[5] So when the practice of cremation decreased in the Roman Empire in the course of the 3rd century, it was not the result of a Christian impact on funerary rites. However, inhumation was the preferred rite for any Christian (Ahrens 2015a, 205), which can explain why cremation was almost completely abandoned when most of the population had adopted the Christian faith from the 4th century onward.

In periods of social, cultural, economic, and/or religious changes, the changes may, for many, create a cognitive crisis, when old norms of life are challenged by new ones. If the changes are felt dramatically, they may be regarded as a threat to the established order and can at times be counteracted by reinventing, renewing, and/or reinforcing old traditions; and old everyday customs can take on new symbolic meanings to signal opposition. One such dramatic change may have been the conversion from a pagan to a Christian society. In some local societies the change may have been gradual and peaceful, in others signalled by returning episodes of violence. At Hierapolis both ancient written and archaeological sources give the impression that the religious change in the town did not happen without confrontations (Huttner 2013, *passim*).

In the apocryphal *Acta Philippi*, perhaps written down some time in the 4th or 5th centuries AD, we are told with what violence the apostles Philip and Bartholomew were hanged from a tree, head down, with iron hooks and nails through their ankles and heels, how Philip's sister and follower, Mariamne, was stripped naked in front of the public, and how an abyss opened and swallowed the Roman proconsul, a viper-cult temple, its priests and 7,000 followers, men, women, and children, all included (Bouvon, Boovier, and Amsler (eds.) 1999, 342–431;[6] D'Andria 2011, 36–8; this volume; cf. also Huttner 2013, 355–71).

The story, even if imaginative and fantastic, contains some basic, real topographical and archaeological foundations. The site is traversed by a tectonic fault line from the depths of which hot water and poisonous gases gush forward here and there. A viper-cult temple can refer both to the temple of the patron and oracular god of Hierapolis, Apollo, and to the sanctuary of Cybele and the god of the Underworld, Pluto, where a curled, sculpted snake has been uncovered. The first temple, as already mentioned (p. 203) was, in around the year 400 AD, violently destroyed and gradually demolished stone by stone, as the round temple of the Plutonium may also have been.

Unfortunately we can only guess with what other kinds of violence Hierapolis had been confronted in the years previous to these actions, and still perhaps for some time continued to be confronted with. According to Amsler (1996), in an article, written before the latest discoveries of the combined Cybele and Pluto sanctuary, the Acts of Philip reveal one level of conflict, which originally existed between Cybele and Apollo, but which in the 5th century was transferred to a conflict between Cybele and St Philip, between paganism and Christianity.

Hierapolis was not the only place where the conversion from pagan to Christian belief caused local conflicts and upheavals (see, for example, Saradi 2008, and the discussion by Bremmer 2014). A classic example is given by the town Apamea (the metropolis of second Syria), in which the fanatic archbishop Marcellus is reported to have destroyed many pagan temples (Theodoret of Cyrrhus, *Historia Ecclesiastica* 5.21; see also Trombley 1993, 123–9), whether referring to actual or partly imagined events.

If we can accept that death, from a societal point of view, is often more important for the living than for the dead (Oestigaard 2015, 374), it should not be dismissed that the family that owned Patrokles' tomb (C311) might have 'reinvented' the use of cremation, a distinctly non-Christian burial practice, in periods of tension in the town between pagans and Christians, to symbolically underline the family's strong adherence to pagan traditions. The family used cremation to create or maintain their own 'death myth', i.e. within the framework of an overriding cosmological and religious scheme and according to the available set of ritual possibilities, it composed and conducted 'the funeral in accordance to: specific causes of death, the ancestors, the spiritual world, or using the deceased as a medium for social outcomes in the reconstruction of society' (Oestigaard 2015, 368). The old traditional cremation rites were thus in late Roman and early Byzantine times at Hierapolis used, not as part of a generally accepted custom of the past, but for a specific societal purpose. The late cremations at Pessinus may have been used in a similar way.

Coins

Another common, but by no means obligatory, pagan Greek practice, was to place one or more coins in the grave. In some cases, the coins were placed in the mouth of the dead, a circumstance in which the coin has often been interpreted as payment to the ferryman Charon, who brought the dead souls across the river Styx to the Underworld, to be joined with the ancestors. The rite, often referred to as 'Charon's obol', was also widely practised in the Roman West even though Charon was not originally part of the Roman religion, and the rite may there have had another religious origin (Stevens 1991, 228), and at times perhaps also another meaning. It has, for example, been suggested that coins found in graves in and around Rome could have been given to the deceased as protection against metaphysical or transcendental forces on their afterlife journey (Ceci 2001, 90–1; on the journey cf., for example, Brandt 2015, 125, 130, 136, 149), alternatively as a provision for the journey or as an offering to the dead.

Coins have been found in many graves in the excavations in the North-East Necropolis, but there seems to be a stronger presence of them in the find material from the 4th and early 5th centuries, than from previous centuries (Indgjerd 2014, 61, fig. 15). It is not known if these coins were placed in the mouth of the dead or otherwise placed in the burials (Indgjerd 2014, 61–2). Two coins from the 4th century found with the cover of cist grave C521 suggest rather that, as a specific part of the burial ceremony, they could have been placed in the grave at the end of the interment and not given specifically to the dead. Cist grave C721, close by the previous one, contained three coins from the 4th and early 5th centuries. Finds from necropoleis in the neighbourhood of Hierapolis (as at Buldan and Laodikeia) confirm that placing coins in graves was common in the late Roman/early Byzantine periods (Ceylan 2000, 78; Şimşek 2011, *passim*). An increase in the amount of coins per burial has also been noted in other areas of the

Mediterranean from the 3rd century AD onwards (Stevens 1991, 225–6). Depositions of multiple coins occurred also in early Christian graves in Greece (Laskaris 2000, 321–3) and in Rome, where Pope Gaius (AD 283–296) is reported to have brought with him to the grave three contemporary Diocletian coins (Migotti 1997, 93, quoting Leclercq 1924, 2185, note 5, who refers to Aringhi 1651, *Roma subterranea*, 1. IV, c. xlviii, t. ii, p. 4263).

The rite of placing coins in burials was still practised and connected with the ferryman Charon in the late 4th-century Athens (Alföldy-Gazdac and Gazdac 2013, 308). The Christian rite of the viaticum, the placing of the eucharist in the mouth of the dead, partially supplanted the rite of the Charon's obol in early Christianity. While the term itself meant provision for the journey (Rush 1941, 91; Grabka 1953, 19; Stevens 1991, 220–1) and the rite still very much resembled the pagan rite in the way it was performed, its meaning had been adapted to the Christian ideas (Grabka 1953, 20–1; Paxton 1996, 32–4). In Christian popular belief the journey of the soul after death could be full of dangers, and provisions for the dead were considered to be necessary (Grabka 1953, 23–6). The coins, just as cross amulets or the eucharist, could help protect the soul on this journey.

The rite was also extensively practised in the mid- and late Byzantine period (Ivison 1993, 216–9; Laskaris 2000, 321–4) just as in the Medieval West (Travaini 2004; Schulze-Dörrlamm 2010, 363–7). At that time the coins in Eastern Mediterranean graves may mostly have served as apotropaic amulets (Ivison 1993, 219–21; Lightfoot, Ivison, *et al.* 2001, 378). Two coins from the 10th century were found accompanied by other mid-Byzantine grave goods in tomb C92, and a coin from the 11th century came to light in neighbouring tomb C311. All three depict a bust of Christ, making their function as icon amulets most likely.

The increased use of coins in graves especially in the 4th and 5th centuries AD in Hierapolis, but also other places (see, for example, Stevens 1991, 224), requires some attention (Table 12.1):[7]

Table 12.1. Hierapolis. Numerical distribution of dated coins according to centuries from the 3rd century BC to the 7th century AD found in respectively tombs in the North Necropolis, as sporadic finds in the North Necropolis, and in the North-East Necropolis. The numbers for the North Necropolis and sporadic finds are taken from Travaglini and Camilleri 2010, 227–9.

	Necropolis/centuries									
BC/AD	3	2	1	/1	2	3	4	5	6	7
North	0	0	1	4	2	3	12	6	3	0
Sporadic N	0	0	0	0	0	0	20	18	0	0
North-East	1	0	0	0	1	4	17	16	2	0
Total	1	0	1	4	3	7	49	40	5	0

A similar, but slightly less pronounced distribution can be observed at Laodikeia (Simsek 2011) (Table 12.2):

Table 12.2. Laodikeia. Numerical distribution of dated coins according to centuries from the 2nd century BC to the 7th century AD found in all tombs excavated and as sporadic finds.

	Finds/centuries								
BC/AD	2	1	/1	2	3	4	5	6	7
All tombs	2	19	13	10	15	36	42	0	19
Sporadic finds	1	1	2	1	3	11	0	0	0
Total	3	20	15	11	18	47	42	0	19

It may be argued that the increased number of coins in the graves is just a result of an augmented minting activity due to increased emissions in bronze by the state in addition to frequent redefinitions of weight standards, shape and value currency systems (Travaglini and Camilleri 2010, 30, quoting Burnett 1987, 131–9). At Hierapolis, for example, coins of the 4th and 5th centuries made up 89% of all coins identified for the period 1st–5th century AD (Travaglini and Camilleri 2010, 25, graphic 5). Since the coins from the necropoleis in Hierapolis are mainly found in large, monumental graves with many depositions over time, it is not possible to read these figures in relation to the total number of graves within each century. At nearby Laodikeia a large number of closed-context burials have been excavated and we may here see if the increase was the result of the number of graves containing coins and not only the result of an increase in numbers of coins in each grave (Table 12.3).

Table 12.3. Laodikeia. Numerical distribution according to centuries of the total number of graves excavated and graves containing coins, followed by the percentage distribution of graves containing coins. The graves are, in principle, organized according to the latest datable object according to the date given to each grave by Şimşek 2011, and the coins allocated to their latest dates.

	Graves/Centuries								
BC/AD	2	1	/1	2	3	4	5	6	7
Graves	6	16	149	87	24	16	31	4	2
Graves w/coins	2	2	15	4	9	14	17	4	2
% Graves/coins	(33)	13	10	5	38	88	55	–	–

The result implies that the high number of coins found in graves from the 3rd to the 5th centuries was the result of an increased use of this custom, not explicitly connected to increased minting activity. While in the preceding centuries the number of graves with coins oscillates between roughly 5% and 15% of the total number of registered graves in each century, in late Antiquity the percentage oscillates between roughly 40% and 90%. Could this increased use of coins in the grave be a reflection of a similar mental attitude as we saw in the use of the late cremations: in periods of change, cognitive conflicts and crises may lead to a reinvention and use of old traditions, giving them a new symbolic meaning? This twist of possible pagan meanings in the use of coins in graves, whether originally intended for Charon, or for other ideological, symbolic, or magical purposes in the world beyond death (cf. Stevens 1991, 228–9), meant that the Christians could easily turn the custom into their own ideology – as they did with the use of the word pollution discussed above (p. 205). However, as has already been observed, to neutralize the symbolic meaning of cremations, the Christians did not find a good answer.

In a transition period, when many, out of convenience, adopted the new religion and rituals, one can perhaps also envisage a third, temporary solution to explain the presence of coins in the graves. Taboos, old customs, and beliefs, classified as folk beliefs (or superstition) (Tarlow 2011, 156–90; 2015, 406–10), are connected to other parts of human rationality than new doctrines and rituals, and are difficult to uproot overnight. This may have caused many individual solutions, not written in decrees, dogmas, or textbooks. Death has always been surrounded with anxiety, both for the moment of death itself and for what happens to the soul afterwards. On the threshold of death some, in a transition period between the use of old traditions and the constitution of new doctrines, may have played a double game: even if buried as a Christian, in a simple Christian grave with a Christian ritual *ad sanctos*, their kin may have equipped the deceased with a coin for Charon (or for other pagan purposes) – just in case…

The many possibilities of interpretation of the uses of the coin in the graves make it difficult to understand the full meaning of this habit in each individual case, but that the coin played an important role in the transition of burial rituals from pagan to Christian, may well be deduced from the sudden reinstitution in the late Roman period of an old custom, which in the Imperial period had often been ignored.

Tombs in a time of recovery

In the mid-7th century Hierapolis was hit by a severe earthquake, which marked the end of the classical city (for archaeological circumstances and coins dating the event, see Travaglini and Camilleri 2010, 41–5). Small signs of economic and social changes (Zaccaria Ruggiu 2012, 423, 430–1) give the impression that the town, by that point, for some time had suffered hardships which made it difficult to mobilize the necessary resources to recover, as it had managed in the past after an earthquake catastrophe (Brandt 2016). The reasons for this fatal situation may have been many, but were not limited to Hierapolis alone: from the second half of the 6th century many classical cities were reduced in size, moved, or they simply perished. The answer may perhaps also be found in the 6th century, when the

Byzantine Empire was ravaged by a concentrated series of high-intensity earthquakes over two generations, by repeated outbreaks of the Justinian plague, by a couple of serious short-term climatic changes, and then topped by the Persian wars in the early 7th century. Together these dramatic natural and human circumstances most likely started a demographic change resulting in a downward population spiral, which would have lasted for a few generations before it stopped and followed by many more before the Byzantine Empire reached its former level of manpower (Brandt 2016; this volume, xx). When the earthquake hit Hierapolis in the mid-7th century the town may thus already have been in an advanced stage of depopulation. Hierapolis was accordingly abandoned except perhaps for some scattered squatting activity (as suggested from the information in D'Andria *et al.* 2005–2006, 383–92; Zaccaria Ruggiu and Cottica 2007, 158–9; Zaccaria Ruggiu 2012, 430–1; Cottica 2012, 457–9, 465).

When a new settlement, or a web of smaller nucleated settlements, started to grow in Hierapolis in the late 8th century AD a completely new urban structure grew out of the ruins of the classical city (on the development of the new town, see Arthur 2008, 87; 2012; Arthur and Bruno 2007; Arthur *et al.* 2012, esp. 580–1), and new modes of burial developed with the new settlement(s). By now inhumation was the only burial practice and simple cist graves (without preserved names) were the dominant grave type, the new modes in Hierapolis were principally connected to the burial context, i.e. the location of the cemeteries. They can at present be listed (with some very approximate dates) as follows:

1. Cemeteries with chapels in old, abandoned church buildings: the cathedral (for recent literature, see Ciotta and Palmucci Quaglino 2002; Arthur 2006; 2012, 268, 280; Arthur and Bruno 2007, 520–3; Peirano 2006; 2008; 2012; Romeo 2007), the church above the theatre (Gullino 2002; Arthur 2006, 152–4; 2008; 2012), and the church of St Philip (D'Andria 2011–2012; Caggia 2014a; 2014b): 9th–12th centuries.
2. A cemetery in previous church cemeteries: the cemetery of the Martyrion (Ahrens and Brandt 2016, 399): 9th–11th centuries(?).
3. Burials in previous monumental Roman tombs: North-East Necropolis (Ahrens and Brandt 2016, 403, 405–9): 8th/9th–13th centuries.
4. A new cemetery around a new church: The church-cum-cemetery north-east of the North Agora (Arthur 2002, 219–21; 2006, 68–9, 118–24; 2008, 87; 2012, 290–2; Arthur and Bruno 2007, 523–8; Arthur *et al.* 2012, 573–6): 10th–11th centuries.
5. Burials in connection with new living quarters within ancient buildings: the old bath building at the bottom of the ceremonial stairs leading to the sanctuary of St Philip (Caggia and Caldarola 2012, 631–2; Caggia 2014a; 2014b, 153–4 and n. 33): 10th–11th centuries(?).

In a period of recovery, whether starting from a re-settlement ex-novo or from the expansion of a surviving scattered settlement, the creation of an identity and a legitimization of its existence would have been essential. The creation of identities lies more in the past than in the present; among the Christians, as it had also been among the pagans, the identity was created by reconnecting with ancient memories and heroes. In the present case, this meant that the saints and martyrs, who had once peopled the bygone town and ancient landscape, were brought back to life in a new light. Foremost among these was, of course, the apostle Philip, whose tomb was still visible, but also other, for us known (Cyriacus and Claudianus: Huttner 2013, 341) and unknown saints and martyrs may have played their role through the revival of their ancient, abandoned churches, whether these were restored to some of the lost splendour or only occupied by small chapels with accompanying burials.

The recreation of an identity was apparently closely linked with another necessity – the need for sacred grounds in which to bury the dead. Here the abandoned churches served a special function, as did possibly the graveyard created on the south-west side of the Martyrion of St Philip soon after it was built in the early 5th century AD. One skeleton, in grave C21, was buried there in the time period AD 895–1020 (according to a 2 sigma calibrated radiocarbon date), that is at the same time as when, in the Martyrion building, two small churches or chapels were active: one lay outside the building along the south-east wall, near the east corner, the other in one of the rooms of the south-western wall, close to the west corner. The two chapels collapsed in a new devastating earthquake by the end of the 10th century, never to be rebuilt.

In the same quake the church of St Philip, restored a couple of centuries previously, collapsed and soon after gave way to burials in the central nave, by now open to the sky, in the close vicinity of the tomb of the saint. A chapel was installed in the north side aisle. The burials in the central nave may well be referred to as a kind of *ad sanctos* ideology (Caggia 2014b, 151–2), even if some consider the concept not appropriate for such late burials (Marinis 2009, 158–9). The graves, in the case of a reborn old burial ideology, would not have been attracted by the physical relics of the saint (relics which had been moved already in the 6th century), rather by the memory of his life – and in which the changed occupation and function of the church moved from a place of sermons to one of burials was not so much to guarantee a safe journey of the dead to the world beyond awaiting resurrection, but to retain the memory of the saint in order to maintain his protective powers both in life and in death. It shall, however, be noted that the graves in the nave were few, only six in number, and marked more clearly than other contemporary tombs. One may, therefore, ask whether the interments here, and in the other two churches with chapel and graves (group 1 above), were not in use for common

people, rather for a select group of the society, whether connected to the social position of the individuals interred or to emerging potent families creating a family sepulchre (Ousterhout 2010, 92; Moore 2013, 84–6) – in the house of their appropriated patron saint or martyr.

As for the graveyard of the Martyrion (Figs. 12.2; 12.13), probably more open and easily accessible for the common people than abandoned churches, the excavations have given no answer to what happened in this mid-Byzantine period. It may still have continued to be used (even without serving chapels), in the same way as the nearby Roman tombs of Attalos (C91) and Eutyches (C92) (Fig. 12.14). In a sampling of 15 skeletons found in both graves for radiocarbon dating (with a 2 sigma calibration) the burials range in date, rather evenly distributed, from the mid-8th to the very late 13th century, dates which comply well with objects found in the tombs. To this shall also be added cist grave C512 from the artificial 4th-century earthen terrace covering tomb C91 (see above, p. 202 and Figs. 12.4, 12.12), a grave radiocarbon-dated to AD 695–970 (Ahrens and Brandt 2016, 412). These tombs may well have been considered part of the Martyrion graveyard, but they could even have been considered as an independent burial area. Here was neither a new nor an old sacred building, which could have defined the surrounding land as sacred. However, the Roman sarcophagus (C310) standing in front of Eutyches' tomb to the east, carrying a standard Roman formula inscription, ends with the two Greek letters Alpha and Omega, a clear reference to its once Christian pedigree (Figs. 12.4; 12.10). If here was once a Christian burial, the ground may have been considered sacred by the mid-Byzantines, and the abandoned tomb buildings close by would thus be well adapted to receive more burials.

Who were buried in the tombs of Attalos and Eutyches in mid-Byzantine times? Analysis of strontium isotopes in their skeletal remains gives some interesting results worth a short consideration (Wong *et al.*, this volume). Strontium isotopes in human bodies reflect the level of strontium content in the soil from which humans and animals take their food. The strontium isotope data of a local population will therefore give similar values; values that differ from these ought to belong to an originally non-local population. According to the analyses the late Roman/early Byzantine population has very stable strontium isotope values; in the mid-Byzantine population the values are very unbalanced (see Wong *et al.*, this volume Fig. 14.4). In the tomb of Eutyches were found five pilgrim badges from three different monasteries in France and one from Rome datable to the late 13th century (Ahrens 2011–2012). The obvious question to ask is whether Eutyches' tomb, for the whole or only for some of its later active sepulchre period, was used as a burial place mainly for non-locals, and in case for non-locals of a large variety of origins? Could some of the burials even have been for 13th-century Turks for a period settling in Hierapolis (Arthur 2012, 297–8)? Such questions,

Fig. 12.14. Hierapolis, North-East Necropolis. Eutyches' tomb (C92): Mid-Byzantine, non-articulated depositions in the space between the benches of the tomb (context B7); north t.l. (archive of the Oslo University Excavations at Hierapolis).

however, can only be answered more confidently when we have substantial isotope data from other burial grounds in the mid-Byzantine Hierapolis.

What characterizes the three burial locations so far discussed is a search for an identity through the use of ancient monuments and a legitimization of their use as sacred grounds for burials. In the church-cum-cemetery (group 4) near the North Agora we find an *ad hoc* solution to a possibly expanding burial problem with no ties to the past, instead it reveals a future solution. If Hierapolis had continued to grow in the mid- and late Byzantine times, the town would have needed to create more burial grounds, grounds which would have had to find their location together with a new church/chapel independent of the town's historical past. The legitimization of the new ground and its identity would then most likely be connected to other saints and martyrs with no historical roots in Hierapolis.

Concluding remarks

In the present presentation, in three stages, covering a period of nearly 1,300 years, an attempt has been made to write the history of entombments at Hierapolis from a view of changes in funerary practices and mental processes.

In the first stage, in the Imperial period, we could observe how part of the population built their tombs away from the busy roads leading into the town on the peaceful hills behind with a magnificent view of the town and the surrounding landscape. The view of the hometown may have played a particular role in the choice of these remote necropoleis. It may further reflect the Greeks' and Romans' taste for the countryside and for villa life, as witnessed in the contemporary pastoral literature. In parallel with this development, the display of wealth and expressions of self-representation and euergetism moved from the monumental tombs to the public space of the town. However, for both areas of burials, along the roads or on the hills, we can observe a repetitive use of standardized monumental tombs and stone sarcophagi placed in family precincts. This mediates an idea of a society in harmony and stability built on a strong family organization, and so do, despite their expressions of anxiety and social preoccupations, the repetitive tomb inscriptions following the same formulas for centuries. This harmony and stability seem to have been broken with the introduction of the Christian religion. This, in the second stage, as we have tried to communicate, due to a changed burial ideology, caused conflicts and counter reactions.

The conflicting burial ideologies can in short be described as a development from pagan, large, monumental, strictly protected family tombs in privately owned, extensive extra mural areas to Christian, simple, individual graves in small defined hallowed areas dedicated to the physical presence of a dead saint or martyr (i.e. burials *ad sanctos*) – outside the town walls. At Hierapolis no late Roman/early Byzantine tomb has yet been found inside the walls. It is a change in which gradually the old family burial traditions cede to burials bonded by a Christian, religious content, but till the end of Hierapolis as a classical city, destroyed in the earthquake in the mid-7th century AD, both burial solutions lived side by side, at times contesting, at times overlapping and supporting each other. In this development from pagan to Christian burials, some pagans, in periods of ideological tensions, may also have 'reinvented' old burial customs, like cremation and the deposit of coins in the graves, as a means to demonstrate their anti-Christian sentiments. The Christians answered by turning the custom of the coins into a burial habit of their own, but appear not to have had a similar answer to the cremation rite.

After the abandonment of the town in the mid-7th century some four to five generations may well have passed before a slow resurrection of the ancient settlement started, possibly in the late 8th century. In this third stage the new settlement needed an identity and a legitimization of its existence, which it tried to do through new burials by establishing a link with the past through the appropriation of old Christian burial grounds and abandoned churches, in the last case, perhaps, occupied by burials of emerging, leading families. This interest in the past can perhaps be read as part of a renewed *ad sanctos* mentality, not based on the physical presence of a dead saint or martyr, but on their memory. The results of the analysis of some strontium isotopes begs the question of whether certain burial modes were reserved for non-locals, like pilgrims, tradesmen, or nomads. As in the pagan world the mid-Byzantine community tied their present life to the past, in this way finding an important social stabilizer in the, by definition, conservative agricultural society. The church-cum-cemetery by the North Agora gave a signal

towards a future, which never materialized due to external political events over which the small town had no control.

The moral of this story is threefold. By looking at the burial practices through the lens of mental processes we have tried to unveil how burial ideologies and belief systems shifted over time, thus revealing more the content of funerary actions than their form (Brandt 2012, ix).

Secondly, among colleagues in the field we have often heard complaints about the chronological compartmentalization of Medieval studies where short-term changes are given much attention, but the long-term ones disappear between one specialized time period and another (Brandt 2012, x). People did not live in time periods, they lived in a continuous cultural stream of actions and events where changes were continuously created and supplanted in periods overlapping each other. In order to understand Ancient and Medieval man we need to look at society as a long continuum, not as stopping places of a train moving from one time period to another. Here may lie one challenge for future studies of Roman and Medieval burial practices, both in Asia Minor and beyond.

Thirdly, it has for long been an accepted 'truth' that cremation, as a burial practice, went out of use in the Mediterranean region in the course of the 3rd century AD, even if examples of later use have been forwarded. The dates established for late cremations at Hierapolis leave open a question as to what criteria have been used for the dating of cremations in the past, on accompanying finds, or on the notion that cremations cannot be late. What is needed in the future is a large-scale study of radiocarbon dates from cremated bodies in every necropolis where these are registered, especially in necropoleis in which burials cannot be dated via other contextual data. We shall then certainly get a different picture of the spread of cremations in time than that which we have today.

Notes

1   Translation: Silvas 2007, 224.
2   Translation by C. Pharr 1952.
3   The dates were provided by Beta Analytic Limited (Miami, Florida).
4   Could there also have been late cremations in the North Necropolis at Hierapolis? See De Bernardi Ferrero 1996, 91 on tomb 1, in which apparently one of four funerary areas inside the tomb was excavated: 'The excavation has brought to light bones belonging to successive disturbed burials, many of which were burnt, as well as ceramic fragments of a wash-basin for domestic use, dating to the Vth–VIth centuries A.D.'? A search in the Hierapolis storerooms confirm both the presence of cremated bones as well as the date to the 5th–6th centuries of the ceramics found in the tomb. In a recent architectural study Filippo Masino and Giorgio Sobrà suggest a date of the tomb to the 4th century (personal communication September 2015).

5   For a late cremation from the necropolis under the church
    of St Peter in Rome, see Sinn 1987, 53, 265–6 no. 714, a
    cinerary marble urn containing, together with the cremated
    bones, a Constantinian coin issued shortly after AD 317;
    the necropolis under St Peter's was abandoned when the
    church was built starting *c*. AD 322, giving an *ante quem*
    date for the urn (we are grateful to Prof. Paolo Liverani for
    this information).
6   The text is composed of two more or less parallel stories;
    'Extrait des voyages du saint et glorieux apôtre Philippe
    depuis l'Acte quinziéme, dans lesquels figure le Martyre' and
    'Martyre du saint et glorieux apôtre Philippe, digne de toute
    louange'.
7   In this and the following tables the coins are allocated
    according to their latest date.

Bibliography

Achim, I. (2015) Churches and graves of the Early Byzantine
    period in *Scythia Minor* and *Moesia Scunda*: The development
    of a Christian topography at the periphery of the Roman Empire.
    In Brandt, Prusac, and Roland (eds.), 287–342.
Ackermann, J. A. (1990) *The villa. Form and ideology of country
    houses*. Princeton, Princeton University Press.
Ahrens, S. (2011) Hierapolis: graver med utsikt. *Klassisk Forum*
    2011.2, 98–108 (expanded English version forthcoming).
Ahrens, S. (2011–2012) Appendice 3. A set of western European
    pilgrim badges from Hierapolis. In D'Andria, 67–75.
Ahrens, S. (2015a) 'Whether by decay or fire consumed...':
    Cremation in Hellenistic and Roman Asia Minor. In Brandt,
    Prusac, and Roland (eds.), 185–222.
Ahrens, S. (2015b) The North-East Necropolis of Hierapolis. In
    Scardozzi (ed.), 77–79.
Ahrens, S. and Brandt, J. R. (2016) Excavations in the North-East
    Necropolis of Hierapolis 2007–2011. In D'Andria, Caggia, and
    Ismaelli (eds.), 395–414.
Alföldy-Gazdac, A. and Gazdac, C. (2013) 'Who pays the
    ferryman?' The testimony of ancient sources on the myth of
    Charon. *Klio* 95, 285–314.
Amsler, F. (1996) The apostle Philip, the viper, the leopard and the
    kid. The masked actors of a religious conflict in Hierapolis of
    Phrygia (Acts of Philip VII–XV and Martyrdom). In *Society of
    Biblical Literature, 1996 Seminar Papers* 35, 432–7. Atlanta,
    Scholars Press.
Anderson, T. (2007) Preliminary osteo-archaeological investigation
    in the North Necropolis. In D'Andria and Caggia (eds.),
    473–93.
Arthur, P. (2002) Hierapolis tra Bisanzio e i turchi. In De Bernardi
    Ferrero (ed.), 217–31.
Arthur, P. (2006) *Byzantine and Turkish Hierapolis (Pamukkale).
    An archaeological guide*. Istanbul, Ege Yayınları.
Arthur, P. (2008) Abitazioni medievali (85); *Agora* Nord – Chiesa
    e cimitero medievale (87); Abitazioni byzantine; Impianto
    produttivo (99); Chiesa sopra il teatro (126). In D'Andria,
    Scardozzi, and Spanò (eds.).
Arthur, P. (2012) Hierapolis of Phrygia: The drawn-out demise
    of an Anatolian city. In N. Christie and A. Augenti (eds.)
    Vrbes Extinctae. Archaeologies of abandoned Classical towns,
    275–305. Farnham and Burlington (VT), Ashgate.

Arthur, P. and Bruno, B. (2007) Hierapolis di Frigia in età
    medioevale (scavi 2001–2003). In D'Andria and Caggia (eds.),
    511–29.
Arthur, P., Bruno, B., Imperiale, M. L., and Tinelli, M. (2012)
    Hierapolis bizantina e turca. In D'Andria, Caggia, and Ismaelli
    (eds.), 565–83.
Berges, D. and Nollé, J. (2000) *Tyana. Archäologisch-historische
    Untersuchungen zum südwestlichen Kappadokien* (Inschriften
    griechischer Städte aus Kleinasien 55.1). Bonn, Habelt.
Bouvon, F., Bouvier, B, and Amsler, F. (eds.) (1999) *Acta Philippi.
    Textus*. Turnhout, Brepols Publishers.
Brandt, J. R. (2012) From sacred space to holy places. The
    christianization of the Roman cityscape: Some reflections.
    Orizzonti. Rassegna di archeologia 13, 151–6.
Brandt, J.R. (2015) Passage to the Underworld. Continuity or
    change in Etruscan funerary ideology and practices (6th–2nd
    centuries BC)? In Brandt, Prusac, and Roland (eds.), 105–83.
Brandt, J. R. (2016) Bysantinsk skjebnetid. Om jordskjelv, pest og
    klimatiske endringer i Anatolia i tidlig bysantinsk tid (300–800
    e.Kr.). *Nicolay arkeologisk tidsskrift* 127, 40–46 (expanded
    English version forthcoming).
Brandt, J. R. (this volume) Introduction. Dead bodies – Live data:
    Some reflections from the sideline, xvi–xxiv.
Brandt, J. R., Prusac, M., and Roland H. (eds.) (2015) *Death and
    changing rituals. Function and meaning in ancient funerary
    practices*. Oxford, Oxbow Books.
Bremmer, J. N. (2014) Religious violence between Greeks,
    Romans, Christians and Jews. In A. C. Geljon and R. Roukema
    (eds.) *Violence in ancient Christianity. Victims and perpetrators*
    (Supplements to Vigiliae Christianae 125), 8–30. Leiden and
    Boston, Brill.
Burnett, A. (1987) *Coinage in the Roman world*. London, Seaby.
Caggia, M. P. (2012) Recenti ricerche sulla collina di San Filippo
    a Hierapolis di Frigia. L'edificio termale di età bizantina. In
    R. D'Andria and K. Mannino (eds.) *Gli allievi raccontano*
    (Atti dell'incontro di studio per i trent'anni della Scuola di
    Specializzazione in Beni Archeologici-Università del Salento),
    171–92. Galatina, Congedo editore.
Caggia, M. P. (2014a) Phrygia Hierapolis'I aziz Philippus tepesi:
    Bisans ve Selçuklu yerleşimleri üzerine gözlemler (IX.–XIV
    yüzyıllar). In D'Andria (ed.) 191–216.
Caggia, M. P. (2014b) La collina di San Filippo a Hierapolis
    di Frigia: osservazioni sulle fasi di occupazione bizantina e
    selgiuchide (IX–XIV sec.). In F. Guizzi (ed.) *Fra il Meandro
    e il Lico. Archeologia e storia in un paesaggio anatolico*.
    Giornata di studio. Sapienza Università di Roma, 30 Marzo
    2012 (Scienze dell'antichita 20.2), 143–61. Rome, Edizioni
    Quasar.
Caggia, M. P. and Caldarola, R. (2012) La collina e il ponte di
    S. Filippo. In D'Andria, Caggia, and Ismaelli (eds.), 601–36.
Ceci, F. (2001) L'interpretazione di monete e chiodi in contesti
    funerari: esempi dal suburbio romano. In M. Heinzelmann,
    J. Ortalli, and P. Fasold (eds.) *Römischer Bestattungsbrauch und
    Beigabensitten in Rom, Norditalien und den Nordwestprovinzen
    von der späten Republik bis in die Kaiserzeit = Culto dei morti
    e costumi funerari romani Roma, Italia settentrionale e province
    nord-occidentali dalla tarda Repubblica all'età imperiale*
    (Internationales Kolloquium, Rom 1.–3. April 1998) Palilia 8,
    87–97. Wiesbaden, Dr Ludwig Reichert Verlag

Ceylan A. (2000) Scavi a Tripolis e nella necropoli di Buldan. In F. D'Andria and F. Silvestrelli (eds.) *Ricerche archeologiche turche nella Valle del Lykos*, 69–88. Galatina, Congedo Editore.

Ciotta, G. and Plamucci Quaglino, L. (2002) La Cattedrale di Hierapolis. In De Bernardi Ferrero (ed), 179–201.

Cottica, D. (2012) Nuovi dati sulle produzioni ceramiche dall'insula 104: i contesti della Casa dell'Inscrizione Dipinta (VII–IX sec. d.C.). In D'Andria, Caggia, and Ismaelli (eds.), 453–68.

D'Andria, F. (2003) *Guida Archeologica. Hierapolis di Frigia (Pamukkale)*. Istanbul, Ege Yayınları.

D'Andria, F. (2007) Le attività della MAIER – Missione Archeologica Italiana a Hierapolis 2000–2003. In D'Andria and Caggia (eds.), 1–46.

D'Andria, F. (2008) Martyrion di San Flilippo (95). In D'Andria, Scardozzi, and Spanò (eds.).

D'Andria, F. (2011) Conversion, crucifixion and celebration. St. Philip's Martyrium at Hierapolis. *Biblical Archaeology Review* 37, 34–46.

D'Andria, F. (2011–2012) Il Santuario e la Tomba dell'apostolo Filippo a Hierapolis di Frigia. *Rendiconti della Pontificia Accademia Romana di Archeologia* 84, 1–75 (including three appendices).

D'Andria, F. (2013) Il *Plutonion* a Hierapolis di Frigia. *Istanbuler Mitteilungen* 63, 157–217.

D'Andria, F. (2014) Havari Philippus'un Phrygia Hierapolis'inde yer alan kutsal alanı ve mezari. In D'Andria (ed.), 101–59.

D'Andria, F. (ed.) (2014) *Cehennem'den cennet'e Hierapolis (Pamukkale). Ploutonion. Aziz Philippus'un mezari ve kutsal alanı*. Istanbul, Ege Yayınları.

D'Andria, F. (this volume) The Sanctuary of St Philip in Hierapolis and the tombs of saints in Anatolian cities, 3–18.

D'Andria, F. and Caggia, M. P. (eds.) (2007) *Hierapolis di Frigia I. Le attività delle campagne di scavo e restauro 2000–2003*. Istanbul, Ege Yayınları.

D'Andria, F., Caggia, M. P., and Ismaelli, T. (eds.) (2012) *Hierapolis di Frigia V. Le attività delle campagne di scavo e restauro 2004–2006*. Istanbul, Ege Yayınları.

D'Andria, F., Caggia, M. P., and Ismaelli, T. (eds.) (2016) *Hierapolis di Frigia VIII.1. Le attività delle campagne di scavo e restauro 2007–2011*. Istanbul, Ege Yayınları.

D'Andria, F. and Gümgüm, G. (2010) Archaeology and architecture in the Martyrion of Hierapolis. In S. Aybek and A. Kazım Ös (eds.) *The land of crossroads. Essays in honour of Recep Meriç* (Metropolis Ionia II), 95–104. Istanbul, Homer Kitabevi Yayınları.

D'Andria, F., Scardozzi, G., and Spanò, A. (eds.) (2008) *Hierapolis di Frigia II. Atlante di Hierapolis di Frigia*. Istanbul, Ege Yayınları.

D'Andria, F., Zaccaria Ruggiu, A., Ritti, T., Bazzana, G. B., and Cacitti, R. (2005–2006) L'iscrizione dipinta con la Preghiera di Manasse a Hierapolis di Frigia. In *Rendiconti della Pontificia Accademia Romana di Archeologia* 78, 348–449.

D'Andria, R. 2012 Il *Martyrion* di San Filippo. Saggi di scavo. In D'Andria, Caggia, and Ismaelli (eds), 635–41.

De Bernardi Ferrero, D. (1996) Excavations and restorations in Hierapolis during 1995. *Kazı Sonuçlari Toplantısı* 18.2, 85–93.

De Bernardi Ferrero, D. (ed.) (2002) *Saggi in onore di Paolo Verzone* (Hierapolis. Scavi e Ricerche 4). Rome, Giorgio Bretschneider Editore.

Fox, S. C. (2012) The burial customs of early Christian Cyprus. A bioarchaeological approach. In M. A. Perry (ed.) *Bioarchaeology and behaviour. The people of the Ancient Near East*, 60–79. Gainesville *et al.*, University Press of Florida.

Gardner, E. A. (1885) Inscriptions copied by Cockerell in Greece II. *Journal of Hellenic Studies* 6, 340–63.

Gonzalez Villaescusa, R. and Lerma, V. (1996) Cristianismo y ciudad, los cemeteries in ambitus murorum. In H. Galinié and E. Zadora-Rio (eds.) *Archéologie di cimetère chrétien. Actes du 2è colloque A.R.C.H.E.A (Orléans, 29 septembre–1er octobre 1994)* (Revue Archéologique du Centre de la France, Supplément 11), 37–44. Tours.

Grabka, G. (1953) Christian viaticum: A study of its cultural background. *Traditio* 9, 1–43.

Gullino, N. (2002) La basilica sopra il teatro. In De Bernardi Ferrero (ed.), 203–16.

Haug, A. (2003) *Die Stadt als Lebensraum. Eine kulturhistorische Analyse zum spätantiken Stadtleben in Norditalien* (Internationale Archäologie 85). Rahden/Westf., Verlag Marie Leidorf.

Hill, D. J. A., Lieng Andreadakis, L. T., and Ahrens, S. (2016) The North-East Necropolis survey 2007–2011: Methods, preliminary results, and representivity. In D'Andria, Caggia, and Ismaelli (eds.), 109–120.

Huttner, U. (2013) *Early Christianity in the Lycus valley* (Ancient Judaism and Early Christianity 85; Early Christianity in Asia Minor (ECAM) 1). Leiden and Boston, Brill.

Indgjerd, H. (2014) *The grave goods of Roman Hierapolis. An analysis of the finds from four multiple burial tombs*. MA diss., University of Oslo.

Ivison, E. A. (1993) *Mortuary practices in Byzantium: An archaeological contribution (c. 950–1453)*. PhD diss., University of Birmingham.

Ivison, E. A. (1996) Burial and urbanism at late Antique and Early Byzantine Corinth (c. AD 400–700). In N. Christie and S. T. Loseby (eds.) *Towns in transition. Urban evolution in late Antiquity and the Early Middle Ages*, 99–125. Aldershot, Scolar Press.

Jacobs, I. (2012) A tale of prosperity. Asia Minor in the Theodosian period. *Byzantion* 82, 113–64.

Jacobs, I. and Waelkens, M. (2014) Sagalassos in the Theodosian age. In I. Jacobs (ed.) *Production and prosperity in the Theodosian age* (Interdisciplinary Studies in Ancient Culture and Religion 14), 91–126. Leeuven and Walpole (MA), Peeters.

Judeich, W. (1898) Inschriften. In C. Humann, C. Cichorius, W. Judeich, and F. Winter, *Altertümer von Hierapolis* (Jahrbuch des Kaiserlich Deutschen Archäologischen Instituts, Ergänzungsheft 4), 67–181. Berlin, Reimer.

Kars, E. A. K. and Kars, H. (2002) *The degradation of bone as an indicator in the deterioration of the European archaeological heritage. Project ENV4-CT98-0712, internal report of the European Commission*. Brussels.

Kiesewetter, H. (1999) Spätbyzantinische Gräber bei der Quellhöhle in der Unterstadt von Troia/Ilion. *Studia Troica* 9, 411–35.

Kiesewetter, H. (this volume) Toothache, back pain, and fatal injuries: What skeletons reveal about life and death at Roman and Byzantine Hierapolis, 265–85.

Laforest, C, Castex, D., and Blaizot, F. (this volume) Tomb 163d in the North Necropolis of Hierapolis in Phrygia. An insight into the funerary gestures and practices of the Jewish Diaspora in Asia Minor in Late Antiquity and the Proto-Byzantine period, 69–84.

Laforest, C., Castex, D., D'Andria, F., and Ahrens, S. (2013) Accès et circulation dans les nécropoles antiques de Hiérapolis (Phrygie, Turquie): L'exemple de la tombe 163D (Nécropole Nord) et de la tombe C92 (Nécropole Est). In *Cahier des thèmes transversaux ArScAn* 11, 2011–2012, 143–50.

Lambrechts, P. (1969) Les fouilles de Pessinonte. La nécropole. *L'antiquité classique* 38, 121–46.

Laskaris, N. G. (2000) *Monuments funéraires paléochretiens (et byzantines) de Grèce*. Athens, Les Editions Historiques Stéfanos D. Basilopoulos.

Leclercq H. (1924) s.v. Anneaux. In F. Chabrol (ed.) *Dictionnaire d'archéologie chrétienne et de liturgie*, vol. I.2, 2185. Paris, Letouzey et Ané.

Leone, A. (2007a) *Changing townscapes in North Africa from Late Antiquity to the Arab conquest*. Bari, Edipuglia.

Leone, A. (2007b) Changing urban landscapes: Burials in North African cities. In Stone and Stirling (eds.), 164–203.

Lightfoot, C., Ivison, E. A., *et al.* (2001) The Amorium project: The 1998 excavation season. *Dumbarton Oaks Papers* 55, 371–99.

Marinis, V. (2009) Tombs and burials in the monastery tou Libos in Constantinople. *Dumbarton Oaks Papers* 63, 147–66.

Meneghini, R. and Santangeli Valenzani, R. (1995) Sepolture intramuranee a Roma tra V e VII secolo d. C. Aggiornamenti e considerazioni. *Archeologia Medievale* 22, 283–90.

Mielsch, H. (1987) *Die römische Villa. Architektur und Lebensform*, Munich, Verlag C. H. Beck.

Migotti, B. (1997) *Evidence for Christianity in Roman southern Pannonia (northern Croatia). A catalogue of finds and sites* (BAR International Series 684). Oxford, Archeopress.

Moore, J. P. (2007) The 'Mausoleum culture' of Africa proconsolaris. In Stone and Stirling (eds.), 75–109.

Moore, S.V. (2013) *A relational approach to mortuary practices within Medieval Byzantine Anatolia*. PhD dissertation, Newcastle University.

Oestigaard. T. (2015) Changing rituals and reinventing tradition: The burnt Viking ship at Myklebostad, Western Norway. In Brandt, Prusac, and Roland (eds.), 359–77.

Okunak, M. (2005) *Hierapolis kuzey nekropolü (159D nolu tümülüs) anıt mezar ve buluntuları*. Unpublished MA dissertation, Pamukkale University, Denizli.

Ousterhout, R. (2010) Remembering the dead in Byzantine Cappadocia: The architectural settings for commemoration. In *Transactions of the State Hermitage Museum LIII: Architecture of Byzantium and Kievan Rus from the 9th to the 12th centuries. Materials of the International Seminar November 17–21, 2009*, 87–98. St. Petersburg, The State Hermitage Publishers.

Paxton, F. S. (1996) *Christianizing death: The creation of a ritual process in Early Medieval Europe*. Ithaca (NY), Cornell University Press (paperback edition of original 1990 publication).

Peirano, D. (2006) La cattedrale di Hierapolis di Frigia: nuove acquisizioni sull'architettura. In F. De Filippi and A. Longhi (eds.) *Architettura e territorio. Internazionalizzazione e ricerca* (Quaderin del Dipartimento Casa-Città 1), 29–32. Turin, Politecnico.

Peirano, D. (2008) Cattedrale (101). In D'Andria, Scardozzi, and Spanò (eds.).

Peirano, D. (2012) Scavi e ricerche nel cantiere del Duomo. In D'Andria, Caggia, and Ismaelli (eds.), 591–600.

Pennacchietti, F. A. (1963) Tre iscrizioni cristiane inedite di Hierapolis Frigia. *Rivista di Archeologia Cristiana*, 39, 131–7.

Pennacchietti, F. A. (1966–1967) Nuove iscrizioni di Hierapolis Frigia. *Atti della Accademia delle Scienze di Torino: II classe di scienze morali, storiche e filologiche* 101, 287–328.

Piepoli L. (2008) Sepolture urbane nell'Apulia tardoantica e altomedievale. Il caso di Herdonia. In G. Volpe and D. Leone (eds.) *Ordona XI. Ricerche archeologiche a Herdonia*, 579–94. Bari, Edipuglia.

Robert, J. and Robert, L. (1967) Bulletin épigraphique. *Revue des études grecques* 80, 453–573.

Romeo, E. (2007) Il progetto Catedrale (Regio III, insula 15, 20). Conoscenza, restauro, ipotesi di valorizzazione. In D'Andria and Caggia (eds.), 495–510.

Ronchetta, D. (2008) Necropoli Nord (59, 61, 65, 67, 71, 73, 75, 79, 81), Nord-Est (91, 93, 97), Est (107, 111, 129, 133), Sud-Est (139), Sud-Ovest (143, 147). In D'Andria, Scardozzi, and Spanò (eds.).

Ronchetta, D. (2012) Necropoli Nord. Indagini nell'area della Porta di Frontino. In D'Andria, Caggia, and Ismaelli (eds.), 495–512.

Ronchetta, D. (2015) Necropoli Nord (59, 61, 65, 67, 71, 73, 75, 79, 81), Nord-Ovest (71), Nord-Est (91, 93, 97), Est (107, 111, 129, 133), Sud-Est (139), Sud-Ovest (143, 147). In Scardozzi (ed.).

Ronchetta, D. (this volume) The South-East Necropolis of Hierapolis in Phrygia: Planning, typologies, and construction techniques, 39–68.

Ronchetta, D. and Mighetto, P. (2007) La Necropoli Nord. Verso il progetto di conoscenza: nuovi dati dalle campagne 2000–2003. In D'Andria and Caggia (eds.), 433–55.

Rush, A. C. (1941) *Death and burial in Christian antiquity*. Washington D.C., The Catholic University of America Press.

Saradi, H. (2008) The Christianization of pagan temples in the Greek hagiographical texts. In Hahn, J., Emmer, S., and Grotter, U. (eds.) *From temple to church. Destruction and renewal of local cultic topography in Late Antiquity* (Religions in the Graeco Roman World 63), 113–34. Leiden and Boston, Brill.

Scardozzi, G. (2006) L'urbanistica di Hierapolis di Frigia: Ricerche topografiche, immagini satellitari e fotografie aeree. In *Archeologia aerea. Studi di aerotopografia archeologica* 2, 83–134. Rome, Libreria dello stato. Istituto poligrafico e zecca dello stato.

Scardozzi, G. (2008) Mura bizantine (90, 93, 105, 107, 127, 129, 135, 139, 143). In D'Andria, Scardozzi, and Spanò (eds.).

Scardozzi, G. (2015) Mura bizantine (90, 93, 105, 107, 127, 129, 135, 139, 143). In Scardozzi (ed.).

Scardozzi G. (ed.) (2015) *Hierapolis di Frigia VII. Atlante di Hierapolis di Frigia*, second edition. Istanbul, Ege Yayınları.

Scheid, J. (2007) Körperbestattung und Verbrennungssitte aus der Sicht der schriftlichen Quellen. In *Körpergräber des 1. – 3. Jahrhunderts in der römischen Welt: internationales Kolloquium Frankfurt am Main, 19. – 20.11.2004* (Schriften des Archäologischen Museums Frankfurt 21), 19–25. Frankfurt, Archäologisches Museum.

Schneider Equini, E. (1972) La necropoli di Hierapolis di Frigia. Contributi allo studio dell'architettura funeraria di età romana in Asia Minore. *Monumenti Antichi. Serie Miscellanea* I.2 (48), 94–142. Rome, Accademia Nazionale dei Lincei.

Schörner, H. (2007) Sepulturae graecae intra urbem. *Untersuchungen zum Phänomen der intraurbanen Bestattungen bei den Griechen* (Boreas Beiheft 9). Möhnesee, Bibliopolis.

Schulze-Dörrlamm, M. (2010) Gräber mit Münzbeigabe im Karolingerreich. *Jahrbuch des Römisch-Germanischen Zentralmuseums Mainz* 57.1, 339–86.

Semeraro, G. (2007) Ricerche archeologiche nel Santuario di Apollo (Regio VII) 2001–2003. In D'Andria and Caggia (eds.), 169–209.

Semeraro, G. (2008) Santuario di Apollo. In D'Andria, Scardozzi, and Spanò (eds.), 117.

Silvas, A. M. (2007) *Gregory of Nyssa: The letters. Introduction, translation and commentary*. Leiden and Boston. Brill.

Şimşek, C. (ed.) (2000) *Laodikeia nekropolü (2004–2010 yılları). Katalog*. Istanbul, Ege Yayınları.

Sinn, F. (1987) *Stadtrömische Marmorurnen* (Beiträge zur Erschliessung hellenistischer und kaiserzeitlicher Skulptur und Arkitektur, vol. 8). Mainz am Rhein, Philip von Zabern.

Staebler, P. D. De (2008) Re-use of carved marble in the city wall. In R. R. R. Smith and J. L. Lenaghan (eds.) *Roman Portraits from Aphrodisias* (exhibition catalogue), 185–99. Istanbul, Yapı Kredi Yayınları.

Stevens, S. T. (1991) Charon's obol and other coins in ancient funerary practice. *Phoenix* 45, 215–29.

Stone, D. L. and Stirling, L. M. (eds.) (2007) *Mortuary landscapes of North Africa*. Toronto, University of Toronto Press.

Travaglini, A. and Camilleri, V. G. (2010) *Hierapolis di Frigia. Le monete. Campagne di scavo 1957–2004* (Hierapolis di Frigia 4). Istanbul, Ege Yayınları.

Tarlow, S. (2011) *Ritual, belief and the dead in early Modern Britain and Ireland*. Cambridge, Cambridge University Press.

Tarlow, S. (2015) Changing beliefs about the dead body in post-Medieval Britain and Ireland. In Brandt, Prusac, and Roland (eds.), 406–15.

Travaini, L. (2004) Saints and sinners: Coins in Medieval Italian graves. *The Numismatic Chronicle* 164, 159–81.

Trombley, F. R. (1993) *Hellenic religion and christianization c. 370–529* (Religions in the Graeco-Roman world, vo. 115/1, Leiden, New York, and Boston, E. J. Brill

Trombley, F. R. (2013) Christianity in Asia Minor: Observations on the epigraphy, In M. A. Sweeney, M. R. Salzman, and W. Adler (eds.) *The Cambridge History of Religions in the Ancient World*, 341–68. Cambridge, Cambridge University Press.

Vanhaverbeke, H. and Waelkens, M. (2002) The Northwestern Necropolis of Hierapolis (Phrygia). The chronological and topographical distribution of the travertine sarcophagi and their way of production. In De Bernardi Ferrero (ed.), 119–45.

Verzone, P. (1960) Il Martyrium ottagono a Hierapolis di Frigia. Relazione preliminare. *Palladio* N. S. 10, 1–20.

Verzone, P. (1961) Relation de l'activité de la mission archéologique Italienne de Hierapolis pour la campagne 1960. *Türk Arkeoloji Dergisi* 11.1, 35–6.

Verzone, P. (1961–1962) Le campagne 1960 e 1961 a Hierapolis di Frigia. *Annuario della Scuola archeologica di Atene e delle missioni italiane in Oriente* 39–40 (N.S. 23–24), 633–45.

Verzone, P. (1965) Le campagne 1962–1964 a Hierapolis di Frigia. *Annuario della Scuola archeologica di Atene e delle missioni italiane in Oriente* 41–42 (N.S. 25–26), 371–89.

Wong, M., Naumann, E., Jaouen, K., and Richards, M. (this volume) Isotopic investigations of human diet and mobility at the site of Hierapolis, Turkey, 228–36.

Yasin, A. M. (2009) *Saints and church spaces in the Late Antique Mediterranean architecture, cult and community*. Cambridge, Cambridge University Press.

Zaccaria Ruggiu, A. (2012) Un quartiere residenziale: l'insula 104. In D'Andria, Caggia, and Ismaelli (eds.), 419–42.

Zaccaria Ruggiu, A. and Cottica, D. (2007) Hierapolis di Frigia fra tarda antichità ed XI secolo: L'apporto dello studio degli spazi domestici nell'*Insula* 104. *Rivista di Archeologia* 31, 139–89.

Zuiderhoek, A. (2009) *The politics of munificence in the Roman empire: Citizens, elites and benefactors in Asia Minor*. Cambridge, Cambridge University Press.

Part II

From death to life: Demography, health, and living conditions

Analysis of DNA in human skeletal material from Hierapolis

Gro Bjørnstad and Erika Hagelberg

Abstract

The analysis of DNA from archaeological bones is a powerful tool to provide information on the origin and migrations of human populations. We analysed mitochondrial DNA from human skeletal remains from the archaeological site of Hierapolis (now Turkey) to shed light on the genetic affinities and relationships of the inhabitants during the Roman and Byzantine eras. Our results indicate that the skeletal remains are in a poor state of preservation and the original bone DNA is highly degraded, but we were able to generate a limited number of informative DNA sequences. We discuss the DNA results in the context of other ancient and present-day data from the area of former Roman Asia Minor.

Keywords: ancient DNA, human genetics, human evolution, mitochondrial DNA

Introduction

The ancient city of Hierapolis was founded in the 3rd century BC in classical Phrygia in south-western Anatolia. During the Byzantine era, Hierapolis became an important pilgrimage centre for the apostle Philip, whose tomb was recently excavated. His tomb had been incorporated into a church in the 5th century (D'Andria 2011–2012).

Hierapolis is surrounded by three large cemetery areas at the east, north, and south of the ancient city. In 2007, a team of archaeologists from the University of Oslo, Norway, started to excavate two reused Roman house tombs with saddle roof, tombs C92 and C91, in what is termed the North-East Necropolis, which is situated near the pilgrimage complex of St Philip (Fig. 13.1). The house tombs bore inscriptions with the name of the original owners, Eutyches and Attalos, respectively. Both tombs contained large numbers of human bones from the Roman and Byzantine eras; the Eutyches tomb contained a minimum of 91 individuals (Kiesewetter, this volume), and the Attalos tomb at least 29 (Henrike Kiesewetter and Helene Russ, personal communication). From the associated archaeological finds it was possible to deduce that the tombs had been used from the 2nd century onwards, and until at least AD 1100 in the case of the Attalos tomb, and approximately AD 1300 in case

of the Eutyches tomb (with a possible exception between the 7th and the 9th centuries). It is highly likely that both tombs were used to bury pilgrims, as a batch of five pilgrim badges from France and Italy were found in tomb C92, as well as a number of small bronze crosses in both tombs (Ahrens 2011–2012). The continual archaeological excavations at the site by teams from the Universities of Lecce and Bordeaux have uncovered several graves under the 5th-century church of St Philip. One of these graves was Roman, and older than the church itself, while others were made of reused church structures in the 11th and 12th centuries, when the church was finally abandoned. The University of Bordeaux archaeologists also excavated a Roman house tomb with saddle roof (T163d) in the so-called North Necropolis. This contained a large number of human bones, representing a minimum of 243 individuals (Laforest *et al.*, this volume). From the inscription at the front of the tomb it was clear it was the property of a Jewish family, and the archaeological finds indicated it was in use from the Augustan period until at least the 5th or 6th century AD.

In late 2010, a project was initiated to investigate the possibility of recovering DNA from the human skeletal remains in Hierapolis. DNA analysis of present-day and past populations represents a powerful tool for exploring

Fig. 13.1. Hierapolis; North-East Necropolis. View of house tombs C91 (bottom left corner) and (C92) (centre) towards the travertine cliffs and the fertile Lykos valley seen from north (archive of the Oslo University Excavations at Hierapolis).

population origins and diversity. DNA from the ancient inhabitants of Hierapolis contains, in principle, information on population structure, health and disease, and burial traditions that can be of great value to complement the information deduced by archaeologists and historians. It was also important to investigate the state of organic preservation of the skeletal remains and evaluate the possibility of undertaking successful genetic analyses.

Samples for DNA analysis were taken from skeletal material from the three above-mentioned house tombs in the North-East and North Necropoleis, as well as four graves under the church of St Philip. In the first instance, we proposed to analyze mitochondrial DNA markers, which represent the maternal lineages, to establish the distribution of these lineages in the city inhabitants during the Roman and Byzantine eras.

Ancient DNA analyses

Ancient DNA is the name given to research on DNA isolated from old biological materials, like skeletal remains from archaeological excavations and museum collections, a field of research which has only developed in the past few decades (Hagelberg *et al.* 2015). The most important concerns of ancient DNA researchers are the degradation of DNA over time, and the possibility of contamination by sources of modern DNA. Ancient DNA can be easily contaminated by DNA from the hands or saliva of people handling bones or performing the genetic analyses, from laboratory consumables that often contain tiny amounts of DNA, and from bacteria and fungi from soil surrounding the skeletal remains. Some of this contamination can interfere with or inhibit the DNA analyses but, more dangerously, it can lead to false results. Studies of human skeletal material can be hampered by the difficulties in distinguishing authentic ancient DNA from the DNA of the archaeologists or geneticists. Ancient DNA researchers adopt stringent

precautions to minimize the risk of contamination (for example, see Hofreiter *et al.* 2001), including physical separation of laboratory facilities for work on ancient and recent biological material, and the use of various controls to help detect contamination.

In past decades, the study of ancient DNA involved extraction of DNA from the source material (bone or another type of tissue), and subsequent analysis using a technique called the polymerase chain reaction, or PCR for short, a method that permits the amplification of a specific, informative piece of DNA from a source containing very little, or highly degraded DNA (Saiki *et al.* 1985). Afterwards, the sequence of the DNA piece would be 'read' using conventional DNA sequencing methods. The result is a series of letters of DNA that contain information pertaining to the individual. The problem with the PCR technique is that it amplifies whatever DNA is present, not just ancient DNA, and it may be impossible to distinguish whether a DNA sequence was original or introduced through contamination. In the case of very old or degraded materials, it was sometimes impossible to filter out the tiny amounts of human DNA from the environment, and the results of the analyses would have been a mixture of ancient and modern sequences, impossible to interpret.

Fortunately, new sequencing techniques, known as next generation sequencing, or NGS, are starting to help overcome some of the difficulties of ancient DNA research. A major advantage of NGS is that it can 'capture' and analyze a high number (hundreds or thousands) of amplified DNA copies from each sample. In ancient biological material, DNA degradation is expected, and the number of original DNA fragments may be extremely small, and potentially overwhelmed by fragments of modern DNA from contamination. The NGS method provides a large mixture of original, damaged ancient sequences, and intact sequences resulting from modern contamination. By scrutinizing the variation observed among the thousands of sequences captured for a given sample, it is possible to evaluate the proportion of ancient sequences (degraded) to modern, intact sequences, using something called a *c*-statistic (Helgason *et al.* 2009), and eventually reconstruct the putative original sequence. In this way, NGS technology is starting to overcome some of the past limitations of ancient DNA research. Unfortunately, this technology was not available to us at the beginning of our analyses, so we had to rely mostly on traditional PCR and sequencing technology.

Mitochondrial DNA

In our genetic analyses of the skeletal remains from Hierapolis we used a combination of traditional ancient DNA techniques, namely PCR and conventional DNA sequencing, and a particular DNA marker called mitochondrial DNA (mtDNA). This marker has been widely used in population and evolutionary studies since the 1980s, for example in the well-known 1987 study on *African Eve* (Cann *et al*. 1987). Animals, including humans, carry the genetic information in chromosomes in the nucleus of the cells, and the DNA of the two parents is shuffled from generation to generation. However, there is a small amount of extra-chromosomal DNA, mtDNA, present in the mitochondria, small structures involved in the energy metabolism of the cell. Scientists discovered that mtDNA appeared to be inherited through the maternal line, without the shuffling process of nuclear DNA, and it has become a favoured marker to study the evolution of maternal lineages.

Of particular value for ancient DNA research is the fact that each animal cell contains many thousand mitochondria, so old bones or museum specimens are likely to have useful amounts of mtDNA, even after hundreds or thousands of years. In short, mtDNA analysis offers a tool to recover informative DNA pieces from archaeological bones, as well as to investigate the spacial and temporal distribution of maternal genetic lineages.

MtDNA diversity worldwide has been well documented and large reference datasets on both living and past human populations are available to researchers. MtDNA sequences in humans exhibit small differences, giving rise to mtDNA variants, called haplotypes. Similar haplotypes are grouped together into haplogroups, denoted with capital letters (A, B, C, and so on), and these have a specific geographical distribution worldwide, reflecting the expansions and migrations of people through time (Fig. 13.2). In Europe and western Eurasia the haplogroups H, I, J, K, T, U, V,

W and X, all deriving from the ancestral N haplogroup, are found in a complex distribution with relatively little geographical structuring. Haplogroup H is by far the most common across Europe with a prevalence of 40–50% (Richards *et al*. 2000). Haplogroups C, D and G, deriving from the other major ancestral haplogroup M, are more prevalent in eastern Eurasia.

Y chromosome markers are the male counterpart to mtDNA, offering information on the paternal history, while autosomal markers (i.e. chromosomal markers) reflect the complex mosaic of the genetic history of both parents and their respective ancestors.

Bone sampling and preservation of the skeletal material

Preparations of bone samples for radiocarbon and DNA analyses

Ancient DNA researchers who work on human material have strict recommendations for avoiding contamination by modern human DNA. In extreme cases, archaeologists wear whole body protective suits during excavations. This was impractical in the hot conditions in Turkey. Instead, the archaeologists and osteologists wore disposable latex gloves during two seasons, 2011 and 2012. However, even this was impractical due to the high ambient temperatures (as high as 40°C+). Gloves alone do not protect the samples from contamination, as DNA is left on everything we touch and can be easily transferred between surfaces,

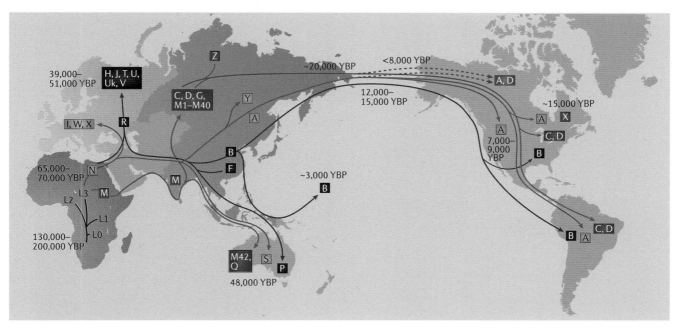

Fig. 13.2. Routes of human migrations and expansions based on mitochondrial DNA (mtDNA) variation. Capital letters represent the mtDNA haplogroups. The migration routes are based on worldwide haplogroup frequencies and the evolutionary relationship between lineages. Reprinted from Stewart and Chinnery (2015) by permission of Macmillan Publishers Ltd (copyright).

even by the gloved hands themselves. In the end, comfort prevailed, and the archaeologists and anthropologists stopped wearing gloves in the later seasons. It was left to the geneticist to decontaminate the bones in the laboratory and carry out proper controls to monitor potential sources of contamination.

After the bones had been examined and recorded by the osteologists, samples were removed for DNA, radiocarbon, and isotope analyses by cutting a slice of bone with a hacksaw, while using gloves, and cleaning the equipment with ethanol between samples. At the end of each excavation season, the samples were exported to the University of Oslo for further processing, with proper approval from the local excavation official and the museum in Hierapolis. Mostly this was a formality and occurred without difficulties, but during the 2012 season the local official only approved for export 28 randomly selected samples, from a total of 243. This was a major setback for the DNA analysis work. The main sampling of the skeletal material from house tombs C92 and C91 was during the 2012 and 2013 seasons, and all samples collected after the 2013 season were approved for export, but they arrived too late to be included in this report.

Skeletal preservation and radiocarbon dating

The preservation of skeletal material, including DNA and collagen, is influenced by temperature, bleaching by the sun, humidity, pH, leaching by water, burning, and microbial infestation (Burger *et al.* 1999). In general, low temperatures are favourable, and retrieval of DNA has been feasible for specimens aged up to 700,000 years given exceptional preservation in permafrost (Orlando *et al.* 2013). Also moderate temperatures, with little seasonal variation, are conducive to good bone preservation, as exemplified by numerous studies on skeletal material found in caves (e.g. Hofreiter *et al.* 2002). The house tombs of the North-East Necropolis of Hierapolis were exposed to high seasonal temperature fluctuations with hot summers and cold winters. The soil composition is probably not favourable for bone preservation, as witnessed by the poor macroscopic preservation of the skeletal material from the house tombs, particularly in the tile graves located close to house tombs C92 and C91. Most of the bones retrieved from the two house tombs were broken into small pieces (Fig. 13.3). The bones of house tomb C92 seemed better preserved, although only two skeletons were reasonably complete, with the

Fig. 13.3. Hierapolis; North-East Necropolis. Skeletal material in situ from Eutyches' tomb (C92) seen from west (archive of the Oslo University Excavations at Hierapolis).

cranium missing or moved for one of them. As mentioned above, the estimated minimum number of skeletons was 91, but only six complete adult femurs were found.

Twenty of the best-preserved bones from house tombs C92 and C91 were selected for radiocarbon dating, based on our subjective evaluation of thickness and appearance of the bones. The size of the bone pieces ranged from 2 to 6.3 g. Typically, 2 g of well-preserved bone yielded between 20 and 100 mg of collagen, adequate for a reliable radiocarbon dating. Dating was done by accelerator radiocarbon dating at Beta Analytic Limited (Miami, Florida). Ten of the bones (eight from C92 and two from C91) produced meaningful dates, which indicated that house tomb C92 had been used at least *c*. between AD 100 and 1200 (unpublished data). The remaining ten bones contained insufficient collagen for dating (0.2–1.0 mg).

DNA analyses of skeletal material from Hierapolis
Extraction, amplification and sequencing of DNA

Once in the ancient DNA laboratory at the University of Oslo, the necessary precautions were taken to avoid contamination by modern DNA. Appropriate protective clothing was worn in the laboratory. The pieces of bone were cleaned by removing the outer surface by means of sandblasting, followed by UV irradiation and the use of bleach to remove DNA from the surfaces of the bones. Appropriate controls were carried out during the various steps of the analyses, including DNA extractions and PCR amplifications.

DNA was extracted and amplified according to a previously published method (Malmström *et al.* 2009). A fragment of DNA 343 base pairs (bp) in length, corresponding to position 16050 to 16392 in the mtDNA reference sequence (Anderson *et al.* 1981) was targeted, using five overlapping fragments. Since ancient DNA is typically broken down into small pieces, the target fragments were short, between 120 and 150 bp, to ensure that even the degraded fragments could be amplified. After amplification, the DNA fragments were identified with a unique barcode, as described previously (Malmström *et al.* 2009), to permit the large piece to be reassembled after sequencing.

The success rate was low, consistent with the presumed poor state of preservation of the bones. Twenty-one samples were sequenced by conventional sequencing at the sequencing facility at the Institute of Biosciences, University of Oslo, and Macrogen (Netherlands). Sequencing of a further 23 samples was attempted at the NGS facility in Oslo. Eight samples were sequenced using both methods. The sequences were assigned to their respective haplogroups using methods described by Vincent Macaulay (stats.gla.ac.uk/~vincent/founder2000/motif.html), and using the mtDNAmanager (mtmanager.yonsei.ac.kr) and Genographic databases (genographic.nationalgeographic.com). The diversity of modern and ancient Anatolian populations was calculated according to Nei (1987).

Preliminary mtDNA results of Hierapolis

Of the 36 individual samples that were submitted for DNA sequencing, only two yielded DNA information in all five targeted fragments (combining the result of NGS and conventional sequencing). Twenty-five samples yielded DNA sequences for three or fewer fragments. The quality of the conventional sequences and the reproducibility and number of captured DNA copies in NGS technology were not good enough to be able to assign the haplogroup of the sequences with any degree of certainly.

The two samples with sequence information in all five targeted fragments were putatively assigned to haplogroups M1 and K respectively. The first was a piece of a cranium of skeleton 108 (DNA number 538, little skeletal information, possibly a female) from house tomb C91, which possessed mutations found in haplogroup M1. The associated archaeological finds indicated that the burials of this context were Byzantine (Kjetil Bortheim, personal communication), but the radiocarbon dating of material from this context failed. The other individual was an immature female, dated to 86–246 AD (skeleton 13, DNA number 41), from the Roman Jewish house tomb T163d, tentatively assigned to haplogroup K.

Several other samples had mtDNA variants that gave indications of haplogroup assignment. A late Byzantine skeleton of an adult male (aged above 30), buried in the crypt under the altar of the St Philip church was probably sub-haplogroup U2. Remains of a wooden board were found under his abdomen and he wore knee-high leather boots (Caroline Laforest, personal communication). From house tomb C92, three individuals also from Byzantine contexts were tentatively assigned to the J or T haplogroups, while another individual from the Roman Jewish house tomb T163d (skeleton 40/DNA number 57) was tentatively assigned to haplogroup M, possibly M1.

The lack of information of the partial sequences was a challenge, particularly for detecting H haplotypes, the most common haplogroup in Turkey and Europe today. No changes, or few changes compared to the reference sequence indicate that the haplotype belongs to haplogroup H. This means that partial sequences exhibiting no differences to the reference sequence were assigned to haplogroup H, in the absence of more complete information from longer DNA sequences.

The present-day inhabitants of Turkey carry a range of European mtDNA sequences, as well as some haplogroups derived from central Asia (Di Benedetto *et al.* 2001, Nasidze *et al.* 2004, Quintana-Murci *et al.* 2004, Schönberg *et al.* 2011). The genetic diversity is attributed to complex genetic exchanges and human movements, as signified by the geographic location of Anatolia at a crossroad between Europe and Asia. Many haplogroups common in Europe today originated in this region and their presence in Anatolia probably reflects this ancient diversity, overlaid with the results of more recent interactions with neighbouring regions. While we emphasize that our mtDNA data from Hierapolis are few in number and limited in scope, and that comparisons with mtDNA frequencies of other geographical areas must perforce be tentative, we can say with confidence that the partial haplogroups in our small Hierapolis sample are of western Eurasian origin (H, JT, K, U) and also found in present-day Turkey, as well as in archaeological bones from Ottoman Ephesos and late Byzantine Sagalassos (Table 13.1).

What do we know about these haplogroups? What we can say for certain is that in any population today we can find a large mixture of different haplogroups, but in certain instances they can give an indication of demography and origins. Some haplogroups are very ancient, but they may have been introduced into a region by recent migrants. Haplogroup U is one of the most ancient haplogroups in Europe, and has been observed in previous studies of ancient DNA, including the analysis of the remains of a 33,000-year-old hunter-gatherer from Central-South European Russia (Krause *et al.* 2010), and Mesolithic European hunter-gatherers from Germany (Bollongino *et al.* 2013), Karelia in Russia (Der Sarkissian *et al.* 2011), and Sweden (Lazaridis *et al.* 2014). The

steppes of Eastern Europe and Central Asia are probably the original geographic location from which dispersal happened during the Stone Age, and again during the Bronze Age. Haplogroup K originated in West Asia between 18,000 and 38,000 years ago and probably spread from the Near East to Europe with early farmers and herders. Haplogroup K was found in approximately 15% of Neolithic samples from Europe, a frequency twice as high as in modern Europeans. The mutation defining haplogroup J is thought to have taken place some 45,000 years ago, probably in West Asia. The mutation defining haplogroup T happened around 29,000 years ago, probably in the East Mediterranean region. Pala *et al.* (2012) suggested that some J and T lineages recolonized Europe from the Near East following the end of the last glaciation. We could speculate that the presence of several individuals possibly belonging to haplogroups J/T may be indicative of pilgrims originating in Europe, as these haplogroups are less frequent in present-day Turkey (Table 13.1).

Two of our Hierapolis individuals of Roman and Byzantine origin may be assigned to M-related haplogroups (haplogroups D, G, M). It should be emphasized that it is not necessarily the individuals themselves who have an exotic origin, but rather their maternal ancestors. M-type haplogroups can indicate eastern connections, although

Table 13.1. Frequency distribution of the mitochondrial DNA haplogroups of present day Turkey, late Byzantine Sagalassos, Türbe in the Artemision of Ottoman Ephesos and Harbour Necropolis of Roman Ephesos.

	Modern Turkey[1]	Late Byzantine Sagalassos[2]	Türbe in the Artemision, Ottoman Ephesos[3]	Harbour Necropolis, Roman Ephesos[4]	Hierapolis
	n=79	n=53	n=14	n=15	n=20
A	0.05 (4)	–	–	–	–
D	0.05 (4)	–	0.07 (1)	0.13 (2)	–
F	0.01 (1)	–	–	–	–
L/M	–	–	–	0.07 (1)	0.10 (2)
MGI	–	–	–	0.07 (1)	0.05 (1)
H	0.37 (29)	0.25 (13)	0.57 (8)	0.20 (3)	0.35 (7)
J/T	0.12 (10)	0.20 (11)	0.07 (1)	0.26 (4)	0.30 (6)
I	0.03 (2)	–	0.14 (2)	0.07 (1)	–
K	0.05 (4)	0.09 (5)	–	–	0.05 (1)
N	0.01 (1)	0.11 (6)	–	0.07 (1)	–
R	0.03 (2)	0.08 (4)	–	–	–
U	0.24 (19)	0.17 (9)	0.14 (2)	0.13 (2)	0.15 (3)
W	–	0.08 (4)	–	–	–
X	0.04 (3)	0.02 (1)	–	–	–

[1]Quintana-Murci *et al.* 2004; Schönberg *et al.* 2011. [2]Ottoni *et al.* 2011. [3]Bjørnstad 2015. [4]Bjørnstad in press.

haplogroup M1 is also found in northern Africa and the Horn of Africa, as well as in the east Mediterranean and Near East. Approximately 4–7% of the present-day population of Turkey has M-related mtDNA haplogroups (Quintana-Murci *et al.* 2004, Schönberg *et al.* 2011). These lineages were neither detected in Palaeolithic nor in Mesolithic Central Europeans (Haak *et al.* 2005, Bramanti 2008), but a similar lineage was found in Neolithic Hungary and possibly in Chalcolithic Spain (Guba *et al.* 2011, Gamba *et al.* 2012) and in North-East Europe 7500 BP (Der Sarkissian *et al.* 2013). M-derived lineages were not observed in an ancient DNA study on Byzantine Sagalassos (Ottoni *et al.* 2011), but haplotypes of the D haplogroup were found in skeletal material of Ottoman and Roman Ephesos (Bjørnstad 2015; in press). It is possible that haplogroup D was introduced to the Anatolian population by the Seljuk Turks or even later, as this haplogroup is now present in about 30% of living Turks (Comas *et al.* 2004).

The Lykos valley where Hierapolis is situated was partially occupied by the Seljuks in 1206 (Arthur 2011), but we cannot rule out that haplogroup D was already present in Anatolia before their arrival, brought in by earlier migrations of people. Moreover, the maritime trading networks of the eastern Mediterranean could have left permanent genetic traces of North African origin in the population of Hierapolis, as shown by the tentative M1-haplotypes. In this context it is worth mentioning that interactions between the inland city of Sagalassos and Egypt and the Levant have been documented through finds of ancient fish bones in Sagalassos, originating from northern Africa and the Near East (Arndt *et al.* 2003).

The diversity in mtDNA haplotypes is high in modern and ancient Anatolian samples (Table 13.2), while Roman Ephesians have the highest number of different haplotypes and Hierapolis has the lowest. The high diversity

in Roman Ephesos is consistent with a harbour city inhabited by peoples of different origins. The relatively low diversity in Hierapolis is probably the result of the comparatively poor quality of the DNA sequences, which limited the number of haplogroups which could be assigned securely. However, it is important to remember that all our ancient DNA data are limited; for example, the Ottoman Ephesos sample is based on 14 skeletons of a small cemetery (Türbe) located in the entrance area of the Artemision.

Future analyses of the skeletal material of Hierapolis

The poor macroscopic preservation of the skeletal material was a challenge for the human osteologists, as well as hampering analyses of stable isotope (Wong *et al.*, this volume) and DNA, and radiocarbon dating. The poor preservation of the bones led to a series of discouraging negative results of our DNA extraction, PCR amplification and sequencing experiments. The haplotype assignments from the skeletal remains of the mid-/late Byzantine pilgrim tomb complexes of the North-East Necropolis and the putative resident Roman and early Byzantine population of the North Necropolis of Hierapolis must perforce be tentative, and the data must be verified by further analyses, including the skeletal material recovered during the 2012 and 2013 archaeological seasons. Another promising avenue would be to use NGS to analyse high-resolution autosomal single nucleotide polymorphisms, a type of analysis that is increasingly preformed on ancient human remains, and which provides information on both maternal and paternal ancestors. Despite the poor preservation of the human skeletal material in Hierapolis, new techniques of skeletal analyses could circumvent some of the limitations posed by DNA degradation, but this must await future studies.

Table 13.2. Haplotype discriminance and haplotype diversity of modern and ancient populations of Turkey based the mitochondrial DNA HVRI region 16050–16365.

		Sample size	Haplotype number	Haplotype discriminance	Haplotype diversity
Turkey[1]	Modern	79	69	0.87	0.99
Sagalassos[2]	Byzantine	51	31	0.61	0.98
Ephesos, Türbe[3]	Ottoman	14	11	0.79	0.93
Ephesos, Harbour Necropolis[4]	Roman	15	14	0.93	0.99
Hierapolis	Roman/Byzantine	20	11	0.55	0.91

haplotype discriminance = number of haplotypes/sample size.

haplotype diversity, $h = \dfrac{(1-\Sigma x^2)n}{n-1}$ (Nei 1987), where n = sample size and x = frequency of each haplotype.

[1]Quintana-Murci *et al.* 2004; Schönberg *et al.* 2011. [2]Ottoni *et al.* 2011. [3]Bjørnstad 2015. [4]Bjørnstad, in press.

Bibliography

Ahrens, S. (2011–2012) Appendice 3. A set of western European pilgrim badges from Hierapolis. In D'Andria, 67–75.

Anderson, S., Bankier, A. T., Barrell, B. G., de Bruijn, M. H. L., Coulson, A. R., Drouin, J., Eperon, I. C., Nierlich D. P., Roe, B. A., Sanger, F., Schreier, P. H., Smith, A. J. H., Staden, R., and Young, I. G. (1981) Sequence and organization of the human mitochondrial genome. *Nature* 290, 457–65.

Arndt, A., Van Neer, W., Hellemans, B., Robben, J., Volckaert, F., and Waelkens, M. (2003) Roman trade relationships at Sagalassos (Turkey) elucidated by ancient DNA of fish remains. *Journal of Archaeological Science* 30, 1095–105.

Arthur, P. (2011) Hierapolis of Phrygia: The drawn-out demise of an Anatolian city. In N. Christie and A. Augenti (eds.) *Vrbes Extinctae. Archaeologies of abandoned Classical towns*, 275–305. Farnham and Burlington (VT), Ashgate.

Bjørnstad, G. (2015) Mitochondrial analyses suggest a maternal European signature of the skeletons from the Türbe. In Ladstätter, S. (ed.) *Die Türbe im Artemision. Ein frühosmanischer Grabbau in Ayasuluk/Selçuk und sein kulturhistorisches Umfeld* (Sonderschriften des Österreichischen Archäologischen Instituts 53), 489–93. Vienna, Österreichisches Archäologisches Institut.

Bjørnstad, G. (in press) Preliminary analyses of DNA in skeletal material from the harbour necropolis suggest Roman Ephesos as a meeting point for east, south, north and west. In M. Steskal (ed.) *Forschungen in der Hafennekropole von Ephesos. Die Grabungen seit 2005 im Bereich des Hafenkanals*. Sonderschriften des Österreichischen Archäologischen Institutes.

Bollongino, R., Nehlich, O., Richards, M. P., Orschiedt, J., Thomas, M. G., Sell, C. Fajkosová, Z. Powell, A., and Burger, J. (2013) 2000 years of parallel societies in Stone Age Central Europe. *Science* 342, 479–81.

Bramanti, B. (2008) Ancient DNA: Genetic analysis of aDNA from sixteen skeletons of the Vedrovice collection. *Anthropologie* 46, 153–60.

Burger, J., Hummel, S., Herrmann, B., and Henke, W. (1999) DNA preservation: A microsatellite-DNA study on ancient skeletal remains. *Electrophoresis* 20, 1722–8.

Cann, R. L., Stoneking, M., and Wilson, A. C. (1987) Mitochondrial DNA and human evolution. *Nature* 325, 31–6.

Comas, D., Plaza, S., Wells, R. S., Yuldaseva, N., Lao, O., Calafell, F., and Bertranpetit, J. (2004) Admixture, migrations, and dispersals in Central Asia: Evidence from maternal DNA lineages. *European Journal of Human Genetics* 12, 495–504.

D'Andria, F. (2011–2012) Il Santuario e la Tomba dell'apostolo Filippo a Hierapolis di Frigia. *Rendiconti della Pontificia Accademia Romana di Archeologia* 84, 1–75 (including three appendices).

Der Sarkissian, C. (2013) Ancient DNA reveals prehistoric gene-flow from Siberia in the complex human population history of North East Europe. *PLoS Genetics* 9, e1003296.

Di Benedetto, G., Ergüven, A., Stenico, M., Castrì, L., Bertorelle, G., Togan, I., and Barbujani, G. (2001) DNA diversity and population admixture in Anatolia. *American Journal of Physical Anthropology* 115, 144–56.

Gamba, C., Tirado, M., Deguilloux, M. F., Pemonge, M. H., Utrilla, P., Edo, M., Molist, M., Rasteiro, R., Chikhi, L., and Arroyo-Pardo, E. (2012) Ancient DNA from an Early Neolithic Iberian population supports a pioneer colonization by first farmers. *Molecular Ecology* 21, 45–56.

Guba, Z., Hadadi, E., Major, A., Furka, T., Juhász, E., Koós, J., Nagy, K., and Zeke, T. (2011) HVS-I polymorphism screening of ancient human mitochondrial DNA provides evidence for N9a discontinuity and East Asian haplogroups in the Neolithic Hungary. *Journal of Human Genetics* 56, 784–96.

Haak, W., Forster, P., Bramanti, B., Matsumura, S., Brandt, G., Tänzer, M., Villems, R., Renfrew, C., Gronenborn, D., Alt, K. W., and Burger, J. (2005) Ancient DNA from the first European farmers in 7500-Year-Old Neolithic sites. *Science* 310, 1016–18.

Hagelberg, E., Hofreiter, M., and Keyser, C. (2015) Ancient DNA: The first three decades. *Philosophical Transactions Royal Society B* 370, 20130371.

Helgason, A., Lalueza-Fox, C., Ghosh, S., Sigurðardóttir, S., Sampietro, M. L., Gigli, E., Baker, A., Bertranpetit, J., Árnadóttir, L., Þorsteinsdottir, U., and Stefánsson, K. (2009) Sequences from first settlers reveal rapid evolution in Icelandic mtDNA pool. *PLoS Genetics* 5, e1000343.

Hofreiter, M., Serre, D., Poinar, H. N., Kuch, M., and Pääbo, S. (2001) Ancient DNA. *Nature Reviews Genetics* 2, 353–9.

Hofreiter, M., Capelli, C., Krings, M., Waits, L., Conard, N., Munzel, S., Rabeder, G., Nagel, D., Paunovic, M., Jambresic, G., Meyer, S., Weiss, G., and Pääbo, S. (2002) Ancient DNA analyses reveal high mitochondrial DNA sequence diversity and parallel morphological evolution of late Pleistocene cave bears. *Molecular Biology and Evolution* 19, 1244–50.

Kiesewetter, H. (this volume) Toothache, back pain, and fatal injuries: What skeletons reveal about life and death at Roman and Byzantine Hierapolis, 268–85.

Krause, J., Fu, Q. M., Good, J. M., Viola, B., Shunkov, M. V., Derevianko, A. P., and Pääbo, S. (2010) The complete mitochondrial DNA genome of an unknown hominin from southern Siberia. *Nature* 464, 894–7.

Laforest, C., Castex, D., and Blaizot, F. (this volume) The grave 163d in the North Necropolis of Hierapolis in Phrygia: An insight into the funerary gestures and practices of the Jewish Diaspora in Asia Minor in late Antiquity and the proto-Byzantine period, 69–84.

Lazaridis, I., Patterson, N., Mittnik, A., Renaud, G., Mallick, S., Kirsanow, K., Sudmant, P. H., Schraiber, J. G., Castellano, S., and Lipson, M., *et al.* (2014) Ancient human genomes suggest three ancestral populations for present-day Europeans. *Nature* 513, 409–413.

Malmström, H., Gilbert, M. T. P., Thomas, M. G., Brandström, M., Storå, J., Molnar, P., Andersen, P. K., Bendixen, C., Holmlund, G., Götherström, A., and Willerslev, E. (2009) Ancient DNA reveals lack of continuity between Neolithic hunter-gatherers and contemporary Scandinavians. *Current Biology* 19, 1758–62.

Nasidze, I., Ling, E. Y. S., Quinque, D., Dupanloup, I., Cordaux, R., Rychkov, S., Naumova, O., Zhukova, O., Sarraf-Zadegan, N., Naderi, G. A., Asgary, S., Sardas, S., Farhus, D. D., Sarkisian, T., Asadov, C., Kerimov, A., and Stoneking, M. (2004) Mitochondrial DNA and Y-chromosome variation in Caucasus. *Annals of Human Genetics* 68, 205–21.

Nei, M. (1987) *Molecular evolutionary genetics.* New York, Colombia University Press.

Orlando, L., Ginolhac, A., Zhang, G. J., Froese, D., Albrechtsen, A., Stiller, M., Schubert, M., Cappellini, E., Petersen, B., Moltke, I., Johnson, P. L., Fumagalli, M., Vilstrup, J. T., Raghavan, M., Korneliussen, T., Malaspinas, A. S., Vogt, J., Szklarczyk, D., Kelstrup, C. D., Vinther, J., Dolocan, A., Stenderup, J., Velazquez, A. M., Cahill, J., Rasmussen, M., Wang, X., Min, J., Zazula, G. D., Seguin-Orlando, A., Mortensen, C., Magnussen, K., Thompson, J. F., Weinstock, J., Gregersen, K., Røed, K. H., Eisenmann, V., Rubin, C. J., Miller, D. C., Antczak, D. F., Bertelsen, M. F., Brunak, S., Al-Rasheid, K. A., Ryder, O., Andersson, L., Mundy, J., Krogh, A., Gilbert, M. T., Kjær, K., Sicheritz-Ponten, T., Jensen, L. J., Olsen, J. V., Hofreiter, M., Nielsen, R., Shapiro, B., Wang, J., and Willerslev, E. (2013) Recalibrating *Equus* evolution using the genome sequence of an early Middle Pleistocene horse. *Nature* 499, 74–81.

Ottoni, C., Ricaut, F. X., Vanderheyden, N., Brucato, N., Waelkens, M., and Decorte, R. (2011) Mitochondrial analysis of a Byzantine population reveals the differential impact of multiple historical events in South Anatolia. *European Journal of Human Genetics* 19, 571–6.

Pala, M., Olivieri, A., Achilli, A., Accetturo, M., Metspalu, E., Reidla, M., Tamm, E., Karmin, M., Reisberg, T., Hooshiar Kashani, B., Perego, U. A., Carossa, V., Gandini, F., Pereira, J. B., Soares, P., Angerhofer, N., Rychkov, S., Al-Zahery, N., Carelli, V., Sanati, M. H., Houshmand, M., Hatina, J., Macaulay, V., Pereira, L., Woodward, S. R., Davies, W., Gamble, C., Baird, D., Semino, O., Villems, R., Torroni, A., and Richards, M. B. (2012) Mitochondrial DNA signals of late glacial recolonization of Europe from near eastern refugia. *American Journal of Human Genetics* 90, 915–24.

Quintana-Murci, L., Chaix, R., Wells, R. S., Behar, D. M., Sayar, H., Scozzari, R., Rengo, C., Al-Zahery, N., Semino, O., Santachiara-Benerecetti, A. S., Coppa, A., Ayub, Q., Mohyuddin, A., Tyler-Smith, C., Qasim Mehdi, S., Torroni, A., and McElreavey, K. (2004) Where West meets East: The complex mtDNA landscape of the Southwest and Central Asian corridor. *American Journal of Human Genetics* 74, 827–45.

Richards, M., Macaulay, V., Hickey, E., Vega, E., Sykes, B., Guida, V., Rengo, C., Sellitto, D., Cruciani, F, Kivisild, T., Villems, R., Thomas, M., Rychkov, S., Rychkov, O., Rychkov, Y., Gölge, M., Dimitrov, D., Hill, E., Bradley, D., Romano, V., Calì, F., Vona, G., Demaine, A., Papiha, S., Triantaphyllidis, C., Stefanescu, G., Hatina, J., Belledi, M., Di Rienzo, A., Novelletto, A., Oppenheim, A., Nørby, S., Al-Zaheri, N., Santachiara-Benerecetti, S., Scozari, R., Torroni, A., and Bandelt, H. J. (2000) Tracing European founder lineages in the Near Eastern mtDNA pool. *American Journal of Human Genetics* 67, 1251–76.

Saiki, R. K., Scharf, S., Faloona, F., Mullis K. B., Horn, G. T., Erlich, H. A., and Arnheim, N. (1985) Enzymatic amplification of beta-globin genomic sequences and restriction site analysis for diagnosis of sickle cell anemia. *Science* 230, 1350–4.

Schönberg, A., Theunert, C., Li, M., Stoneking, M., and Nasidze, I. (2011) High-throughput sequencing of complete human mtDNA genomes from the Caucasus and West Asia: High diversity and demographic inferences. *European Journal of Human Genetics* 19, 988–94.

Stewart J. B. and Chinnery P. F. (2015) The dynamics of mitochondrial DNA heteroplasmy: Implications for human health and disease. *Nature Reviews Genetics* 16, 530–42.

Wong, M., Naumann, E., Jaouen, K., and Richards, M. (this volume) Isotopic investigations of human diet and mobility at the site of Hierapolis, Turkey, 228–36.

Isotopic investigations of human diet and mobility at the site of Hierapolis, Turkey

Megan Wong, Elise Naumann, Klervia Jaouen, and Michael Richards

Abstract

Isotopic analyses of human remains at the site of Hierapolis, Turkey, have been used to reconstruct diet and mobility patterns in the Roman (= Roman/early Byzantine, 1st–7th centuries AD) and Byzantine (= mid-Byzantine, 9th–13th centuries AD) periods. Results of carbon and nitrogen isotope analysis indicate that individuals at Hierapolis subsisted on a diet mainly composed of C_3 plants and terrestrial animals with more dietary variation in the Byzantine period than in the Roman period. This trend of variation continues when examining mobility at Hierapolis as well. While local baseline strontium isotope analysis is ongoing, preliminary results of strontium isotope signatures of individuals from Roman and Byzantine periods indicate increased mobility of possibly non-local individuals into the site during the Byzantine period. There is increased variation in strontium isotope values during this time period (values between 0.70718 and 0.71065) and this is contrasted by a close clustering of sample values from Roman individuals (values between 0.70814 and 0.70827). Future research into the regional strontium isotope signatures of Hierapolis will allow for a greater understanding of the mobility of local and non-local individuals.

Keywords: Byzantine, diet, isotopes, mobility, necropolis, Roman.

Introduction

The use of isotopic analyses of human remains to reconstruct past subsistence (carbon and nitrogen isotope measurements of bone collagen) and mobility (strontium isotope analysis of teeth) are established methodologies in archaeological research (e.g. van der Merwe and Vogel 1978; Schoeninger and Deniro 1984; Richards *et al.* 2000; Choy *et al.* 2010). This paper presents isotopic analyses of human skeletal remains recovered from the North-East Necropolis at the site of Hierapolis with the aim of characterizing diet, and possible differences in subsistence methods through time, and to assess the degree of human mobility at the site.

Mobility and diet at Hierapolis are especially interesting as the influence and role of the site transformed throughout its history, from the Hellenistic (3rd century BC) to late Byzantine periods (14th century AD).

The ancient city of Hierapolis is located in south-western Turkey just 15 km north of Denizli and directly adjacent to the modern village of Pamukkale (see Fig. 14.1, Leucci *et al.* 2013). Hierapolis was initially founded in the Hellenistic period around 200 BC and, as part of the Pergamene kingdom, it was ceded to Roman control in 133 BC. Throughout both these periods, Hierapolis was a thriving city, and during the Roman period it came to be an especially important Asian city that linked Anatolia to the Mediterranean (Leucci *et al.* 2013; Nuzzo *et al.* 2009).

In the course of the Roman Empire's conversion to Christianity, Hierapolis transformed into a Christian city and became an important site for followers of the Christian faith. A church and a martyrion were constructed to commemorate the death and martyrdom of the Apostle Philip and Hierapolis was subsequently considered a pilgrimage site of significance during this time period (D'Andria

Fig. 14.1. Map of Turkey with site of Hierapolis shown.

2011–2012; Arthur 2011; D'Andria and Gümgüm 2010; Nuzzo *et al.* 2009). Following violent seismic activity in the mid-7th century AD, Hierapolis was abandoned for some generations and re-emerged on a more modest level only in the 8th/9th centuries AD. While occupation and pilgrimages to the site continued throughout the mid-Byzantine period, the city came to its eventual demise sometime in the 14th century (Leuccie *et al.* 2013).

Skeletal samples from the North-East Necropolis

Samples for isotopic analysis were taken from individuals interred in the North-East Necropolis. The necropolis has been under continuous excavation by the Norwegian team led by Prof. J. Rasmus Brandt since 2007. Before this time, little archaeological work on this part of the city had been conducted; therefore, the tombs and sarcophagi in this necropolis had yet to be mapped thoroughly and excavated. The individuals interred in this area date to both the Roman and Byzantine periods. For clarity, this paper uses the term Roman to refer to Roman/early Byzantine periods (1st–7th centuries AD) and the term Byzantine to refer to the mid-Byzantine period (9th–13th centuries AD). One of the tombs in this portion of the necropolis, specifically tomb C92, yielded impressive finds of a few Byzantine crosses and also five pilgrim badges. These badges are from Rome and various regions in France (Ahrens 2011–2012) and they reinforce the concept that Hierapolis was an important pilgrimage site and add insight to the possible sphere of mobility seen at this site.

For carbon and nitrogen analysis, the fragmented human remains from Hierapolis were recovered from tombs with a complex stratigraphy, and thus the context information for each individual sample is limited (Kiesewetter, this volume; Laforest 2015). The material has been exposed to a warm and dry climate – conditions that are not favourable in terms of collagen preservation. Skeletal remains that are interred in regions with higher temperatures tend to degrade at a more rapid pace than those buried in more temperate environments (van Klinken 1999). Hot climates negatively impact bone preservation because bone hydrolysis occurs faster at higher temperatures and as bone hydrolysis increases the bone becomes more vulnerable to chemical alteration and degradation (von Endt and Ortner 1984). Therefore, most of the skeletal samples from Hierapolis, experiencing seasonal average temperatures of over 40°C, had significant collagen degradation.

Teeth were also sampled from a total of 27 individuals and were prepared for strontium isotope analysis. These samples came from both the Byzantine and Roman periods and from four different tombs. Teeth samples taken from tombs C91 and C92 are from individuals dating to the Byzantine era and the remaining samples are Roman from tomb T163d and sarcophagus C309.

Reconstruction of diet and mobility

Strontium isotope analysis for assessment of human mobility

The use of strontium isotopes ([87]Sr and [86]Sr) is a well-established methodology for analyzing past human mobility and migration patterns through the analysis of human tissues (Bentley 2006; Cox and Sealy 1997; Ezzo *et al.* 1997; Knudson *et al.* 2004; Price *et al.* 2000). This method is based on the concept that these isotopes can be used as geochemical signatures acquired through the consumption of water and plants that have incorporated strontium from the local bedrock (Bentley 2006). The ratio of [87]Sr/[86]Sr will vary depending on the mineral composition of the bedrock, rubidium (Rb) content, as well as its age. Over time, rubidium, specifically [87]Rb with a half-life of 4.88×10^{10} y, will decay and form [87]Sr; consequently, older rock formations will display ratios weighted higher in [87]Sr, than formations formed in later time periods.

Additional factors, such as mineral weathering and erosion of the underlying bedrock, sea spray, and marine-derived precipitation, can lead to a difference between the [87]Sr/[86]Sr ratio in the bedrock and the actual bioavailable [87]Sr/[86]Sr signature that would be reflected in our analysis; therefore, it is important to establish local strontium baselines in order to accurately interpret results within the local environment (Bentley 2006). To construct a local baseline, plants and snail shells are collected to determine the range of [87]Sr/[86]Sr ratios available for absorption by humans. Baseline materials for the Hierapolis site and surrounding areas are currently being analyzed and will be discussed in future publications.

To assess the [87]Sr/[86]Sr signatures of humans, human dental enamel is usually the tissue that is analyzed. Dental enamel is largely resistant to diagenetic alteration in the burial environment and is therefore a good representation of the biologically available strontium in the food and drink consumed during the time of tooth mineralization (Budd *et al.* 2000; Montgomery *et al.* 2000; Trickett *et al.* 2003). Tooth formation begins *in utero* and the crowns of permanent dentition are formed by approximately seven years old, the exception to this timeline is the third molar crown, which forms slightly later (usually between the ages of 9.9 and 12.6 years of age). Accordingly, the [87]Sr/[86]Sr ratios of permanent dentition are a good representation of the location of an individual's childhood.

Carbon and nitrogen analysis for assessment of human diet

Isotope analyses on human remains have increasingly become an integrated part of archaeological research where human remains are present. Still, at present, few isotope analyses have been conducted on human remains from Turkey. In the present study, human remains were analysed for $\delta^{13}C$, $\delta^{15}N$ and δ [87]Sr/[86]Sr-values. The value of $\delta^{13}C$ in bone collagen normally varies between −6 and −22‰ in mammals (Pollard and Heron 2008). Due to various factors, such as the difference in stable carbon isotope values between marine bicarbonate and atmospheric CO_2, as well as different carbon fractionation processes in C_3 and C_4 photosynthetic pathway plants (as shown in Propstmeier *et al.*, in this volume, p. 241), marine organisms and C_4-plants like millet, have higher carbon isotope ratios than terrestrial organisms and C_3 plants like wheat and rye. The ratio of the isotope ^{15}N to ^{14}N increases with each trophic level in the food chain, and is thus more abundant in animals compared to plants. As the food chains in marine ecosystems are longer, marine organisms are often elevated in $\delta^{15}N$ compared to organisms from terrestrial ecosystems. Bone collagen carbon isotope values of $\leq -19‰$ are commonly perceived as reflecting a diet based mainly on C_3 plants and terrestrial animals (Fuller *et al.* 2012). A recent study (Fuller *et al.* 2012) presented an isotopic reconstruction of human and animal diet from Classical Hellenistic to Byzantine periods in Turkey, concluding that the protein intake of humans mainly derived from C_3 plants and animals feeding on C_3 plants. This was also the conclusion from an isotopic study of Greek Byzantine diet (Bourbou *et al.* 2011; cf. also Propstmeier *et al.*, this volume).

Different bone and tooth elements from the human remains represent different periods of the individuals' lifetime; isotopic values in teeth are integrated through consumption of food during the time of tooth formation, and reflect diet in childhood and adolescence (Sealy 1995). Bone, on the other hand, continually turns over, meaning the isotopic values represent an average diet of the last years before death. This time span may be between 5 and 25 years, depending on a number of different factors such as sex and age (Hedges *et al.* 2007). The samples from Hierapolis presented in this paper are from both teeth and bone, thus representing different periods of the individuals' lifetime.

Methods

Strontium sample preparation and processing

Samples were processed for strontium isotope analysis at the clean laboratory and MC-ICP-MS facility at the Department of Human Evolution, Max Planck Institute for Evolutionary Anthropology (Leipzig, Germany) using the ion exchange method as outlined in Deniel and Pin (2001). Surface contaminants were removed from teeth using a Dremel rotary tool equipped with a sonicated diamond saw. A small piece of enamel from the tooth crown was removed using the Dremel tool with a diamond edge attachment. Enamel was then separated from any adjoining dentine under magnification using the Dremel tool and then cleaned by ultrasonication in deionized water; during this process the water was changed multiple times. Enamel samples were then transferred to the clean laboratory and rinsed in ultrapure acetone and dried. Samples weighing ~10–20 mg were placed in separate Teflon beakers and digested at 120°C in 1 ml of 14.3 M HNO_3. A hot plate was then used to evaporate the solution containing the enamel sample and, once dry, was mixed with 1 ml of 3 M HNO_3. Clean, preconditioned 2 ml columns filled with clean Sr-spec resin (EiChrom, Darien, IL) were used to purify the strontium from the enamel solutions. Each sample was reloaded into its respective column three times. The resin containing the strontium was washed twice with 3 M HNO_3, after which the strontium was eluted from the resin using ultrapure deionized water into a clean Teflon tube and evaporated on a hot plate until the sample was dry. Each sample was re-dissolved in 3 M HNO_3 and measured on a Thermo Fisher Neptune MC-ICP-MS instrument. Samples were externally corrected to the standard SRM_987 (Strontium carbonate, NIST, USA) and measured parallel to the external standard SRM_1486 (Bone Meal, NIST, USA), as well as one beaker blank per batch.

Carbon and nitrogen sample preparation and processing

All samples were prepared at the department of Archaeology, Conservation, and History, University of Oslo, as described in Brown *et al.* (1988), modified by Richards and Hedges (1999). Solid or drilled bone samples of 200–500 mg were demineralized in 0.5 M HCl at 4°C over several days, gelatinized in a pH3 solution at 70°C for 48 h. The solution was filtrated using an ezee-filter and then through an additional ultrafiltration step to isolate protein fragments of >30K. Collagen samples were then freeze-dried and analysed in a Flash EA mass spectrometer at the Max Planck Institute for Human Evolution in Leipzig. All results are reported as a ratio of δ relative to the standards VPDB for ^{13}C and AIR for ^{15}N, in parts per thousand (‰).

Results and discussion
Strontium isotope ratios

Strontium isotope ratios from the 27 individuals are presented in Table 14.1 and are characterized by more variation during the Byzantine period than in the Roman period (see Fig. 14.2). The values of the 19 Byzantine

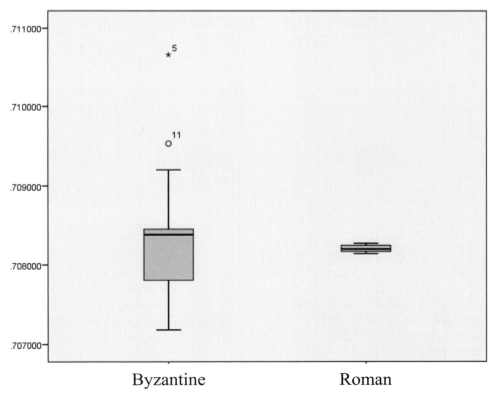

Fig. 14.2. Hierapolis. Variation between Sr values in Roman and Byzantine periods. A box represents the 25th–75th percentiles (with the median as a bold vertical line) and whiskers show the 10th–90th percentiles.

Table 14.1. Hierapolis. $^{87}Sr/^{86}Sr$ values for twenty-seven humans. R = Roman/Early Byzantine Period (1st – 7th century AD); B=Middle Byzantine Period (9th–13th century AD).

Sample No.	Grave context	Find No.	Skeleton No.	Age	Sex	Period	$^{87/86}Sr$ Enamel	$^{87/86}Sr$ Dentine
H7s	C92:B07	F0120	29	25–35	F?	B	0.709201	0.708537
H485	C92:B07	F1407	58	3–5	?	B	0.707182	0.708097
H4x	C92:B07	F0179	22	20–30	F	B	0.708158	0.708253
H2x	C92:B07	F0185	37	30–40	M	B	0.708286	0.708315
H474	C92:B07	F1851	78	30–40	M	B	0.710657	not enough material
H398	C92:B07	F0238	32	25–35	M?	B	0.707489	0.707924
H9s	C92:B07	F0238	32	25–35	M?	B	0.707501	0.708107
H10s	C92:B07	F0268	33	35–45	F	B	0.708206	0.708264
H5x	C92:B07	F0371	39	50–60	F	B	0.708803	0.708803
H3s	C92:B07		23	30–40	F?	B	0.708439	0.708359
H10x	C92:B07			no info	no info	B	0.709534	0.708550
H11x	C92:B07		NB H11x & H11 not identical	no info	no info	B	0.708429	0.708351
H8x	C92:B07		13a	40–60	M	B	0.707306	0.708106
H9x	C92:B07		13b			B	0.707620	0.708079
H504	C91:B93	F3085	101	15–25	?	B	0.708382	not enough material
H467	C91:B93	F1900	69	17–20	F	B	0.707988	0.708282
H316	C309: B26 (sarcophagus).	F0564b	41b	5 to 6	?	R	0.708272	0.708315
H314	C309: B26 (sarcophagus).	F0600				R	0.708252	0.708322
H21s	C92:B07	F0089				B	0.708461	0.708383
H16s	C92:B07			4–5?	no info	B	0.708411	0.708388
H18s	C92:B07			6–8?	no info	B	0.708443	0.708347
H65	T163d: US76		22	20–49	F	R	0.708198	0.708195
H21	T163d: US83		25	20–49	F	R	0.708189	0.708304
H64	T163d: US172		43	20–49	F	R	0.708140	0.708243
H108	T163d: US59		59	3–7	not det	R	0.708146	0.708297
H11	T163d: US31		6	20–30	F	R	0.708238	0.708270
H541	T163d: US261		71	30–59	not det	R	0.708196	0.708236

individuals varied between 0.70718 and 0.71065 while all of the Roman individuals (*n=8*) group around 0.7080 (see Fig. 14.3). The majority of these values from both time periods (*n=18*) cluster tightly between 0.7080 and 0.7085.

While local strontium isotope baselines values for Hierapolis are still being determined, a possible interpretation of these preliminary results could be that the local baseline for bioavailable strontium falls between 0.7080 and 0.7085 (see Fig. 14.4) and that the Byzantine individuals with values outside of this range are non-locals. There are currently no discernable patterns between these possibly non-local individuals. Tooth samples are from both males and females and four of these individuals are above the possible local range while five fall below; potentially indicating diversity in the regional locations of non-local visitors to Hierapolis. The only possible trend is that the outlying individuals are all from tomb C92 with the exception of two samples from tomb C91. It is interesting to note that the pilgrimage badges from Rome and different regions in France were found in tomb C92 possibly reinforcing the idea that non-local individuals were interred there.

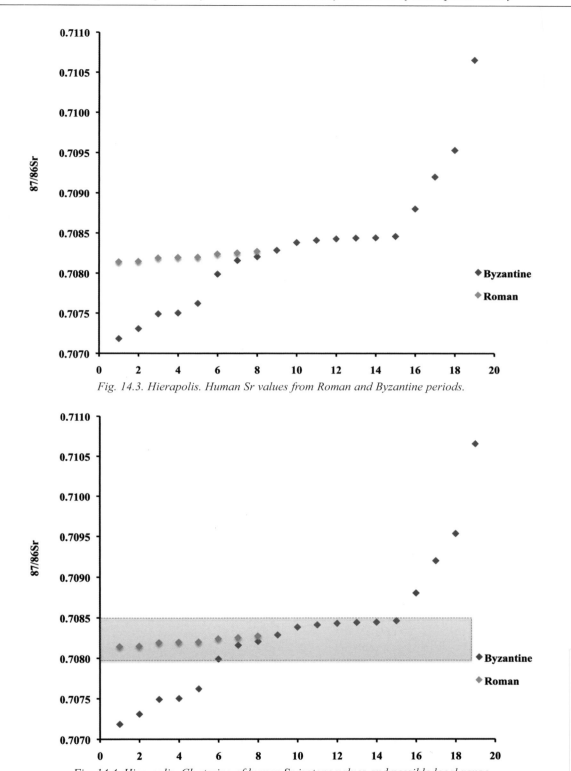

Fig. 14.3. Hierapolis. Human Sr values from Roman and Byzantine periods.

Fig. 14.4. Hierapolis. Clustering of human Sr isotope values and possible local range.

Carbon and nitrogen isotope ratios

The challenging preservation conditions are evident from the available collagen samples. From a total of 53 human and 19 animal samples, only seven human and two animal samples provided sufficient collagen with a C/N ratio within the required range of 2.9–3.6 (DeNiro 1985). Commonly, samples with <1% collagen preserved are excluded (Ambrose *et al.* 1997). However, we have included two samples of

<1% collagen as the sample procedure involved an extra ultrafiltration step to isolate protein fragments of >30K. As only seven human samples were viable for this analysis, two samples not run for this study are included to increase our sample size (see below). All results from human and animal samples are listed in Table 14.2.

The $\delta^{13}C$ results from humans range from −20.5 to −18.9‰ with an average of −19.3‰ (S.D. 0,5) and $\delta^{15}N$ results range from +8.9 to +11.1‰ with an average of +9.8‰ (S.D. 0.8) (see Fig. 14.5). $\delta^{13}C$ results from animal samples range from −20.1 to −20.3‰, and $\delta^{15}N$ from +3.7 to +5.1‰ (see Fig. 14.5). No average or S.D. is estimated for the animal bones as only two results are present.

Results show that the seven humans from Hierapolis all consumed a diet mainly consisting of C_3 plants and products from terrestrial animals, and correspond well with previously published results from Turkey (Fuller *et al.* 2012). Six of the seven humans have slightly elevated $\delta^{13}C$-values compared to the animal bones, suggesting that most humans had a modest intake of C_4 plants in their diet.

The $\delta^{15}N$-values indicate a considerable emphasis on foods from terrestrial animals. The $\delta^{15}N$-values vary to a certain degree, implying a various degree of plant foods as dietary component for the analysed individuals. Individual 389, from the mid-Byzantine burial in tomb C92 is distinctly different from the others; and shows depleted $\delta^{13}C$-values compared to the other human samples, and also compared to the two animal samples. The results are consistent with an individual consuming exclusively C_3 plants and animals

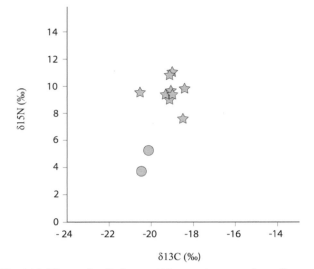

Fig. 14.5. Hierapolis. Carbon and Nitrogen isotope values of human (star) and animal (circle) samples.

feeding only on C_3 plants. The $\delta^{13}C$ depletion of individual 389 may thus reflect a difference on feeding strategies for domestic animals. Individual 460 from the same Byzantine burial shows elevated $\delta^{15}N$ values, and thus both individuals from the Byzantine period differ to a certain degree from the Roman samples. Additional $\delta^{13}C$ and $\delta^{15}N$-analyses of two Byzantine individuals (F2294 and F1412, see Table 14.2)

Table 14.2. Hierapolis. $\delta^{13}C$ and $\delta^{15}N$ values from seven human and two animal samples from Hierapolis. R=Roman/Early Byzantine Period (1st–7th century AD); B=Middle Byzantine Period (9th–13th century AD).

Sample No.	Grave & context	Find No.	Skeleton No.	Species	Element	Age	Sex	Period	$\delta^{13}C$	$\delta^{15}N$	%C	%N	C:N	% Coll
389	C92:B07	F75	30*	*H. Sapiens*	Tooth	30–40*	M*	B*	-20.5	9.6	43.2	14.2	3.5	6.5
460	C92:B07	F1740		*H. Sapiens*	Tooth	40–50*	M*	B*	-19.0	11.1	42.9	13.9	3.6	12.2
450	C322:G09	F2342		*H. Sapiens*	Tooth			R*	-19.2	8.9	43.7	14.5	3.5	9.9
35#	T163d: US55		14	*H. Sapiens*	Femur	20–39		R	-19.2	10.6	40.0	14.1	3.3	1.2
41#	T163d: US49		13	*H. Sapiens*	Femur	Ca. 15		R	-18.9	9.7	35.1	12.2	3.4	0.7
106#	Ha2011: US242		6	*H. Sapiens*	Humerus	Adult		R	-19.4	9.6	36.6	12.3	3.5	0.4
107#	Ha2011: US242		5	*H. Sapiens*	Ulna	Adult		R	-19.1	9.2	45.3	15.9	3.3	2.4
F2294*	C91			*H. Sapiens*	Femur			B	-18.5	7.6				
F1412*	C91			*H. Sapiens*	Femur			B	-18.4	9.9				
F 1095–87 a	C91	F1095		*Ovis/Capra*	Bone			B	-20.3	3.7	31.4	10.5	3.5	1.90
374–21	T163d			Hare/Rabbit	Bone			R	-20.1	5.1	42.1	14.3	3.4	11.4

Results from Gro Bjørnstad (F2294, F1412) and Henrike Kiesewetter, personal communications. #Results from Laforest 2015.

from burial C91 (unpublished data) confirm this pattern; the individuals had values of −18.4, and +9.9 and −18.5, and +7.6 respectively. The Byzantine samples also differ from each other, suggesting different places of origin. The results are interesting in comparison with the $^{87}Sr/^{86}Sr$-values, implying a high frequency of non-local individuals amongst the Byzantine individuals. The dietary analyses support this conclusion, suggesting indeed that the Byzantine population of Hierapolis consisted of a significant part of non-local individuals. While the two individuals from burial C92 presented in this paper were represented only by teeth, indicating childhood diet, individuals F2294 and F1412 were sampled from bone, representing an average of consumed isotopic values during the last years before death. Thus, at least for these two, differing dietary values may indicate a recent arrival at Hierapolis.

A limited amount of information may be extracted from the two samples from animal bones, though the difference in $\delta^{15}N$-values is interesting. Both animals are classified as herbivores, and thus the elevated $\delta^{15}N$-level in the rabbit/hare sample may indicate that fertilizers were used in cultivation for fodder production for certain species or periods – a strategy that would also affect $\delta^{15}N$-levels in humans (cf. also Propstmeier *et al.*, this volume).

As sex determination was only achievable for two male individuals, it is not possible to discuss any dietary patterns related to sex, but we notice that most $\delta^{13}C$ and $\delta^{15}N$ values fall within a narrow range suggesting a diet consistent of similar amounts of different food components. For the few individuals where age determination was possible, no pattern is evident for the different age groups. However, as the number of individuals is low, and osteological information limited, we make no conclusive remarks regarding social dietary patterning.

Conclusion

These first human strontium isotopic results from our study show that there was a wider range of strontium isotopic values during the Byzantine period than during the Roman period. Until baseline values can be established it can only be hypothesized that strontium values that fall outside of the 0.7080 and 0.7085 range are possibly non-local individuals, strontium signatures from more distant geologies. Baseline samples for Hierapolis and the surrounding region have been collected and are currently under analysis. Once the local geology can be characterized, a more thorough investigation of human strontium values can be conducted and will allow for a greater understanding of human mobility at Hierapolis.

Dietary results from this research indicate that individuals at Hierapolis subsisted mainly on a diet comprised of C_3 plants and terrestrial animals. Conclusions from carbon and nitrogen analysis also seem to support preliminary results from strontium analysis, as more variation in diet can be seen in the Byzantine individuals possibly indicating differences due to regionally distinct diets. Conversely, Roman individuals had a more homogenous subsistence pattern implying no differences in these individuals' region of origin. The slightly elevated $\delta^{15}N$-values in the two herbivorous fauna samples may indicate a reliance on fertilizers in the cultivation process and this process would also affect the $\delta^{15}N$-values of humans. However, given the relatively small sample sizes for both humans and fauna these results need to be investigated further, with more robust sample sizes, to substantiate these results.

Bibliography

Ahrens, S. (2011–2012) Appendix 3. A set of western European pilgrim badges from Hierapolis. In D'Andria, 67–75.

Ambrose, S. H., Butler, B. M., Hanson, D. B., Hunter-Anderson, R. L., and Krueger, H. W. (1997) Stable isotope analysis of human diet in the Marianas Archipelago, Western Pacific. *American Journal of Physical Anthropology* 104, 343–61.

Arthur, P. (2011) Hierapolis of Phrygia: The drawn-out demise of an Anatolian city. In N. Christie and A. Augenti (eds.) Vrbes Extinctae. *Archaeologies of abandoned Classical towns*, 275–305. Farnham and Burlington (VT), Ashgate.

Bentley, A. R. (2006) Isotopes from the earth to the archaeological skeleton: A review. *Journal of Archaeological Method and Theory* 13.3, 135–87.

Bourbou, C., Fuller, B. T., Garvie-Lok, S. J., and Richards, M. P. (2011) Reconstruction the diets of Greek Byzantine populations (6th – 15th centuries AD) using carbon and nitrogen stable isotope ratios. *American Journal of Physical Anthropology* 146, 569–81.

Brown, T. A., Nelson, D. E., Vogel, J. S., and Southon, J. R. (1988) Improved collagen extraction by modified Longin method. *Radiocarbon* 30.2, 171–7.

Budd, P., Montgomery, J., Barreiro, B., and Thomas, R. G. (2000) Differential diagenesis of strontium in archaeological human dental tissues. *Applied Geochemistry* 15, 687–94.

Choy, K., Jean, O. R., Fuller, B. T., and Richards, M. P. (2010) Isotopic evidence of dietary variations and weaning practices in the Gaya cemetery at Yeanri, Gimhae, South Korea. *American Journal of Physical Anthropology* 142, 74–84.

Cox, G. and Sealy, J. (1997) Investigating identity and life histories: Isotopic analysis and historical documentation of slave skeletons found on the Cape Town foreshore, South Africa. *International Journal of Historical Archaeology* 1, 207–24.

D'Andria, F. (2011–2012) Il Santuario e la Tomba dell'apostolo Filippo a Hierapolis di Frigia. *Rendiconti della Pontificia Accademia Romana di Archeologia* 84, 1–75 (including three appendices).

D'Andria, F. and Gümgüm, G. (2010) Archaeology and architecture in the Martyrion of Hierapolis. In S. Aybek and A. Kazım Ös (eds.) *The land of crossroads. Essays in honour of Recep Meriç* (Metropolis Ionia II), 95–104. Istanbul, Homer Kitabevi Yayınları.

Deniel, C. and Pin, C. (2001) Single-stage method for the simultaneous isolation of lead and strontium from silicate samples for isotopic measurements. *Analytica Chimica Acta* 426, 95–103.

DeNiro, M. J. (1985) Postmortem preservation and alteration of in vivo bone collagen isotope ratios in relation to paleodietary reconstruction. *Nature* 317, 806–9.

Ezzo, J. A., Johnson, C. M., and Price, T. D. (1997) Analytical perspective on prehistoric migration: A case study from east-central Arizona. *Journal of Archaeological Science* 24, 447–66.

Fuller, B. T., Cupere, B. D., Marinova, E., Van Neer, W., Waelkens, M., and Richards, M. P. (2012) Isotopic reconstruction of human diet during the Classical-Hellenistic, Imperial, and Byzantine periods at Sagalassos, Turkey. *American Journal of Physical Anthropology* 149, 157–71.

Hedges, R. E. M., Clement, J. G., Thomas, D. L., and O'Connell, T. C. (2007) Collagen turnover in the adult femoral mid-shaft: modeled from anthropogenic radiocarbon tracer measurements. *American Journal of Physical Anthropology* 133, 808–16.

Kiesewetter, H. (this volume) Toothache, back pain, and fatal injuries: What skeletons reveal about life and death at Roman and Byzantine Hierapolis, 268–85.

Klinken, G. J. van (1999) Bone collagen quality indicators for palaeodietary and radiocarbon measurements. *Journal of Archaeological Science* 26, 687–95.

Knudson, K. J., Price, T. D., Buikstra, J. E., and Blom, D. E. (2004) The use of strontium isotope analysis to investigate Tiwanaku migration and mortuary ritual in Bolivia and Peru. *Archaeometry* 46, 5–18.

Laforest, C. (2015) *La sépulture collective 163d de la nécropole nord de Hiérapolis (Phrygie, Turquie, période augusteenne – VIIe s. de notre ère): fouille et enregistrement des dépôts, gestes et pratiques funéraires, recrutement.* PhD dissertation, University of Bordeaux.

Leucci, G., DiGiacomo, G., Ditaranto, I., Miccoli, I., and Scardozzi, G. (2013) Integrated ground-penetrating radar and archeological surveys in the ancient city of Hierapolis of Phrygia (Turkey). *Archaeological Prospection* 20, 285–301.

Merwe N. J. van der and Vogel J. C. (1978) 13C content of human collagen as a measure of prehistoric diet in woodland North America. *Nature* 276, 815–6.

Montgomery, J., Budd, P., and Evans, J. (2000) Reconstructing the lifetime movements of ancient people: A Neolithic case study from southern England. *European Journal of Archaeology* vol. 3, 370–85.

Nuzzo, L., Leucci, G., and Negri, S. (2009) GPR, ERT and magnetic investigations inside the martyrium of St Philip, Hierapolis, Turkey. *Archaeological Prospection* 16, 177–92.

Pollard, A. M. and Heron, C. (2008) *Archaeological chemistry.* 2nd edition. Cambridge, The Royal Society of Chemistry.

Price, T. D., Manzanilla, L., and Middleton, W. H. (2000) Residential mobility at Teotihuacan: A preliminary study using strontium isotopes. *Journal of Archaeological Science* 27, 903–14.

Propstmeier, J., Nehlich, O., Richards, M. P., Grupe, G., Müldner, G. H., and Teegen, W.-R. (this volume) Diet in Roman Pergamon: Preliminary results using stable isotope (C, N, S), osteoarchaeological and historical data, 237–49.

Richards, M. P. and Hedges, R. E. M. (1999) Stable isotope evidence for similarities in the types of marine foods used by Late Mesolithic humans at the sites along the Atlantic coast of Europe. *Journal of Archaeological Science* 26, 717–22.

Richards, M. P., Pettitt, P. B., Trinkaus, E., Smith, F. H., Paunovic, M., and Karavanic, I. (2000) Neanderthal diet at Vindija and Neanderthal predation: The evidence from stable isotopes. *Proceedings of National Academy of Sciences USA* 97, 7663–6.

Schoeninger, M. J. and DeNiro, M. J. (1984) Nitrogen and carbon isotopic composition of bone collagen from marine and terrestrial animals. *Geochimica et Cosmochimica Acta* 48, 625–39.

Sealy, J., Armstrong, R., and Schrire, C. (1995) Beyond lifetime averages: Tracing life histories through isotopic analysis of different calcified tissues from archeological human skeletons. *Antiquity* 69, 290–300.

Trickett, M. A., Budd P., Montgomery, J., and Evans, J. (2003) An assessment of solubility profiling as a decontamination procedure for the 87Sr/86 Sr analysis of archaeological human skeletal tissue. *Applied Geochemistry* 18, 653–58.

von Endt, D. W. and Ortner, D. J. (1984) Experimental effects of bone size and temperature on bone diagenesis. *Journal of Archaeological Science* 11, 247–53.

Diet in Roman Pergamon: Preliminary results using stable isotope (C, N, S), osteoarchaeological and historical data

Johanna Propstmeier, Olaf Nehlich, Michael P. Richards,
Gisela Grupe, Gundula H. Müldner, and Wolf-Rüdiger Teegen

Abstract

Stable isotope analysis of bone collagen can offer important clues towards the reconstruction of past lives, as it provides information about past human diets, and subsistence of past populations. In this paper the living conditions of the ancient populations of Pergamon (today Prov. İzmir, Turkey) in Roman and late Byzantine times are in focus.

By the use of carbon (C) isotope ratios the amounts of C_3- vs. C_4-plant food and marine vs. terrestrial food can be determined. Further information on the protein sources can be achieved with the nitrogen (N) isotope value, such as the amount of animal vs. plant protein. The input of marine or freshwater food sources can be achieved with the measurements of the sulphur (S) isotope value. The results of the carbon, nitrogen and sulphur isotope analysis of the analyzed Roman and late Byzantine inhabitants of Pergamon indicate a diet based on C_3-plants and terrestrial animals with no input of marine food.

In addition, results of the palaeopathological analysis of the jaws and teeth of the sampled individuals are presented. The results are discussed in the light of the medical writings by Galen from Pergamon (2nd century AD). The palaeopathological data is consistent with a plant-based diet with a remarkably high carbohydrate consumption and a certain amount of (animal) protein.

Keywords: dental diseases, diet reconstruction, Pergamon, Roman period, stable isotope analysis (C, N, S).

Introduction

The ancient city Pergamon (in Greek) or Pergamum (in Latin) is located at the northern rim of the modern city Bergama, İzmir Province (Turkey) in western Asia Minor, approximately 30 km from the Aegean Sea. The Hellenistic city is on a promontory north of the ancient river Kaikos (today Bakırçay). The hilltop with its acropolis is famous for Hellenistic architecture and art (Radt 1999). At some distance from the Hellenistic city contemporaneous monumental tumuli are located, where the rulers and other high ranked people were buried (Kelp in Otten *et al.* 2011).

Settlement in the lower city began mostly only in Roman times. The Roman lower city is located at the foot of the acropolis hill under the northern part of the modern city, both north and south of the river Kaikos (Radt 1999; Pirson 2011; 2014). The Roman burial grounds were mostly discovered around the ancient lower city under the modern city of Bergama (cf. Pirson 2014, 104 fig. 3).

Two burial grounds were identified on the flanks of the promontory (see Teegen, this volume, fig. 16.2). During the construction work for a cable car station in 2007 a Roman necropolis was discovered quite unexpectedly. The so-called South-East Necropolis is located just outside the city walls of Eumenes II (197–159 BC). Foundations of several funerary monuments were excavated by the German Archaeological Institute at Istanbul in close cooperation with the Museum

Bergama. Around the monuments, generally containing multiple burials, several other burials were discovered, including both inhumations and some cremations. The excavations were continued in the summers of 2011 and 2013–14. In particular, one multiple burial with at least 11 individuals is worth mentioning (see Teegen, this volume, fig. 16.3).The burials are dated archaeologically to the 1st to 4th century AD (personal communication by S. Japp and U. Kelp, September 2014). Recent radiocarbon dating of human bone samples, mainly from the 2014 excavations in the South-East Necropolis revealed a calibrated time range from 40 BC to 415 AD (2 sigma probability) (GrA 62665–71, 62677–8, 63897; van der Plicht 2015).

The second cemetery was discovered in 2010 on the northern slope of the promontory (Pirson 2011). The so-called North Cemetery dates to the late Byzantine period. In 2011 eight burials were excavated (Pirson 2012).

Furthermore, human burials from two sites in the cemeteries of Elaia, the ancient port of Pergamon, were sampled (Teegen 2013, 142–3). The burials date to the Hellenistic and Roman periods (Pirson 2013, 130–1).

This paper presents the first results regarding the diet in Roman and Byzantine Pergamon. The effects of food on the body can not only be shown from a palaeopathological point of view, but also using stable isotope data. Both will be shown and discussed in this article. The combination of the elements carbon, nitrogen and sulphur will create a more detailed and accurate understanding of the dietary practices of ancient populations.

Historical accounts of Roman and Byzantine diet refer to grain as the base of the diet (Grant 2000; Teall 1959). But dry legumes and other plants must also have played an important role. Analysis of the stable carbon isotope ratios can determine the amounts of C_3- vs. C_4-plant food. Archaeozoological studies show that the favoured consumed terrestrial animals were sheep or goat, pig and to a lesser extent, cattle (see below). The analysis of the stable carbon and nitrogen isotope ratios give information about the average amount of animal food in the general diet. As mentioned above, Pergamon is situated some 30 km from the Mediterranean Sea. Archaeozoological investigations revealed fish bones and shells from the excavations in the ancient city and the South-East Necropolis (see below). Therefore, the question has to be raised whether and to which extent marine food was part of the diet. This can be addressed by analysis of the stable nitrogen and sulphur isotope ratios.

[JP, WRT]

Diet and dental diseases

Diet

Diet is fundamental to all humans and all animals as well. To study them we have a variety of approaches. The archaeological approach deals not only with vessels and other objects for storage, preparation, cooking and serving of food and drink, but also with the food and drink itself: animal bones, plant remains and burnt residues on pots are particularly important. Liquids can also leave traces in the ceramics which can be determined using e.g. gas chromatographic methods and/or stable isotope analysis (cf. Craig *et al.* 2011). Other sources include ancient images as well as written texts.

Archaeozoological analyses of animal bones from Roman and Byzantine Pergamon were carried out by the late Joachim Boessneck and Angela von den Driesch in the 1980s (Boessneck and von den Driesch 1985; von den Driesch 2008). Recently, Michael MacKinnon (2011) has published the animal bones from the eastern slope of Pergamon. All investigations have shown ovicaprines, pigs and cattle to be main sources for animal protein (Fig. 15.1). In some samples, however, other species can be of particular importance; in sample PE08-So-04-14 the frequency of molluscs is nearly 15%, in sample PE10-Ar-04 the frequency of tortoise is around 8% (cf. Fig. 15.1).

The eminent ancient physician and surgeon Galen of Pergamon (approx. 129–216/7 AD) mentioned in his book *De Alimentorum Facultatibus* that pig meat is the most nutritious (Grant 2000). Chicken is the most common fowl. To Galen it seemed, together with pheasant, to be the best fowl meat, for digestion and nutrition (Grant 2000). Archaeozoological investigations revealed, until now, mostly domestic fowl (Boessneck and von den Driesch 1985; von den Driesch 2008; MacKinnon 2011; personal observations by WRT).

Marine and aquatic food was also consumed. Fish bones have rarely been found. They are, however, present in small numbers (von den Driesch 2008; MacKinnon 2011; personal observations by WRT in 2012–14). In the Roman necropolis fish bones were laid on plates in grave contexts. Furthermore, in some amphora used as the burial containers for neonates, small fish bones were found upon flotation of the contents (Teegen 2015). This is an indication of the transport of fish or fish products like *garum* (Roman fish sauce). The species are still unidentified.

Better known are molluscs like *Cardium* spec., which are often found in the upper city. More difficult is their dating. Galen mentioned molluscs as being a laxative for the stomach (Grant 2000). He, however, preferred oysters. They were only sometimes found in Pergamon (personal observation by WRT).

Until now we have had no archaeobotanical data for human nutrition from Pergamon. Cisterns were discovered during the 'Stadtgrabung', yielding several animal bones (Boessneck and von den Driesch 1985). Samples for archaeobotany were, however, never taken nor analysed.

In this context, Galens' treatise *De Alimentorum Facultatibus*, in which he also discussed food plants and their properties, is quite important. He mentioned in

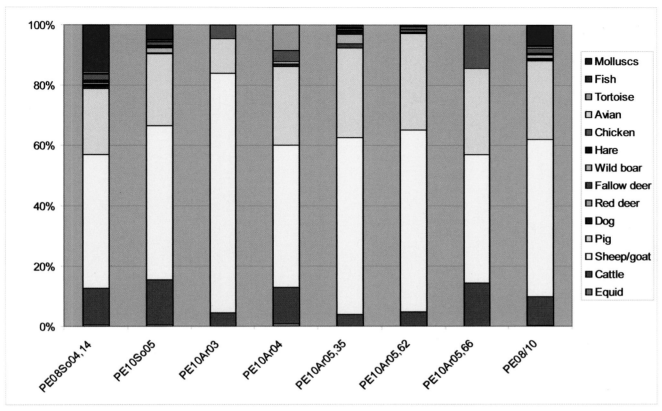

Fig. 15.1. Pergamon. Animal bones (N=2522) from recent excavations in the late Hellenistic and Roman city. Data from MacKinnon 2011 (Graphics: W.-R. Teegen).

particular barley soup with beans as forming the daily diet for gladiators, boxers, ringers and other heavyweight athletes (Grant 2000). Barley or wheat soups are – together with bread – the most common diet for the lower classes. Beans were in Antiquity and the Middle Ages – and are still today – an important source of plant protein in the Mediterranean diet.

As we will see shortly, the plant-based diet was common for most Roman and Byzantine Pergameneans analysed, so far, for stable isotopes.

Dental Diseases

The dental diseases present in the samples from Pergamon are dental caries, abscesses, dental calculus, parodontopathies and intra vitam tooth loss. Furthermore, teeth were probably used as tools (see Teegen, this volume, Fig. 16.7).

Dental caries (Fig. 15.2) was a common finding, mainly in the molar region. The defects could have destroyed half of the tooth crown or more. Dental caries is an infectious disease, mainly caused by *Streptococcus mutans* and other *Streptococcaceae* or other acid uric bacteria in an acidic environment (pH<7) (Schroeder 1997). The presence of carbohydrates promotes these bacteria. Indirectly,

caries defects are indicators of a cariogenic diet, rich in carbohydrates.

As we have seen bread and soup (pulses) were the common daily food source for most Romans at Pergamon and elsewhere. Honey was used as a sweetener. Whether dates were eaten regularly remains an open question. They are, however, quite common in Roman sites from the eastern Mediterranean, and are mentioned by Galen as a sweet. All these sweets caused prolific dental caries, which often lead to abscesses at the tooth root and intra vitam tooth loss. This was also observed.

Another indicator of a specific diet is dental calculus. It develops mainly in a basic environment with a pH>7. It is often associated with a protein-rich diet; this view has also been discussed in the scientific literature (Lieverse 1999). Dental calculus was regularly observed in the inhumation burials from Pergamon. It is also, like today, an indicator of inadequate dental hygiene. The bacteria in the dental plaque can cause an inflammation of the gingival tissue (gingivitis) and then of the alveolar bone (Schroeder 1997; Raitapuro-Murray *et al.* 2014). But also the dental calculus, solidified dental plaque, can develop gingivitis. All this leads to a reduction of the alveolar rim (parodontitis) and in some cases to intra vital tooth loss.

Fig. 15.2. Pergamon, South-East Necropolis. Adult individual (PE11So11_031 #1) with parodontitis, caries and dental abscesses (Photo: W.-R. Teegen).

The kind of dental abrasion can also give some insight into diet. Meat consumers have often a special abrasion of the front teeth. Bread and pulse consumers exhibit an enlarged abrasion of the molars due to grinding (Smith 1984). Also severe abrasion of the teeth, caused, for example, by stone grind in the flour, could have led to abscesses and intra vitam tooth loss.

Analysing the teeth of the ancient Pergameneans leads to the assumption that carbohydrates were the basis of the daily diet (cf. also Wong *et al.*, Kiesewetter, and Nováček *et al.*, all this volume). Protein sources should have been present, but to a lesser degree. This view was supported by stable isotope analysis (see below).

In addition, the presence of numerous periodic disorders in the odontogenesis in terms of transverse linear enamel hypoplasia on the dental crowns of the individuals from the Roman and late Byzantine necropolis was found (for details see W.-R. Teegen, this volume).

[WRT]

Stable isotope analysis[1]

Analysis of stable carbon (δ^{13}C), nitrogen (δ^{15}N), and sulphur (δ^{34}S) isotopes in bone collagen can offer important insights into the reconstruction of diet and therefore into the subsistence, but also the mobility of past populations. Isotopes are atoms of the same atomic number with the same number of protons but a different number of neutrons, thus they differ in mass.

The isotopes of nitrogen and carbon of the bone collagen reflect the origins and the amount of plant and animal proteins. In addition, sulphur isotope analysis of archaeological material can be used to answer questions related to marine versus terrestrial diets and the mobility of people. The ratio of stable carbon, nitrogen, and sulphur isotopes is expressed as the relation between the lighter and the heavier isotopes (here ^{13}C/^{12}C, ^{15}N/^{14}N, ^{34}S/^{32}S) of the analysed sample divided by the same isotope relation in an international standard (e.g. Nehlich 2015, 2). It is expressed in delta (δ) notation in parts per thousands (‰), for example:

$$\delta^{13}C \text{ (in ‰)} = ((^{13}C/^{12}C)_{sample} / (^{13}C/^{12}C)_{standard}) - 1$$

The δ^{15}N-value is a good indicator of the trophic level on which the investigated individual stands. There is a gradual enrichment, on average 3.4‰, of the heavy isotope from one trophic level to the next (Grupe *et al.* 2012). With that it is possible to evaluate if the person has a mainly herbivorous, omnivorous, or carnivorous diet. Herbivores show the lowest values, carnivores the highest.

By use of carbon isotope ratios the amounts of C_3- vs. C_4-plant food and marine vs. terrestrial food can be determined. Plants use different types of photosynthesis (C_3- and C_4-plants), which results in different $\delta^{13}C$ values. C_3-plants show values between $-22‰$ and $-37‰$ and C_4-plants between $-10‰$ and $-13‰$ (Ambrose and Norr 1993). The $\delta^{13}C$-value provides information about C_3- and C_4-plants in the diet of humans and animals. It is also possible to distinguish terrestrial from limnic and marine habitat. In limnic biospheres the atmospheric CO_2 serves as the carbon source of the plants, which leads to values between -37 and $-27‰$. Saltwater plants use the dissolved bicarbonate, which is about 7‰ more positive, resulting in values between -18 and $-16‰$ (Coltrain *et al.* 2004). In water habitats the $\delta^{15}N$-value is in general higher, the result of longer food chains.

As the $\delta^{34}S$ value in the available food for humans ultimately depends on the $\delta^{34}S$ values of the ecosystem, the consumption of fresh or saltwater fish can be determined (Nehlich *et al.* 2011).

Additionally, sulphur isotope ratios reflect the origin of food consumed during the last 10–20 years of life of an individual (Nehlich *et al.* 2011; Oelze *et al.* 2012). Plants gain most of their sulphur from the soil and the $\delta^{34}S$ signal of the soil is derived from the local bedrock, thus collagen $\delta^{34}S$ values of animals and humans can be used to identify the region in which an individual normally resides and therefore identify migrants (Nehlich *et al.* 2014; Nehlich 2015).

The combination of the $\delta^{34}S$ data with the $\delta^{13}C$ and $\delta^{15}N$ values creates a more detailed and accurate understanding of the dietary practices of these populations as well as providing hitherto unknown information about migratory individuals (Nehlich 2015, 13).

Material and methods

Material

For her master's thesis (Propstmeier 2012; 2013) at the Ludwig-Maximilian-University in Munich (Supervisor: Prof. Gisela Grupe), 70 (66 human and four animal) bone samples were analysed for stable isotopes of the elements C and N. Furthermore, the question was raised as to whether there was a marine component in their diet, since the results revealed an elevated content of stable carbon and nitrogen isotopes. Thus, the second set of further 61 human and 15 animal bones, and the 70 samples mentioned above were measured for their sulphur, carbon, and nitrogen values in the spring 2013 at the University of British Columbia in Vancouver (supervisors: Prof. M. P. Richards and Dr O. Nehlich).

Altogether 56 samples from the South-East Necropolis were analysed for stable isotopes, including three individuals from the 2007 campaign, ten individuals from single burials and 43 samples (minimum number of individuals [MNI] sampled = 10!) from the multiple burial (PE11-So11-Gr15) from the 2011 campaign.

Eight burials with 13 individuals from the North Cemetery were excavated in 2011, but only nine individuals could be sampled. For comparison, one individual from a Hellenistic monumental tumulus found in Elaia was also sampled.

In total 66 human bones and four animal bones were analysed at the Ludwig-Maximilian-University in Munich (Germany) for carbon and nitrogen isotopes.

For the sulphur, carbon, and nitrogen isotope analysis to be processed in the laboratories of the University of British Columbia in Vancouver (Canada) and the Max Planck Institute for Evolutionary Anthropology (Leipzig, Germany), 76 samples (including 15 animal bones) were collected during the 2012 campaign. Sixty-nine samples were taken from the Roman South-East Necropolis at Pergamon (campaigns 2007 and 2011). The other seven samples are from Roman burials in Elaia, excavated in 2012.

Furthermore, all 70 samples studied at Munich were (re-)analysed for sulphur, carbon and nitrogen isotopes in Vancouver and Leipzig.

The total number of bone samples measured is 146. Unfortunately not all results from the two sets are available yet: only from 62 humans and 10 animals for the sulphur isotopes and 55 humans and eight animals for the carbon and nitrogen isotopes. Due to organizational reasons the carbon and nitrogen isotopes were measured separately from the sulphur isotopes. At the moment, not all datasets (C, N, S) are complete.

C- AND N-ISOTOPE ANALYSIS AT THE LUDWIG-MAXIMILIAN-UNIVERSITY

For the stable isotope analysis of the light elements carbon and nitrogen the collagen-gelatine extraction was carried out in the BioCenter of the Ludwig-Maximilian-University. 500 mg of bone powder sample was demineralized in 5 ml of 1M HCl for 20 minutes and then washed in distilled water by centrifuging at 3,000 rpm for 5 minutes.

The pellet was incubated in 5 ml of 0.125 M NaOH for 20 hours. Another washing step in distilled water followed until a pH-value of 5.5 was reached. The samples were incubated in 5 ml of 0.001M HCl (pH 3) for 10–17 hours in a hot water bath at 90°C, filtered (with a pore size of 5 μm) and lyophilized for 3–5 days at −50°C. The amount of extracted gelatine was weighed and placed in relation to the weight of the whole bone sample, which will be referred to as collagen weight proportion in per cent (weight %). Finally 0.5 mg of each sample was measured for $\delta^{13}C$- and $\delta^{15}N$-isotopes in a mass spectrometer of the type ThermoFinnigan Delta Plus.

In total, 66 human and four faunal samples of the ancient city of Pergamon were analysed for carbon and nitrogen isotopes for the master's thesis.

C-, N-, AND S-ISOTOPE ANALYSIS AT THE
UNIVERSITY OF BRITISH COLUMBIA IN VANCOUVER

Approximately a 0.5 g piece of each bone was demineralized with 0.5M HCl at 4°C for 48 hours. This step was repeated with fresh 0.5M HCl again for 48 hours. Depending on the bone sample, this step had to be repeated again until the bone was fully demineralized. After that the acid was removed from the bone samples and rinsed with purified water three times. The remaining collagen was denatured with HCl (pH3) at 75°C for 48 hours. An Ezee filter® was used to filter the sample fluid. Each sample was filled into a 30kDa Pall Microsep centrifuge filter tube and centrifuged at 4000rpm for 10 minutes. This purification process was done until the sample volume was less than 1 ml. The purified collagen had to be lyophilized for 48 hours, before measuring it in the mass spectrometer. Each sample was weighed and 4 mg put into tin capsules which were analysed on an ElementarVario Micro Cube elemental analyser coupled to an Isoprime 100 mass spectrometer (Elementar Americas Inc., New Jersey).

The second set of samples (N=146) was analysed for carbon, nitrogen, and sulphur isotopes.

Results

To use the results of the bone samples they have to fulfil certain quality criteria. Besides the weight percentage (see above), the most important criteria is the C/N ratio of 2.9 to 3.6 (van Klinken 1999), which is the result of the molar proportion of carbon to nitrogen considering the atomic mass. The amount of sulphur from archaeological bone collagen should fall into a similar range as modern bone collagen, which ranges between 0.15 and 0.35% (Nehlich and Richards 2009).

$\delta^{13}C$ and $\delta^{15}N$ results processed at the LMU in Munich

From the first set of 70 samples 19 (28.8%) had to be excluded due to the quality criteria (C/N, weight%); six of nine late Byzantine individuals (66.6%), the only Hellenistic individual and 12 of 56 Roman samples (21.4%). In Figure 15.3 the $\delta^{13}C$- and $\delta^{15}N$-values of all the human samples are shown.

As the sample number (n=46) of the multiple burial PE11-So11-Gr15 does not represent the actual individual number, the most represented bone, the left tibia (n=7) was used to determine the minimum number of individuals. This results in a total number of individuals measured (n=20), i.e. 10 Roman individuals of single burials, 3 late Byzantine individuals, and the 7 individuals of multiple burials (Fig. 15.4).

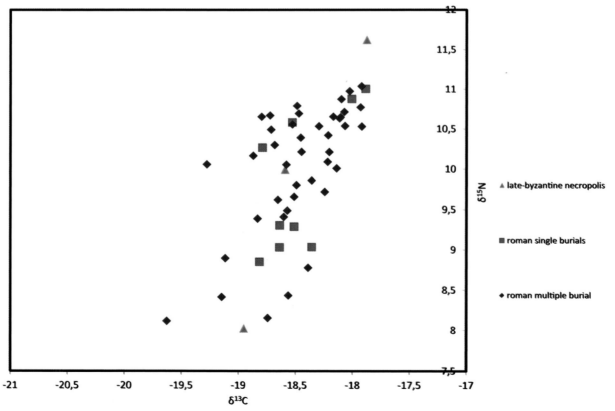

Fig. 15.3. Pergamon, Roman South-East Necropolis and late Byzantine North Cemetery. $\delta^{13}C$ and $\delta^{15}N$ values (data from Propstmeier 2012) (Graphics: J. Propstmeier).

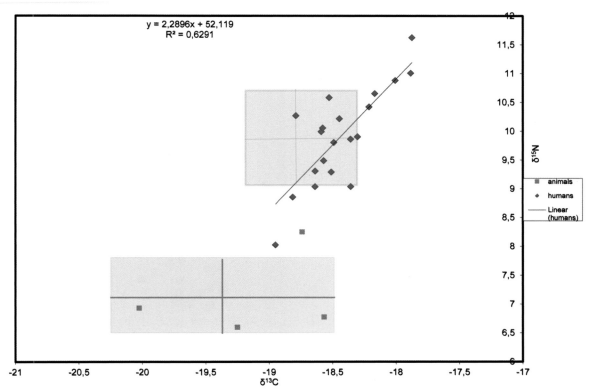

Fig. 15.4. Pergamon, Roman South-East Necropolis and late Byzantine North Cemetery. $\delta^{13}C$ and $\delta^{15}N$ isotope values with the mean of the four animals and the humans (n=20) (data from Propstmeier 2012) (Graphics: J. Propstmeier).

The resulting mean of the minimum number of human samples is for $\delta^{13}C$ −18.4±0.3‰ and for $\delta^{15}N$ 9.9±0.8‰ (Fig. 15.4). The $\delta^{13}C$-mean value of the herbivorous animals is −19.3±0.6‰ and the $\delta^{15}N$-mean is 6.8±0.1‰. The carnivorous dog lies in between the omnivorous humans and the herbivorous animals with a $\delta^{15}N$-value of 8.2‰ and a $\delta^{13}C$-value of −18.8‰. This indicates a diet based on C_3-plants and C_3-plant-consuming terrestrial animals with some individuals having higher $\delta^{15}N$-values.

[JP, GG]

$\delta^{34}S$, $\delta^{13}C$ and $\delta^{15}N$ results processed at the UBC in Vancouver

Only half of the samples (n=72/146) have so far been measured for sulphur. The human and animal samples that were measured fulfil the quality criteria for the $\delta^{34}S$-values (%S), as they range from 0.2–0.3% (Nehlich and Richards 2009). The herbivorous animals (n=8, cattle, pig, equid, and sheep) have a mean $\delta^{34}S$-value of 6.8±1.1‰, the ratios range from 5.1‰ (cattle) to 8.3‰ (cattle). The median of the herbivorous is 6.8‰. The two carnivorous dogs show $\delta^{34}S$-values of 6.2‰ and 7.1‰ with a mean of 6.7±0.4‰ (Fig. 15.5).

Sulphur isotope values of all humans analysed so far (n=62) range from 2.6‰ to 14.5‰ with a mean of 6.9±2.0‰ and a median of 6.8‰ (Fig. 15.5).

As mentioned above, not only sulphur isotope, but also carbon and nitrogen isotope values were determined. As seven of the human samples have not fulfilled the quality criteria (C/N, weight%), only 55 samples were used for interpretation. The $\delta^{15}N$- and $\delta^{13}C$-values of the humans have a mean value for nitrogen of 9.9±0.9‰ and for carbon of −18.9 ±0.6‰. The herbivores (n=6) have a mean $\delta^{15}N$ of 5.2±0.9‰ and $\delta^{13}C$ of −20.1±0.7‰ (Fig. 15.6).

The two dogs show $\delta^{15}N$- and $\delta^{13}C$-values of 7.8±0.7‰ and −18.9±0.4‰ respectively, ranging in between the herbivores and the humans (Fig. 15.6).

[JP, ON, MPR]

Interpretation

$\delta^{13}C$- and $\delta^{15}N$-values processed at the LMU in Munich

The herbivores, sheep and cattle show $\delta^{13}C$- and $\delta^{15}N$-values that indicate a diet based on C_3-plants. The $\delta^{15}N$-value of the pig is, with 6.6‰, even lower than the herbivores and refers to a plant-based diet. Meat was probably too valuable to feed to pigs or to throw away. The mainly carnivorous dog has a $\delta^{15}N$-value of 8.2‰ and falls in between the herbivorous animals and the omnivorous

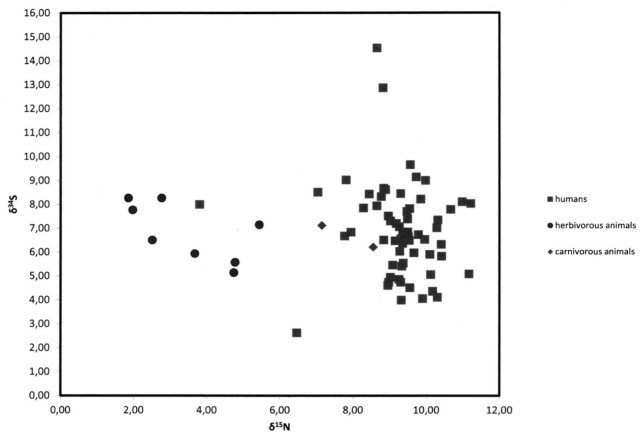

Fig. 15.5. Pergamon, Roman South-East Necropolis and late Byzantine North Cemetery. $\delta^{15}N$ and $\delta^{34}S$ isotope values for humans, herbivorous and carnivorous animals (Graphics: J. Propstmeier).

humans; indicating again, that meat was probably rare and was not fed to animals like dogs and pigs.

The mean value of the humans is −18.4±0.3‰ for $\delta^{13}C$ and 9.9±0.8‰ for $\delta^{15}N$ showing a diet based on C_3-plants and C_3-plant-consuming animals. There is a significant positive correlation between the N- and the C- isotopes (Pearson 0.629, p<0.01) of the humans. Marine input increases both the C- and the N-values, which could indicate a marine input in some of the individuals of Pergamon. In several publications a positive correlation between the $\delta^{13}C$- and $\delta^{15}N$-values were explained by a marine input and to some extent proven by an oligo element analysis (Giorgi *et al.* 2005). Freshwater fish were probably not consumed. Eating freshwater fish could result in more negative $\delta^{13}C$-values. As Vika and Theodoropoulou (2012) have recently shown for the Eastern Mediterranean, freshwater and marine fish are often indistinguishable in their $\delta^{13}C$- and $\delta^{15}N$-values.

The $\delta^{15}N$-value of the above-mentioned faunal species lies in the average range of 6.8‰. Humans, which have consumed these animals, must have values, if one assumes a fractionation factor of 3‰, higher than 9.8‰. All samples

having lower values must have had a mainly herbivorous diet (Fig. 15.7, in green). One individual even has nitrogen values lower than the dog. Reasons for this could be a different population stratum or that this person lived in a different time; so warlike conflicts, bad harvests, or epidemics could have occurred and some individuals may have temporarily suffered from an insufficient food supply.

The humans who have higher values than 9.8‰ had a mixed diet of plants and animals (Fig. 15.7, in blue). Some individuals even have much higher nitrogen values than the standard deviation of the mean value (Fig. 15.4). This seems to indicate a marine input in their diet; they have higher C-values as well (Fig. 15.7, in red). The sulphur isotope analysis could, however, not confirm this result (see below).

Summarizing the results, the diet of the investigated individuals was based on C_3-plants such as grain and vegetables like beans, as well as C_3-plant-consuming terrestrial animals. These findings are supported by the results of the archaeozoological investigation and the dental diseases found in the cemeteries studied (see above).

[JP, GG]

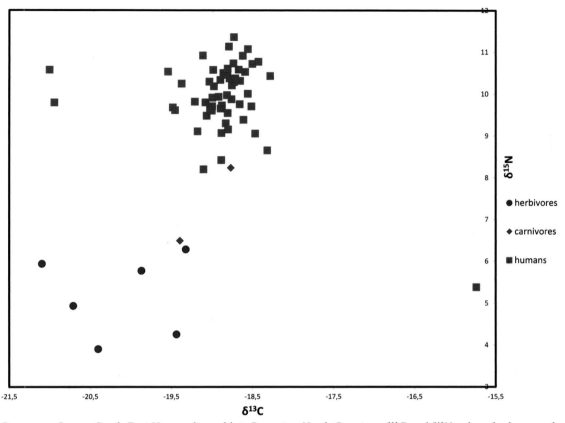

Fig. 15.6. Pergamon, Roman South-East Necropolis and late Byzantine North Cemetery. $\delta^{13}C$ and $\delta^{15}N$ values for humans, herbivorous and carnivorous animals (Graphics: J. Propstmeier).

The early Byzantine warrior

In 2006 an exceptional early Byzantine burial of a young male individual and several grave goods (including weapons) was discovered in Pergamon (Otten *et al.* 2011). Stable isotope analysis of the elements nitrogen, carbon, strontium, and oxygen was carried out by G. Müldner, J. Evans, and A. Lamb (in Otten *et al.* 2011) at the University of Reading. For comparison the skeletons of a Roman adult male and a late Byzantine child were analysed as well (see Fig. 15.8).

Both of the last mentioned individuals showed similar $\delta^{13}C$- and $\delta^{15}N$-values to the individuals analysed in Munich. But the early Byzantine male had an elevated $\delta^{13}C$ values in contrast to the other individuals which indicates a diet with different plants (C_4-plants). To our present knowledge, C_4-food plants were not or only rarely used in Pergamon. For the early Byzantine warrior a different origin can be assumed (Fig. 15.3), where C_4-plants played a more major role in the daily diet. A precise point of origin could not be determined by the results of the strontium and oxygen stable isotope analysis as the data were consistent with the wider East Mediterranean area (Müldner *et al.* in Otten *et al.* 2011).

[GHM]

$\delta^{34}S$-, $\delta^{15}N$- and $\delta^{13}C$-values processed at the UBC in Vancouver

Preliminary results of the sulphur, carbon, and nitrogen isotope analysis of humans and animals of Pergamon will be presented here. As mentioned above, not all results for the 146 samples studied were available yet for presentation (for sulphur: humans n=62, animals n=10; for carbon and nitrogen: human n=55, animals n=8).

The $\delta^{34}S$-values of the herbivores (n=8) have a rather small range (5.2‰ to 8.3‰) with a mean of 6.8±1.1‰. As the trophic level shift of sulphur isotope ratios between diet and consumers' tissue is low, around −1.0‰, the $\delta^{34}S$ range of the humans can be expected to be in the $\delta^{34}S$ range of the animals. Humans with solely terrestrial-based diets would have $\delta^{34}S$-values within the range of the $\delta^{34}S$-values of the terrestrial animals (or within 1‰, due to the small fractionation effect). The $\delta^{34}S$ range of the humans is 2.6‰ to 14.5‰ with a mean of 6.9±2.0‰ and fall into the range of terrestrial animals, representing the local variety and dietary variability. This suggests that the main dietary resource consisted of terrestrial domesticated animals; there is no indication of a marine input. But there are also a few humans that do not fall into this range (Fig. 15.5). It can

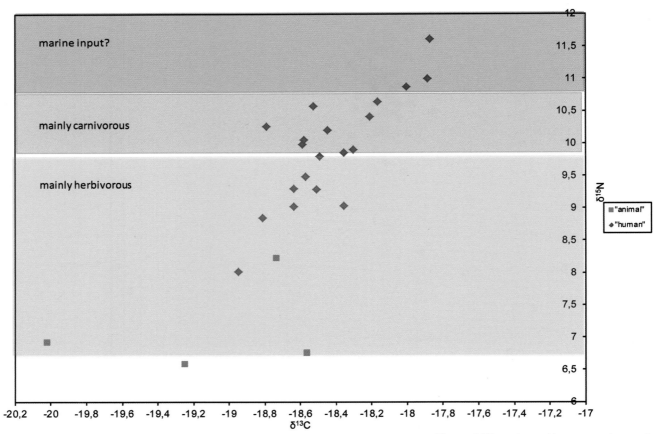

Fig. 15.7. Pergamon, Roman South-East Necropolis and late Byzantine North Cemetery. δ¹³C and δ¹⁵N values of humans and animals. Green: mostly herbivorous diet, blue: omnivorous diet, red: probable marine input (Graphics: J. Propstmeier).

be concluded that either the investigated animals do not represent the full range of biologically available $\delta^{34}S$ values or some individuals migrated to Pergamon. A marine input can also be excluded; in this case, higher $\delta^{13}C$-values would be expected.

The two mainly carnivorous dogs show values that are also placed within the range of the terrestrial animals.

The herbivores (n=6) show a $\delta^{13}C$-mean value of −20.1±0.6‰ and a slightly lower $\delta^{15}N$ value of 5.2±0.8‰ than the herbivorous animals analysed in the first set at the LMU in Munich. The two dogs have a $\delta^{15}N$ mean value of 7.8±0.6‰ and a $\delta^{13}C$ mean value of −18.9±0.4‰. This places the dogs again in between the herbivores and the humans.

The $\delta^{15}N$ and $\delta^{13}C$ mean values of the humans of the second dataset are quite similar to the mean values of the samples of the first dataset. Most of the human values (mean value of 9.9±0.9‰ for $\delta^{15}N$ and −18.9±0.5‰ for $\delta^{13}C$) fall within the standard deviation of the mean (Fig. 15.6). This again indicates a diet based on C_3-plants and C_3-plant-consuming animals.

Two individuals from the Roman necropolis show very low carbon values, the lowest one is a child between 7 and 12 years of age. Another individual shows very high carbon values, indicating a diet based on C_4-plants, like the above-mentioned early Byzantine warrior (Otten *et al.* 2011). It has very low nitrogen values, indicating that this person had a mainly herbivorous diet (Fig. 15.7), as he/she is in the same range as the herbivorous animals. It cannot be excluded, that these outliers come from a different place of origin with different food traditions. Again the $\delta^{34}S$ values for these individuals are not yet available for interpretation.

These outliers should be studied more in detail. Bad harvests have to be taken into consideration; such events were described by Galen for the 2nd century AD. In general, the alimentary status of the individuals investigated was not bad. As described by Teegen (this volume), enamel hypoplasias are quite common in Pergamon, both in the Roman and Byzantine period. Severe forms are very rare. At the moment, there are no indications of severe malnourishment. It seems more likely that the enamel defects were mainly caused by diseases in childhood.

It was surprising that there was not a significant amount of marine food consumption by individuals at this site close to the Mediterranean Sea.

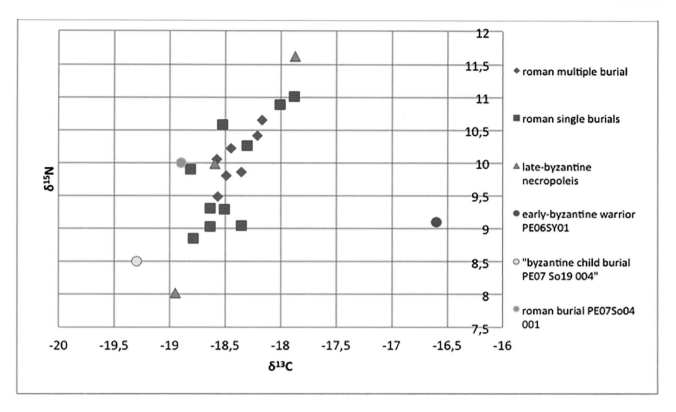

Fig. 15.8. Pergamon. δ¹³C and δ¹⁵N values of the Roman South-East Necropolis and the late Byzantine North Cemetery in comparison to the Early Byzantine warrior burial (PE06-Sy01) a Late Byzantine child burial (PE07-So19-004) and the Roman adult (PE07-So04-001) (data from Müldner in Otten et al. 2011) (Graphics: J. Propstmeier).

As there are still samples to evaluate, they will help to gain an even better and broader insight into the diet and the living conditions of the populations of Pergamon.

[JP, ON, MPR]

Conclusion

The results of the carbon, nitrogen, and sulphur isotope analysis of the Roman and late Byzantine skeletons from Pergamon indicate a diet based on C_3-plants and terrestrial animals with no input of marine food. Similar observations were made for Sagalassos (Fuller *et al.* 2012), one of the few sites from Asia Minor where analyses of stable isotopes for diet reconstruction were carried out (for other new investigations see the relevant contributions in this volume). This is in accordance with the archaeozoological investigations at Pergamon, where only rarely fish bones were observed (Boessneck and von den Driesch 1985; von den Driesch 2008; MacKinnon 2011; personal observations by WRT during the 2012–14 campaigns). The same can be assumed for the Byzantine period (cf. Kroll 2010; 2012). The sometimes quite high

amount of molluscs (cf. Fig. 15.1), which can also be observed in Byzantine contexts (personal observations by WRT in Priene and Pergamon), probably left no, or only slight, traces in the isotopic values. This needs, however, further investigation.

The palaeopathological study of the jaws and teeth indicates a more plant-based diet with a remarkably high carbohydrate consumption and a certain amount of (animal) protein. The present study underlines the importance of sulphur isotope analysis for the investigation of possible marine food intake.

[WRT, JP]

Note

1 The stable isotope analyses were part of the project 'Anthropological and palaeopathological investigations in the Roman South-East Necropolis of Pergamon' (head W.-R. Teegen) in the framework of two projects 'The Roman South-East Necropolis of Pergamon', supported by the Gerda Henkel Foundation 2011–14 (principal investigator F. Pirson). All raw data will be published in the final publication of the South-East Necropolis, to be edited by U. Kelp and W.-R. Teegen.

Acknowledgements

We are grateful to the Pergamon excavation of the German Archaeological Institute (director Professor Felix Pirson), the Gerda Henkel-Foundation and the Ludwig-Maximilians-University Munich, and the Natural Sciences and Engineering Research Council of Canada (NSERC) for support. The Turkish Ministry of Culture and Tourism (Ankara) kindly granted all necessary permits. We are further grateful to the anonymous reviewer for helpful comments.

Bibliography

Ambrose, S. and Norr, L. (1993) Experimental evidence for the relationship of the carbon isotope ratios of whole diet and dietary protein to those of bone collagen and carbonate. In J. Lambert and G. Grupe (eds.) *Prehistoric human bone: Archaeology at the molecular level*, 1–37. Berlin, Heidelberg, and New York, Springer.

Boessneck, J. and von den Driesch, A. (1985) *Knochenfunde aus Zisternen in Pergamon.* Munich, Institut für Paläoanatomie, Domestikationsforschung und Geschichte der Tiermedizin der Universität München.

Coltrain, J., Harris, J., Cerling, T., Ehleringer, J., Dearing, M., Ward, J., and Allen, J. (2004) Rancho La Brea stable isotope biogeochemistry and its implications for the palaeoecology of Late Pleistocene, coastal southern California. *Palaeogeography, Palaeoclimatology, Palaeoecology* 205, 199–219.

Craig, O. E., Steele, V. J., Fischer, A., Hartz, S., Andersen, S. H., Donohoe, P., Glykou, A., Saul, H., Jones, D. M., Koch, E., and Heron, C. P. (2011) Ancient lipids reveal continuity in culinary practices across the transition to agriculture in Northern Europe. *Proceedings of the National Academy of Sciences* 108, 17910–5.

Driesch, A. von den (2008) Tierreste aus dem Podiensaal. In H. Schwarzer, *Das Gebäude mit dem Podiensaal in der Stadtgrabung von Pergamon. Studien zu sakralen Banketträumen mit Liegepodien in der Antike* (Die Stadtgrabung. Teil 4. Altertümer von Pergamon XV.4), 309–13. Berlin and New York, de Gruyter.

Fuller, B. T., Cupere, B., de Marinova, E., Van Neer, W., Waelkens, M., and Richards, M. P. (2012) Isotopic reconstruction of human diet and animal husbandry practices during the Classical-Hellenistic, Imperial, and Byzantine periods at Sagalassos, Turkey. *American Journal of Physical Anthropology* 149.2, 157–71.

Giorgi, F., Bartoli, F., Iacumin, P., and Mallegni, F. (2005) Oligoelements and isotopic geochemistry: A multidisciplinary approach to the reconstruction of the palaeodiet. *Human Evolution* 20, 55–82.

Grant, M. (2000) *Galen on food and diet*. London and New York, Routledge.

Grupe, G., Christiansen, K., Schröder, I., and Wittwer-Backofen, U. (2012) *Anthropologie. Ein einführendes Lehrbuch* (2nd ed.). Berlin, Heidelberg, and New York, Springer.

Kiesewetter, H. (this volume), Toothache, back pain, and fatal injuries: What skeletons reveal about life and death at Roman and Byzantine Hierapolis, 268–85.

Klinken, G. J. van (1999) Bone collagen quality indicators for palaeodietary and radiocarbon measurements. *Journal of Archaeological Science* 26, 687–95.

Kroll, H. (2010) *Tiere im Byzantinischen Reich. Archäozoologische Forschungen im Überblick* (Monographien des Römisch-Germanischen Zentralmuseums 87). Mainz, Römisch-Germanisches Zentralmuseum.

Kroll, H. (2012) Animals in the Byzantine Empire: An overview of the archaeozoological evidence. *Archeologia Medievale* 39, 93–121.

Lieverse, A. R. (1999) Diet and the aetiology of dental calculus. *International Journal of Osteoarchaeology* 9, 219–32.

MacKinnon, M. R. (2011) Animal use at Hellenistic Pergamon: Evidence from zooarchaeological analyses. *Archäologischer Anzeiger* 2011.2, 193–8.

Nehlich, O., and Richards, M. P. (2009) Establishing collagen quality criteria for sulphur isotope analysis of archaeological bone collagen. *Archaeological and Anthropological Sciences* 1.1, 59–75.

Nehlich, O. (2015) The application of sulphur isotope analyses in archaeological research: A review. *Earth-Science Reviews* 142, 1–17.

Nehlich, O., Fuller, B., Jay, M., Mora, A., Nicholson, R., Smith, C., and Richards, M. P. (2011) Application of sulphur isotope ratios to examine weaning patterns and freshwater fish consumption in Roman Oxfordshire, UK. *Geochimica et Cosmochimica Acta* 75, 4963–7.

Nehlich, O., Oelze, V., Jay, M., Conrad, M., Stäuble, H., Teegen, W.-R., and Richards, M. P. (2014) Sulphur isotope ratios of multi-period archaeological skeletal remains from central Germany: A dietary and mobility study. *Anthropologie* (Brno) 52, 15–33.

Nováček, J., Scheelen, K., and Schultz, M. (this volume) The wrestler from Ephesus: Osteobiography of a man from the Roman period based on his anthropological and palaeopathological record, 318–38.

Oelze, V., Koch, J., Kupke, K., Nehlich, O., Zäuner, S., Wahl, J., Weise, S., Rieckhoff, S., and Richards, M. P. (2012) Multi-isotopic analysis reveals individual mobility and diet at the Early Iron Age monumental tumulus of Magdalenenberg, Germany. *American Journal of Physical Anthropology* 148, 406–21.

Otten, T., Evans, J., Lamb, A., Müldner, G., Pirson, A., and Teegen, W.-R. (2011) Ein frühbyzantinisches Waffengrab aus Pergamon. Interpretationsmöglichkeiten aus archäologischer und naturwissenschaftlicher Sicht. *Istanbuler Mitteilungen* 61, 347–422.

Pirson, F. (2011) Pergamon – Bericht über die Arbeiten in der Kampagne 2010. *Archäologischer Anzeiger* 2011.2, 81–212.

Pirson, F. (2012) Pergamon – Bericht über die Arbeiten in der Kampagne 2011. *Archäologischer Anzeiger* 2012.2, 175–274.

Pirson, F. (2013) Pergamon – Bericht über die Arbeiten in der Kampagne 2012. *Archäologischer Anzeiger* 2013.2, 79–164.

Pirson, F. (2014) Pergamon – Bericht über die Arbeiten in der Kampagne 2013. *Archäologischer Anzeiger* 2014.2, 101–76.

Plicht, J. van der (2015) *Unpublished Reports CIO/433-2015/PWL and CIO/582-2015/PWL.* 07.04.2015 and 21.09.2015.

Propstmeier, J. (2012) *Die Lebensbedingungen in Pergamon: Nahrungsrekonstruktion mit Hilfe stabiler Stickstoff- und Kohlenstoffisotope einer römischen und spätbyzantinischen Nekropole*. Unpublished Master's thesis, University of Munich.

Propstmeier, J. (2013) Analyse stabiler Isotope zur Ernährungsrekonstruktion. *Archäologischer Anzeiger* 2013.2, 141–42.

Radt, W. (1999) *Pergamon. Geschichte und Bauten einer antiken Metropole*. Darmstadt, Wissenschaftliche Buchgesellschaft.

Raitapuro-Murray, T., Molleson, T. I., and Hughes, F. J. (2014) The prevalence of periodontal disease in a Romano-British population *c.* 200–400 AD. *British Dental Journal* 217.8, 459–66.

Schroeder, H. E. (1997) *Pathobiologie oraler Strukturen*, 3rd ed. Basel, Karger.

Smith, B. H. (1984) Patterns of molar wear in hunter-gatherers and agriculturalists. *American Journal of Physical Anthropology* 63, 39–56.

Teall, J. L. (1959) The grain supply of the Byzantine Empire, 330–1025. *Dumbarton Oaks Papers* 13, 87–139.

Teegen, W.-R. (2013) Die anthropologisch-paläopathologischen Untersuchungen 2012. *Archäologischer Anzeiger* 2013.2, 138–43.

Teegen, W.-R. (2015) Die anthropologisch-paläopathologischen Untersuchungen in Pergamon 2014. *Archäologischer Anzeiger* 2015.2, 158–63.

Teegen, W.-R. (this volume) Pergamon – Kyme – Priene: Health and disease from the Roman to the late Byzantine period in different locations of Asia Minor, 250–67.

Vika, E., and Theodoropoulou, T. (2012) Re-investigating fish consumption in Greek antiquity: Results from δ 13 C and δ 15 N analysis from fish bone collagen. *Journal of Archaeological Science* 39.5, 1618–27.

Wong, M., Nauman, E., Jaouen, K., and Richard, M. (this volume) Isotopic investigations of human diet and mobility at the site of Hierapolis, Turkey, 228–36.

Pergamon – Kyme – Priene: Health and disease from the Roman to the late Byzantine period in different locations of Asia Minor

Wolf-Rüdiger Teegen

Abstract

This paper focuses on the preliminary results from the 2007, 2011, and 2013 excavations in the Roman South-East Necropolis and the Byzantine North Cemetery at Pergamon. Furthermore, some singular cases from the Byzantine cemeteries at Kyme Eolica (severe infant diseases) and Priene (alterations of the spine and malformations) will be presented. Diet and dental diseases from Pergamon will mainly be discussed in the paper given by Johanna Propstmeier and co-workers (this volume).

The individuals from the Roman necropolis, probably belonging to the middle class, were relatively healthy. The following diseases were observed: frontal and maxillary sinusitis, mastoiditis, inflammatory processes of the skull and its veins, linear enamel hypoplasia, root hypoplasia, Harris' lines, degenerative joint diseases (body joints and vertebral column), Schmorl's nodes, button osteoma, and fibro-osseous tumour. More affected were probably the individuals of lower status from the Byzantine North Cemetery. Special medical interventions like trepanations have, so far, not been observed.

The infants from Kyme Eolica suffered from haemorrhagic and inflammatory processes of the skull and its veins, such as, for example, meningitis. Furthermore, a birth trauma in the form of a haematoma on the external lamina of the skull was observed.

The individuals from Priene showed a similar spectrum of diseases as the people from Pergamon. Furthermore, several fractures of the postcranial skeleton were observed. Three out of 12 cases of spondylolysis of the 5th lumbar vertebra are present. There is one female skeleton with syndromic sagittal craniosynostosis, malformation of the temporal lobe and the Fossa cranii posterior, shortening of the long bones on the left side and twins (at least) who all died in the 8th to 9th lunar months.

Most individuals from Pergamon and Priene showed more or less extended enamel hypoplasia, mostly on the permanent teeth. These findings indicate a high level of stress between one and seven years of age. People of different levels of social status were affected. This also means a high level of disease occurred in the early years of the children across different social settings.

Keywords: Byzantine period, Kyme, medical history, mortality, palaeodemography, palaeopathology, Pergamon, Priene, Roman period.

Introduction

Health is surely the most important consideration for everyone. This is true not only today, but in particular in ancient times, where medical support was scarce or non-existent. The only sources for analysing the health status for ancient populations are their skeletal remains, either from inhumations (cf. Fig. 16.3) or from cremations (cf. Fig. 16.10).

As is well known, human skeletal remains are first-class 'bio-historic documents', which can offer insights into the life history of ancient people. By analyzing skeletal remains, it can be possible to reconstruct the 'osteobiography' of an individual or of an entire population. Therefore, human skeletons should be carefully excavated and studied. Still today, in many Mediterranean regions, however, skeletal remains are generally given less attention than the accompanying grave goods. This should change in the future.

The topic of this paper is the health of selected Roman and Byzantine cemetery populations from western Asia Minor (Fig. 16.1). The human skeletons from Roman and late Byzantine Pergamon, late Byzantine Kyme and Priene were recently studied from an osteoarchaeological and palaeopathological point of view. This paper gives a preliminary overview of the most important findings. Data from all cemeteries will be published as monographs.

The text presents mainly the oral presentation at the Fredrikstad conference in 2013 with references added.

Some recent discoveries made during the 2014 campaign in Pergamon were also included in the current paper.

Methods

Sex and age of juveniles and adults were determined using the recommendations of the European Association of Anthropologists (Ferembach *et al.* 1980; Rösing *et al.* 2007). Furthermore, the closure of the maxillary sutures was used for age determination (Mann *et al.* 1991).

The sex of the subadults was determined according to Schutkowski (1990; 1993). The age of foetuses, infants, and children was determined either using the long bone length (for foetuses by Kósa 1978, for infants and children by Schmid and Künle 1958, Stloukal and Hanáková 1978, Schaefer *et al.* 2009) and/or the development status of the teeth (Ubelaker 1989).

Metrics were recorded according to Martin 1928 and Bräuer 1988, and body height was calculated using Pearson's 1899 formulae, corrected by Rösing 1988. Epigenetic traits

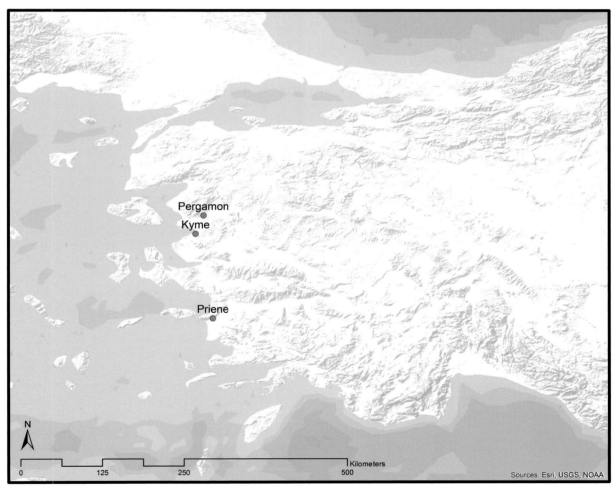

Fig. 16.1. Asia Minor. Localization of Pergamon, Kyme and Priene.

were registered using Hauser and De Stefano (1989) as a reference for the skull, Alt (1997) for the teeth, and Wiltschke-Schrotta 1988 for the postcranial skeleton.

For the analysis of pathological alterations several well-known papers and textbooks were used, mainly by Michael Schultz (1988; 1993; 2001; 2003), the late Don Ortner (2003), and others (e.g. Fornaciari and Giuffra 2009; Lewis 2009; Waldron 2001; Walker 2012).

All bone surfaces were checked macroscopically and microscopically for the presence of pathological alterations and Harris lines (cf. Ameen *et al.* 2005). Some x-rays were carried out by radiologist Dr İdris Yavuzyılmaz from Bergama (Prov. İzmir). For Priene, CTs and x-rays were taken at the hospital in Söke (Prov. Aydın).

This paper focuses on the preliminary results from the 2007, 2011, and 2013–14 excavations in the Roman South-East Necropolis and the Byzantine North Cemetery at Pergamon (Pirson 2012). Furthermore, I will present some singular cases from the (late) Byzantine cemeteries at Priene (alterations of the spine and malformations) and at Kyme (severe infant diseases and infant mortality).

Some preliminary results are published for Pergamon in the annual excavation reports (Teegen 2011d; 2012a; 2013; 2014; 2015), for Priene in the annual meeting reports of the American Association of Physical Anthropology (Teegen 2010; 2011c) and for Kyme in a booklet about the Italo-Turkish excavations and an excavation report (Teegen 2012b; forthcoming).

Pergamon

Material

During the construction work for the cable car at Bergama a Roman necropolis was discovered quite unexpectedly. Foundations of several funeral monuments were excavated by the German Archaeological Institute in close cooperation with the Museum of Bergama in spring 2007 (Mania 2008; Pirson 2008). Around the monuments, generally containing multiple burials, several other burials were discovered, including both inhumations and some cremations. The excavations were continued in summer 2011, 2013, and 2014 (Pirson 2012; 2013; 2014). This was made possible thanks to the generous support of the Gerda Henkel Foundation (Düsseldorf). From the analysis of the ceramics and other findings from the burials, they can be dated to the 1st to 4th century AD (personal communication by S. Japp and U. Kelp, 2014). A first test of radiocarbon dates of selected human skeletons yielded a calibrated time span between 40 BC and 415 AD (GrA62665–71, 62677–8, 63897; van der Plicht 2015).

The human skeletal remains were investigated by the undersigned from an anthropological and palaeopathological point of view. There are more than 50 burials and eight funerary monuments with 12 burial shafts. The graves contain more than 100 individuals, including approximately 25 cremations. Furthermore, an early Byzantine warrior burial (Teegen 2011b) and 13 individuals from the late Byzantine North Cemetery were studied. The location of these burials are shown in Fig. 16.2.

Results

Multiple burials are quite common in the Roman South-East Necropolis (cf. Fig. 16.3). But we can sometimes observe multiple burials in the Byzantine cemeteries: there are up to three individuals in the North Cemetery from Pergamon, or up to four individuals in Kyme and Priene (see below). Generally, the human remains from the different burials are mixed up. This makes a reconstruction of the single individuals quite difficult. This is particularly complicated, when people of the same age or the same sex were buried together. The cremations also sometimes contain the remains of multiple individuals.

PALAEODEMOGRAPHY

In the Roman burials from the South-East Necropolis, both males and females are present. Their ages range from foetuses up to individuals of 60-plus years of age. The majority, however, died below 40 years of age. The same is true for the adults from the Pergamean late Byzantine North Cemetery and the late Byzantine cemeteries at Priene and Kyme (see below; Fig. 16.21). The average lifespan for females is approximately 10 years shorter than for males.

The burial chambers in the funerary monuments contained generally more than one individual. Burial 15, discovered in 2011, contained at least 11 individuals, mostly adults (Fig. 16.3). But it included also the left vertebral arch of the 2nd cervical vertebra of a late foetus or a neonate. The burial in funerary monument 7, excavated in 2014 contained also a minimum of 11 individuals: at least six adults, one child and three late foetuses or neonates and fragments of a cremated individual. From an archaeological point of view, the foetuses or neonates could not be linked to any of the females buried in the same tomb. Therefore, it remains unclear, if they were the remains of foetal/neonatal burials or of one or more pregnant females. Their skeletal remains are of particular interest. There is one secure case from the South-East Necropolis and a questionable case from the North Cemetery: in grave monument 6 a 25–35-year-old female skeleton (PE07 So4 006) was excavated. During the examination of the bones the left tibia of a foetus was found between the pelvis and the lumbar vertebrae. Age determination of foetuses is possible using their long bone length. The tibia belonged to a foetus in the seventh gestational month (Teegen 2009). In the Byzantine North Cemetery, a foetus of the same age can probably also be linked to a young female. Unfortunately, this burial was plundered by grave robbers. The context is, therefore, not totally clear. Nevertheless, these cases clearly show the dangers of pregnancy and giving birth in Antiquity

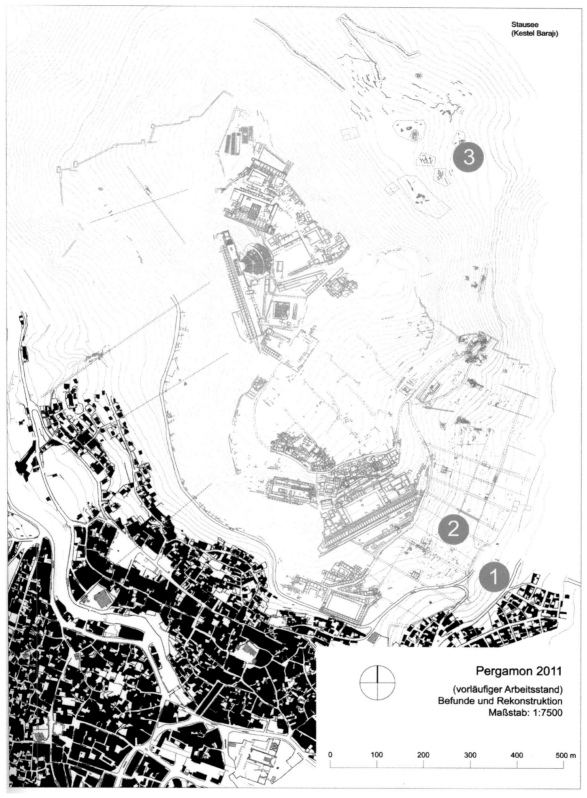

Fig. 16.2. Pergamon. Localization of the cemeteries mentioned in text. 1. Roman South-East Necropolis; 2. Early Byzantine warrior burial; 3. Late Byzantine North Cemetery. Modified after Pirson 2012.

Fig. 16.3. Pergamon. Roman South-East Necropolis. Multiple burial (PE11-So-11-Gr15) with at least 11 individuals.

and beyond. They can also explain the early mortality of adult females.

PALAEOPATHOLOGY

Traces of several diseases could be found analysing the bone finds. This included diseases of the teeth and jaws, the skull, the respiratory system, joints (vertebrae and postcranial bones), trauma, tumours, and malformation.

The dental diseases present in the samples from Pergamon are dental caries, abscesses, dental calculus, parodontopathies and intra vitam tooth loss. They are discussed in the paper by Propstmeier *et al.*(this volume). Furthermore, teeth were probably used as tools.

Enamel hypoplasia

Unspecific stress markers like linear transverse enamel hypoplasias are quite common (Fig. 16.4). They were the result of a growth disruption during tooth development (Hillson 2014 with extensive bibliography). Severe cases also cause growth disruption on the underlying dentine tissue. This could sometimes be observed in cremations from Pergamon, where the enamel of the tooth crown was not preserved (Fig. 16.5).

The onset of enamel hypoplasias can be determined by their distance to the dento-enamel junction, using either metrics (cf. Rose *et al.* 1985, 294 Fig. 9.4; Hillson 2014, 179–81) or counting the number of pericymata (Hillson 2014, 182–4). The latter is more precise, but requires, however, the application of microscopic techniques. This is quite difficult in the field.

Enamel hypoplasia of the deciduous dentition develops between the late foetal stages and approximately two years of age. In the permanent dentition it develops mainly between the first and the eleventh year of age. Further information can be derived by observing the hypoplastic defects in the root dentine (Teegen 2004; cf. Fig. 16.4). By analyzing the third molar we can also collect information about tooth

growth disruption between 12 and approximately18 years of age, when the root is fully grown.

Not only in Pergamon, but also in the other sites studied, enamel hypoplasias were rarely found in the crowns of deciduous teeth. In one case from Pergamonan enamel defect was found at the crown tips of a deciduous molar (Fig. 16.6). Simon Hillson (2014, fig. 3.13) has charted the development of the deciduous teeth around birth. This means that the enamel defect on Fig. 16.6 was caused by a growth

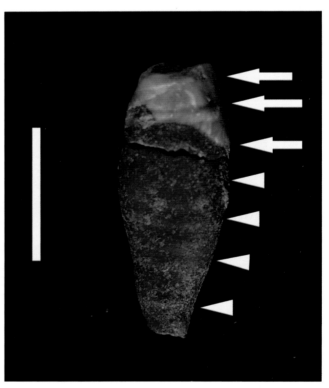

Fig. 16.4. Pergamon. Roman South-East Necropolis. Canine with enamel and root hypoplasias (PE11-So-11-044-Gr24). Scale 1 cm.

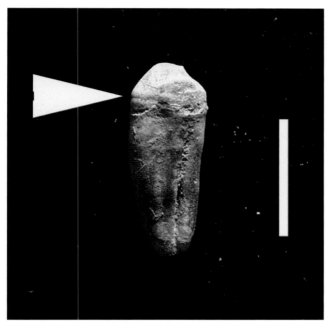

Fig. 16.5. Pergamon. Roman South-East Necropolis. Dentine hypoplasia in a cremated premolar (PE11-So-11-041-1s). Scale 1 cm.

disturbance during the late phase of pregnancy or around birth.

In the samples studied, enamel hypoplasias were primarily observed in the permanent teeth. In Pergamon they mainly developed between two and seven years, with a peak between three and four years. There were only a few observations below two years of age. Tooth abrasion in later life has to be taken into consideration.

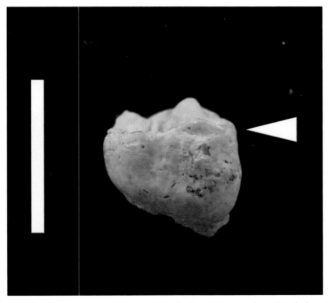

Fig. 16.6. Pergamon. Roman South-East Necropolis. 'Birth line' in a deciduous molar of a small infant (PE07-So04-017, KF 147). Scale 1 cm.

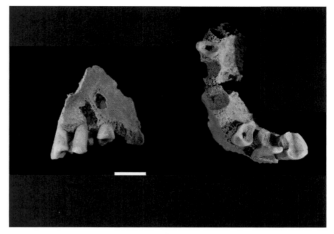

Fig. 16.7. Pergamon. Roman South-East Necropolis. Teeth as tools (?) in the upper jaw of an adult individual (PE11-So-11-031_#1). Scale 1 cm.

Teeth as tools

The last point regarding the teeth is the observation of unusual abrasion. The individual PE11-So-11-031_#1 exhibits a groove with a diameter of approximately 5 mm on the occlusal surface of an incisive tooth, running from the buccal to the palatal side of the tooth (Fig. 16.7). Normal chewing cannot produce this kind of abrasion. This individual probably held something between his teeth, regularly and over a long period. The traces are quite characteristic of using teeth as tools (cf. Alt and Pichler 1998).

In the Roman population from the South-East Necropolis diseases of the internal lamina of the skull (Schultz 1993; 2001) are rare. This is true for infants and adults alike and stands in contrast to Byzantine burials from Pergamon itself (Schultz 1989a; Schultz and Schmidt-Schultz 1994;

Fig. 16.8. Pergamon. Roman South-East Necropolis. Organised epidural hematoma on the internal lamina of the skull of a young female (PE14-Ar-02-084).

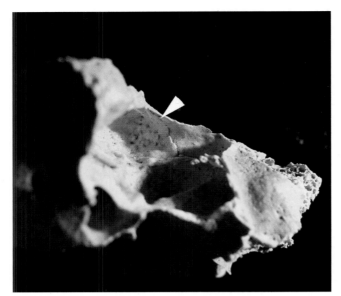

Fig. 16.9. Pergamon. Roman South-East Necropolis. Maxillary sinusitis (PE07-So04-003).

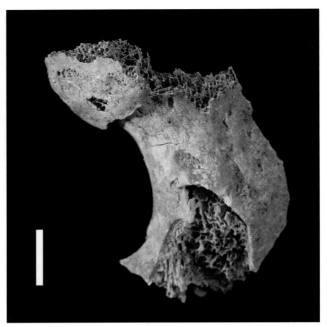

Fig. 16.10. Pergamon. Roman South-East Necropolis. Degenerative joint disease of the femoral head of a cremated adult (PE11-So-10-047-Gr26). Scale 1 cm.

Teegen 2014) or Byzantine Kyme (see below). One of these rare cases is a skull fragment of a young adult female. Here, several tree-shaped areas of new bone formation and serpent-like vessel impressions are present (Fig. 16.8). They are very likely the result of an organized epidural haematoma (Schultz 1993; 2001; 2003). This could have been caused by an accident or interpersonal violence.

Respiratory diseases

The most common infection in the osteoarchaeological record is sinusitis, the inflammation of the pneumatic cavities in the skull (cf. Schultz 1993; Aufderheide and Rodriguez-Martín 1998, 257). They are quite common in all cemetery groups and therefore in all social classes. The frequency of maxillary sinusitis (Fig. 16.9) is generally higher than all other forms due to dental abscesses in the upper jaw. Sinusitis at Pergamon was probably at least partially caused by rainy winters, with storms and poor heating systems as well as allergic reactions to the fumes. It is interesting to note that even people of rank, like the old man buried in the monumental İlyastepe tumulus near Pergamon, suffered from this type of infection (Teegen 2011a).

Other diseases of the skull were rarely observed in adults. However, in one case, a traumatic injury of the external lamina of the skull was present, which will be presented later (cf. Fig. 16.13).

Joint diseases

Degenerative joint diseases (cf. Schultz 1988, 481–7; Aufderheide and Rodriguez-Martín 1998, 93–7) were mainly observed in the spinal column rather than in the body joints. The femoral head of an adult from the

cremation burial PE11-So-10-047-Gr26 (Fig. 16.10) shows the characteristic new bone formation at the rim and the fungus-like structure. In general, the degree of degenerative joint diseases is quite low in the Roman population studied. This probably correlates with the middle class appearance of the South-East Necropolis.

Due to poor preservation, the degree of joint diseases (cf. Schultz 1988, figs. 170–1) in the individuals from the Byzantine North Cemetery could rarely be determined.

Degenerative changes of the vertebral joints are more common. Sometimes there are also Schmorl's nodes present. In one adult individual (PE11-So-11-053-Gr27) at least three thoracic vertebrae are ankylosed due to

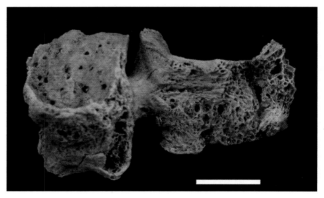

Fig. 16.11. Pergamon. Roman South-East Necropolis. Ankylosis of the anterior ligament of three thoracic vertebrae (PE11-So-11-053-Gr27). Scale 1 cm.

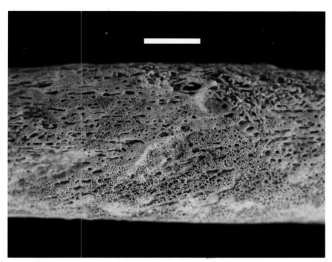

Fig. 16.12. Pergamon. Late Byzantine North Cemetery. Burial 4 with osteomyelitic process of both tibiae. Fine plaque-like structures indicating a process which was still on-going at the time of death. Scale 1 cm.

Fig. 16.13. Pergamon. Roman South-East Necropolis. Sharp trauma on the skull of an adult male (PE11-So-11-044-Gr24).

ossification of the anterior ligaments (Fig. 16.11). Here, an ankylosing spondylitis seems likely (cf. Aufderheide and Rodriguez-Martín 1998, fig. 6.14).

Unspecific infections

Remarkable is a case of osteomyelitis from the Byzantine North Cemetery (Fig. 16.12). It is very likely that it was caused by an infectious disease. Both tibia shafts are enlarged and exhibit an anomalous surface structure. Unfortunately, both tibiae are quite badly preserved and broken. This makes it easily possible to see the internal structure of the shafts. They show several layers of newly built bone. This is a quite typical reaction to an inflammatory process.

Histological analyses of a bone sample were carried out by Michael Schultz from Göttingen University; the results of which will be published elsewhere. The macroscopic image of the outer layer of the tibia also shows some fine plaque-like structures, indicating a process which was still ongoing at the time of death (Fig. 16.11). Probably, the spread of infection was also the cause of death, for example, by a general sepsis.

Trauma

Traumas both of the skull and the postcranial skeleton are rare in the Romans from Pergamon. There are two cases of cranial trauma (cf. Fig. 16.13) and several rib fractures. A well-healed fracture of a tibia probably indicates some kind of medical intervention.

Tumours

Today, tumours are quite a common cause of death. In Antiquity, the contrary was the case, mainly due to the lower life expectancy. But nevertheless, tumours can be regularly

found in the archaeological record. From Roman Pergamon, at least three probable cases are known. One case seems to be a fibro-osseous tumour in the frontal sinus (Fig. 16.14). This kind of benign tumour often develops after chronic inflammatory processes of the sinus.

The other case is a little more problematic. In the left metatus acusticus externus of an adult male from the multiple burial 15 (cf. Fig. 16.3) (PE11-So-11-Gr15) an external auditory exostosis (EAE) was discovered (Fig. 16.15). The form of the alteration is clear and it can easily be described as EAE. Here, however, it is not clear if it is a benign tumour or the result of taking cold baths or an epigenetic trait (Torus auditivus). There

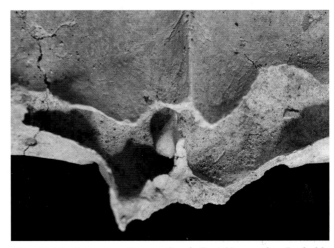

Fig. 16.14. Pergamon. Roman South-East Necropolis. Probable fibro-osseous tumour (Ø ca. 5 mm) in the left frontal sinus (PE11-So-11-053-Gr27).

is also one case of a slight torus auditivus, observed in the burials from the 2007 rescue excavation in the same South-East Cemetery. When we consider the outcome of this phenomenon, it is clear that these two people could have had difficulties in hearing.

As D. Campillo and J. Baxarias (2008) have shown, a cholesteatoma is a quite likely interpretation for the case of the adult male PE11-So-11-Gr15 (Fig. 16.15). Computer tomography has, however, still to be performed. The presence of EAE on only one side makes it not very likely that this alteration was the result of cold baths.

Sometimes tumours can be found in cremated bones. This is the case in the adult individual PE11-So-10-044-4 (Teegen 2013, 139, fig. 63).

Malformations

Studying the last skeletons from the 2007 rescue excavation in summer 2013, a quite interesting case was discovered. Burial PE07-So-09-008 contained the skeleton of a young female. Her skull is deformed (Fig. 16.16) and the mastoid processes are of different dimensions. The latter could indicate the presence of a torticollis. Analyzing the sutures of the skull, a premature craniosynostosis seems likely. Both alterations can be found together (Kutterer and Alt 2008). Nearby, another deformed skull was found, belonging to a middle-aged individual (PE07-So-09-009). Deformation by soil pressure seems not very likely. Due to its more advanced age, the cause for the deformation could not clearly be ascertained.

People with premature craniosynostosis and torticollis have a quite different physical appearance. There is one unusual Roman helmet from the Xanten area in the province of Germania inferior (von Prittwitz und Gaffron *et al.* 1991), which was especially designed for an officer with

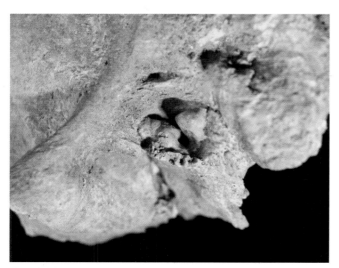

Fig. 16.15. Pergamon. Roman South-East Necropolis. External auditory exostosis (PE11-So-11-Gr15_Skull).

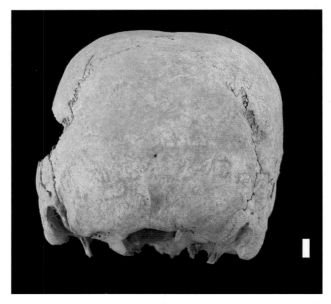

Fig. 16.16. Pergamon. Roman South-East Necropolis. Torticollis and skull deformation (Female, 21-25 years; PE07-So09-008). Scale 1 cm.

torticollis. People with skull deformation due to premature craniosynostosis have a normal IQ, but sometimes learning disorders (Magge *et al.* 2002).

Priene

The ancient city Priene (Turkey; Fig. 16.1) was excavated by German archaeologists between 1895 and 1898 (Wiegand and Schrader 1904). Since 1998, teams from the Universities of Frankfurt and Bonn, with support from the German Archaeological Institute at Istanbul, have continued the excavations (Raeck 2008).

During the excavations by the University of Frankfurt (2005–10) in the former sanctuary of the Egyptian Gods (AEG) (Hennemeyer 2005; Raeck 2008; 2009), a late Byzantine cemetery was discovered. It probably dates to the 13th/14th century AD. Radiocarbon dates are still missing. Furthermore, some small grave groups were excavated throughout the ancient city (Fig. 16.17); several late Byzantine burials were discovered: in the so-called Agora chapel (AK) three burials with at least six individuals were excavated. Nearby, in the so-called south complex (SK), two other burials were found and excavated. Outside the city near the street to the acropolis (Teloneia) burials of a previously unknown Byzantine cemetery were discovered. In the Hellenistic to Roman cemetery, by the main gate, some dispersed human bones were also found.

Results

The anthropological and palaeopathological analysis of the late Byzantine skeletons from Priene revealed a minimal

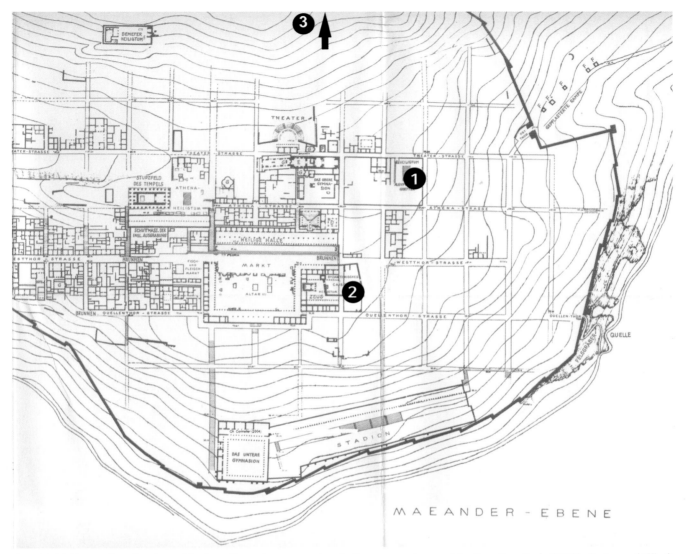

Fig. 16.17. Priene. Localization of the Late Byzantine cemeteries: 1. Former sanctuary of the Egyptian Gods (AEG); 2. Agora (AK); 3. Teloneia. Modified after Wiegand and Schrader 1904.

number of 50 individuals from the AEG, AK and SK sites. The inhumations from Priene are similarly preserved as those from Pergamon.

Spondylolysis

Degenerative joint diseases of the spine and of the large body joints are common in late Byzantine Priene. Furthermore, there are three cases of spondylolysis (Fig. 16.18). Two cases of spondylolysis of the 5th lumbar vertebra (L5) from the AEG area and one from the SK area were discovered. The frequency of three out of 12 individuals (25%) with a preserved 5th lumbar vertebra is quite high. This approximates the frequency found in Inuit populations (Ortner 2003, 147). In modern populations of the Old World the frequency lies between 2.3% and 10%, predominantly

males (Walker 2012, 116). In Priene only males between 20 and 50 years of age are affected. In only one case is the vertebral arch still preserved (Fig. 16.18). In one individual the L5 has slipped approximately 5 mm ventrally. This indicates a slight spondylolisthesis.

All individuals with spondylolysis show severe degenerative joint disease (DJD) of the lumbar vertebral bodies, indicating heavy physical loading and probably lower back pain. It is striking that the individuals buried inside the Agora chapel, who probably belonged to a higher social class, show neither signs of spondylolysis nor severe DJD.

Due to severe DJD in the present cases, it seems more likely that spondylolysis was due to physical stress of the lumbar and sacral vertebrae, probably at a young age. A genetic origin, can, however, not fully be excluded.

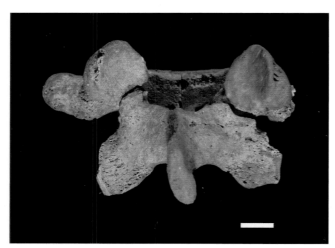

Fig. 16.18. Priene. Late Byzantine Cemetery in the former sanctuary of the Egyptian Gods (AEG). Spondylolysis of the 5th lumbar vertebra (AEG 5.15/2). Scale 1 cm.

CONGENITAL ALTERATIONS

The most unusual case is a young female with syndromic sagittal craniosynostosis and twins. During the 2010 campaign at Priene, a somewhat disturbed burial was excavated at the periphery of the late Byzantine Cemetery in the former sanctuary of the Egyptian Gods (Fig. 16.19). The skeleton was nearly complete, only the lower leg and foot bones are not preserved due to a disturbance of the burial (Fig. 16.19).

The analysis revealed a young adult female of about 20–25 years of age with a calculated body height of 148±3 cm

according to Pearson's (1899) formulae, corrected by Rösing (1988). She died after a miscarriage of at least twins approximately 8 to 9 lunar months after conception. Her robust skull with a clear scaphocephalic appearance shows premature sagittal craniosynostosis (Fig. 16.20). The cranial index of 74.4 is in the dolicocephalic range. Furthermore, the skull is slightly deformed due to a closure of the right occipito-mastoid suture. There are also indications of a slight torticollis. The CT scan shows thickening of the left cranial vault of up to 10 mm. The right temporal and sphenoid bones are also heavily thickened and deformed. Their structure indicates a temporal lobe malformation. The internal lamina of the occipital bone shows an abnormal structure without marked areas of the occipital poles and the posterior cranial fossae. This is a strong indication of modifications of the outer form of the brain and the cerebellum. The frontal teeth of the upper and lower jaw are protruding, the lower jaw shows a trema of 15 mm width, giving her a singular appearance. Multiple linear enamel hypoplasias are present. Furthermore, there is a shortening of the limbs, but only on the left side. Due to this association of malformations the present case should be called a syndromic sagittal craniosynostosis. It is the only case of its kind in the late Byzantine population of Priene. Torticollis and cranial deformation was, however, also recorded in two burials nearby.

Kyme

During recent years the Italian archaeological excavation at Kyme (MIKE) has focused its work on the agora of the

Fig. 16.19. Priene. Late Byzantine Cemetery in the former sanctuary of the Egyptian Gods, burial 39. Young female with cranial deformation and premature closure of the sagittal suture.

Fig. 16.20. Priene. Late Byzantine Cemetery in the former sanctuary of the Egyptian Gods, burial 39. Premature closure of the sagittal suture. Scale 1 cm.

ancient city (La Marca and Mancuso 2012). Above the ancient structures, an early Christian basilica and other ecclesiastical buildings were found. Around the basilica is a Byzantine cemetery with 44 burials. Thirty of them contained bones of at least 49 humans.

Results

The skeletal material from Kyme is very well preserved – much better than in Pergamon and Priene. This is due to the location of the tombs on the ancient agora with its structures made of marble.

The age distribution is quite different from all other sites. In the Agora Cemetery burials of neonates, infants, and children are most common. They have a much higher frequency than in both Pergamon and Priene. Furthermore, due to the excellent state of preservation, slight pathological changes can also be discovered.

MORTALITY

The age distribution of the Kyme sample is striking (Fig. 16.21: 48% of the individuals were below 12 months of age, and the major part were neonates (and a foetus = 30%). Only 26% of the sample belonged to adults. Similar patterns were observed by Arzu Demirel (this volume) for the Byzantine infants around the lower city church at Amorium.

The Byzantine Cemetery at Kyme offers a rare insight into infant mortality patterns. As we have seen before in the burials of pregnant females, pregnancy and

giving birth were high risk periods for mother and child. Between four and six months of age the probability of dying was less. The death risk increases at the end of the first year of life. This probably corresponds with an early weaning period.

PALAEOPATHOLOGY

Unfortunately, the risky time of the first year of life is not reflected in the enamel hypoplasia ratios. As we have seen for Pergamon, individuals with enamel hypoplasia below two years of age were rarely found. One of them was found in burial 41. This small infant shows a severe enamel defect of the upper right deciduous first incisor (tooth 51) and root hypoplasia (Fig 16.22). The enamel defect developed *in utero,* the root hypoplasias around nine and 12 months post partum (Ferembach *et al.* 1980; Ubelaker 1989).

The same individual from burial 41 shows new bone formation on the internal lamina of the parietal bone (Fig. 16.23). This was the outcome of an inflammatory process, which was, however, not survived. The root hypoplasias possibly represent a growth disturbance caused by the inflammatory process of the internal lamina of the skull.

INFLAMMATORY PROCESSES OF THE INTERNAL LAMINA OF THE SKULL

Another small infant from burial 40 suffered from an inflammatory process in the skull base. Here, the ala minor ossis sphenoidalis shows clear signs of an inflammation (Fig. 16.24).

The 12–24-month-old infant from burial 29 showed severe pathological alterations on the internal lamina of the skull vault (Teegen 2012b, 18 figs. 4–7). Here, an inflammation of the sinus sagittalis superior and adjacent areas is present. Both occipital poles show signs of increased intracranial pressure. Furthermore, signs of an inflammatory process were also present in their impressions digitatae.

We can conclude that the infants from burials 29, 40, and 41 suffered from meningitis. This disease caused high fever, severe headaches, opisthotonus and was very likely the cause of death. Meningitis was not uncommon during the Byzantine period and was also observed by Michael Schultz (1989a), for example, at Pergamon or in Bogazköy (Schultz 1989b).

A neonate from burial 2 has a haematoma on the external lamina of the skull vault (Fig. 16.25). These are the remnants of a so-called kephal haematoma, which often develops during birth. As recent medico-legal findings show, they can reach considerable dimensions.

CRIBRA ORBITALIA

In the burial population from the Agora Cemetery at Kyme there is only one case of severe cribra orbitalia (grade III

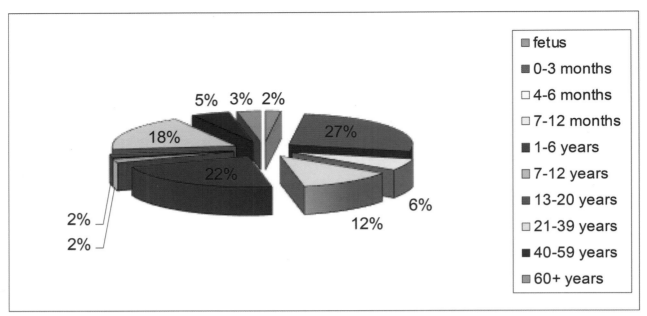

Fig. 16.21. Kyme. Byzantine Agora Cemetery. Age distribution of the buried individuals (N=49).

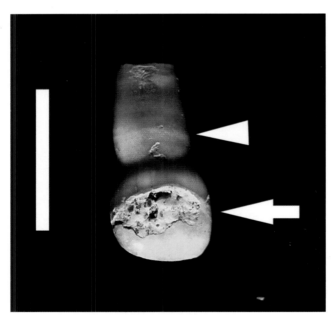

Fig. 16.22. Kyme. Byzantine Agora Cemetery, burial 41. Small infant with enamel defect of the upper right deciduous first incisor (tooth 51; arrow) and root hypoplasia (arrow head). Scale 1 cm.

Fig. 16.23. Kyme. Byzantine Agora Cemetery, burial 41. Small infant with new bone formation following an inflammatory process on the internal lamina of the parietal bone. Scale 1 cm.

according to Hengen 1971). It belongs to a child aged between nine and 13 years. Traditionally it is believed that cribra orbitalia is a sign of anaemia (Steinbock 1976; Martin *et al.* 1985). Several diseases and alterations can cause a porotic or spongy orbita, e.g. tumours, inflammatory processes, anaemia, but also post-mortem erosion (Schultz 1986; Götz 1989). An exact diagnosis is, however, only possible by microscopic analysis of thin sections (Schultz 1993; 2001; 2003).

SPONDYLOLYSIS

Also one of the few adult burials showed a spondylolysis of the 5th lumbar vertebra. Together with the degenerative joint diseases of the large joints and the vertebral joints this is an indication of heavy physical loading since childhood (see above). Furthermore, there is one juvenile with Morbus Scheuermannt, exhibiting Schmorl's nodes in early years.

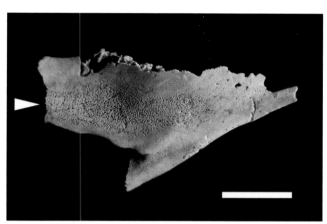

Fig. 16.24. Kyme. Byzantine Agora Cemetery, burial 40. Small infant with inflammatory process on the Ala minor ossissphenoidalis. Scale 1 cm.

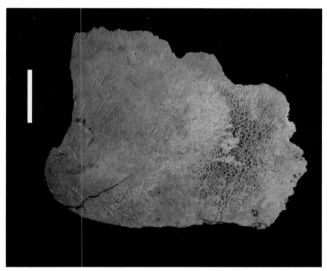

Fig. 16.25. Kyme. Byzantine Agora Cemetery, burial 2. Small infant with hematoma on the external lamina of the skull vault. Scale 1 cm.

Short discussion

The people buried in the Roman South-East Necropolis at Pergamon can be interpreted as being some kind of 'middle class'. This is indicated by the funerary monuments and also some luxurious grave goods, such as a funerary bed (Mania 2008). The more or less good health of these people correlates with their status.

In contrast, it is quite difficult to determine the social setting of the Byzantine burials, either from Pergamon, Kyme or Priene. Due to its remote location on the northern slopes of Pergamon's acropolis hill, the North Cemetery seems to be relatively poor. For Priene and Kyme it is likely that they were also some kind of Byzantine 'middle class', like shopkeepers. They were neither poor peasants nor rich aristocrats. In contrast

to their Roman counterparts, the Byzantine families exhibited signs of heavy physical loading.

It is difficult to determine when people died. The only reliable sources are grave inscriptions and other epigraphic sources. Walter Scheidel (1996) and others have shown that, in the Mediterranean and Egypt, mortality peaked in the summer months. The same can be presumed for Asia Minor and the cities mentioned in this paper. The graves themselves only rarely reveal the season of burial.

Studying the Roman and Byzantine populations from Pergamon, Kyme and Priene, differences between both time periods can be recognized (see also Kiesewetter, this volume). In particular, the degree of degenerative joint disease is quite low in the Roman sample and higher in the Byzantines. Furthermore it has to be noted that the cases of spondylolysis – mentioned above – were only observed in Byzantine Priene and Kyme and not in the sample from Roman Pergamon. The latter is not only larger but also contained more lumbar vertebrae than the populations mentioned before.

Also the presence of Schmorl's nodes, protruding of the intervertebral disk in the adjacent vertebral bodies (Walker 2012, 179–80, figs. 284–5), in juveniles indicates heavy physical loading. Schmorl's nodes in juveniles can be an indication of Scheuermann's disease (Ortner 2003, 464). They have, until now, only been observed in Byzantine juvenile skeletons from Priene and recently also from Pergamon (Teegen 2014). The onset of Schmorl's nodes in Romans from the South-East Necropolis begins generally only in the age range from the late 30s or early 40s. This is quite normal for prehistoric or historic populations (Aufderheide and Rodríguez-Martín 1998, 97).

Both, the presence or absence of spondylolysis and Morbus Scheuermann probably hints at different lifestyles and working conditions in Roman and Byzantine times in the three sites from Asia Minor.

In all three populations enamel (and root) hypoplasias are quite frequent. This indicates a lot of stress mostly between (one) two and seven years of age. In some cases (cf. Figs. 16.6 and 16.22) stress began before or during birth. This seems, however, to be the exception. In part, the enamel hypoplasis were probably caused during weaning and/or by (infectious) diseases in childhood; in some cases from Kyme, a connection between inflammatory processes on the internal lamina of the skull and the onset of enamel and root hypoplasias seems quite probable. There are, however, a wide range of causes for enamel hypoplasias (Jälevik and Norén 2000).

It is quite interesting to observe Galen's notes regarding episodes of famine in the rural populations of the Roman Empire during the second half of the 2nd century AD (Schlange-Schöningen 2003). He mentioned poor peasants eating acorns and non-edible plant stuff. In times of famine the main efforts of the public administration was concentrated on procuring wheat for the urban population thus preventing riots. This could also result in different frequencies of enamel hypoplasia in urban and rural populations. In fact, in a recent paper, Saskia Hin (2014) argued for a lower rate of enamel hypoplasia in the *urbs* Rome, in confrontation with the rural parts of the *campagna* or rural Italy in general. For Pergamon and its hinterland the rural populations are still missing. In the city itself, enamel defects are, however, quite common.

In the palaeopathological literature we can often find the statement that enamel hypoplasia is an indicator for the lower social classes (Sweeny *et al.* 1971; Schultz 1982). This is, however, not true. High frequencies of enamel hypoplasia can regularly be observed in high rank burials – at least from the Iron Age onwards (cf. Little *et al.* 1992; Teegen and Schultz 2009, 19) – also in Hellenistic Pergamon (Teegen 2011a, 157). In this context it has to be considered that children of status were often brought up by a wet nurse. This could also lead to infections during breast-feeding and in consequence also to enamel hypoplasias. Due to these findings, enamel hypoplasia or other stress markers cannot be used as an indicator of higher or lower social status.

'Food is a total social fact' – with these words the eminent French sociologist and anthropologist Marcel Mauss concluded his main work *Le don* (Mauss 1923–4) in the English translation (Mauss 1954). This also includes the cultural sphere of diet and its social status. The consumption of meat (Purcell 2003) and fish (Wilkins 1993) is also a good status indicator in Antiquity. The composition of animal bone assemblages can give important clues on the social status of its consumers. Piglets, lambs, and calves together with domestic fowl will preferably be found in contexts of higher social status. Meat consumption can be determined by analysing stable isotopes of carbon and nitrogen (for details see the contributions by Propstmeier *et al.* and Wong *et al.*, both this volume). Analysing skeletons of known status, as for example, the late Medieval and Renaissance family of the Medici in Florence (Italy), using stable isotopes, showed this very clearly (Fornaciari 2008).

Roman and Byzantine skeletons from Pergamon were analyzed for stable isotopes by G. H. Müldner (in Otten *et al.* 2011) and J. Propstmeier (2012; 2013; Propstmeier *et al.*, this volume). These analyses reveal individuals with different diets (more plant or more meat oriented). Further studies are, however, needed to analyze whether this image also reflects different social status.

Medical history and palaeopathology

There is a strong bias in the epigraphical evidence regarding ancient medicine. Ancient medicine is generally seen through the eyes of its principal leaders (eminent doctors),

who were dependent on their rich benefactors or patients, respectively (Nutton 2012). Furthermore, there are interested laymen, mostly belonging to the upper class, who could afford to spend time on such a hobby. Therefore, we know quite a lot about medicine available for those who could afford it. In contrast, we know very little about the medical problems of the middle classes and nearly nothing about those of the poor.

Galen himself was quite typical (Schlange-Schöningen 2003). He came from a wealthy family from Pergamon and belonged to the municipal aristocracy. Later he was surgeon to the gladiators, an extremely well-paid and acclaimed occupation – similar to some sports doctors today. Being surgeon to gladiators means also gaining spectacular insight into the anatomy and pathology of the human body and the skeleton. In Rome he was, at least at times, the personal doctor of five emperors, and an acclaimed author in the fields of medicine and philosophy. His writing influenced western medicine for more than one and a half millennia.

Palaeopathology gives us an invaluable insight into everyday life and diseases which often – or mostly – are not mentioned in the medical texts. They were, however, quite important for these ancient and Medieval populations. Furthermore, palaeopathology can evaluate health status in different social settings (Pitts and Griffin 2012), where often no textual evidence is present. Our investigations also contribute to the history of diseases itself as do also ancient images of diseased people, a collection of which can be found in the much-acclaimed book by Mrko Grmek and Danielle Gourevitch (1998).

Acknowledgments

I am most grateful to the Pergamon excavation of the German Archaeological Institute (director Prof. Felix Pirson), the Priene excavation (Prof. Wulf Raeck), the Kyme excavation (Prof. Antonio La Marca) and their teams, my collaborators Johanna Propstmeier MSc, Saskia Wunsch BA, and Sabrina Kutscher BA, the Gerda Henkel-Foundation, the Deutsche Forschungsgemeinschaft and last but not least the Ludwig-Maximilians-Universität München for support. The Turkish Ministry of Culture and Tourism (Ankara) kindly granted all necessary permits. I am further grateful to Dr Henrike Kiesewetter for reading my paper. The helpful comments of an anonymous reviewer are gratefully acknowledged. All remaining errors are of course my own. All figures are made by myself and are published here with courtesy by respectively the German Archaeological Institute at Istanbul (Pergamon excavations) (Figs. 16.1–16), the University of Frankfurt (Priene excavations) (Figs. 16.17–20), and the Italian archaeological excavation at Kyme (MIKE) (Figs. 16.21–25).

Bibliography

Alt, K. W. (1997) *Odontologische Verwandtschaftsanalyse. Individuelle Charakteristika der Zähne in ihrer Bedeutung für Anthropologie, Archäologie und Rechtsmedizin.* Stuttgart and Jena, Gustav Fischer Verlag.

Alt, K. W. and Pichler, S. L. (1998) Artificial modifications of human teeth. In K. W. Alt, F. W. Rösing, and M. Teschler-Nicola (eds.) *Dental anthropology. Fundamentals, limits, and prospects,* 387–415. Vienna, Springer.

Ameen, S., Staub, L., Ulrich, S., Vock, P., Ballmer, F., and Anderson, S. E. (2005) Harris lines of the tibia across centuries: A comparison of two populations, medieval and contemporary in Central Europe. *Skeletal Radiology* 34, 279–84.

Aufderheide, A. C. and Rodriguez-Martín, C. (1998) *The Cambridge encyclopedia of human paleopathology.* Cambridge, Cambridge University Press.

Bräuer, G. (1988) Osteometrie. In Knußmann (ed.), 160–232.

Campillo, D. and Baxarias, J. (eds.) (2008) *Quaranta anys de paleopatologia en el Museu d'Arqueologia de Catalunya* (Museu d'arqueologia de Catalunya Barcelona, Monografies 12). Barcelona, Museu d'arqueologia de Catalunya.

Demirel, F. A. (this volume) Infant and child skeletons from the Lower City Church at Byzantine Amorium, 306–17.

Ferembach, D., Schwidetzky, I., and Stloukal, M. (1980) Recommendations for age and sex diagnosis of skeletons. *Journal of Human Evolution* 9, 517–49.

Fornaciari, G. (2008) Food and disease at the Renaissance courts of Naples and Florence: A paleonutritional study. *Appetite* 51.1, 10–14.

Fornaciari, G. and Giuffra, V. (2009) *Lezioni di paleopatologia,* Genova, ed. ECIG.

Gilbert, R. I. and Mielke, J. H. (eds.) (1985) *The analysis of prehistoric diets* (Studies in Archaeology). Orlando, San Diego, and New York, Academic Press.

Götz, W. (1989) Histologische Untersuchungen an Cribra orbitalia, ein Beitrag zur Paläopathologie des Orbitadaches. Unpublished Master's thesis, University of Göttingen.

Grmek, M. D. and Gourevitch, D. (1998) *Les maladies dans l'art antique.* Paris, Broccard.

Hauser, G. and De Stefano, G. F. (1989) *Epigenetic variants of the human skull.* Stuttgart, Schweizerbartsche Verlagsbuchhandlung.

Hengen, O. P. (1971) Cribra orbitalia: Pathogenesis and probable etiology. *Homo* 22, 57–75.

Hennemeyer, A. (2005) Das Heiligtum der Ägyptischen Götter in Priene. In A. Hoffmann (ed.), *Ägyptische Kulte und ihre Heiligtümer im Osten des Römischen Reichs* (Byzas 1), 139–53. Istanbul, Ege Yayınları.

Hillson, S. (2014) *Tooth development in human evolution and bioarchaeology.* Cambridge and New York, Cambridge University Press.

Hin, S. (2014) *The first healthy metropolis in Europe's history? Urban-rural differences in health status in ancient Rome.* http://iussp.org/sites/default/files/event_call_for_papers/Hin_HealthyMetropolis.pdf (visited 07.01.2015).

Jälevik, B. and Norén, J. G. (2000) Enamel hypomineralization of permanent first molars: A morphological study and survey of possible aetiological factors. *International Journal of Paediatric Dentistry* 10, 278–89.

Kiesewetter, H. (this volume) Toothache, back pain, and fatal injuries: What skeletons reveal about life and death at Roman and Byzantine Hierapolis, 268–85.

Knußmann, R. (ed.) (1988) *Anthropologie* (Handbuch der vergleichenden Biologie des Menschen 1,1), 160–232. Stuttgart and New York, Gustav Fischer Verlag.

Kósa, F. (1978) Identifikation der Feten durch Skeletuntersuchungen. In H. Hunger and D. Leopold (eds.), *Identifikation,* 211–41. Leipzig, Barth.

Kutterer, A., and Alt, K. W. (2008) Cranial deformations in an Iron Age population from Münsingen-Rain, Switzerland. *International Journal of Osteoarchaeology* 18, 392–406.

La Marca, A., and Mancuso, S. (eds.) (2012) *Catalogo della Mostra Fotografica 'Scavi archeologici a Kyme d'Eolide (Turchia)'* (Centro Direzionale BCC Mediocrati, maggio 2012). Arcavacata di Rende, Centro Editoriale e Librario dell'Università della Calabria.

Lewis, M. E. (2009) *The bioarchaeology of children. Perspectives from biological and forensic anthropology* (Cambridge Studies in Biological and Evolutionary Anthropology 50). Cambridge and New York, Cambridge University Press.

Little, B. J., Lanphear, K. M., and Owsley, D. W. (1992) Mortuary display and status in a 19th-century Anglo-American cemetery in Manassas, Virginia. *American Antiquity* 57, 397–418.

Magge, S. N., Westerveld, M., Pruzinsky, T., and Persing, J. A. (2002) Long-term neuropsychological effects of sagittal craniosynostosis on child development. *Journal of Craniofacial Surgery* 13.1, 99–104.

Mania, U. (2008) Die Südostnekropole. *Archäologischer Anzeiger* 2008.2, 112–8.

Mann, R. W., Jantz, R. L., Bass, W. M., and Willey, P. S. (1991) Maxillary suture obliteration: A visual method for estimating skeletal age. *Journal of Forensic Science* 36.3, 781–91.

Martin, D. L., Goodman, A. H., and Armelagos, G. J. (1985) Skeletal pathologies as indicators of quality and quantity of diet. In Gilbert and Mielke (eds.), 227–79.

Martin, R. (1928) *Lehrbuch der Anthropologie in systematischer Darstellung. Bd. 2: Kraniologie, Osteologie.* Jena, Gustav Fischer Verlag.

Mauss, M. (1923–4) Essai sur le don. *L'Année Sociologique* N.S. 1, 30–186.

Mauss, M. (1954) *The gift: Forms and functions of exchange in archiac societies.* New York, Norton.

Nutton, V. (2012) *Ancient medicine.* Second ed. (Sciences of Antiquity Series). London and New York, Routledge.

Ortner, D. J. (2003) *Identification of pathological conditions in human skeletal remains* (2nd ed.). San Diego, Academic Press.

Otten, T., Evans, J., Lamb, A., Müldner, G., Pirson, A., and Teegen, W.-R. (2011) Ein frühbyzantinisches Waffengrab aus Pergamon. Interpretationsmöglichkeiten aus archäologischer und naturwissenschaftlicher Sicht. *Istanbuler Mitteilungen* 61, 347–422.

Pearson, K. (1899) Mathematical contributions to the theory of evolution. V. On the reconstruction of the stature of prehistoric races. *Philosophical Transactions of the Royal Society of London A* 192, 169–244.

Pirson, F. (2008) Pergamon – Bericht über die Arbeiten in der Kampagne 2007. *Archäologischer Anzeiger* 2008.2, 83–155.

Pirson, F. (2012) Pergamon – Bericht über die Arbeiten in der Kampagne 2011. *Archäologischer Anzeiger* 2012.2, 175–274.

Pirson, F. (2013) Pergamon – Bericht über die Arbeiten in der Kampagne 2012. *Archäologischer Anzeiger* 2013.2, 79–164.

Pirson, F. (2014) Pergamon – Bericht über die Arbeiten in der Kampagne 2013. *Archäologischer Anzeiger* 2014.2, 101–76.

Pirson, F., Japp, S., Kelp, U., Nováček, J., Schultz, M., Stappmanns, V., Teegen, W.-R., and Wirsching, A. (2011) Der Tumulus auf dem İlyastepe und die pergamenischen Grabhügel. *Istanbuler Mitteilungen* 61, 117–203.

Pitts, M. and Griffin, R. (2012) Exploring health and social well-being in Late Roman Britain: An intercemetery approach. *American Journal of Archaeology* 116.2, 253–76.

Plicht, J. van der (2015) *Unpublished Reports CIO/433-2015/PWL and CIO/582-2015/PWL.* 07.04.2015 and 21.09.2015.

Prittwitz und Gaffron, H.-H. von, Spiering, B., and Eggert, G. (1991) Der Reiterhelm des Tortikollis. *Bonner Jahrbücher* 191, 225–46.

Propstmeier, J. (2012) Die Lebensbedingungen in Pergamon: Nahrungsrekonstruktion mit Hilfe stabiler Stickstoff- und Kohlenstoffisotope einer römischen und spätbyzantinischen Nekropole. Unpublished Master's thesis, University of Munich.

Propstmeier, J. (2013) Analyse stabiler Isotope zur Ernährungsrekonstruktion. *Archäologischer Anzeiger* 2013.2, 141–2.

Propstmeier, J., Nehlich, O., Richards, M. P., Grupe, G., Müldner, G. H., and Teegen, W.-R. (this volume) Diet in Roman Pergamon: Preliminary results using stable isotope (C, N, S), osteoarchaeological and historical data, 237–49.

Purcell, N. (2003) The way we used to eat: Diet, community, and history at Rome. *American Journal of Philology* 124.3, 329–58.

Raeck, W. (2008) 2007 Yılı Priene Çaslismalari/Die Arbeiten in Priene im Jahre 2007. In *30. Kazı Sonuçları Toplantısı. 1. Cilt. 26–30 Mayis 2008 Ankara*, 33–52. Ankara, T. C. Kültür Bakanliği.

Raeck, W. (2009) Urbanistische Veränderung und archäologischerBefund in Priene. In A. Matthaei and M. Zimmermann (eds.), *Stadtbilder im Hellenismus*, 307–21. Berlin, Verlag Antike.

Rösing, F. W. (1988) Körperhöhenrekonstruktion aus Skelettmaßen. In Knußmann (ed.), 586–600.

Rösing, F. W., Graw, M., Marré, B., Ritz-Timme, S., Rothschild, M. A., Rötzscher, K., Schmeling, A., Schröder, I., and Geserick, G. (2007) Recommendations for the forensic diagnosis of sex and age from skeletons. *Homo* 58, 75–89.

Rose, J. C., Condon, K. W., and Goodman, A. H. (1985) Diet and dentition: Developmental disturbances. In Gilbert and Mielke (eds.), 281–305.

Schaefer, M., Scheuer, L. and Black, S. M. (2009) *Juvenile osteology. A laboratory and field manual.* Amsterdam, Elsevier.

Scheidel, W. (1996) Seasonal mortality in the Roman Empire. In W. Scheidel, *Measuring sex, age and death in the Roman Empire. Explorations in ancient demography* (Journal of Roman Archaeology, Supplementary Series 21), 139–63. Ann Arbor, Journal of Roman Archaeology.

Scheuer, L. and Black, S. (2000) *Developmental juvenile osteology.* San Diego, Academic Press.

Schlange-Schöningen, H. (2003) *Die römische Gesellschaft bei Galen: Biographie und Sozialgeschichte* (Untersuchungen zur antiken Literatur und Geschichte 65). Berlin and New York, Walter de Gruyter.

Schmid, F. and Künle, A. (1958) Das Längenwachstum der langen Röhrenknochen in bezug auf Körperlänge und Lebensalter. *Fortschritte Röntgenstrahlen* [Röfo] 89.3, 350–6.

Schultz, M. (1982) Umwelt und Krankheit des vor- und frühgeschichtlichen Menschen. In *Kindlers Enzyklopädie. Der Mensch Bd. 2*, 259–312. Munich, Kindler.

Schultz, M. (1986) *Die mikroskopische Untersuchung prähistorischer Skeletfunde. Anwendung und Aussagemöglichkeiten der differentialdiagnostischen Untersuchung in der Paläopathologie* (Archäologie und Museum 6). Liestal, Amt für Museen und Archäologie des Kantons Basel-Land.

Schultz, M. (1988) Paläopathologische Diagnostik. In Knußmann (ed.), 480–96.

Schultz, M. (1989a) Osteologische Untersuchungen an den spätmittelalterlichen Skeleten von Pergamon – Ein vorläufiger Bericht. In *IV. Arkeometri Sonuçları Toplantısı (Ankara 23–7 Mayis 1988)*, 111–8. Ankara, T. C. Kültür Bakanliği.

Schultz, M. (1989b) Nachweis äußerer Lebensbedingungen an den Skeleten der frühmittelalterlichen Bevölkerung von Bogazkale/Hattussa. In *IV. Arkeometri Sonuçları Toplantısı (Ankara 23–7 Mayis 1988)*, 119–20. Ankara, T. C. Kültür Bakanliği.

Schultz, M. (1993) *Vestiges of non-specific inflammations of the skull in prehistoric and historic populations. A contribution to palaeopathology* (Anthropologische Beiträge 4A/B). Aesch BL, Anthropologisches Forschungsinstitut Aesch and Anthropologische Gesellschaft Basel.

Schultz, M. (2001) Paleohistopathology of bone: A new approach to the study of ancient diseases. *Yearbook of Physical Anthropology* 44, 106–47.

Schultz, M. (2003) Light microscopic analysis in skeletal paleopathology. In D. Ortner, 73–108.

Schultz, M. and Schmidt-Schultz, T. H. (1994) Krankheiten des Kindesalters in der mittelalterlichen Population von Pergamon. Ergebnisse einer paläopathologischen Untersuchung. *Istanbuler Mitteilungen* 44, 181–201.

Schultz, M. and Schmidt-Schultz, T. H. (2004) 'Der Bogenschütze von Pergamon'. Die paläopathologisch-biographische Rekonstruktion einer interessanten spätbyzantinischen Bestattung. *Istanbuler Mitteilungen* 54, 243–56.

Schutkowski, H. (1990) Zur Geschlechtsdiagnose von Kinderskeletten. Morphognostische, metrische und diskriminanzanalytische Untersuchungen. Unpublished PhD thesis, University of Göttingen.

Schutkowski, H. (1993) Sex determination of infant and juvenile skeletons: I. Morphognostic features. *American Journal of Physical Anthropology* 90, 199–205.

Steinbock, T. R. (1976) *Paleopathological diagnosis and interpretation. Bone diseases in ancient human populations.* Springfield (Ill.), Charles C. Thomas.

Stloukal, M. and Hanáková, H. (1978) Die Länge der Langknochen altslawischer Bevölkerungen – Unter besonderer Berücksichtigung von Wachstumsfragen. *Homo* 29, 53–69.

Sweeney, E. A., Saffir, J. A., and De Leon, R. (1971) Linear hypolasia of deciduous incisor teeth in malnourished children. *American Journal of Clinical Nutrition* 24, 29–31.

Teegen, W.-R. (2004) Hypoplasia of the tooth root: A new unspecific stress marker in human and animal paleopathology [abstract]. *American Journal of Physical Anthropology, Supplement* 38, 193.

Teegen, W.-R. (2009) Burials of a pregnant woman and neonates from the Roman South-East Necropolis of Pergamon (Bergama, Prov. Izmir, Turkey): A preliminary report. Poster, Réncontres autour de la mort des tout-petits. Mortalité foetale et infantile. 3–4 décembre 2009, Musée d'Archéologie Nationale Saint-Germain-en-Laye. http://gaaf.emonsite.com/pages/rencontres/rencontre-autour-des-tout-petits-2009/pre-actes.html (visited 03.08.2013).

Teegen, W.-R. (2010) Spondylolysis in Late Byzantine Priene (Turkey) [abstract]. *American Journal of Physical Anthropology, Supplement* 50, 228–9.

Teegen, W.-R. (2011a) Die menschlichen Skelettreste. In Pirson *et al.*, 146–65.

Teegen, W.-R. (2011b) Der anthropologisch-paläopathologische Befund. In Otten *et al.*, 369–93.

Teegen, W.-R. (2011c) A female skeleton with syndromic sagittal craniosynostosis from Late Byzantine Priene (Turkey) – a CT investigation [abstract]. *American Journal of Physical Anthropology, Supplement* 52, 292.

Teegen, W.-R. (2011d) Die anthropologisch-paläopathologischen Untersuchungen. *Archäologischer Anzeiger* 2011.2, 186–8.

Teegen, W.-R. (2012a) Die anthropologisch-paläopathologischen Untersuchungen 2011. *Archäologischer Anzeiger* 2012.2, 255–8.

Teegen, W.-R. (2012b) La Tomba 29 nell'area dell'agorà: un caso studio paleopatologico. In La Marca and Mancuso (eds.), 18.

Teegen, W.-R. (2013) Die anthropologisch-paläopathologischen Untersuchungen 2012. *Archäologischer Anzeiger* 2013.2, 138–43.

Teegen, W.-R. (2014) Die anthropologisch-paläopathologischen Untersuchungen in Pergamon 2013. *Archäologischer Anzeiger* 2014.2, 152–56.

Teegen, W.-R. (2015) Die anthropologisch-paläopathologischen Untersuchungen in Pergamon 2014. *Archäologischer Anzeiger* 2015.2, 158–63.

Teegen, W.-R (forthcoming) Gli uomini della necropolis bizantina sull'agora di Kyme Rapporto preliminare su gli studi antropologici e paleopatologici. In La Marca (ed.), Kyme Eolica VI (submitted, not yet published).

Teegen, W.-R. and Schultz, M. (2009) Eine slawische Burg und ihre 'fürstlichen' Bewohner: Starigard/Oldenburg 10. Jh. In L. Clemens and S. Schmidt (eds.) *Sozialgeschichte der mittelalterlichen Burg* (Interdisziplinärer Dialog zwischen Archäologie und Geschichte 1), 13–24. Trier, Kliomedia.

Ubelaker, D. H. (1989) *Human skeletal remains: Excavation, analysis, interpretation* (2nd ed.) (Manuals on Archaeology 2). Washington D.C., Taraxum.

Waldron, T. (2001) *Palaeopathology* (Manuals in Archaeology). Cambridge, Cambridge University Press.

Walker, D. (2012) *Disease in London, 1st–19th centuries. An illustrated guide to diagnosis* (MOLA Monograph 56). London, Museum of London.

Wiegand, T. and Schrader, H. (1904) *Priene. Ergebnisse der Ausgrabungen und Untersuchungen in den Jahren 1895–8.* Berlin, Georg Reimer.

Wilkins, J. (1993) Social status and fish in Greece and Rome. *Food, Culture and History* 1, 191–203.

Wiltschke-Schrotta, K. (1988) Das frühbronzezeitliche Gräberfeld von Franzhausen I. Analyse der morphologischen Merkmale mit besonderer Berücksichtigung der epigenetischen Varianten. Unpublished PhD thesis, University of Vienna.

Wong, M., Naumann, E., Jaouen, K., and Richards, M. (this volume) Isotopic investigations of human diet and mobility at the site of Hierapolis, Turkey, 306–17.

Toothache, back pain, and fatal injuries: What skeletons reveal about life and death at Roman and Byzantine Hierapolis

Henrike Kiesewetter

Abstract

At the ancient site of Hierapolis the University of Oslo has undertaken intense archaeological fieldwork in the North-East Necropolis from 2007 to 2014. The present study focuses on the examination of human bones recovered from two neighbouring Roman saddle-roofed house tombs of the late 1st/early 2nd century AD, which have been completely excavated. One of the Roman tombs was extensively reused as a grave in the mid-Byzantine period. Minimum numbers of about 100 Roman – some of them cremated – and 60 Byzantine individuals offer a special opportunity for diachronic osteological and palaeopathological analyses at Hierapolis.

The frequency of dental diseases, signs of deficiencies, degenerative joint disease, and fracture patterns in Roman and Byzantine skeletons are compared in order to draw conclusions about the general well-being, health status, and living conditions of the people. The palaeopathological investigations indicate that the health of the Romans buried in the house tombs was better than that of the mid-Byzantines reusing these grave structures. The Romans showed less signs of degenerative diseases, infectious diseases, dental diseases, malnutrition, and childhood stress factors compared to the Byzantines. Demographic analyses point to a higher life expectancy in the Roman population. However, palaeopathological and palaeodemographical analyses are limited by the difficult find situation with very fragmentary and scattered bones.

Keywords: cremation, Hierapolis, late Roman, mid-Byzantine, osteology, palaeopathology, Roman.

Introduction: The tombs and the burials

Hierapolis is a very well-known Hellenistic-Roman-Byzantine site in Phrygia, Turkey. Three large burial grounds, the North, North-East, and South necropoleis enclose the ancient town, which lies on a plateau bordered to the west by a steep slope towards the Lykos river valley. The immense size of the cemeteries and the quantity and preservation of funeral monuments excavated is unique. Based on an estimation of the number of deaths during the centuries at Hierapolis, it seems that every deceased citizen could have found a burial place in one of these necropoleis (Ahrens, this volume). Thus, the site of Hierapolis holds a huge amount of funerary data and offers a special opportunity for diachronic osteological studies of the human remains.

The North-East Necropolis with more than 750 registered tombs and sarcophagi was laid out in the Roman Imperial period, but partly reused in Byzantine times. It is located on the steep hillside behind the ancient town with a beautiful view facing the Lykos river valley. The Norwegian team completely excavated three Roman saddle-roofed house tombs, three sarcophagi, a Roman cist grave, and several Byzantine tile graves. While the burial monuments are quite well preserved, in many cases the preservation of the human bones found within the graves was comparatively poor, sometimes almost no skeletal material remained. Erosion, soil conditions, earthquakes, grave robberies, and extensive reuse of the graves explain why many skeletons are so badly preserved (Ahrens and Brandt 2016; Selsvold *et al.* 2012, 15).

For the present osteological investigations two Roman house tombs, C92 and C311, densely packed with human bones, were chosen. Inscriptions reveal the names of the tomb owners and even the profession of one of them: Eutyches (C92) and Patrokles the younger, who was a purple dyer (C311). The funerary buildings were constructed in the 1st/2nd century AD and are situated close to the later founded martyrion building of St Philip, the apostle. They were built on the same terrace and share a wall (Figs. 17.1–2). Probably due to an earthquake Patrokles' tomb partially collapsed (possibly in the 6th/7th century AD) and all objects and bones found inside the tomb seem to be Roman. There is no evidence of later reuse of this funerary building. The still intact house tomb of Eutyches on the other hand was intensely utilized for burials in the mid-Byzantine period (Ahrens *et al.* 2013, 14; Bortheim *et al.* 2014, 82; Ahrens and Brandt 2016, 405–6; Wenn *et al.*, this volume, 209–11).

The Byzantine layers in tomb C92 were densely filled with human remains (Fig. 17.3). At first glance, it seemed to have been a Byzantine 'mass grave' with more or less contemporaneous inhumations, after, for example, an epidemic outbreak. Yet, finds of coins and cross pendants among the human remains date from the 9th to the 12th century AD and argue for a continuous and steady reuse of the funeral monument and against a 'mass grave'. Radiocarbon dates from the human bones confirmed the archaeological stratigraphy and a history of several centuries of inhumations in the mid-Byzantine period. Five pilgrim badges from the 13th century discovered in top layers of the tomb were found without clear connection to any of the burials (Ahrens 2011–2012).

Archaeological investigations at Hierapolis reveal that a complete change of the ancient town took place beginning in the early 5th century AD with the transition of the city

Fig. 17.2. Hierapolis. Roman house tombs of Eutyches and Patrokles, seen from south.

from pagan to Christian. The centrally placed sanctuaries of Apollo and Pluto were gradually dismantled and eliminated (D'Andria 2013, 175–7) and a new religious centre grew up around the tomb of Philip the apostle, a sanctuary eventually composed of a bath building, a church, and a martyrion building (D'Andria 2011–2012). After an earthquake in the mid-7th century Hierapolis never fully recovered and appears to have been abandoned. However, traces of new life of a rural character can be discerned already by the 8th century followed by more intense activity in the following two centuries, possibly including the creation of a fortified *kastron* (Arthur 2012, 280–4). A new destructive earthquake around AD 1000 and the first appearance of Turkoman tribes a century later caused a set-back to the settlement, which in the 12th century was reduced to a small rather insignificant settlement, as reported in 1190 by Frederick Barbarossa, when he, on his crusade, passed through the area on his way towards Jerusalem (Arthur 2012, 285–97; see also D'Andria 2010; 2011–2012). Other pilgrims, however, followed in his footsteps at least until the end of the following century (Ahrens 2011–2012).

The human remains discovered in the two house tombs C92 and C311 thus provide the possibility for a diachronic comparison between the Roman population of the urban city and the mid-Byzantine population of the rural town of Hierapolis. Together with the archaeological information and interpretation, conclusions on the quality of life in Antiquity can be drawn from these data.

Material and methods

Chronology of burials and condition of the human remains

It is estimated that over 100,000 human bones and bone fragments were recovered from the Roman house tombs of Eutyches (C92) and Patrokles (C311). Only two partly

Fig. 17.1. Hierapolis. Roman house tomb of Eutyches (C92) surrounded by sarcophagi and the neighbouring collapsed Roman house tomb of Patrokles (C311), aerial view 2009.

Fig. 17.3. Hierapolis. Byzantine layers with human remains inside Eutyches' tomb (C92).

articulated skeletons (SN 32 and SN 38) were discovered in the Byzantine strata of Eutyches' tomb. Some vertebrae were still connected, in a few cases the spine with ribs was almost completely preserved and occasionally even in articulation with pelvic bones. Some bones were exposed in one-joint-articulation, for example, pelvis and femur. All other bones were disarticulated and found scattered within the tomb, many were fragmented and in a poor state of preservation (Figs. 17.3–4). Due to the commingled nature of the human remains, it was impossible to identify single individuals beside the two individuals found in half-articulation. Only 12 boxes with small bone fragments from tomb C311, excavated at the end of the 2014 campaign, remain to be examined and could not be included in the present study.

Radiocarbon dates from the human bones reveal rather continuous interments in Patrokles' grave (C311) from the 1st until the 4th and even 5th/6th centuries AD. While

Patrokles' tomb was used as a burial place from the Imperial Roman period, when it was built, until it collapsed in the late Roman period, the neighbouring Eutyches' tomb was intensely reused in Byzantine times. Most of the human remains were found in the upper layers and date mainly to the mid-Byzantine period. Radiocarbon dates range from the 8th to the 13th century AD, indicating that the tomb was reused for at least 500 years. Bones from the bottom of tomb C92 reveal radiocarbon dates from the 1st to the 4th century AD. It seems that during the 5th, 6th, and 7th centuries AD the tomb was not in use (although coins from the 5th century AD were discovered in the tomb; see Ahrens and Brandt 2016, 405–6) and a hiatus of about 300 years divided the two burial phases. Due to extensive reuse and consequent disturbances in Eutyches' tomb, mixed strata containing bones from both periods were found. All layers, revealing both Roman as well as Byzantine radiocarbon

Fig. 17.4. Hierapolis. Fragmented human remains from Eutyches' tomb (C92).

dates, were excluded from the present analyses. Only one well-preserved skull (SN 97, B52), discovered in mixed layers, and radiocarbon dated to the 10th century AD, was included in the Byzantine sample. Since a more precise dating of the different layers and consequently for most of the bones was not possible, in the present osteological analyses Roman/late Roman remains were compared with mid-Byzantine remains. Very few human remains unearthed at the bottom of Eutyches' tomb and dating to the Roman period were burnt. Quite unexpected, therefore, was the high percentage of cremated remains found in the house tomb of Patrokles. Calculating minimum numbers of individuals, about a third of the burials from Patrokles' tomb were cremations. Most of the burnt bones showed a dark brown to black colour (charred bones) indicating a burning temperature of about 400° Celsius. Some bones were white or cream-coloured displaying parabolic heat cracks and shrinking up to about 25% (cremated bones), which is typical for burning temperatures of about 800° Celsius (Wahl 2007). The burnt bones were unusually well preserved (Fig. 17.5), even a completely preserved charred skull of a woman aged between 20 and 30 years was discovered. Radiocarbon dates from the 5th/6th centuries AD obtained from the burnt bones suggest that cremation at Hierapolis was also practised in the late Roman period (see Wenn *et al.*, this volume).

A representative sample of 4,000 bones and fragments from the Byzantine layers of tomb C92 was analyzed: most frequently ribs (18%) and vertebrae (18%), followed by teeth (17%), hand (11%) and foot bones (11%) were excavated. Long bones and long bone fragments were more or less equally represented: humerus, radius, ulna, and tibia (2% each), femur (3%), and fibula (1%). More bones and bone fragments were present from the more fragile skeletal parts like crania (6%), pelvis (4%), and shoulder (3%) (numbers calculated by the author from unpublished data

taken in 2010–2011 by Helene Russ and in 2009 by Silje Jeanette Hofstad Strømslund). No indications of reburials in the Byzantine period were found. In reinterments the dead body would have been buried elsewhere and after exhuming the skeleton the bones would have been transferred to the Roman house tomb. If this had been the case, one would expect to find fewer small and fragile bones, like finger and foot bones, ribs, and vertebrae, and comparatively more long bones and skulls.

Osteological and palaeopathological methods

Before studying the human remains samples for genetic and isotope analyses were taken with gloves from all intact crania and from well-preserved long bones. Then bones were cleaned and whenever possible broken crania were restored. As mentioned above, it was impossible to identify whole skeletons. Since there were such a vast number of bone fragments, anthropological analyses in 2012 first concentrated on the examination of reasonably well-preserved skulls and mandibles. Age estimation for adult crania was made using the degree of suture closure (Meindl and Lovejoy 1985); sex estimation was made according to cranial morphology (Buikstra and Ubelaker 1994). Skulls of children and juveniles were aged according to common standards (Scheuer and Black 2000) with an emphasis on age estimation by dentition.

In the following campaigns in 2013/14 information gathered from postcranial elements was added to the results gained from cranial analyses. Examination of postcranial remains was adapted to the special find situation and only better preserved bones, including burnt bones, were analyzed. Sex and adult age determination of the postcranium was restricted to pelvic bones (Buikstra and Ubelaker 1994; Lovejoy *et al.* 1985). All other postcranial remains were divided into adult remains with closed epiphyses and subadult remains with open epiphyses or epiphyses in fusion. For the infants, children, and juveniles age was determined following recommendations by Scheuer and Black (2000).

In order to estimate a minimum number of adult individuals (MNI) buried in tomb C92 adult skulls (with more than half preserved – to avoid double counting) and adult mandibles were registered. Analyses of examined postcranial elements did not provide higher MNIs for the adults. For the subadult remains MNI was estimated by counting distinct cranial and postcranial elements according to age. Additionally, for tomb C311 the MNI for burnt and unburnt remains was determined by counting the parts of the skull where the inner ear is located (internal acoustic meatus), which due to its dense osseous structure is frequently preserved in cremations (Wahl 1982). Due to a tight schedule in 2014, 12 boxes with human bones from tomb C311 remain to be examined in the future.

Fig. 17.5. Hierapolis. Adult anklebones (talus), unburnt, charred (black), and cremated (white with shrinking), all from Patrokles' tomb (Roman).

Fig. 17.6. Hierapolis. Burnt mandible with ante-mortem tooth loss and teeth partly destroyed by fire (Roman).

Each bone was examined for any evidence of pathological alteration. A special focus was placed on signs of dental diseases, nutritional deficiencies, infections, degenerative joint disease, and signs of injuries. All teeth were scored for carious lesions, calculus deposition, and enamel hypoplasia. The jawbones were analyzed for periodontal disease, abscess formation, and ante-mortem tooth loss. Skulls were examined for porosities of the cranial vault (porotic hyperostosis) and pitting of the orbits (cribra orbitalia) according to established standards (Stuart-Macadam 1991). Degenerative joint disease and degenerative disease of the spinal column was documented only for adult bones with fused epiphyses. All joints were studied for evidence of arthritic changes and scored for severity. Due to the fragmentary nature of the human remains, proximal and distal articulations of a bone were analyzed separately and the number of diagnostic articulation surfaces and affected articulation surfaces was counted. For vertebrae osteophytic lipping and Schmorl's nodes were documented independently for superior and inferior surfaces. All traumatic lesions of the bone were recorded and trauma type, causes of injury, and state of healing were described in detail.

Palaeodemography

Palaeodemographic analyses reveal that deceased of all age groups, neonates, children, juveniles, and adults, females and males were buried in the Roman house tombs of Eutyches (C92) and Patrokles (C311). This is also the case for the time of Byzantine reuse of tomb C92. Even the remains of foetuses were found among the bones. For the foetal bones it remains uncertain whether they belonged to premature infants or originated from deceased pregnant mothers. In Roman layers from Eutyches' and

Patrokles' tombs altogether a minimum of 99 individuals (MNI) were unearthed, 27 (27%) of them died before reaching the age of 15 years. Due to the fractured nature, especially of the burnt bones in the Roman sample it was, in many cases, not possible to distinguish between individuals over 15 and individuals over 20. In the Byzantine strata of Eutyches' tomb MNI is 61 with 19 (31%) individuals younger than 15 years and 23 (38%) younger than 20 years (Tables 17.1–2).

For Patrokles' tomb (C311) a minimum number of 69 individuals – 45 unburnt burials and 24 cremations – was recorded by counting the internal acoustic meatuses. They all date to the Roman and late Roman period. Table 17.1 shows the age-at-death distribution of the individuals buried in the tomb. Age-at-death for the subaadult individuals was determined by long bone measurements and closure of epiphyses. Only eight adult skulls are comparatively well preserved: one burnt skull of a woman, who died at an age of about 20–30 years, and seven unburnt crania, one of the males reached an age of 60–80 years. Twenty-three skulls and skull fragments show female features, 25 reveal male features. Only seven reasonably well-preserved pelvic bones were found in Patrokles' tomb, all showing female features. Small fragments from the pelvic bones, which might be usable for sexing, remain to be examined.

In Eutyches' tomb (C92) a minimum number of 30 individuals were discovered in Roman layers and at least 61 individuals were exposed in a mid-Byzantine context. The age-at-death distribution for the Roman and Byzantine individuals unearthed in the tomb is given in Table 17.2. The Roman bones from the bottom layers of the tomb were very fragmentary and only five adult skulls and ten adult mandibles were almost completely preserved. The preservation of bones in Byzantine layers

Table 17.1. Hierapolis. Age-at-death distribution of Roman human remains from house tomb C311 presented as minimum numbers of individuals (MNI).

Age groups (years)	Unburnt (MNI)	Burnt (MNI)	Burnt and Unburnt (MNI)
Foetuses	2	0	2
Neonates/infants (0–1)	3	1	4
Young children (1–7)	2	1	3
Children (8–14)	1	1	2
Adults and juveniles (>15)	37	21	58
All individuals	45	24	69

Table 17.2. Hierapolis. Age-at-death distribution of human remains from house tomb C92 presented as minimum numbers of individuals (MNI) for Romans and Byzantines separately.

Age groups (years)	Roman (MNI)	Byzantine (MNI)
Foetuses	3	1
Neonates/infants (0–1)	4	4
Young children (1–7)	5	7
Children (8–14)	4	7
Juveniles (15–20)	4	4
All adults	10	38
All individuals	30	61

was better: 38 reasonably well-preserved adult crania and 18 complete adult mandibles were found. Twelve more or less complete pelvic bones were discovered in the Byzantine layers, four show female eight show male features.

It was possible to determine age and sex of 51 adult skulls; for 38 Byzantine skulls from tomb C92 and for 13 Roman skulls, five from tomb C92, and eight from tomb C311, one of which was burnt. Results are listed in Table 17.3. In the Roman population three females and nine males could be identified. These numbers are obviously too small to calculate a sex-ratio for the Romans. However, most striking is the fact that two males of the Roman population

reached an age beyond 60 years, while no individual of the Byzantine population survived after 60 years. For the Byzantine population 15 females and 20 males were found (sex ratio 1.3). If all individuals with undeterminable sex were female the sex-ratio would have been 1.1, reflecting a 'natural' population. Two thirds of the Byzantine females buried in the tomb died young and did not reach an age of 40 years. However, it needs to be considered that these data originate from a relatively small number of individuals. But the data may at least indicate a general trend of the demographic composition of the deceased buried in the two house tombs. The results also indicate that there was no selection of individuals who were buried in the tombs.

Stress and strain: Palaeopathology

Dental diseases

Examination of dentition and dental diseases provides information on diet, general health status, dental hygiene, and dental treatment. Tooth decay caused by caries is usually related to a diet high in sugar and carbohydrates and is increased by low standards of oral hygiene, but genetic predisposition also influences the formation of dental cavities. Another indicator of neglected oral hygiene is the excessive formation of dental calculus. Periodontal disease with loss of the bony support of the teeth may arise from chronic inflammation of the gums (gingivitis), malposition of the teeth, and old age (Strohm and Alt 1998). Tooth loss is the final stage of periodontal disease. Alternatively, tooth extraction – performed as dental treatment for toothache – will obviously be a reason for ante-mortem tooth loss. In order to obtain information on the general health status of an individual permanent teeth not only of adults but also of children were scored for enamel hypoplasia. Decreased enamel thickness (hypoplasia) is visible on the tooth crowns as lines and provides a chronological memory of childhood stress such as malnutrition or severe infection.

In the present study, these dental indicators of living standards were compared for the Roman and mid-Byzantine period. Teeth were available for the adult Byzantine population from 23 upper and 18 lower jaws and for the adult Roman population from 11 upper jaws and 35 mandibles and mandibular fragments. However, in many cases the

Table 17.3. Hierapolis. Age-at-death and sex distribution of adult skulls from house tombs C92 and C311; Romans and Byzantines listed separately.

Adult age groups (years)	Roman females (N)	Roman males (N)	Roman sex? (N)	Byzantine females (N)	Byzantine males (N)	Byzantine sex? (N)
20–39	2	5	1	10	12	2
40–59	1	2		5	8	1
60–80	0	2		0	0	0

jaws and dentition were not completely preserved and especially in burnt remains teeth were often destroyed by fire (Fig. 17.6). These facts unfortunately limited the number of observable teeth. Table 17.4 presents the prevalence of dental diseases for both periods. While teeth were examined for carious lesions, calculus, and hypoplasia, tooth sockets (alveoles) were examined for ante-mortem tooth loss, abscess formation, and periodontal disease.

The most common dental disease found in the Roman material was periodontal disease (33%), while in the mid-Byzantine material calculus formation (32.2%) was most commonly diagnosed. The differences between the two populations are illustrated in Fig. 17.7. Except for periodontal disease the oral health appeared to have been better in the adult Roman population. The higher frequencies of periodontal disease in the Romans might be explained by older age at death of the Romans. Ante-mortem tooth loss was probably, in most cases, associated with severe carious lesions of the teeth and subsequent extraction. A higher prevalence of carious lesions in the Byzantine sample correlates with a higher rate of tooth loss.

A most serious condition reveals a maxilla found in Eutyches' tomb, in the Byzantine period, of an individual between 50 and 60 years of age, possibly male. A dental root abscesses destroyed the jawbone (apical cyst) as seen in Fig. 17.8. Infection on the tip of the dental root leading to abscess formation and resorption of the bone might have had severe consequences on the general health of the individual with a risk of general infection, strong pain, and a reduced ability to eat.

Signs of childhood stress was also less frequent in the Roman population (3.6% of the teeth show enamel hypoplasia). In the Byzantine population, 6.5% of the teeth revealed enamel hypoplasia. Counted by dentition, not by single tooth, 10.5% of the Byzantines had at least one affected tooth indicating malnutrition or other diseases causing severe physical stress during childhood. Table 17.5 gives information on the age of defect formation in childhood. Two Byzantine males and a female were suffering from childhood stress. While the males had one stress episode, the female underwent, at the age of one year and again at the age of five, a period

of severe stress. Also one of the Roman skeletons, which could not be sexed, revealed an enamel defect formed at the age of about three years. In all cases the enamel defect showed mild expressions with thin horizontal lines on the teeth.

The number of Roman jaws suitable for sexing was insufficient to give sex-related analyses. For the Byzantine period results of dental examination are also listed for females and males separately (Table 17.6). However, due to sometimes poor preservation, not all jaws could be sexed. Especially mandibles with a considerable amount of ante-mortem tooth loss could not be sexed due to morphological changes caused by the tooth loss. Therefore, the sexed sample shows less tooth loss. Females revealed less carious lesions, calculus formation, and periodontal disease. Influencing factors could be the younger age of the female sample, different diet with less sugar and carbohydrates, and better oral hygiene.

Infectious diseases and deficiencies

All bones were examined for signs of deficiencies and infection. Signs of rickets and osteomalacia caused by severe Vitamin D deficiency were neither found in the Roman nor in the Byzantine sample. This is not so much surprising for a Mediterranean site like Hierapolis, since Vitamin D is produced in the skin under the influence of ultraviolet radiation of the sunlight. Also quite predictable in the Mediterranean area with its typical vegetation is the lack of evidence of scurvy caused by Vitamin C deficiency observed in the skeletal material. Signs commonly related to iron-deficiency-anaemia, however, were observed on the skulls of several skeletons. In these cases, the eye-sockets revealed small round lesions of the bone called orbital pitting or cribra orbitalia. Fig. 17.9 shows a male skull from Byzantine layers in tomb C92 displaying typical lesions for cribra orbitalia in the left orbit (SN 97). Other skulls exhibited pitting on the external surface of the cranium called porotic hyperostosis. These lesions are thought to be caused as a result of malnutrition, anaemia, or infection. Some authors think that haemolytic and megaloblastic anaemia are most likely responsible for porotic changes of the skull (Walker *et al.* 2009).

Table 17.4. Hierapolis. Incidence of dental diseases for adults, Roman versus Byzantine population (tombs C92 and C311);

Period	Alveoles exam. (N)	Teeth exam. (N)	Ante-mortem tooth loss (%)	Caries (%)	Calculus (%)	Periodontal disease (%)	Apical cyst (%)	Enamel hypoplasia (%)*
Roman	239	187	15.1	8.5	23.5	33.0	0.9	3.6
Byzantine	548	240	18.6	10.0	32.2	28.9	1.8	6.5

*enamel hypoplasia counted by single tooth (not by dentition).

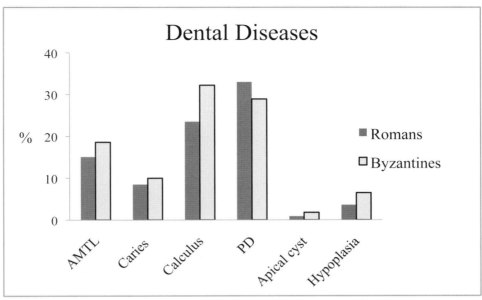

Fig. 17.7. Hierapolis. Distribution of dental disease, Roman versus Byzantine population (AMTL: ante-mortem tooth loss; PD: periodontal disease).

In the Mediterranean area porotic hyperostosis of the skull and orbits (cribra orbitalia) are often discussed as being associated with chronic infection, for example, with malaria (Angel 1966; Stuart-Macadam 1992).

Table 17.7 demonstrates the incidence of cranial porotic hyperostosis and Table 17.8 shows the frequency of cribra orbitalia. None of the Roman skulls showed signs of porotic hyperostosis and the only Roman individual revealing cribra orbitalia was an infant that died at an age between six months and one year. The prevalence of porotic hyperostosis in the Byzantine population was 9.1%, the prevalence of cribra orbitalia 22.2%. Table 17.9 shows the distribution of cribra orbitalia within the Byzantine population. One

of the Byzantine individuals of unknown sex displaying orbital lesions was a child aged between three and five years. All other individuals examined for cribra orbitalia were adults. Two females, one male, and two adult individuals of unknown sex revealed the lesions. Obviously, the small numbers do not allow any statistical interpretation regarding the distribution between sexes.

Age- and activity-related diseases of joints and spinal columns

One of the most common diseases detectable on ancient human bones is the degenerative joint disease (DJD) also known as osteoarthritis. DJD is caused by wear and tear of

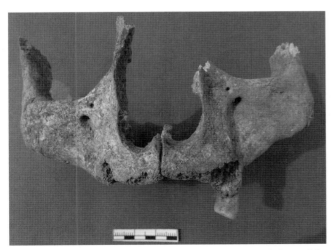

Fig. 17.8. Hierapolis. Maxilla with ante-mortem tooth loss of the upper incisors and apical cyst on the right side (Byzantine, tomb C92).

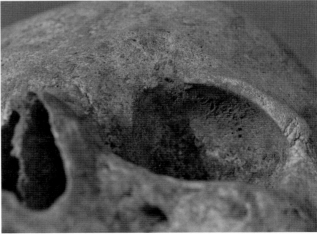

Fig. 17.9. Hierapolis. Byzantine male skull with cribra orbitalia in the left orbit (tomb C92, SN 97).

Table 17.5. Hierapolis. Enamel hypoplasia detected on skulls from tombs C92 and C311 revealing age of defect formation.

Individuals	Sex	Age of defect formation (years)	Age at death (years)
Skull 79 (C92, Byzant.)	Female	1 and 5	20–30
Skull 37 (C92, Byzant.)	Male	2	30–40
Skull 32 (C92, Byzant.)	Male	4–5	25–35
BN 3778 (C311, Roman)	?	2–4	adult

Table 17.6. Hierapolis. Incidence of dental diseases for adults in the Byzantine population (tomb C92), females versus males.

Sex	Alveoles exam. (N)	Teeth exam. (N)	Ante-mortem tooth loss (%)	Caries (%)	Calculus (%)	Periodontal disease (%)
Byzantine females	153	72	11.1	6.9	18.3	16.1
Byzantine males	172	63	9.9	15.9	38.8	34.2

bones and joints and subsequent degradation of the joints and is not caused by inflammation (even though the name osteoarthritis might suggest this). On the surface of the joints new bone outgrowths called marginal osteophytes become visible. The disease reflects mechanical stress on the joints increasing with heavy physical labour, injury, overweight, and age. Symptoms of osteoarthritis are joint pain, tenderness, stiffness, and atrophy of muscles. The analyses of distribution of osteoarthritic joints in ancient skeletons help to reconstruct past activities and occupational specialization (e.g. Rogers and Waldron 1995; Hollimon 2011; Roberts and Manchester 2001).

Table 17.7. Hierapolis. Incidence of cranial porotic hyperostosis: Roman versus Byzantine population (tombs C92 and C311).

Period	Skulls examined (N)	Porotic hyperostosis (n)	Porotic hyperostosis (%)
Roman	18	0	0.0
Byzantine	33	3	9.1

Table 17.8. Hierapolis. Incidence of cribra orbitalia: Roman versus Byzantine population (tombs C92 and C311).

Period	Orbits examined (N)	Cribra orbitalia (n)	Cribra orbitalia (%)
Roman	11	1	9.1
Byzantine	27	6	22.2

Table 17.9. Hierapolis. Incidence of cribra orbitalia: Byzantine females versus males (tomb C92).

Sex	Orbits examined (N)	Cribra orbitalia (n)
Byzantine females	12	2
Byzantine males	12	1
Byzantine sex?	3	3

In the present study DJD was subdivided into slight osteoarthritic changes with osteophytic formation of less than 1 mm and moderate to severe degenerative changes with osteophytic formation larger than 1 mm. Table 17.10 shows the prevalence of DJD for the Roman and Byzantine population. As mentioned above proximal and distal joints of a bone were counted separately due to the fragmentary condition of the remains. The mid-Byzantine bones show a prevalence of osteoarthritis of 16.6%; 3.3% bones displayed moderate to severe changes. DJD was less frequent in the Roman population with 10.7% bones affected; 2.1% bones revealed moderate to severe changes. No distinct differences in arthritic changes could be found between the right and left side of the skeleton (Table 17.11).

Tables 17.12 and 17.13 show the distribution of DJD among joint complexes for the Roman and Byzantine population respectively. The results are illustrated in Fig. 17.10. Except for the wrist, where no arthritic changes were diagnosed in the Byzantine population, the Romans showed fewer changes in all other joint complexes. Hip (24.5%) and knee (23.5%) were the locations most affected by degeneration in the Byzantines; knee (20.0%) and shoulder (18.2%) were most affected by degeneration in the Romans. Especially a right shoulder (see Fig. 17.11) found in the Roman layers of tomb C311 reveals severe arthritic changes of the joint. These changes of the right

Table 17.10. Hierapolis. Distribution of degenerative joint disease (DJD): Roman versus Byzantine population (tombs C92 and C311), subdivided into slight degeneration with osteophytic formation <1 mm and moderate to severe degeneration with osteophytic formation >1 mm.

Period	Bones exam. (N)*	DJD (%)	<1mm (n)	<1mm (%)	>1mm (n)	>1mm (%)
Roman	140	10.7	12	8.6	3	2.1
Byzantine	211	16.6	28	13.3	7	3.3

* proximal and distal surfaces of bones counted separately

Table 17.11. Hierapolis. Distribution of degenerative joint disease examined on the right side versus left side.

Period	Bones right/ left (N)	DJD right (n)	DJD right (%)	DJD left (n)	DJD left (%)
Roman	80/60	9	11.3	6	10.0
Byzantine	111/100	18	16.2	17	17.0

Table 17.12. Hierapolis. Distribution of degenerative joint disease among joint complexes in the Roman population (tombs C92 and C311), subdivided in slight degeneration with osteophytic formation <1 mm and moderate to severe degeneration with osteophytic formation >1 mm.

Joint	Bones exam. (N)	DJD (%)	<1 mm (n)	<1 mm (%)	>1 mm (n)	>1 mm (%)
Shoulder	11	18.2	1	9.1	1	9.1
Elbow	20	10.0	2	10.0	0	0.0
Wrist	15	7.1	1	7.1	0	0.0
Hip	19	15.8	3	15.8	0	0.0
Knee	5	20.0	1	20.0	0	0.0
Ankle	34	2.9	1	2.9	0	0.0

Table 17.13. Hierapolis. Distribution of degenerative joint disease among joint complexes in the Byzantine population (tomb C92), subdivided in slight degeneration with osteophytic formation <1 mm and moderate to severe degeneration with osteophytic formation >1 mm.

Joint	Bones exam. (N)	DJD (%)	<1 mm (n)	<1 mm (%)	>1 mm (n)	>1 mm (%)
Shoulder	28	21.5	5	17.9	1	3.6
Elbow	53	11.3	6	11.3	0	0.0
Wrist	13	0.0	0	0.0	0	0.0
Hip	49	24.5	7	14.3	5	10.2
Knee	17	23.5	3	17.6	1	5.9
Ankle	19	5.3	1	5.3	0	0.0

shoulder joint might be associated with athletic exercise like spear-throwing. Since sexing and aging of the single bones or bone fragments was impossible or inadequate, differences in DJD frequencies between sexes and age groups could not be analyzed.

Degenerative changes of the spine can be seen as changes of the small joints between vertebrae (articular facets), called spinal osteoarthrosis, and as changes of the superior and inferior surfaces of the vertebral bodies, called osteophytosis. In the present material vertebral bodies were much better preserved than the small articular facets, which were often incomplete or missing. Therefore, investigations focused on the analyses of osteophytosis. Results were recorded for the cervical, thoracic, and lumbar

spine separately. Degenerative changes of the spine usually increase from the cervical to the lumbar sections. Generally, by the third decade a large proportion of individuals develop osteophytosis, by the fifth decade the condition affects almost every individual (Roberts and Manchester 2001). As an example of degenerative disease of the spine, Fig. 17.12 shows thoracic vertebrae with severe changes and osteophytic lipping of about 5 mm found in Byzantine layers in tomb C92.

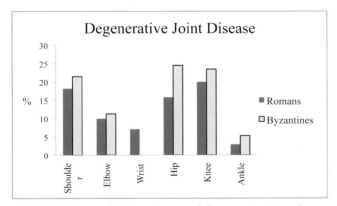

Fig. 17.10. Hierapolis. Distribution of degenerative joint disease (all lesions slight, moderate, and severe together), Roman versus Byzantine population.

Fig. 17.11. Hierapolis. Scapula, lateral view, showing osteophytes at glenoid cavity border FN 4550 (Roman, tomb C311).

For the Roman population 251 vertebral surfaces and for the Byzantine population 755 surfaces could be examined. The results are listed in Tables 17.14 and 17.15 and shown in Fig. 17.13, displaying the distribution of degenerative changes along the spine. The highest prevalence of osteohytosis (71.1%) is recorded for the lumbar spine in the Roman population; 60% of the lumbar vertebrae in the Roman sample even revealed moderate to severe changes with osteophytic lipping of more than 1 mm. The high frequency of lumbar degeneration could be explained by the old age of the individuals examined. For the Byzantine population the prevalence of lumbar degeneration was

34.6%. For the cervical and thoracic spine prevalence of degeneration was about 25% and 35% respectively and no marked difference was found between the Byzantine and Roman sample.

In addition, vertebral bodies were examined for typical indentations called Schmorl's nodes. They are associated with degeneration of the intervertebral discs and are quite common, especially in the lower thoracic and lumbar regions of the spine. Schmorl's nodes can occur idiopathically or are attributed to trauma, heavy lifting, and age, but are also reported to be commonly found in elite athletes, caused by great force placed on the lower spine (Waldron 2009). Tables 17.16 and 17.17 show the results for both populations. Again, the highest frequency of Schmorl's nodes (26.1%) was found on lumbar vertebrae of the Roman sample.

We have information on sex and age for only two spines belonging to the partly articulated Byzantine skeletons found in tomb C92: skeleton 32 (FN 234), a male, 25 to 35 years old, with spine, ribs, and pelvis found in articulation, reveals fusion of the 2nd and 3rd cervical vertebra, scoliosis of the cervical spine, and moderate osteophytosis of the thoracic vertebrae. Skeleton 38 (FN 235), a male, 20 to 30 years old, with shoulder girdle, spine, pelvis, and right leg discovered in articulation, displays Schmorl's nodes and moderate osteophytosis of the lumbar vertebrae.

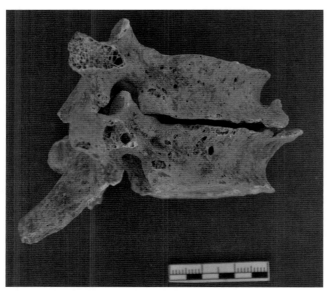

Fig. 17.12. Hierapolis. Thoracic vertebrae with severe changes and osteophytic lipping of about 5 mm (Byzantine, tomb C92).

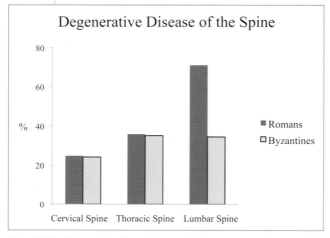

Fig. 17.13. Hierapolis. Distribution of degenerative disease of the spine (all lesions slight, moderate, and severe together), Roman versus Byzantine population.

Table 17.14. Hierapolis. Distribution of osteophytes among sections of the spine in the Roman population (tombs C92 and C311), subdivided in slight degeneration with osteophytic formation <1 mm and moderate to severe degeneration with osteophytic formation >1 mm.

Spine	Vertebral surfaces (N)	Osteophytes (n)	Osteophytes (%)	<1 mm (%)	>1 mm (%)
Cervical	117	29	24.8	9.4	15.4
Thoracic	89	32	36.0	14.6	21.4
Lumbar	45	32	71.1	11.1	60.0

Table 17.15. Hierapolis. Distribution of osteophytes among sections of the spine in the Byzantine population (tomb C92), subdivided in slight degeneration with osteophytic formation <1 mm and moderate to severe degeneration with osteophytic formation >1 mm.

Spine	Vertebral surfaces (N)	Osteophytes (n)	Osteophytes (%)	<1 mm (%)	>1 mm (%)
Cervical	215	52	24.2	14.4	9.8
Thoracic	358	126	35.2	18.2	17.0
Lumbar	182	63	34.6	14.3	20.3

Table 17.16. Hierapolis. Distribution of Schmorl's nodes among sections of the spine in the Roman population (tombs C92 and C311).

Spine	Vertebral surfaces (N)	Schmorl's nodes (n)	Schmorl's nodes (%)
Cervical	147	2	1.4
Thoracic	122	16	13.1
Lumbar	69	18	26.1

Table 17.17. Hierapolis. Distribution of Schmorl's nodes among sections of the spine in the Byzantine population (tomb C92).

Spine	Vertebral surfaces (N)	Schmorl's nodes (n)	Schmorl's nodes (%)
Cervical	227	0	0
Thoracic	409	51	12.5
Lumbar	210	22	10.5

Signs of injury: Fractures

The pattern and location of bone fractures help to draw conclusions on the type of injury and traumatization and provides information concerning the occurrence of accidents and/or violence in a population (Larsen 1999). Well-healed fractures are also an indication for some kind of medical treatment. Due to the fragmentary state of the present material (including many burnt bones from the Roman layers) investigations were somewhat restricted, especially as only few bones were completely preserved. Signs of healed fractures were detectable on only six postcranial bones of almost 2,000 bones and bone fragments examined.

In the Byzantine sample five healed fractures of the post-cranium were found:

- Healed mid-shaft fracture of right clavicle (adult, sex unknown; FN161, C92); see Fig. 17.14
- Healed fracture of left clavicle (15–25 years, sex unknown; FN1812, C92)
- Healed fracture of left transverse process of first thoracic vertebra (adult, sex unknown; FN366, C92)
- Oblique shaft fracture of left tibia, well aligned and healed with callus formation (adult, sex unknown; FN1272, C92)
- Healed transverse fracture of distal shaft of left fibula (adult, sex unknown; FN1853, C92)

In the Roman sample one fracture was recorded:

- Healed rib fracture (adult, sex unknown; FN3127, C92)

The prevalence of fractures in the Roman and Byzantine samples is shown in Tables 17.18 and 17.19. As mentioned above, since only a few bones were completely preserved and thus investigations were limited, these percentages give only a rough orientation on incidences of injuries. Nevertheless, the fractures observed offer clues to the type and nature of traumatization. Clavicle fractures are often caused by a fall onto the shoulder or an outstretched arm. A direct blow to the shoulder can also result in a clavicle fracture. This may happen in athletic competition or fighting. Fractures of tibia and fibula are often connected and are usually associated with falls or direct blows to the lower leg. Oblique fractures of the tibia shaft usually occur when there are twisting forces, which are often found in sport injuries (for example, while running into another player during a ball game). Force to the chest like a fall or attack can cause rib fractures. Fractures of the thoracic spine are typically caused by falls from heights or accidents (see fracture causes according to Fischer, *OrthoInfo*). Similar fracture patterns, as diagnosed in the present material, are recorded, for example, from a mid-Byzantine urban population from the 11th century AD Kastella in Central Crete (Bourbou 2009).

From 51 fairly well-preserved adult crania found in the tombs C92 and C311 only one exhibits a fracture. The skull was unearthed in the mid-Byzantine layer of house tomb C92 (SN70). It was the cranium of a man, who died at an age of about 30–40 years. A sharp edged injury of 25 mm was discovered on the left parietal bone (see Fig. 17.15). The wound displays a straight profile and results from a cut with a sharp edged weapon, e.g. a sword or an axe. Both, the position of the wound above the hat brim line and on the left side of the head are typical criteria for homicidal blows (Kremer *et al.* 2008). The location of the fracture indicates that the stroke to the head came from above. Either the assaulter was sitting on horseback, or the victim was kneeling, or the attacker used a long handled weapon (polearm). A face-to-face fight with a right-handed

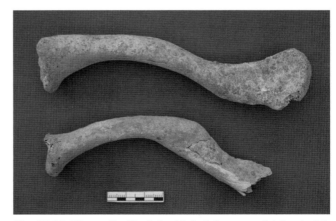

Fig. 17.14. Hierapolis. Healed mid-shaft fracture of right clavicle (FN161, Byzantine, tomb C92); for comparison normal clavicle above.

Table 17.18. Hieraplis. Distribution of fractures in the Roman adult population (tombs C92 and C311)

Bone examined	Diagnostic bones (N)*	Fractured (n)	Fractured (%)
Skull	13	0	–
Clavicle	6	0	–
Vertebra	21	0	–
Rib	9	1	11.1
Humerus	3	0	–
Radius	3	0	–
Ulna	4	0	–
Femur	6	0	–
Tibia	4	0	–
Fibula	0	0	–

* more than 80% of bone preserved

Table 17.19. Hierapolis. Distribution of fractures in the Byzantine adult population (tomb C92)

Bone examined	Diagnostic bones (N)*	Fractured (n)	Fractured (%)
Skull	38	1	2.6
Clavicle	33	2	6.1
Vertebra	221	1	0.5
Rib	1	0	–
Humerus	17	0	–
Radius	10	0	–
Ulna	9	0	–
Femur	17	0	–
Tibia	18	1	5.6
Fibula	15	1	6.7

* more than 80% of bone preserved.

opponent armed with a long-handled battle-axe could also explain the traumatic event. Those long handled battle-axes can be seen in many Byzantine illustrations; for example, on the 11th century mosaic in the Nea Moni monastery on the island of Chios showing Varangian guards in ceremonial costumes (http://www.byzantinemilitary.blogspot.it), or in a miniature of the 14th century held and digitized by the British Library, showing Philip Augustus arriving in Palestine (http://www.bl.uk/catalogues/illuminatedmanuscripts/searchSimple.asp). The head injury reveals first signs of healing with some new bone formation and remodeling of the wound (Fig. 17.16). It requires at least some days or even weeks to show these visible changes. However, no rounding and smoothing of the margins of the wound were noticed, which takes six months and more to form

(Steyn and İşcan 2000; Waldron 2009). Therefore, it seems that the man survived the head injury only for several days or a few weeks (Ahrens *et al.* 2012).

Discussion

The aim of the present study was to compare the health status, general wellbeing, and living conditions of the people buried in Roman house tombs at Hierapolis and of the people reusing these tomb structures in mid-Byzantine times. Human remains from two completely excavated Roman tombs, one of which had been intensely reused in mid-Byzantine times, have been analyzed. There was a large amount of human bones from both periods available for the investigations. The great challenge, however, was the

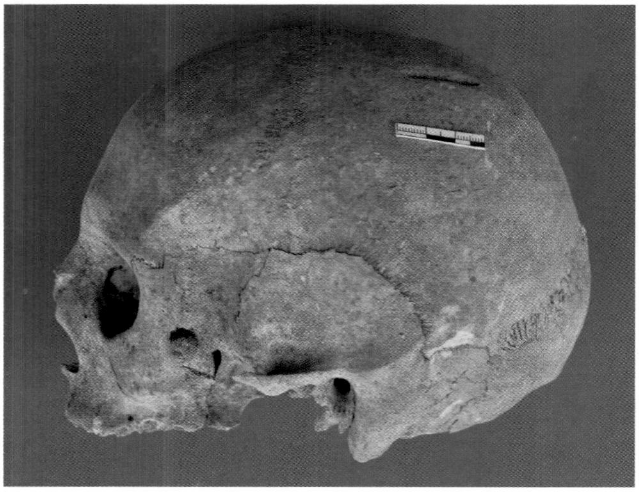

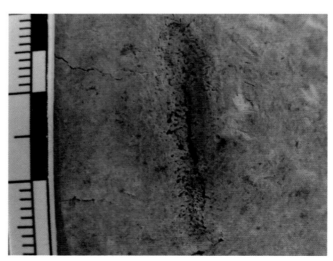

Fig. 17.15. Hierapolis. Male skull displaying a sharp-edged injury on the left parietal bone (SN70; Byzantine, tomb C92).

Fig. 17.16. Hierapolis. Head injury, close up view (SN70), revealing first signs of healing.

fragmentary preservation of the skeletal material and the find situation with almost no bones in articulation. The condition of the bones and the lack of complete skeletons restricted osteological analyses. Dealing with the comingled remains, it was particularly difficult to evaluate the demographic profile of the populations. Paleopathological examinations were also limited by the fragmentary nature of the skeletal material. Especially interactions with regard to sex and gender could only be observed on cranial bones, but not on postcranial remains. Of course, sex estimation from the adult cranium alone is a problem; osteologists claim accuracy for sex identification from the skull of about 80–92% (Mays and Cox 2000). Furthermore, it was impossible to link pathological findings on postcranial bones to a specific adult age group, e.g. individuals aged 20–40, individuals aged 40–60 and so on. For example, it was not possible to relate degeneration of the spine to sex or age. These facts obviously limit the interpretation of the palaeopathological results. However, it was still possible to examine many of the bones and compare the results.

The house tombs of Eutyches and Patrokles not only provided skeletal material from the two different periods, they also revealed completely different burial customs for the Roman and Byzantine times. Funeral rites in Imperial Roman times were quite elaborate and luxurious with the construction of a family tomb, the decoration of the tomb, for example, with theatre masks, garlands of flowers, and so on. Also the cremations found in Patrokles' tomb surely required detailed planning and organization. Particular mortuary behaviours (cremation versus inhumation) were probably chosen according to individual preference and had nothing to do with differences in social status. As mentioned by others (Ahrens 2015; Ubelaker and Rife 2007), cremations and simple inhumations were often found side by side in Roman Asia Minor. Yet, it was always believed that in the 2nd/3rd century inhumations completely replaced cremations (Arthur 2006; Ahrens 2015). Now, the radiocarbon dates from burnt bones sampled in 2014 shed a new light on funeral practice at Hierapolis. It seems that we have some evidence of very late cremations dating to the late Roman period (see also Wenn *et al.,* this volume).

In contrast to Roman mortuary rites, the reburials in the mid-Byzantine period did not require a lot of time, energy, or money. What at first glance seemed to be a Byzantine 'mass grave' was in fact a continuous reuse of the tomb structure over several hundred years. The dead corpses were simply interred in the existing Roman funeral monument. Yet, the simple burials do not necessarily reflect poverty and a low social status of the deceased. They rather express a change in religious beliefs. The main reason for reusing Eutyches' tomb may have been the similarity and the physical closeness to the tomb of Philip the Apostle (Ahrens and Brandt 2016, 413). Philip the Apostle was said to be buried in a similar Roman house tomb – originally part of the present North-East Necropolis – that was later embedded into a basilica (D'Andria 2011–2012). With the find of the Christian pilgrim badges from Western Europe (Ahrens 2011–2012) and new results from strontium analyses (Wong *et al.,* this volume) the Roman tomb of Eutyches may also have been the final resting place of some pilgrims in the 13th century AD.

Besides difficulties discussed above, the present palaeodemographic data reveal at least a trend of the demographic composition of the people buried in the tombs. It has been shown that – in Roman as well as in Byzantine times – not only older children and adults, but also young children and neonates were interred in the tombs. These finds suggest that every member of the society had 'access' to the house tombs and was worth being buried there. Even neonates and infants were not excluded and seem to have been considered as full members of the family in the pagan as well as in the Christian community. Determination of adult age-at-death was possible for 51 crania (13 Roman and 38 Byzantine skulls) and determination of sex was possible for 47 adult crania (12 Roman and 35 Byzantine skulls). The data show that all age groups were represented and women as well as men found their place in the tombs. In summary, all demographic results suggest that individuals buried in the tombs – in Roman and Byzantine times – were not selected but rather reflect a 'natural' population.

The palaeodemographic results also reveal differences between the Roman and Byzantine population. In Roman layers 27% of the individuals unearthed were neonates, children, and juveniles younger than 15 years, while in Byzantine layers 31% of examined individuals were younger than 15. Furthermore, it is remarkable that two Roman men reached an age beyond 60, while none of the Byzantines survived 60 years. In fact, two-thirds of the Byzantine women did not even reach the age of 40. Although all these results are based on a comparatively small number of individuals and calculated from minimum numbers of individuals the data may at least indicate a general trend of the demographic composition. It seems that life expectancy was higher in the Roman period. Reaching old age in Antiquity was indeed an achievement and can be related to good health, sufficient nutrition, and privileged living conditions for the Romans.

Good health and living conditions for the Roman period are also reflected by the palaeopathological investigations. Signs of infections, anaemia, and malnutrition, such as porotic hyperostosis of the skull and porotic pitting of the orbits (cribra orbitalia) occurred less frequent in the Roman remains (frequency of cribra orbitalia: 9.1% in Romans, 22.2% in Byzantines, see Tables 17.7 and 17.8). A special indicator of childhood stress, such as malnutrition and infection, is the occurrence of enamel hypoplasia. Lower rates were found in the Roman sample (3.6% versus 6.5% in the Byzantine sample). The result indicates comparatively better health conditions for the children in the Roman period.

Comparison of frequencies of degenerative joint disease implies, that the Byzantine population was more exposed to physical stress and strain from daily occupations. The distribution of degenerative disease of joints (Fig. 17.10) reveals higher incidences in the Byzantine compared to the Roman population. Degenerative alterations were diagnosed on 10.7% Roman and on 16.6% Byzantine bones (Table 17.10). In the Roman sample only the lumbar spine showed severe degeneration and Schmorl's nodes, which could be related to occupational heavy lifting, but also to old age of the individuals. Schmorl's node may also occur as idiopathic (or genetic) growth disturbance of the vertebrae. The pronounced degeneration of the lumbar spine in the Roman sample would correlate with an average older age of the Roman population as implied by the demographic data.

Healed fractures were observed on one Roman and five Byzantine bones. They were discovered only in adults; however, for all healed fractures the traumatic events could have taken place during childhood or adolescence. All postcranial fractures observed were most likely caused by accidents, falls, or assaults. No specific bone fracture related to battlefield injury was discovered on the postcranial skeleton. Pathological analyses of the fractures showed that all were well aligned and healed without signs of severe complications, such as non-union, or osteomyelitis. These finds indicate the ability to provide some kind of treatment and care. For example, successful healing of a fractured lower limb, which was diagnosed twice in the Byzantine sample, requires immobilization of the patient for several weeks. Yet, no evidence of specific medical treatment like surgery performed on the bones was discovered. While the fractures detected on the postcranial skeleton were mainly related to accidents, a Byzantine male skull revealed an injury that demonstrates clear evidence of violence and is a typical war wound. The victim survived even this severe injury for several days, maybe weeks. Surely, he benefited from intense care and support. In the end, however, all efforts were in vain and the patient died.

Examination of Roman and Byzantine dentition suggests a better oral hygiene by the Romans. Roman teeth show less calculus formation, less tooth decay, and less ante-mortem tooth loss suggesting better dental care (Fig. 17.7). As already reported by Bourbou (2010) from Byzantine skeletons from Crete, cleaning of the teeth was apparently not a 'common activity' of the Byzantine population. Evidence of dental treatment can be found for the Roman as well as the Byzantine period. The correlation between carious lesions and ante-mortem tooth loss in both populations indicates that toothache was treated by tooth extraction (Künzl 2002). It seems that toothache caused by severe caries was a common complaint for the Byzantine population, but also affected the Roman population. Only periodontal disease – commonly linked to older age – was more frequently diagnosed in the Roman sample.

The present osteological results from the Hierapolis North-East Necropolis – comparatively better general health and presumably higher life expectancy in the Roman versus mid-Byzantine skeletal sample – are not surprising, they rather confirm other studies. Angel (1972) had already reported a higher life expectancy in Imperial Roman than in mid-Byzantine times from the Eastern Mediterranean area. He presented data from Greek cemeteries showing that longevity of adults noticeably decreased for both sexes from the Imperial Roman to the Byzantine period. He also found that death ratios for children increased in Byzantine times. Demographic results from the Hierapolis mid-Byzantine sample – 38% of the examined individuals died before reaching adulthood – correspond with analyses from 13 early and mid-Byzantine Greek skeletal populations. From 1,081 individuals examined by Bourbou (2013), 41.5% were in the non-adult age group (see also Demirel, this volume).

Reasons for child mortality could have been numerous diseases. Some are listed in paediatric textbooks of the Byzantine era as, for example, diarrhoea, cholera, worms, fever, and infection of tonsils and throat (Talbot 2009). A strong correlation between malnutrition, poor hygiene, infection, and childhood mortality is reported by many authors (e.g. Larsen 1999; Bourbou 2013). One of the skeletal markers indicative of infection and malnutrition in childhood is enamel hyperplasia of permanent teeth. While in the Hierapolis Roman sample only 3.6% of teeth show these hypoplastic defects, 6.5% of the Byzantine teeth examined and 10.5% of the adult individuals reveal enamel hypoplasia. Enamel hypoplasia is also frequently recorded from other Byzantine sites. From Cretan samples of the 7th to 12th century AD a prevalence of 6.1% enamel hypoplasia in adult individuals has been recorded (Bourbou 2010); 9.6% of the individuals from mid-Byzantine Korytiani in West Greece (10th/11th century AD; Papageorgopoulou and Xirotiris 2009) reveal enamel hypoplasia, and in the early Byzantine sample from the Sanctuary at the Isthmus of Corinth the prevalence is as high as 42.86% (Rife 2012). High frequencies of enamel defects are also found in the late Byzantine period. At Troy the frequency in the late Byzantine population is 48% (Kiesewetter, unpublished) and children of the late Byzantine period show even 90% hypoplasia at Ephesus (Schultz 1989) and 67% enamel lesions at Pergamon (Schultz and Schmidt-Schultz 1994).

Other skeletal markers commonly discussed as signs of chronic infection and malnutrition – of children and adults – are the iron-deficiency related alterations of the orbit (cribra orbitalia) and the skull (porotic hyperostosis). Again, at Hierapolis an increase of these pathologies is found from the Roman (9.1% cribra orbitalia, 0% porotic hyperostosis) to the mid-Byzantine period (22.2% cribra orbitalia, 9.1% porotic hyperostosis). From early Byzantine sites an incidence of cribra orbitalia around 13% and of porotic hyperostosis from 5% to 8% is recorded (e.g. Rife 2010; Bourbou and Tsilipakou 2009). At Troy individuals from the late Byzantine period (13th century AD) revealed also cribra orbitalia (Kiesewetter 2014); of 102 individuals 26% were affected (Kiesewetter, unpublished).

In the 1st centuries AD, living conditions at Roman Hierapolis were probably, like in other Imperial Eastern Mediterranean cities (e.g. at Pergamon, South-East Necropolis, Teegen in Pirson 2013; Teegen, this volume), much better than in Rome or in the Northern provinces (Angel 1972). The ancient city of Rome is described as a 'death-trap', with current epidemics, baths that may have concentrated the pathogens, lead poisoning, and higher death rates (Hopkins 2009). Also in the Northern provinces living standards were presumably harsh compared to Hierapolis. For example, at a Roman settlement (vicus) in Southern Germany (Stettfeld) higher frequencies of degenerative disease of the spine and of dental diseases (including periodontal disease, ante-mortem tooth loss, calculus, abscesses of the root) were diagnosed (Wahl and Kokabi 1988). This quality of life in the Eastern Mediterranean cities seemed to change in Byzantine times. One of the reasons for the worsening living conditions in the mid-Byzantine period was, according to Angel, the man-produced destruction of natural resources (deforestation, erosion, soil-exhaustion). Additional factors responsible for a decline of living conditions may at Hierapolis be related to the 7th-century earthquake, but perhaps also to the Persian wars (AD 603–28), and to the plague of Justinian that haunted the Byzantine Empire (AD 542–750) (Vasold 2003; Brandt 2016; this volume). Taken together they may have caused a demographic drop and it seems that the population at Hierapolis never have fully recovered from these events (D'Andria 2010; Arthur 2012).

To summarize, the human remains from the North-East Necropolis at Hierapolis examined in the present study enlarge the picture of life and death in Roman/late Roman and mid-Byzantine times, which we already have from archaeological finds. Well-being and the quality of everyday life of a population depends on several factors, such as nutritional status, physical workload, medical treatment, dental health, hygiene, and childhood diseases. The results from the present palaeopathological analyses of Roman and mid-Byzantine bones allow a multiple-attribute measurement of general health for both populations – besides all limitations caused by the fractured skeletal material which has been discussed above. It has been shown that general health was better in the Roman sample. The Roman skeletons revealed less signs of degeneration caused by physical workload, less signs of infection, of childhood diseases, and better dental care. While one of the Byzantine skulls showed clear evidence of a war injury, no indication of battlefield wounds was found in the Roman sample. Furthermore, demographic investigations suggest a higher life expectancy and somewhat lower child-mortality for the Romans.

Concluding, osteological analyses of the human burials from the house tombs of Eutyches and Patrokles show that the change of Hierapolis from a prosperous urban city of the Roman Imperial period to a more rural mid-Byzantine town erected on the ancient ruins is also reflected in the health status of its inhabitants. With the decline of the city, life expectancy and general health of the population decreased. Yet, due to the limitations of osteological analyses caused by the nature of the skeletal material – first of all the incomplete and sometimes poor preservation of the commingled remains – the present study can only provide provisional results and conclusions. Moreover, it needs to be considered that the present results come from the examination of human bones from two house tombs, while a minimum of some 600 tombs and sarcophagi are reported from the Hierapolis North-East Necropolis (Hill et al. 2016, 112).

Acknowledgements

I would like to thank Prof. Dr J. Rasmus Brandt and Prof. Dr Francesco D'Andria for the opportunity to work at Hierapolis and for their support, interest, and all the organization of the excavation. Many thanks go to Dr Sven Ahrens, who introduced me to the site and its archaeology in 2010. I am also grateful to the whole Norwegian excavation team, especially to Helene Russ, for her great help examining and discussing the bones, to Dr Gro Bjørnstad and Prof. Dr Erika Hagelberg for many discussions and for organizing the C14 dates. Camilla Wenn and Kjetil Bortheim have been carefully excavating all the human remains for many years. Megan Wong helped sorting the bones in 2013, Caroline Fisker, Linda Grytan, and Anette Eriksen-Sand helped cleaning the bones in 2014.

Photographs courtesy *Missione archeologica italiana a Hierapolis in Frigia* (Fig. 17.1) and the Oslo University Excavation Project at Hierapolis (Figs. 17.2–6, 8–9, 11–12, 14–16).

Bibliography

Ahrens, S. (2011–2012) A set of western European pilgrim badges from Hierapolis of Phrygia. *Rendiconti della Pontificia Accademia Romana di Archeologia* 84, 67–75.

Ahrens, S. (2015) 'Whether by decay or fire consumed ...': Cremation in Hellenistic and Roman Asia Minor. In J. R. Brandt, M. Prusac, and H. Roland (eds.) *Death and changing rituals. Function and meaning in ancient funerary practices* (Studies in Funerary Archaeology 7), 185–222. Oxford, Oxbow Books.

Ahrens (this volume) Social status and tomb monuments in Hierapolis and Roman Asia Minor, 131–48.

Ahrens, S. and Brandt J. R. (2016) Excavations in the North-East Necropolis of Hierapolis 2007–2010. In D'Andria, Caggia, and Ismaeli (eds.), 395–414.

Ahrens, S., Bjørnstad, G., Bortheim, K., Kiesewetter, H., Russ, H., Selsvold, I. and Wenn C. C. (2013) Hierapolis 2012 – Excavations and analyses. *Nicolay Arkeologisk Tidsskrift* 120, 13–22.

Angel, J. L. (1966) Porotic hyperostosis, anemias, malarias, and marshes in the prehistoric Eastern Mediterranean. *Science* 153 no. 3737, 760–3.

Angel, J. L. (1972) Ecology and population in the Eastern Mediterranean. *World Archaeology* 4, 88–105.

Arthur, P. (2006) *Byzantine and Turkish Hierapolis (Pamukkale). An archaeological guide.* Istanbul, Ege Yayınları.

Arthur, P. (2012) Hierapolis of Phrygia: The drawn-out demise of an Anatolian city. In N. Christie and A. Augenti (eds.) Urbes Extinctae, *Archaeologies of abandoned Classical towns*, 275–305. Farnham, Ashgate.

Bortheim, K., Cappelletto, E., Hill, D., Kiesewetter, H., Russ, H., Selsvold, I., Wenn, C. C. and Wong, M. (2014) Hierapolis 2013 – Revisiting old finds and procuring new ones. *Nicolay Arkeologisk Tidsskrift* 123, 78–84.

Bourbou, C. (2009) Patterns of trauma in a Medieval urban population (11th century AD) from central Crete. In Schepartz, Fox and Bourbou (eds.), 111–20.

Bourbou, C. (2010) *Health and disease in Byzantine Crete (7th – 12th centuries AD).* Farnham, Ashgate.

Bourbou, C. (2013) 'Hide and Seek': The bioarchaeology of children in Byzantine Greece. In Sioumpara, E. and Psaroudakis, K. (eds.) *Themelion, 24 papers in Honor of Professor Petros Themelis from his students and colleagues*, 465–83. Athens, Society of Messenian Archaeological Studies.

Bourbou, C. and Tsilipakou, A. (2009) Investigating the human past of Greece during the 6th – 7th centuries AD. In Schepartz, Fox, and Bourbou (eds.), 121–36.

Brandt, J. R. (2016) Bysantinsk skjebnetid. Om jordskjelv, pest og klimatiske endringer i Anatolia i tidlig bysantinsk tid (300–800 e.Kr.). *Nicolay arkeologisk tidsskrift* 127, 40–46 (expanded English version forthcoming).

Brandt, J. R. (this volume) Introduction. Dead bodies – Live data: Some reflections from the sideline, xvi–xxiv.

Buikstra, J. E. and Ubelaker, D. H. 1994 *Standards for data collection from human skeletal remains. Proceedings of a seminar at the The Field Museum of Natural History* (Arkansas Archaeological Survey Research Series 44). Fayettville, Arkansas Archeological Society.

D'Andria, F. (2010) *Hierapolis of Phrygia (Pamukkale). An archaeological guide.* Istanbul, Ege Yayınları.

D'Andria, F. (2011–2012) Il Santuario e la tomba dell'apostolo Filippo a Hierapolis di Frigia. *Rendiconti della Pontificia Accademia Romana di Archeologia* 84, 1–52.

D'Andria, F. (2013) Il *Ploutonion* a Hierapolis di Frigia. *Istanbuler Mitteilungen* 63, 157–217.

D'Andria, F., Caggia, M. P., and Ismaelli, T. (eds.) *Hierapolis di Frigia VII. Le attività delle campagne di scavo e restauro 2007–2011.* Istanbul, Ege Yayınları.

Demirel, F. A. (this volume) Infant and child skeletons from the Lower City Church at Byzantine Amorium, 306–17.

Fischer, S. J. (ed.) *OrthoInfo. American Academy of Orthopaedic Surgeons*; http://orthoinfo.aaos.org

Hill, D. J. A., Lieng Andreadakis, L. T. and Ahrens S. (2016) The North-East Necropolis survey 2007–2011: methods, preliminary results and representativity. In D'Andria, Caggia, and Ismaelli (eds.), 109–120.

Hollimon, S. E. (2011) Sex and gender in bioarchaeological research: Theory, method, and interpretation. In S. C. Agarwal and B. A. Glencross (eds.) *Social Bioarchaeology*, 149–82. Chichester, Wiley-Blackwell.

Hopkins, K. (2009) The political economy of the Roman Empire. In I. Morris and W. Scheidel (eds.) *The dynamics of ancient empires: State power from Assyria to Byzantium* (Oxford Studies in Early Empires), 178–204. Oxford and New York, Oxford University Press.

Kiesewetter, H. (2014) Paläoanthropologische Untersuchungen in Troia. In E. Pernicka, C. B. Rose, and P. Jablonka (eds.) *Troia 1987–2012: Grabungen und Forschungen I, Forschungsgeschichte, Methoden und Landschaft* (Studia Troica Monographien 5), 610–42. Bonn, Habelt.

Kremer, C., Racette, S., Dionne, C. A., and Sauvageau, A. (2008) Discrimination of falls and blows in blunt head trauma: Systematic study of the hat brim line rule in relation to skull fractures. *Journal of Forensic Sciences* 53.3, 716–9.

Künzl, E. (2002) *Medizin in der Antike. Aus einer Welt ohne Narkose und Aspirin.* Stuttgart, Theiss Verlag.

Larsen, C. S. (1999) *Bioarchaeology. Interpreting behavior from the human skeleton* (Cambridge Studies in Biological Anthropology). Cambridge, Cambridge University Press.

Lovejoy, C. O., Mindle, R. S., Przybeck, T. R., and Mensforth R. (1985) Chronological metamorphosis of the auricular surface of the ilium: A new method for determination of adult skeletal age at death. *American Journal of Physical Anthropology* 68, 15–28.

Mays, S. and Cox, M. (2000) Sex determination in skeletal remains. In M. Cox and S. Mays (eds.) *Human osteology in archaeology and forensic science*, 117–30. London, Greenwich Medical Media Ltd.

Meindl, R. S. and Lovejoy, C. O. (1985) Ectocranial suture closure. A revised method for the determination of skeletal age at death based on the lateral-anterior sutures. *American Journal of Physical Anthropology* 68, 57–66.

Papageorgopoulou, C. and Xirotiris, N. I. (2009) Anthropological research on a Byzantine population from Korytiani, West Greece. In Schepartz, Fox, and Bourbou (eds.), 193–221.

Pirson, F. (2013) Pergamon – Bericht über die Arbeiten in der Kampagne 2012 – Die anthropologisch-paläopathologischen Untersuchungen 2012 by Teegen, W.-R. *Archäologischer Anzeiger* 2013.2, 138–41.

Rife, J. L. (2012) *The Roman and Byzantine graves and human remains* (Isthmia XI). Princeton, The American School of Classical Studies at Athens.

Roberts, C., and Manchester, K. (2001) *The archaeology of disease.* Sparkford, Sutton Publishing.

Rogers, J. and Waldron, T. (1995) *A field guide to joint disease in archaeology.* Chichester, Wiley and Sons Ltd.

Schepartz, L. A., Fox, S. C., and Bourbou C. (eds.) (2009) *New directions in the skeletal biology of Greece* (Hesperia Supplement 43). Princeton, The American School of Classical Studies at Athens.

Scheuer, L. and Black, S. (2000) *Developmental juvenile osteology.* San Diego, Academic Press.

Schultz, M. (1989) Ergebnisse osteologischer Untersuchungen an mittelalterlichen Kinderskeletten unter besonderer Berücksichtigung anatolischer Populationen. *Anthropologischer Anzeiger* 47, 39–50.

Schultz, M. and Schmidt-Schultz, T. H. (1994) Krankheiten des Kindesalters in der mittelalterlichen Population von Pergamon. *Istanbuler Mitteilungen* 44, 181–201.

Selsvold, I., Solli, L. and Wenn, C. C. (2012) Surveys and Saints – Hierapolis 2011. *Nicolay Arkeologisk Tidsskrift* 117, 13–22.

Steyn, M. and İşcan, M. Y. (2000) Bone pathology and antemortem trauma in forensic cases. In J. A. Siegel, P. J. Saukko, and G. C. Knupfer (eds.) *Encyclopedia of forensic sciences*, 217–27. San Diego, Academic Press.

Strohm, T. F. and Alt, K. W. (1998) Periodontal diseases – etiology, classification, and diagnosis. In K.W. Alt, F. W. Rösing, and M. Teschler-Nicola (eds.) *Dental anthropology. Fundamentals, limits, and prospects*, 227–46. Vienna, Springer Verlag.

Stuart-Macadam, P. (1991) Anaemia in Roman Britain: Poundbury Camp. In H. Bush and M. Zvelebil (eds.) *Health in past societies. Biocultural interpretations of human skeletal remains in archaeological contexts* (British Archaeological Reports, International Series 567), 101–13. Oxford, Tempus Reparatum.

Stuart-Macadam, P. (1992) Porotic hyperostosis: A new perspective. *American Journal of Physical Anthropology* 87, 39–47.

Talbot, A.-M. (2009) The death and commemoration of Byzantine children. In A. Papaconstantinou and A.-M. Talbot (eds.) *Becoming Byzantine. Children and childhood in Byzantium* (Dumbarton Oaks Byzantine Symposia and Colloquia, April 28–30, 2006), 283–307. Harvard, Dumbarton Oaks.

Teegen, W.-R. (this volume) Pergamon – Kyme – Priene: Health and disease from the Roman to the Late Byzantine period in different locations of Asia Minor, 250–67.

Ubelaker, D. H. and Rife, J. L. (2007) The practice of cremation in the Roman-era cemetery at Kenchreai, Greece. *Bioarchaeology of the Near East* 1, 35–57.

Vasold, M. (2003) *Die Pest. Ende eines Mythos*. Stuttgart, Theiss Verlag.

Wahl, J. (1982) Leichenbranduntersuchungen. Ein Überblick über die Bearbeitungs- und Aussagemöglichkeiten von Brandgräbern. *Prähistorische Zeitschrift*, 57, 1–125.

Wahl, J. (2007) Investigations on pre-Roman and Roman cremation remains from Southwestern Germany: Results, potentialities, and limits. In C. W. Schmidt and S. A. Symes (eds.) *The analysis of burned human remains*, 145–61. London, Academic Press.

Wahl, J. and Kokabi, M. (1988) *Das römische Gräberfeld von Stettfeld I. Osteologische Untersuchung der Knochenreste aus dem Gräberfeld*. Stuttgart, Theiss Verlag.

Waldron, T. (2009) *Palaeopathology*. Cambridge, Cambridge University Press.

Walker, P. L., Bathurst, R. R., Richman, R., Gjerdrum, T., and Andrushko, V. A. (2009) The causes of porotic hyperostosis and cribra orbitalia: A reappraisal of the iron-deficiency-anemia hypothesis. *American Journal of Physical Anthropology* 139.2, 109–25.

Wenn, C. C., Ahrens, S., and Brandt, J. R. (this volume) Romans, Christians, and pilgrims at Hierapolis in Phrygia. Changes in funerary practices and mental processes, 196–216.

Wong, M., Naumann, E., Jaouen, K., and Richards, M. (this volume) Isotopic investigations of human diet and mobility at the site of Hierapolis, Turkey, 228–36.

Health and disease of infants and children in Byzantine Anatolia between AD 600 and 1350

Michael Schultz and Tyede H. Schmidt-Schultz

Abstract

Infant and child skeletons (n = 333) from Byzantine cemeteries in Anatolia, Turkey (Arslantepe, Boğazkale, Pergamon, and Ephesos), dating from the 7th to the 14th century AD were examined using macroscopy, radiology, low-power microscopy, light microscopy including polarization microscopy, scanning-electron microscopy and proteomics. The skeletons were studied with respect to frequencies of deficiency and inflammatory diseases (morbidity) and demographic parameters (mortality). For a comparable evaluation, disease profiles were established to characterize the quality of the individual population health. The results revealed very poor living conditions for the infants and children who lived in the final period of the Byzantine Empire (c. 13th/14th century AD), whereas the children who lived in the early period of the Empire (c. 7th – 10th century AD), which was the time when Turkish tribes had not significantly attacked Byzantium, showed evidence of better living conditions. Apparently, politically induced factors, such as war and privation, had a long-term influence on the health of the infants and children of the late Byzantine period.

Keywords: Arslantepe, Boğazkale, Byzantine period, Ephesos, morbidity, mortality, palaeopathology, Pergamon.

Introduction

Archaeological skeletal remains, mummies and bog bodies represent biohistorical documents, which are able to give very precise and detailed information on the living conditions of past population groups, such as nutrition, housing and working situations, geographic and climatic factors, and hygienic and sanitary conditions, as well as diseases and disabilities (Schultz 1982; Schultz *et al.* 2006).

Until the 1980s, skeletal remains of infants and children were only of minor interest in physical anthropology. This was also the case in the relatively new research field of palaeopathology, when subadult skeletons, particularly of foetuses and infants, were frequently neglected.

Current bioarchaeology, especially palaeopathology, deals with the biosocial background of prehistoric and early historical diseases and the quality of living conditions of past populations (Carli-Thiele and Schultz 2001; Lewis 2009; Larsen 1997; Schultz 1982). It is well known that inadequate living conditions cause diseases. Thus, if we know which ancient diseases were prevalent, we should be able to reconstruct, of course, within limits, the living conditions of ancient populations (Schultz 1982; 2001a; Schultz *et al.* 2008a). Infants, children, and the elderly people are the weakest groups within a human community mainly because, in the young, their immune systems have not completely developed or, in the elderly, have already started to decrease (Schultz *et al.* 1998; Schultz 2001a; Schultz and Schmidt-Schultz 2008a).

The Byzantine Empire, in its glorious early period (AD 476 – *c.* 630), did not only rule Anatolia, Greece, and southern parts of the Balkan Peninsula, but also most of the Near East, some small parts of the western Middle East, parts of North Africa including Egypt, parts of the Hispanic Peninsula, middle and lower Italy, Sicily, Sardinia and Corsica. The decline starts slowly (after AD 1071), mainly caused by Turkish tribes (e.g. the Seljuks, the Turks of the

Emirates of Karasi and Aydın, and later the Ottoman Turks) which broke through the border in the west and started to reduce the Byzantine state to the territory of Constantinople and smaller parts of Western Anatolia and Thrace (*c.* AD 1300/1350). Thus, the question arises, if it can be supposed that the decline of the empire and the consequential political and economic changes might have affected the health situation of the Byzantine people living in Anatolia.

Materials

For this study, the skeletal remains of infants and children from six Christian burial places, which were situated at four Byzantine Anatolian settlements dating from the 7th to the 14th century AD, were available (Fig. 18.1). Two cemeteries date from the early Byzantine period (Arslantepe and Boğazkale), three from the late Byzantine period (Pergamon and Ephesos) and one just after the Turkish conquest

(Pergamon). The Byzantine settlements of Arslantepe (about 1103 AD) and Boğazkale (about 1075 AD) were conquered by the Turkmenian Dynasty of the Danishmend, who were of Turko-Persian origin and who ruled north-central and parts of eastern Asia Minor in the 11th and 12th centuries AD. However, after 1174–76, the territory of the Danishmend was taken over by the Seljuks who were also horse people and represented a branch of the Oghuz-Turks. They had founded the Sultanate of Rum in the second half of the 11th century which persisted till 1307. After the decline of the Sultanate of Rum, small emirates filled the vacuum of former Seljuk power. Two of these lordships were the Emirates of Karasi (Beylik of the Karesioğlu) founded in 1297 and of Aydın (Beylik of the Aydınoğlu) established in 1308. These dynasties also stemmed from a branch of the Oghuz-Turks, as the Seljuks, and lasted till 1390 when the Ottomans took over the rule. Thus, these Turkish dynasties conquered Pergamon (Karasi) and Ephesos (Aydın).

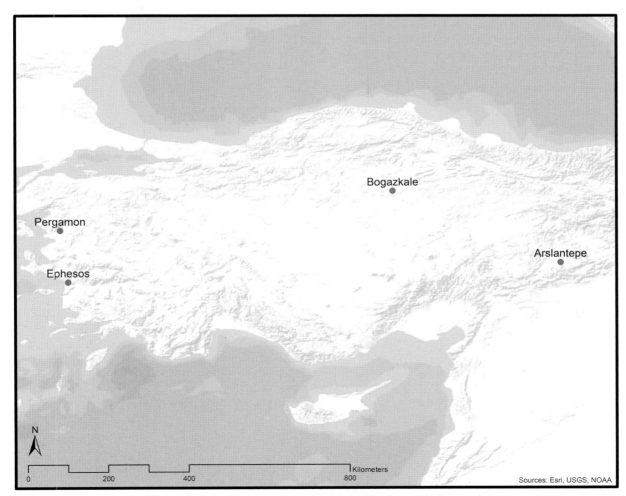

Fig. 18.1. Anatolia. Map with the position of the four sites discussed.

Arslantepe

The site Arslantepe, near Malatya in central eastern Anatolia, dates to the early Byzantine period (*c*. AD 600–800) (Frangipane 1997; 1999). The infant/child population (n = 97) from the Byzantine cemetery on the top of the settlement hill (Tepe) belonged to a relatively wealthy community, whose settlement was situated at the crossroads of important ancient trade routes (Frangipane 1997; 1999). The skeletal remains are in excellent condition and were studied by the authors in the years 1995, 1997, and 1999.

Boğazkale

The site Boğazkale, near Çorum in central Anatolia, dates to the early Byzantine period (*c*. AD 600–900) (Neve 1990). The infant/child population (n = 77) from the Byzantine cemetery, which is situated next to a small church (Neve 1996, 81, Fig. 229) in the upper city of the antique ruins of the Hittite capital of Ḫattuša, belongs to a relatively poor rural community of farmers and shepherds whose settlement was situated in the central Anatolian highland where farming was difficult and cattle breeding inefficient. Therefore, sheep and goat farming was carried out. The well preserved infant/child skeletons were examined by the authors in the years 1985, 1987, and 1990 after preparation by M. Brandt (UMG).

Pergamon

From Pergamon, present-day Bergama in north-western Anatolia, infant/child skeletons from two neighbouring Medieval cemeteries in the upper city (Radt 1989; 1995; 1999), were analyzed. The preservation of the skeletons (n = 69) was good or fair to middling. Five individuals were not included in this study because they belonged to the Roman or the early Byzantine period. The remaining infants and children (n = 64) from the two cemeteries belonged to a relatively poor urban community (cf. Rheidt 1991). Their skeletal remains were studied by the authors in the years 1986–89, 1993, 1995–96, and 1999 (Schultz and Schmidt-Schultz 1995) and can be split into two chronological sub-groups.

The first group (n = 49) dates to the late Byzantine period (*c*. AD 1200–1315). The fortified settlement was more or less under permanent siege by the Turks of the Emirate of Karasi and, finally, conquered around AD 1315. This group can be divided into two further sub-groups: the first half of the 13th century (n = 23) and second half of the 13th century (n = 19). Only seven individuals cannot be distributed with certainty to one of the two sub-groups, but belong generically to the 13th century.

The second group (n = 15) dates to the early Turkish period (*c*. AD 1315–1350). After the conquest of the city, only a small group of people survived living and working as slaves of the Turkish conquerors. Apparently, the survivors obtained permission from the local Turkish administration to bury their dead in the traditional Christian cemeteries of the Byzantine period and use the adjacent chapels for services.

Ephesos

Ephesos, situated close to the modern city of Selçuk, lies in southwestern Anatolia. From this site two burial areas, St Mary's Church and the Byzantine Palace, which yielded the skeletal remains of 95 infants and children dating to the late Byzantine period, were examined. The authors studied the skeletal remains of the infants and children from the Byzantine Palace in 2010 and from St Mary's Church in 2010–2011. A small sample of subadult skeletons from St Mary's Church had already been examined preliminarily by the first author at the Anthropological Laboratory of Berna Alpagut at the University of Ankara in 1987 (Schultz 1989a).

St Mary's Church, the main and famous basilica of the Byzantine city of Ephesos, served as a funerary church, at the latest, from the 7th century to at least the 14th century AD. The first information on the excavation of the cemetery was provided by Stefan Karwiese (1989). The skeletons studied (n = 55) are relatively well-preserved and date probably to the final period of the Byzantine reign in southwestern Anatolia (*c*. AD 1100–1310), as suggested by Sabine Ladstätter (2010; 2011). The city was besieged by the Seljuks and later, after AD 1304, conquered by the Beylik of the Aydınoğlu.

The Byzantine Palace, originally the domicile of an important civil servant or even the bishop, dates to late Antiquity and/or the early Byzantine period. This interesting building was abandoned, at the latest, after the 10th century AD and, apparently in the 14th century AD the palace ruins served as a funerary area (Pülz 2010; 2011; 2015). In a small cemetery, the skeletal remains of 40 infants and children were excavated, whose skeletons are relatively well preserved.

Methods and techniques

Age determination was carried out using the criteria suggested by Ferembach *et al.* 1979; Johnston 1962; Kósa 1978; Rösing *et al.* 2007; Scheuer and Black 2000; Schmid and Künle 1958; Stloukal and Hanáková 1978; Szilvássy 1988; and Ubelaker 1978. Sex determination has not yet been conducted.

For the demographic analysis of the mortality rates, the individuals were distributed between the relevant age groups: 1) Fetus-Newborn, 2) Infans-Ia (birth to end of the second year), 3) Infans-Ib (beginning of the third to the end of the sixth year) and 4) Infans-II (beginning of the seventh to the end of the fourteenth year). If an individual matches two or more age groups, it was distributed proportionally to these groups. Thus, fraction numbers might appear.

The mortality rates of the infants and children from Arslantepe cannot be presented yet as they will be revealed in a forthcoming doctoral thesis.

For the palaeopathological investigation, the infant and child skeletons (n = 333) from these four Byzantine settlements were basically investigated on the excavation site using the techniques of macroscopy and low-power microscopy (Carli-Thiele 2001; Lewis 2009; Schultz 1988a). If these techniques could not help to establish a reliable diagnosis, selected examples were taken and, permission was obtained from the Turkish Antiquity Service or the responsible archaeological museum, sent by the responsible excavator to the Department of Anatomy of the University Medical School Göttingen. At the department, additional investigations were carried out, such as radiology, light microscopy including polarization microscopy (Schultz 1988b; 1993; 2001; 2003; 2012), scanning-electron microscopy (Schultz 1988b) and proteomics (Schmidt-Schultz and Schultz 2004).

For this study, six diseases occurring relatively frequently in the Middle Ages were chosen: diseases which had significant influence on the quality of life and could

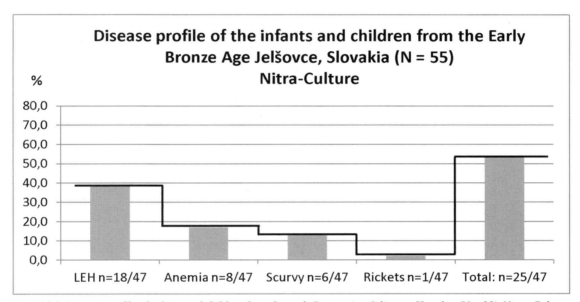

Fig. 18.2. Disease profile of infants and children from the early Bronze Age Jelšovce, Slovakia (N = 55). Nitra-Culture.

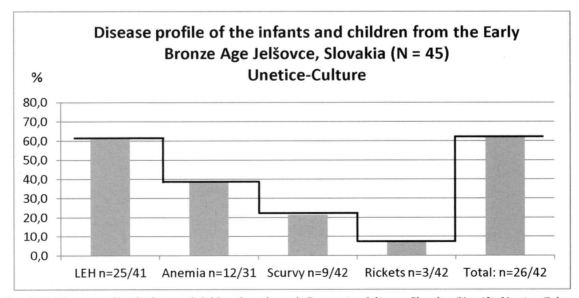

Fig. 18.3. Disease profile of infants and children from the early Bronze Age Jelšovce, Slovakia (N = 45). Unetice-Culture.

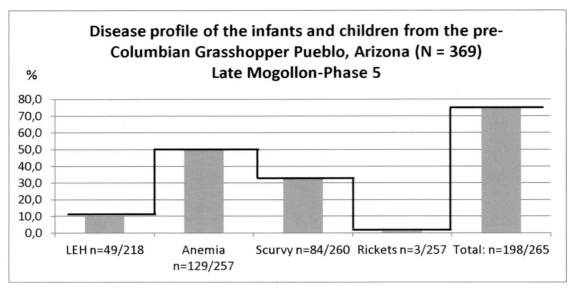

Fig. 18.4. Disease profile of infants and children from the pre-Columbian Grasshopper Pueblo, Arizona (= 369). Late Mogollon-Phase 5.

directly or indirectly provoke premature death. Thus, three deficiency diseases were selected: rickets (chronic vitamin D deficiency), scurvy (chronic vitamin C deficiency) and anaemia. Anaemia can be provoked by iron deficiency (e.g. El-Najjar *et al.* 1976) or lack of essential amino acids, such as tryptophan (e.g. Schultz 1982; 2001b), and in subadults also by malaria (Schultz 1990). However, anaemia might also occur as a consequence of parasitic diseases (e.g. Reinhard 1992; Larsen and Sering 2000) or of genetically caused factors, for instance thalassemia (e.g. Ascenzi *et al.* 1991). Furthermore, three inflammatory/infectious diseases were selected: osteomyelitis (chronic inflammation of bones), otitis media (inflammatory middle ear diseases), and meningeal reactions, for instance, bacterial meningitis (inflammatory-haemorrhagic meningeal diseases). Thus, the skeletal remains were investigated for the presence of vestiges of these diseases in all of the six populations (Arslantepe: one population; Boğazkale: one population; Pergamon: two populations; and Ephesos: two populations). The purpose was to calculate the frequencies of these diseases and to create disease or morbidity profiles ('disease curves') (Schultz and Schmidt-Schultz 2014; cf. Schultz *et al.* 1998), which characterize the morbidity of each population or population group. A disease profile is a plateau-line in a diagram connecting the top of the columns which represent the particular disease frequencies. The relation of the particular height of the columns to each other (= single frequencies) characterizes the profile (Figs. 18.2–4). As the examples demonstrate, the profiles (black lines) are very similar (Figs. 18.2 and 3) or completely different (Fig. 18.4). The two examples from the early Bronze Age settlement at Jelšovce (Slovakia) (Figs. 18.2 and 3) and from the pre-Columbian Grasshopper Pueblo (Arizona, USA)

(Fig. 18.4) show clear differences. The skeletal remains of the two early Bronze Age infant and child populations were excavated in the Nitra Valley at Jelšovce (Slovakia). The subadults of the Nitra-Culture date approximately to the period 2200–2000 BC, the subadults from the Unetice-Culture date approximately to 1900–1700 BC. The two populations were not related genetically and expressed different material cultures. They lived at the same place; however, there was a time gap between the two populations of at least one hundred years. Interestingly, they have the same disease profile (Figs. 18.2 and 3). Of course, the disease frequencies are different: the comparison shows that all frequencies calculated for the subadults of the Nitra-Culture are lower than in the subadults of the Unetice-Culture. The probable cause for the same disease profiles is the same biotope which, apparently, had a greater influence on human health in prehistoric times than human culture (Schultz *et al.* 1998). For comparison purposes, a pre-Columbian Pueblo population from the North American Southwest is presented (Fig. 18.4). This example shows a completely different disease profile.

On the basis of these disease profiles, the health status of populations can be compared to estimate the individual disease stress of each population (Schultz *et al.* 1998). A comparable procedure can be conducted using mortality profiles, which characterize the mortality in the various age groups of subadults.

Prevalence is a characteristic factor which tells us how many individuals of a group or a population of a defined size suffered from a certain disease at a particular time. In palaeopathology, a population comprises, as a rule, a time period of several centuries. Thus, the authors avoid using the term 'prevalence' and instead employ the term 'frequency'.

Results

Case study as palaeopathology

As a rule, the detailed examination of archaeological skeletons offers the possibility – of course within certain limits – to establish a partial biography (palaeobiography), sometimes also called 'osteobiography' of the deceased individual (Schultz 2011). As an example, a case from late Byzantine Pergamon is presented which illustrates the aetological context of various diseases (Case PE 02.10.81 – KA 67 – Burial 14/ Ind. C /PERG86#18/, cf. Schultz and Schmidt-Schultz 1995).

This child lived in the first half of the 13th century AD and died within his individual age range of 3¾ and 6¼ years. Apparently, the child suffered, for a short while before his death, from extremely pronounced chronic scurvy and, additionally, from chronic anaemia. The chronic vitamin C deficiency induced considerable bleeding in the region of the alveolar margins of the upper and lower jaws, the interior of some tooth sockets and the hard gum. Due to the chronic nutrient deficiencies, the immune system decreased. Probably, this precipitated the occurrence of the osteomyelitic process of the skull vault. As the process penetrated through the skull vault, it caused an inflammation of the meninges (pachymeningitis/meningitis) which, probably, was the cause of death.

This disease history vividly demonstrates the health problems Byzantine children generally suffered from during the period of the decline of the Empire. The described case is not an isolated instance. It is the norm in these uncertain and unstable times when the facilities and the infrastructure of the regular system collapsed.

Morbidity

The study of the aetiology and, particularly the epidemiology of the six selected diseases in the skeletal remains of Byzantine infants and children, might provide new insights on the interpretation of health in past populations (cf. Schultz 1984; 1988/1989; 1990; 2001b; Schultz and Schmidt-Schultz 1995; Carli-Thiele 1996; Kreutz 1997; Schultz *et al.* 1998; Lewis 2002). Thus, the nature, cause, spread and frequency of deficiency and inflammatory/ infectious diseases observed in the six subadult populations described here, might add something to our knowledge on the morbidity and the mortality of Medieval subadult populations in Asia Minor.

Arslantepe (Fig. 18.5)

From the early Byzantine period Arslantepe, which represents apparently a small urban community, skeletal remains of a total of 97 infants and children were available for the palaeopathological analysis. However, due to the sometimes fragmentary preservation, not all individuals could be used for this analysis. Thus, to diagnose rickets, 48 individuals were suitable. Only one individual suffered from chronic rickets: n = 1/48 (2.1%). Characteristic lesions of scurvy are exhibited in six out of 56 individuals: n = 6/56 (10.7%). Vestiges of anaemia are observable in seven individuals: n = 7/56 (12.5%). No case of osteomyelitis (n = 0/56; 0.0%) and otitis media (n = 0/48; 0.0%) could be diagnosed. Features due to meningeal reactions are seen in seven out of 60 individuals: n = 7/60 (11.7%).

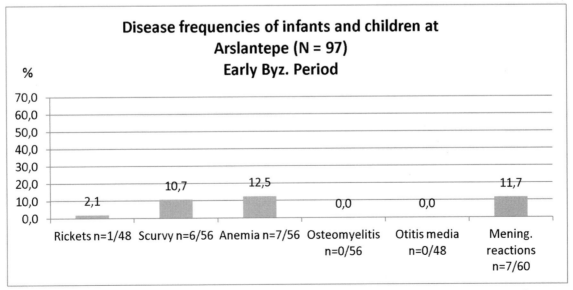

Fig. 18.5. Arslantepe. Disease frequencies of infants and children (N = 97). Early Byzantine period.

Boğazkale (Fig. 18.6)

From the early Byzantine village community at Boğazkale, skeletal remains of a total of 77 infants and children were available for the palaeopathological investigation. However, in this population, not all individuals could be integrated into this study because of fragmentary preservation of some skeletons. The results of recent investigations, for instance the microscopic analysis of thin-ground sections prepared from bone samples of the infants and children from Boğazkale, state that the disease frequency of scurvy and anaemia has to be revised in line with the description presented in previous publications (cf. Schultz 1986; 1989a; 1989b; Schultz and Schmidt-Schultz 1995).

No case of rickets (n = 0/64, 0.0%) and of otitis media (n = 0/64) could be diagnosed. However, 18 individuals exhibited vestiges of chronic scurvy: n = 18/64 (28.1%). Lesions due to anaemia are found in 17 individuals: n = 17/64 (26.6%). Only a low rate of osteomyelitis is stated (n = 3/59; 5.1%). Vestiges caused by meningeal reactions are observed in 16 individuals: n = 16/64 (25.0%).

Pergamon (Figs. 18.7–9)

From the late Byzantine/early Turkish city of Pergamon, skeletal remains of a total of 64 infants and children were available for the palaeopathological study. Unfortunately, in Pergamon the poor preservation of the skeletal remains, worse than in the other two described populations of Arslantepe and

Boğazkale reduced considerably the total individual number. Furthermore, due to the dating and the historical-political circumstances, the total population has to be divided into two groups: the population before (A) and after (B) the conquest of the city in approximately 1315 AD.

First we should look at the total population of 64 individuals which consists of individuals of the late Byzantine period and the early Turkish period (Fig. 18.7). Rickets cannot be diagnosed (n = 0/41; 0.0%). Scurvy shows a relatively high frequency: n = 10/41 (24.4%). Also anaemia is present with the same frequency: n = 11/45 (24.4%). Osteomyelitis (n = 4/40; 10.0%) and otitis media (n = 4/16; 25%) can be diagnosed, particularly the latter with a relatively high frequency. Characteristic changes due to meningeal reactions are seen in a remarkably high disease frequency: n = 21/43 (48.8%).

Late Byzantine population before the conquest of the city (Fig. 18.8)

When we only consider the individuals interred before the conquest around 1315 AD, this late Byzantine population (n = 49) demonstrates a very similar distribution of disease frequencies and an almost identical disease profile (Fig. 18.8) as the total population described before (Fig. 18.7). However, in five of the six diseases selected for this study, the disease frequencies are clearly higher: rickets (n = 0/29; 0.0%), scurvy (n = 8/29; 27.6%), anaemia (n = 8/31; 25.8%), osteomyelitis (n = 4/28; 14.3%), otitis media (n = 4/14; 28.6% and meningeal reactions (n = 19/31; 61.3%).

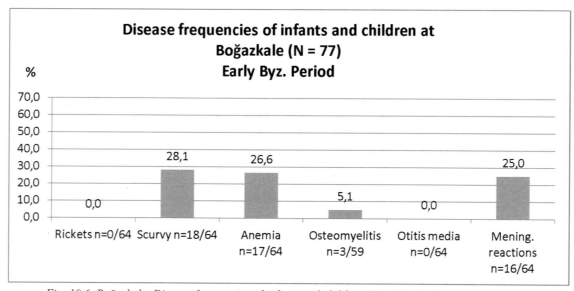

Fig. 18.6. Boğazkale. Disease frequencies of infants and children (N = 77). Early Byzantine period.

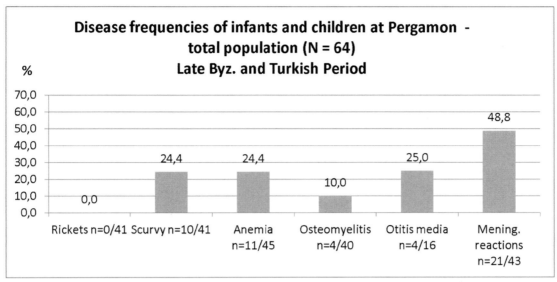

Fig. 18.7. Pergamon. Disease frequencies of infants and children (N = 64). Late Byzantine and Turkish periods.

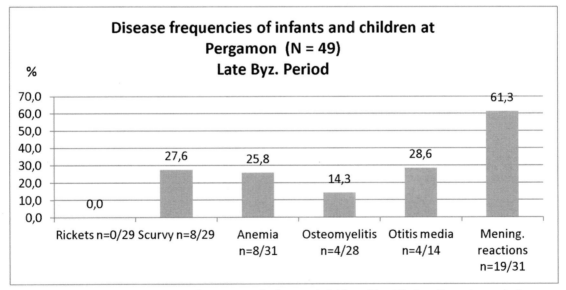

Fig. 18.8. Pergamon. Disease frequencies of infants and children (N = 49). Late Byzantine period.

THE BYZANTINE POPULATION AFTER THE CONQUEST BY THE TURKS (Fig. 18.9)

The group of the individuals interred after the conquest of the city (n = 15) show a completely different disease profile and different disease frequencies. No cases of rickets (n = 0/12; 0.0%) osteomyelitis (n = 0/12; 0.0%) and otitis media (n = 0/2; 0.0%) are observed. The frequencies of scurvy (n = 2/12; 16.7%), anaemia (n = 3/14; 21.4%) and meningeal reactions (n = 2/12; 16.7%) are not very high.

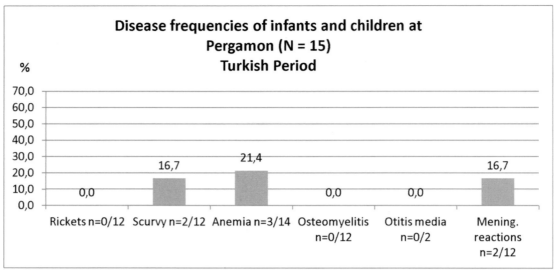

Fig. 18.9. Pergamon. Disease frequencies of infants and children (N = 15). Early Turkish period.

Ephesos (Figs. 18.10–12)

From the late Byzantine period of the important city of Ephesos, relatively well-preserved skeletal remains of a total of 95 infants and children were available for the palaeopathological analysis. These human remains, which apparently date from the final period of Byzantine rule in this city (14th century AD), were recovered from the area of St Mary's Church and a small burial area in the former Byzantine Palace.

Let us begin with the total population of 95 individuals (Fig. 18.10). The disease profile is similar to that of the late Byzantine population of Pergamon (Fig. 18.8). However, the frequencies are lower, with the exception of scurvy. Scurvy at Ephesos shows a relatively high

frequency: n = 18/53 (34.0%). Vestiges of rickets cannot be detected (n = 0/47; 0.0%). Anaemia appeared moderately: n = 12/53 (22.6%). Interestingly, no case of osteomyelitis was diagnosed (n = 0/49; 0.0%). Changes due to otitis media are seen in an unexceptional frequency (n = 8/38; 21.1%), whereas vestiges caused by meningeal reactions are observable with a high frequency: n = 20/39 (51.3%).

St Mary's church (Fig. 18.11)

The relatively large group of infants and children (n = 55) buried in the area of St Mary's Church show a higher disease frequency than the total population, with the exception of otitis media. Of course, as in the total population, no cases of

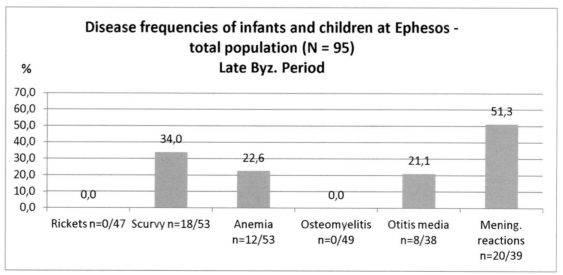

Fig. 18.10. Ephesos. Disease frequencies of infants and children – total population (N = 95). Late Byzantine period.

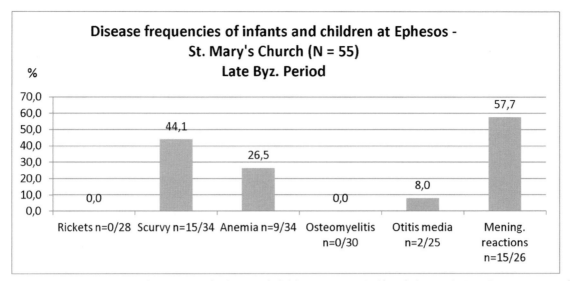

Fig. 18.11. Ephesos. Disease frequencies of infants and children – St Mary's Church (N = 55). Late Byzantine period.

rickets (n = 0/28; 0.0%) and osteomyelitis (n = 0/30; 0.0%) were diagnosed. The frequency of otitis media (n = 2/25; 8.0%) is strikingly low for Byzantine populations of this time, whereas the frequencies of scurvy (n = 15/34; 44.1%) and anaemia (n = 9/34; 26.5%) are distinctly higher. The frequency of observed vestiges due to meningeal reactions is exorbitant (n = 15/26; 57.7%).

THE BYZANTINE PALACE (Fig. 18.12)

The human skeletal remains of 40 infants and children allow a tendency on the health situation of this relatively small group to be observed. Similar to the group from St Mary's Church, no cases of rickets (n = 0/19; 0.0%) and osteomyelitis

(n = 0/19; 0.0%) were diagnosable. Characteristic lesions due to scurvy occur with a relatively low frequency: n = 3/19 (15.8%). Vestiges of anaemia are visible in only three individuals: n = 3/19 (15.8%). Otitis media shows the highest frequency of all diseases studied at Ephesos: n = 6/13; 46.2%). Morphological features caused by meningeal reactions are observable in five out of 13 individuals: n = 5/13 (38.5%).

Mortality

It is well known that morbidity influences mortality. Thus, also in Byzantine Anatolia, the mortality of infants and children reflects the living conditions considerably.

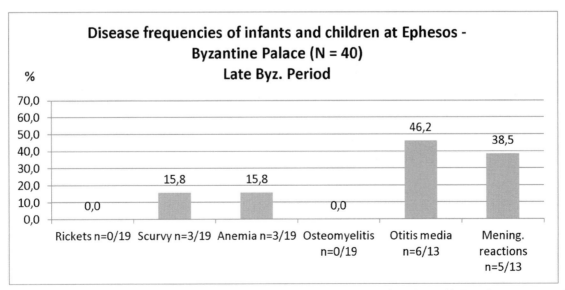

Fig. 18.12. Ephesos. Disease frequencies of infants and children – Byzantine Palace (N = 40). Late Byzantine period.

However, as this article focuses on the diseases, only some short information is given on the mortality of the infants and children whose diseases are presented here.

Boğazkale (Fig. 18.13)

In this rural mountain-village community, mortality shows its highest frequency in the age group Infans-Ib in which 30 of the total of 77 children died (39.0%). Whereas the mortality rate of foetuses and newborns is with only two available individuals rather low (2.6%). In the age group Infans-Ia, 20 infants of the total child population died (26.0%). A similar frequency is observed in the older

children who represent age group Infans-II in which the remaining 25 children died (32.5%).

Pergamon

LATE BYZANTINE PERIOD (Figs 18.14–16)

As expected, in the late Byzantine infant and child population from Pergamon dating from the entire 13th century, mortality is most frequent in the young infants (Fig. 18.14): 21 of the 49 individuals died in the age group Infans-Ia (42.9%). Thus, causes of early death, which are as a rule typical of awkward living conditions, characterize this

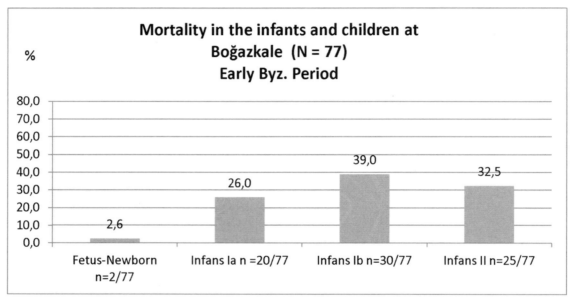

Fig. 18.13. Boğazkale. Mortality in the infants and children (N = 77). Early Byzantine period.

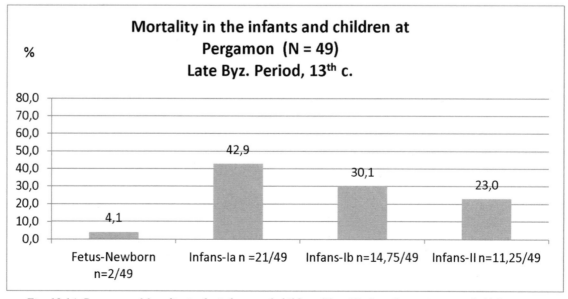

Fig. 18.14. Pergamon. Mortality in the infants and children (N = 49). Late Byzantine period, 13th century.

population (cf. Schultz 1989a; 1989c). Apart from that, the mortality rates of the total late Byzantine child population from Pergamon are similar to Boğazkale: In the age group Fetus-Newborn two out of 49 individuals were observed (4.1%); 14.75 out of 49 individuals died in the age group Infans-Ib (30.1%) and 11.25 out of 49 individuals died in the age group Infans-II (23.0%).

Considering the mortality of the infants and children in the first (Fig. 18.15) versus the second half of the 13th century (Fig. 18.16) which is, indeed, the proximate time before the conquest of the city, it appears that the mortality rates within the various age groups are similar in both sub-periods, however, with some exceptions: in the first half of the 13th century no individual died in the age group Fetus-Newborn (n = 0/23; 0.0%), whereas in the second half of the century, only a few deceased individuals were observed in this age group (n = 2/19; 10.5%). Furthermore, in the age group Infans-Ib, relatively fewer children died (n = 6.25/23; 27.2%) in the first half of the 13th century than in the second half (n = 6/19; 31.6%). However, as the sample is very small, there is not really a difference. A similar situation is found in the age group Infans-II. Only in the age group Infans-Ia is there a seemingly greater difference in the mortality: In the first half of the 13th century the number of the infants who died is larger (n = 11/23; 47.8%) than in the second half

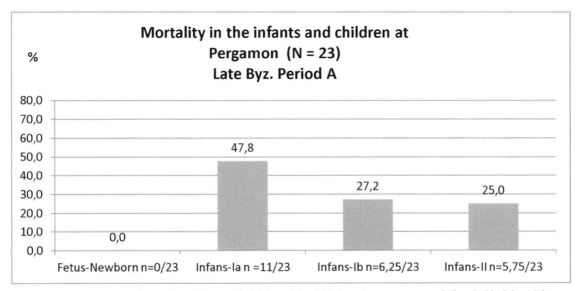

Fig. 18.15. Pergamon. Mortality in the infants and children (N = 23). Late Byzantine period, first half of the 13th century.

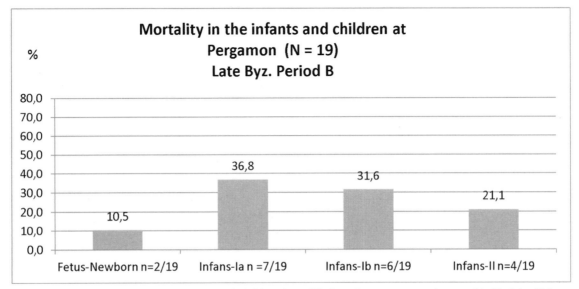

Fig. 18.16. Pergamon. Mortality in the infants and children (N = 19). Late Byzantine period, second half of the 13th century.

of the century (n = 7/19; 36.8%). Thus, living conditions at Pergamon were probably very similar throughout the whole 13th century.

EARLY TURKISH PERIOD (Fig. 18.17)

A completely different situation was apparently present after the conquest of the Byzantine city. The small group of surviving infants and children demonstrated a completely different pattern in their mortality. The lowest mortality is found in the age groups Fetus-Newborn and Infans-Ia: No foetus or newborn could be observed (n = 0/15; 0.0%) and only one infant out of 15 died in the age group Infans-Ia (6.7%).

In the age group Infans-Ib, 4.5 out of 15 children died (30.0%). The highest mortality struck the elder children of age group Infans-II with 9.5 out of 15 (63.3%).

Ephesos (Figs. 18.18–20)

The study of the mortality of the total infant and child population from Ephesos which combines the sub-groups from St Mary's Church and the Byzantine Palace (Fig. 18.18), shows that in this late Byzantine population it was apparently a similar situation as in contemporaneous Pergamon. The highest mortality is found in the age

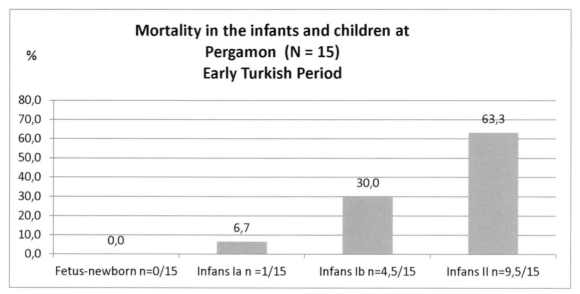

Fig. 18.17. Pergamon. Mortality in the infants and children (N = 15). Early Turkish period.

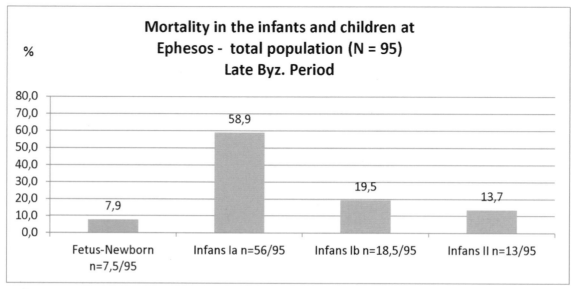

Fig. 18.18. Ephesos. Mortality in the infants and children – total population (N = 95). Late Byzantine period.

group of Infans-Ia, with 56 individuals out of 95 (58.9%), whereas the lowest mortality is observed in the age groups Fetus-Newborn (n = 7.5/95; 7.9%) and Infans-II (n = 13/95; 13.7%). The mortality in age group Infans-Ib is only slightly higher (n = 18.5/95; 19.5%).

St Mary's church (Fig. 18.19)

Almost no deceased individuals were observed in the age group Fetus-Newborn (n = 0.5/95; 09%). Of course, similar to the total population, the mortality in the age group Infans-Ia is, with 26.5 out of 55 infants, rather high (48.2%). In the remaining two age groups Infans-Ib

(n = 15/55; 27.3%) and Infans-II the mortality is moderate (n = 13/55; 23.6%).

Byzantine palace (Fig. 18.20)

The mortality of the individuals interred in the Byzantine Palace is, indeed, different than the mortality of the individuals from St Mary's Church. There is a considerable number of dead individuals in the age group Fetus-Newborn II (n = 7/40; 17.5%). In the Byzantine Palace as well as in the burial grounds of St Mary's Church, the mortality of the infants is highest in the age group Infans-Ia. For this age group, the mortality (n = 29.5/40; 73.8%) is even

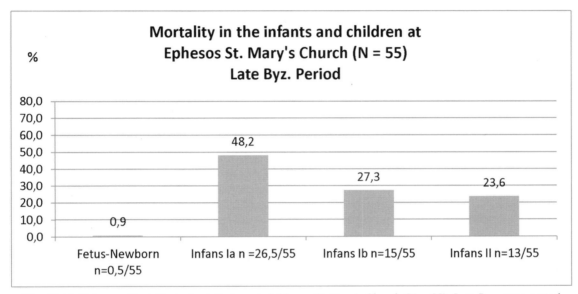

Fig. 18.19. Ephesos. Mortality in the infants and children – St Mary's Church (N = 55). Late Byzantine period.

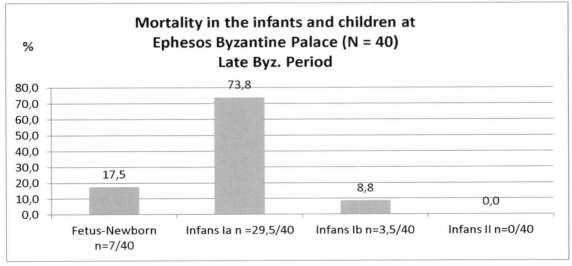

Fig. 18.20. Ephesos. Mortality in the infants and children – Byzantine Palace (N = 40). Late Byzantine period.

higher than in St Mary's Church. The mortality of the children assigned to age group Infans-Ib is low (n = 3.5/40; 8.8%). In age group Infans-II, no dead child was observed (n = 0/40; 0.0%).

Discussion
General remarks

The study of health and disease in prehistoric and early historical infant and child populations is a fascinating field dealing with extremely informative findings, which provide insights into the living conditions of ancient populations. As already mentioned, the nature and frequency of diseases of the infant and child age are very reliable indicators for the quality of ancient living conditions, also for the whole community (Carli-Thiele and Schultz 2001; Schultz 1994, Schultz *et al.* 1998). As the skeleton of infants and children is in an active growing stage, bony vestiges caused by diseases are, as a rule, faster established and relatively stronger expressed in their morphology (for example, changes due to inflammatory meningeal reactions), whereas, in adults, in many cases, macroscopically visible changes first become manifest in the bony tissue only after a considerable time (for example, vestiges of meningeal reactions). On the other hand, because the skeleton of subadults is growing relatively fast, these changes will not be stable and will only be diagnosable for a certain time because of the remodelling process which might reduce or cut vestiges of pathological processes away.

Bone changes in the skeleton of subadults, particularly in infants and young children, are frequently morphologically so differentiated and manifold that, at first sight, it seems difficult to establish a reliable diagnosis using only macroscopic techniques. However, using microscopy, morphological structures due to pathological processes might be reliably differentiated and diagnosed. As an example, one might think of the morphological status display of *Cribra orbitalia* which does not represent an independent disease, but is, however, the morphological feature of several different diseases, such as anaemia, scurvy, and the sequel of inflammatory processes of the paranasal sinuses or the lacrimal gland, or even rickets (Schultz 1990; 2001; 2012).

Because of the richness of different morphological features in the bones of diseased subadults and the given aetological context, comparatively many diseases are diagnosable. This enables us to establish individual courses of diseases which lead to biographic information on a deceased individual, which we call palaeobiography or 'osteobiography' (Nováček *et al.*, this volume; Schultz 2011). The sum of palaeobiographies helps in reconstructing the living environment of past people.

The authors act on the assumption that the disease status of infants and children mirrors the health situation of the whole population because infants and children belong to the weakest groups within a human community.

As this study is the first which deals exclusively with diseases diagnosed in the skeletal remains of subadults from Byzantine Anatolia, the results presented here should be regarded as a first attempt to characterize the health situation in a part of Medieval Anatolia. The authors are aware that in several cases the used sample sizes are relatively small. However, this lies in the nature of such analyses. Archaeological skeletal remains are frequently incomplete and affected by diagenesis, which makes it sometimes impossible to analyze large sample sizes. Therefore, no statistical tests were conducted.

Perhaps, the question might rise, why are the skeletons of subadults so important for archaeological research, particularly in Byzantine Anatolia? In all honesty, up to now, we know relatively little about the health situation of Byzantine populations. Of course, we have written records available which tell us about the rise and the decline of the Byzantine Empire. We know a lot about the historical changes which probably influenced the life and the health of the people in the rural and the urban areas of Anatolia. However, this information tells us nothing substantial about the individual health status of the people who lived in the country and the cities of this huge Empire. Thus, we have only scarce or even no information on the individual fate, the health status and the living conditions of the common people who very probably suffered from the political and economic changes that came along with the decline of Byzantine power. Indeed, we have some thousands of skeletons excavated from Byzantine cemeteries, which might tell us much more about the living conditions during the constantly changing history of the Empire. However, in general, the results of scientific records published on the physical anthropology and the palaeopathology of Byzantine populations are, up to now, relatively meagre in relation to a comparative consideration of the health situation, including a comprehensive observation of the aetiology of the diseases observed in various areas of Anatolia. Of course, this assumption is valid not only for Byzantine populations, but also for many other prehistoric and historical populations in general. Certainly, there is information available dealing with the nature and the frequencies of trauma, fractures, dental caries, osteoarthritis, etc. in Medieval Anatolia (e.g. Otten *et al.* 2011, Pirson *et al.* 2011), but this information does not really exploit the capacity of the richness of the skeletal remains available in museum and university collections of this country.

As a rule, one would expect that people who live in different cultures and represent different biological groups would show a different range of disease patterns and different disease frequencies. Of course, this has happened occasionally in the past. However, as palaeopathological studies have shown, this is not necessarily the case (Schultz *et al.* 1998; Schultz and Schmidt-Schultz 2014). Apparently, the influence of the natural biotope is a crucial factor for the outbreak of diseases and sometimes a stronger initiating cause than genetic and cultural factors. Using disease profiles, this behaviour can be verified (Schultz and Schmidt-Schultz 2014). However, as Byzantine Anatolia apparently is demonstrating, the outbreak and the occurrence of diseases might also be provoked by human-induced factors such as a consequence of war, general collapse of economic systems such as trade and agricultural supply, which opens the floodgates to famine and epidemics. These circumstances cause a higher morbidity rate, which, of course, provokes an increase in the mortality rate.

The state of health of infants and children in Byzantine Anatolia

Arslantepe

At Arslantepe, all three selected deficiency diseases could be diagnosed with strikingly low frequencies: rickets 2.1%, scurvy 10.7% and anaemia 12.5% (Fig. 18.5). Particularly, the presence of chronic vitamin D deficiency is unusual in geographic areas, such as southeastern Anatolia, where sunshine is frequent because vitamin D3 (cholecalciferol) is produced by the human organism from a preform (previtamin D2) in the human skin by ultraviolet radiation during sunlight exposure. However, the presence of one case of rickets in this population can be explained by the fact that this child was probably seriously sick and had to stay for a long time, for instance, many weeks, inside the typical houses of that time which had only small windows in order to keep sunlight and heat out, as is usual for this geographic area. Thus, the child was isolated from the necessary UV light and due to his disease was probably unable to eat in a proper way. This might be an explanation for the existence of one case of rickets at Arslantepe.

Also the three selected infectious diseases show a remarkably low frequency: osteomyelitis 0.0%, otitis media 0.0% and meningeal reactions 11.7%. Thus, caused by the morbidity rate, it stands to reason that the people from Arslantepe belonged to a relatively well-to-do population which possessed some economic wealth and seemed to be relatively healthy. This assumption is apparently supported by the archaeological findings which described Arslantepe as a probably fortified settlement on an important trading route.

Boğazkale

On the basis of new results provided by the microscopic analysis, new data for the frequencies of anaemia and scurvy in the early Byzantine population from Boğazkale are now available (Fig. 18.6). The infant and child population from this village situated in the central Anatolian highland belongs to a relatively poor rural community of farmers and shepherds. The economic situation might be reflected in the relatively high frequencies of scurvy (28.1%) and

anaemia (26.6%). As we know, anaemia might be caused not only by iron deficiency (e.g. El-Najjar et al. 1976; Ortner 2003) or lack of essential amino acids, for instance tryptophan (Schultz 1982; 2001), but also by parasites, for instance vermicular diseases (Reinhard 1992; Larsen and Sering 2000), which seems very probable in this case, or protozoosis, such as malaria (Schultz 1990). It is striking that vestiges of pathological meningeal reactions were frequently diagnosed in the infants and children from Boğazkale. It cannot be excluded that this relatively high frequency of 25.0% might be influenced by sheep and goat farming, which also might be responsible for the occurrence of anaemia via parasitic disease. The frequencies of osteomyelitis (5.1%) and otitis media (0.0%) are very low and come near to the frequencies observed in the relatively wealthy population of Arslantepe (Fig. 18.5). It is interesting that the populations from Arslantepe and Boğazkale lived in the early Byzantine period, at a time in which the Empire could guarantee peace, order, and supply the necessities indispensable to life. It is obvious that the infant and child population from Arslantepe were blessed with better health, probably due to more favourable living conditions than the subadults from Boğazkale; who belonged to a poor rural population (cf. Schultz 1986; 1989a).

The highest mortality in the Boğazkale infant and child population is found in the age group Infans-Ib (39.0%) (Fig. 18.13). In the age group Infans-II, 32.5% of the subadults died. Thus, most of the subadults died after the second year of life (71.5%). The relatively high frequencies of scurvy and anaemia in the children older than two years and the low mortality before the second year of life suggest the conclusion that in early Byzantine Boğazkale infants were probably well-protected from diseases by breast-feeding.

Pergamon

The total infant and child population from Pergamon is relatively small (n = 69 individuals). There is a very small group dating from the Roman or early Byzantine period (n = 5) which is not included into this study, a large group dating from the late Byzantine period (n = 49) and a smaller group dating from the early Turkish period (n = 15). The group from the 13th century AD can be divided into two chronological sub-groups, one from the first half (n = 23), the other from the second half of the 13th century (n = 19). A small group of seven individuals could not be allocated to the two sub-groups, but all belong to the 13th century.

The evaluation of the six diseases provides interesting insights into the health situation of the population from Pergamon (Fig. 18.7). The disease profile of the total population (n = 64) is characterized by the high frequency of vestiges of meningeal reactions, for instance, bacterial meningitis (48.8%). As expected and similar to the population from Boğazkale (Fig.18.6), rickets cannot be diagnosed (0.0%). The frequencies of scurvy (24.4%) and anaemia (24.4%) show the same level as seen in the infant and child population from Boğazkale (Fig. 18.6), however, there is a distinct difference to the population from Arslantepe (Fig. 18.5). In comparison to Arslantepe, osteomyelitis (10.0%) was observed, however, in a higher frequency than at Boğazkale. Otitis media can be diagnosed with a relatively high frequency (25%).

All infants and children dating from the 13th century were compiled into one group (n = 49) because of the small sample sizes. The disease frequencies in this late Byzantine population (Fig. 18.8) reflect the data already presented in the total population (Fig. 18.7). With the exception of rickets, all frequencies are higher in the late Byzantine population than in the total population. Particularly striking is the extraordinarily high frequency of meningeal reactions (61.3%) (Fig. 18.8). This result gives us a slight indication as to the existence of possible epidemics due to insufficient hygienic and sanitary conditions, which probably were present during the final period of the siege of the city, which lasted many months (Radt 1999). Otitis media holds the second place in the frequencies (28.6%) (Fig. 18.8). A high prevalence of otitis media within a population, for instance today in the countries of the so-called Third World, is always an indicator of inadequate living conditions. This general assessment is also valid for the pre-antibiotic times (Schultz *et al.* 2008b). Thus, the high frequency of inflammatory diseases in the middle ear region emphasizes the assumption that late Byzantine Pergamon represented an unhealthy place. The causes for these poor living conditions are probably due to the political conditions, which brought war and privation over the Anatolian people of the 13th century.

For the evaluation of mortality, an attempt was made to see if there are differences between the mortality of the subadults who died in the first half and in the second half of the 13th century, the time just before the conquest of the city by the Turks. Thus, there was the suspicion that living conditions were worse during the last decades before the conquest in approximately 1315 AD. However, there is no significant difference between the mortality in the first (Fig. 18.15) and the second half of the 13th century (Fig. 18.16). In the first half of the 13th century no individual died in age group Fetus-Newborn, whereas in the second half of this century two individuals died (10.5%). However, the number of individuals is too small to draw further conclusions. The highest mortality in both sub-groups of the 13th century, as well as in the whole late Byzantine infant and child population from Pergamon, is found in age group Infans-Ia (Figs. 18.14–16).

The Byzantine people who survived the siege and the conquest lived as slaves of the Turkish victors. Astonishingly, the health situation of these few infants and children who survived (n = 15) was much better than of the subadults who lived during the late Byzantine period. No vestiges of rickets, osteomyelitis, or otitis media were diagnosed (0.0%) and the frequencies of scurvy (16.7%), anaemia (21.4%) and meningeal reactions (16.7%) were relatively low to moderate (Fig. 18.9). Apparently, the Turkish lords cared for their slaves and nourished them in a proper way to maintain their workforce. At first sight, the mortality pattern of this infant and child population looks completely different from all other populations presented here (Fig. 18.17): 63.3% of the subadults died in the age group Infans-II. This is a remarkable frequency. Of course, this population is very small and this might affect the statistics. However, perhaps there is a possible explanation for this unusual mortality pattern. If we concede that these individuals were slaves and that the older children had to work, perhaps even very hard, this might be a cause of early death.

Ephesos

The populations from Ephesos date, like Pergamon, from the final period of the Byzantine Empire in western Anatolia. The disease profiles of the total infant and child population from Ephesos (Fig. 18.10) and of the age-equivalent population interred in the burial areas of St Mary's Church (Fig. 18.11), are very similar and strongly resemble the disease profile of the subadults from late Byzantine Pergamon (Fig. 18.8). In the subadult population from St Mary's Church, meningeal reactions (57.7%) and scurvy (44.1%) show the highest frequencies (Fig. 18.11). Thus, it can be supposed that the living conditions at Pergamon and Ephesos during the 13th century AD were apparently the same. Neither in the subadult population from St Mary's Church nor in the subadults from the Byzantine Palace, were vestiges of osteomyelitis found (0.0%).

Astonishingly, the infant and child population from the Byzantine Palace presents a completely different result (Fig. 18.12). Here, otitis media (46.2%) and meningeal reactions (38.5%) are the most frequent diseases, whereas scurvy (15.8%) and anaemia (15.8%) show, with the exception of the subadults from Arslantepe (Fig. 18.5), the lowest frequency of all studied Byzantine populations. Thus, we can carefully hypothesize that the infants and children buried in the area of the former Byzantine Palace did not belong to a socially low-ranking group. In any case, these two populations are particular in several aspects which are discussed later. Summarized it can be said that the infant and child population from St Mary's Church present, with the exception of otitis media, the highest disease frequencies of the two subadult populations from Ephesos.

Also the mortality of the total infant and child population from St Mary's Church and the Byzantine Palace (Fig. 18.18) resembles the disease profile of the late Byzantine population of the 13th century from Pergamon (Fig. 18.14). However, the frequency of deceased infants in the age group Infans-Ia at Ephesos is higher than in Pergamon. It is striking, that individually considered, the two late Byzantine infant and child populations from the St Mary's Church and the Byzantine Palace show a completely different mortality pattern to that observed at Pergamon. However, there is one striking similarity: as in the late Byzantine population from Pergamon, in both the population from St Mary's Church (48.2%) (Fig. 18.19) and from that of the Byzantine Palace (73.8%) (Fig. 18.20), the highest mortality rate was found in the young infants of the age group Infans-Ia. In particular, the very high mortality of almost 74% of the infants of group I in the population from the Byzantine Palace is striking.

Furthermore, there are two major demographic differences between the two populations from Ephesos. In the population from St Mary's Church, only 0.9% of the individuals are found in the age group Fetus-Newborn (Fig. 18.19), whereas in the population from the Byzantine Palace 17.5% of the individuals match this age group (Fig. 18.20). In the Byzantine Palace, no child could be assigned to the age group Infans-II, whereas in St Mary's Church 23.6% of the children belonged to this age group (Figs. 18.19–20).

The pattern of the diseases and of the mortality of the two subadult populations from Ephesos are exceptions within the populations presented here. Excepting the subadult population from Boğazkale, in all late Byzantine infant and child populations presented in this study, the mortality in age group Infans-Ia has the highest frequency. This is particularly visible in the two populations from Ephesos. Here, the population from the Byzantine Palace presents an enormously high mortality of 73.8%. Also the population from St Mary's Church has, with 48.2%, a very high mortality in this age group. There are various causes for such a high mortality at this young age. For instance, one and in our opinion a very probable cause might be a shortened period of breast-feeding, which means that the weaning period had started before the end of the second year of the life of the infants. In this case, the lack of breast-milk could decrease the already not yet efficient immune system. The non-efficient immune system would provoke an increased disposition for the susceptibility to infectious diseases, which might be fatal in this young age and explain the high mortality rate in the age group Infans-Ia. Otitis media (St Mary's Church 8.0%, Byzantine Palace 46.2%) and meningeal reactions (St Mary's Church 57.7%, Byzantine Palace 38.5%), such as meningitis, which were diagnosed in both subadult populations in very high frequencies, could be the causing factors for the increase of the mortality rate (Figs. 18.11–12). For the rest, according to the frequencies of the deficiency diseases, the infants and children from the Byzantine Palace seemed to be relatively healthy.

Summary

As expected, the results revealed poor to very poor living conditions in the infants and children who lived in the time of the decline of the Byzantine Empire (*c.* 13th/14th century AD), whereas the subadults who lived in the time when the Byzantine Empire was still powerful (*c.* 7th – 10th century AD), showed, as a rule, evidence of better living conditions. Subadults who lived after the conquest had apparently better living conditions, however, the mortality rate was high in the elder children. This might be associated with child labour.

Acknowledgments

The authors would like to thank many colleagues for their help, information and support, particularly Wolfgang Radt, former director of the German excavations at Pergamon, German Archaeological Institute (DAI) at Istanbul (Turkey), Klaus Rheidt, chair and head of the Department of Research in Architecture at the Brandenburg University of Cottbus (Germany), Peter Neve, former director of the German excavations at Boğazkale, German Archaeological Institute (DAI) at Istanbul, field office Ankara (Turkey), Marcella Frangipane, director of the Italian excavations at Arslantepe, Italian Archaeological Mission of Oriental Anatolia of the University of Rome 'La Sapienza' (Italy), Sabine Ladstätter, director of the Austrian excavations at Ephesos, Austrian Archaeological Institute (ÖAI) at Vienna (Austria), and Andreas Pülz, director of the recent Austrian excavations of the Byzantine Palace at Ephesos, Austrian Academy of Science (ÖAW) and the Austrian Archaeological Institute (ÖAI) at Vienna (Austria). The authors thank Berna Alpagut, formerly head of the Anthropological Laboratory, University of Ankara (Turkey), for support during the examination of skeletal remains in 1987. For the preparation of the thin-ground sections, the authors thank Michael Brandt and for the preparation of samples for scanning-electron microscopy Ingrid Hettwer-Steeger, both Department of Anatomy and Embryology, University Medical School Göttingen (Germany). For help in the finalization of the figures, the authors thank Jan Nováček, Department of Anatomy, University Medical School Göttingen (Germany).

Bibliography

Ascenzi, A., Bellelli, A., Brunori, M., Citro, G., Ippoliti, R., Lendaro, E., and Zito, R. (1991) Diagnosis of thalassemia in ancient bones: Problems and prospects in pathology. In D. Ortner and A. Aufderheide (eds.) *Human paleopathology. Current synthesis and future options*, 73–5. Washington, D.C., Smithsonian Institution Press.

Carli-Thiele, P. (1996) Spuren von Mangelerkrankungen an steinzeitlichen Kinderskeleten – Vestiges of Deficiency Diseases in Stone Age Child Skeletons. In M. Schultz (ed.) *Advances in Paleopathology and Osteoarchaeology I*, 13–267. Göttingen, Verlag Erich Goltze.

Carli-Thiele, P. and Schultz, M. (2001) Wechselwirkungen zwischen Mangel- und Infektionskrankheiten des Kindesalters bei neolithischen Populationen. In A. Lippert, M. Schultz, S. Shennan and M. Teschler-Nicola (eds.) *Mensch und Umwelt während des Neolithikums und der Frühbronzezeit in Mitteleuropa*, 273–85. Rahden, Westfalen.

El-Najjar, M. Y., Ryan, D. J. Turner II, C. G., and Lozoff, B. (1976) The etiology of porotic hyperostosis among the prehistoric and historic Anasazi Indians of Southwestern United States. *American Journal of Physical Anthropology* 44, 477–87.

Ferembach, D., Schwidetzky, I., and Stloukal, M. (1980) Recommendations for age and sex diagnosis of skeletons. *Journal of Human Evolution* 9, 517–49.

Johnston, F. E. (1962) Growth of the long bones of infants and young children at Indian Knoll. *American Journal of Physical Anthropology* 20, 249–54.

Karwiese, S. (1989) *Die Marienkirche in Ephesos – Erster vorläufiger Grabungsbericht 1984–6*. Wien, Verlag der Österreichischen Akademie der Wissenschaften.

Kósa, F. (1978) Identifikation der Feten durch Skelettuntersuchungen. In H. Hunger and D. Leopold (eds.) *Identifikation*, 211–41. Leipzig, Barth.

Kreutz, K. (1997) Ätiologie und Epidemiologie von Erkrankungen des Kindesalters bei der bajuwarischen Population von Straubing (Niederbayern). In M. Schultz (ed.) *Beiträge zur Paläopathologie – Contributions to paleopathology*, I. 1–159; II.1–273. Göttingen, Cuvillier Verlag.

Larsen, C. S. (1997) *Bioarchaeology: Interpreting Behavior from the Human Skeleton*. Cambridge, Cambridge University Press.

Larsen, C. and Sering, L. (2000) Inferring iron deficiency anemia from human skeletal remains: The case of Georgia Bight. In P. Lambert (ed.) *Bioarchaeological studies in life in the age of agriculture*. Tularosa, 116–33. University of Alabama Press.

Lewis, M. E. (2002) *Urbanisation and child health in Medieval and post-Medieval England* British Archaeological Series, British Series no. 339). Oxford, Archaeopress.

Lewis, M. E. (2009) *The bioarchaeology of children. Perspectives from biological and forensic anthropology* (Cambridge Studies in Biological and Evolutionary Anthropology 50). Cambridge, Cambridge University Press.

Neve, P. (1996) *Ḫattuša Stadt der Götter und Tempel – Neue Ausgrabungen in der Hauptstadt der Hethiter*. Mainz, Verlag Philipp von Zabern.

Ortner, D. J. (2003) *Identification of pathological conditions in human skeletal remains*, second ed. San Diego, Academic Press.

Pülz, A. (2015) web page of the ÖAW (http://www.oeaw.ac.at/antike/index.php?id=67#c169)

Radt, W. (1989) Vorbericht über die Kampagne 1988. *Archäologischer Anzeiger* 1989, 387–412.

Radt, W. (1990) Pergamon – Vorbericht über die Kampagne 1989. *Archäologischer Anzeiger* 1990, 397–424.

Reinhard, K. J. (1992) Patterns of diet, parasitism and anemia in prehistoric West North America. In P. Stuart-Macadam and S. Kent (eds.) *Diet, demography, and disease: Changing perspectives on anemia*, 219–58. New York, Aldine de Gruyter.

Rheidt, K. (1991) *Die Stadtgrabung, Teil 2: Die byzantinische Wohnstadt* (Altertümer von Pergamon XV.2). Berlin and New York, W. De Gruyter.

Rösing, F. W., Graw, M., Marré, B., Ritz-Timme, S., Rothschild, M. A., Rötzscher, K., Schmeling, A., Schröder, I., and Geserick, G. (2007) Recommendations for the forensic diagnosis of sex and age from skeletons. *Homo* 58, 75–89.

Scheuer, L. and Black, S. (2000) *Developmental juvenile osteology*. San Diego, Academic Press.

Schmid, F. and Künle, A. (1958) Das Längenwachstum der langen Röhrenknochen in Bezug auf Körperlänge und Lebensalter. *Fortschritte Röntgenstrahlen* [Röfo] 89, 350–6.

Schmidt-Schultz, T. H. and Schultz, M. (2004) Bone protects proteins over thousands of years: extraction, analysis, and interpretation of extracellular matrix proteins in archaeological skeletal remains. *American Journal of Physical Anthropology* 123, 30–9.

Schultz, M. (1982) Umwelt und Krankheit des vor- und frühgeschichtlichen Menschen. In H. Wendt and N. Loacker (eds.) *Kindlers Enzyklopädie. Der Mensch*, Vol. 2, 259–312. München, Kindler Verlag.

Schultz, M. (1984) The diseases in a series of children's skeletons from Ikiztepe, Turkey. In V. Capecci and E. Rabino Massa (eds.) *Proceedings of the 5th European meeting of the Paleopathology Association*, 321–5. Siena, University of Siena.

Schultz, M. (1986) Der Gesundheitszustand der frühmittelalterlichen Bevölkerung von Boğazkale/Ḫattuşa. In *IV. Araştırma Sonuçları Toplantısı (Ankara 26–30 Mayis 1986)*, 401–9. Ankara, T. C. Kültür Bakanliği.

Schultz, M. (1988a) Paläopathologische Diagnostik. In R. Knussmann (ed.) *Anthropologie. Handbuch der vergleichenden Biologie des Menschen*, Vol. I, 1, *Wesen und Methoden der Anthropologie*, 480–96. Stuttgart and New York, Fischer Verlag.

Schultz, M. (1988b) Methoden der Licht- und Elektronenmikroskopie. In R. Knussmann (ed.) *Anthropologie. Handbuch der vergleichenden Biologie des Menschen*, Vol. I, 1, *Wesen und Methoden der Anthropologie*, 698–730. Stuttgart and New York, Fischer Verlag.

Schultz, M. (1988/1989) Erkrankungen des Kindesalters bei der frühbronzezeitlichen Population von Hainburg/Niederösterreich. *Mitteilung der Anthropologischen Gesellschaft Wien* 118/119, 369–80.

Schultz, M. (1989a) Nachweis äußerer Lebensbedingungen an den Skeleten der frühmittelalterlichen Bevölkerung von Bogazkale/Hattussa. In *IV. Arkeometri Sonuçları Toplantısı (Ankara 23–7 Mayis 1988)*, 119–20. Ankara, T. C. Kültür Bakanliği.

Schultz, M. (1989b) Osteologische Untersuchungen an den spätmittelalterlichen Skeleten von Pergamon – Ein vorläufiger Bericht. In *IV. Arkeometri Sonuçları Toplantısı (Ankara 23–7 Mayis 1988)*, 111–8. Ankara, T. C. Kültür Bakanliği.

Schultz, M. (1989c) Ergebnisse osteologischer Untersuchungen an mittelalterlichen Kinderskeletten unter besonderer Berücksichtigung anatolischer Populationen. *Anthropologischer Anzeiger* 47, 39–50.

Schultz, M. (1990) Erkrankungen des Kindesalters bei der frühbronzezeitlichen Population vom İkiztepe (Türkei). In F. M. Andraschko and W. R. Teegen (eds.) *Gedenkschrift für Jürgen Driehaus*, 83–90. Mainz, Verlag Philipp von Zabern.

Schultz, M. (1993) *Vestiges of non-specific inflammations of the skull in prehistoric and historic populations. A contribution to palaeopathology* (Anthropologische Beiträge 4A/B). Aesch BL, Anthropologisches Forschungsinstitut Aesch and Anthropologische Gesellschaft Basel.

Schultz, M. (1994) Leben, Krankheit und Tod. Skelettfunde als Spiegel der Lebensbedingungen. In A. Jockenhövel and W. Kubach (eds.) *Bronzezeit in Deutschland* (Archäologie in Deutschland), special issue, 15–7.

Schultz, M. (2001a) Krankheit und Tod im Kindesalter bei bronzezeitlichen Populationen. In A. Lippert, M. Schultz, S. Shennan, and M. Teschler-Nicola (eds.) *Mensch und Umwelt während des Neolithikums und der Frühbronzezeit in Mitteleuropa*, 287–305. Rahden/Westfalen, Verlag Marie Leidorf.

Schultz, M. (2001b) Paleohistopathology of bone: A new approach to the study of ancient diseases. *Yearbook of Physical Anthropology* 44, 106–47.

Schultz, M. (2003) Light microscopic analysis in skeletal paleopathology. In D. J. Ortner (ed.) *Identification of pathological conditions in human skeletal remains* (2nd ed.), 73–108. San Diego, Academic Press.

Schultz, M. (2011) Paläobiographik. In G. Jüttemann (ed.) *Biographische Diagnostik*, 222–36. Berlin and Bremen, Lengerich.

Schultz, M. (2012) Light microscopic analysis of macerated pathologically changed bone. In C. Crowder and S. Stout (eds.) *Bone histology. An anthropological perspective*, 253–95. New York, London, Tokyo, and Boca Raton, CRC Press.

Schultz, M. and Schmidt-Schultz, T. H. (1995) Krankheiten des Kindesalters in der mittelalterlichen Population von Pergamon. *Istanbuler Mitteilungen des Deutschen Archäologischen Instituts* 44, 181–201.

Schultz, M. and Schmidt-Schultz, T. H. (2014) The role of deficiency diseases in infancy and childhood of Bronze Age populations. In L. Milano (ed.) *Paleonutrition and food practices in the Ancient Near East towards a multidisciplinary approach* (History of the Ancient Near East/Monographs XIV), 25–42. Padova, S.A.R.G.O.N. Editrice e Libreria.

Schultz, M., Schmidt-Schultz, T. H. and Kreutz, K. (1998) Ergebnisse der paläopathologischen Untersuchung an den frühbronzezeitlichen Kinderskeletten von Jelšovce (Slowakische Republik). In B. Hänsel (ed.) *Mensch und Umwelt in der Bronzezeit Europas – Man and Environment in Bronze Age Europe*, 77–90. Kiel, Oetker-Voges Verlag.

Schultz, M., Schmidt-Schultz, T. H,. Gresky, J,. Kreutz, K. and Berner, M. (2006) Morbidity and mortality in the Late PPNB populations from Basta and Ba'ja (Jordan). In M. Faerman, L.K. Horwitz, T. Kahana and U. Zilberman (eds.) *Faces from the past: Diachronic patterns in the biology of human populations from the Eastern Mediterranean* (British Archaeological Reports, International Series 1603), 82–99. Oxford, Archaeopress.

Schultz, M., Timme, U., Hilgers, R., Schmidt-Schultz, T. H. (2008a) Preliminary results of the bioarchaeological and sociobiological investigation on the infants and children from Grasshopper Pueblo, Arizona. In: A. L. W. Stodder (ed.) *Reanalysis and Reinterpretation in Southwestern Bioarchaeology* (Anthropological Research Papers No. 59), 127–40. Tuscon, Arizona State University.

Schultz, M., Timme, U., Hilgers, R., and Schmidt-Schultz, H. T. (2008b) Die Krankheiten der Kinder des Grasshopper Pueblo (Arizona) – Ergebnisse paläopathologisch-bioarchäologischer Untersuchungen. In J. Piek and T. Terberger (eds.) *Traumatologische und pathologische Veränderungen an prähistorischen und historischen Skelettresten – Diagnose, Ursachen und Kontext*, 137–60. Rahden/Westfalen, Verlag Marie Leidorf.

Stloukal, M. and Hanáková, H. (1978) Die Länge der Längsknochen altslawischer Bevölkerungen – Unter besonderer Berücksichtigung von Wachstumsfragen. *Homo* 29, 53–69.

Szilvássy, J. (1988) Altersdiagnose am Skelett. In R. Knussmann (ed.) *Anthropologie: Handbuch der vergleichenden Biologie des Menschen*, Vol. I, 1, Wesen und Methoden der Anthropologie, 421–43. Stuttgart and New York, Fischer Verlag.

Ubelaker, D. H. (1989) *Human skeletal remains: Excavation, analysis, interpretation*, (2nd ed.) (Manuals on Archaeology 2). Washington D.C., Taraxum.

Infant and child skeletons from the Lower City Church at Byzantine Amorium

F. Arzu Demirel

Abstract

Amorium is a Byzantine city situated within the Emirdağ district of the province of Afyonkarahisar. Excavations in the 2007, 2008, and 2009 seasons yielded many tombs containing infant and child skeletons forming part of a cemetery datable to the 10th and 11th centuries. Many are multiple burials and were located principally to the north of the main church in an area called A20 situated to the east of the baptistery. Skeletal remains of 128 individuals from 36 tombs were analysed to reveal the demography and health status of the population. Among them 49.2% of the individuals died prenatally, at birth, or soon after birth. The results of this study indicate that the unfavourable geographic conditions of Amorium's location might have produced inadequate hygienic conditions and the malnourishment of the mothers, which in turn caused infections and complications during pregnancy and birth leading to premature and stillbirths. Those infants that did survive the critical neonatal and postnatal periods were generally able to carry on their lives into puberty and adulthood.

Keywords: Amorium, Byzantine funeral rites, infant mortality, infant skeletal population, mid-Byzantine, perinatal period, post-neonatal period.

Introduction

Amorium is a Byzantine city situated within the Emirdağ district of the province of Afyonkarahisar (Fig. 19.1). Systematic excavations on the site have shown that the Byzantine city was settled between 5th to 11th centuries. Amorium was one of the few surviving cities in Anatolia during the so-called Dark Ages and the importance of the city also comes from being the home town of a Byzantine Imperial dynasty known as the Amorians (Lightfoot and Lightfoot 2007). Excavations in the Lower City Church in the 2007, 2008, and 2009 seasons yielded many undisturbed tombs containing 128 infant and child skeletons, forming part of a cemetery datable to the 10th and 11th centuries (Fig. 19.2). Many tombs contained multiple burials and were located principally to the north of the main church in an area called A20 situated to the east of the baptistery (Fig. 19.3). The first basilica of the church was destroyed by a fire during the Arab siege in AD 838 (Ivison 2012, 65) and rebuilt in

the late 9th–early 10th centuries (Ivison 2010, 328–38). The excavations in A20 revealed a large bed for making plaster, indicating that this area was used as a construction yard during the reconstruction of the church complex, probably in the late 9th century. Further excavations showed that, following the second construction period of the church, this area was used as a cemetery, right up to the late 11th century (Lightfoot *et al.* 2009). Most of the burials were simple pits, but some of the tombs were lined and covered by tile and stone fragments.

Cemeteries that include an area reserved for infant and child burials are quite rare in the Byzantine world, and so this cemetery deserves special attention (Lightfoot *et al.* 2009). The skeletons are all orientated in a west/east direction, in other words the head is to the west, the feet to the east, which is the common custom for Christian burials. During the excavation of the skeletons, in most cases it was only possible to trace the deposition of the uppermost skeleton, since the remains of the earlier burials had been pushed to

Fig. 19.1. Amorium. Geographical location (www.afyonweb.com).

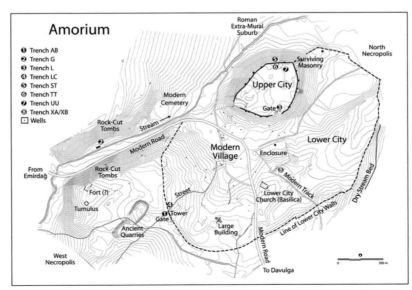

Fig. 19.2. Amorium. City plan (courtesy of the Amorium Excavations Project).

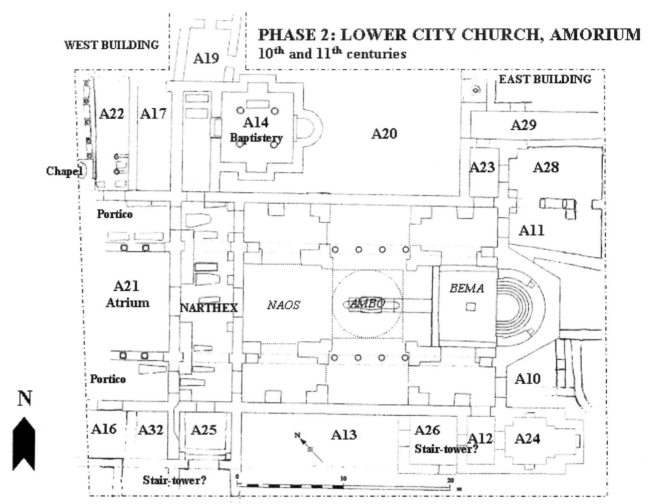

Fig. 19.3. Amorium. Plan of the Lower City Church and A20 area. Plan by Benjamin Arubas and Eric Ivision. (courtesy of the Amorium Excavations Project).

the sides of the tombs and lost their anatomical positions. Owing to the reuse of the tombs during the Byzantine period and the fragile nature of infant and child bones the skeletal remains are very fragmented and the general preservation condition is moderate to poor. In this study the aim was to ascertain the demographic structure and health status of this population as well as the reasons behind the special nature of this unusual cemetery.

Methods

Dental development is recognized as the most reliable method for aging infant and child remains since it has been accepted that teeth are less affected by exterior factors than other criteria for subadult remains (El-Nofely and İşcan 1989; Lewis 2007a; Smith 1991; White and Folkens 2005). However, in this population most of the teeth were lost post-

mortem, and so it was not possible to determine the age of most of the individuals according to dental development. Hence, the age determinations are based on the long bone diaphysial developments, which is another commonly used criterion for aging of subadult skeletal groups. Therefore, age determinations are referenced according to Scheuer and Black 2000, 2004; Kósa 1989; and Ubelaker 1989 for the long bone developments of the subadult remains; Buikstra and Ubelaker 1994 for the long bone and dental development of the child remains. Ortner 2003; Roberts and Manchester 1995; Aufderheide and Rodriguez-Martin 2008 are referred to for the pathological lesions. Terms of perinatal, neonatal, and post-neonatal differ in references and these are referred to as defined by Lewis (2007a). So, the term perinatal is assigned to around birth meaning from 24 weeks in uterus to seven postnatal days, neonatal from birth to 27 postnatal days, and post-neonatal from 28 days up to one year of age.

Demographic Structure

The distribution of infant and child remains according to tombs in A20 area Fig. 19.4 shows that only 14 tombs are single, and the remaining 22 tombs contained multiple burials. The difference between the death ratio of the infants and children is quite remarkable. The ages of the individuals range between 22–24 week-old foetuses to approximately eight years of age. The most remarkable feature of the study group is the great number of deaths in the earlier periods of life. Out of 128 individuals 107 died as early as 22 weeks in uterus up to one year of age; this makes 83.6% of the whole population. It was not possible to determine the age of nine individuals owing to the poor preservation conditions and inadequate number of the skeletal remains. Between the ages of one and two, mortality ratios decrease strikingly to as low as 3.9%, which can be followed in Figs. 19.5–6.

There is no pathology and no anomalies are recorded that are certain indicators of the cause of death, but some lesions probably connected with infection (Grave 14, left tibia, neonatal) (Fig. 19.7) were observed. A probable case of rickets was observed on both sides of the tibia of a one- or two-year-old infant, but no certain diagnosis was possible owing to the lack of the remaining part of the skeleton (Fig. 19.8). Also, a very rare congenital case recorded here was an extra bone development on the *pars lateralis* (Fig. 19.9). Detailed study of the pathological conditions of these skeletal remains is in progress.

Discussion

Pregnancy and delivery were thought by the Byzantines to be critical periods for both mother and child, since complications during pregnancy or childbirth are frequently attested (Bourbou 2010, 104) and high infant mortality is mentioned for Byzantine populations in many texts (e.g. Bourbou 2004; 2010; Dennis 2002; Talbot 2009). Despite the fact that 128 individuals composing this population may not be representative of infant and child mortality over 200 years, still the preponderance of individuals that lost their lives at the very early stages of life within this group is remarkable.

The infant mortality ratios for some Anatolian Byzantine, earlier and contemporary societies are given here in Table 19.1. Among these, some data provided here is only within the subadult group along with Amorium while the data for the other sites gives the mortality rates within the population. Infant remains in skeletal records are sometimes biased due to their fragile nature and their representation sometimes depends on the demographic profile of the society, but the available evidence suggests that infant mortality was often high in most ancient societies, whilst some performed better individually such as some Roman, mid- and later Byzantine sites as well as some Medieval sites which can be followed in

Table 19.1. The single Roman site Börükçü has a considerably lower ratio of infant death (11.1%) than the sites of the early Byzantine period (mortality rates fluctuating between 14.3% and 23.9%). According to these figures the mortality of children seems to worsen in late Antiquity, but then improves slightly in the Mid-Byzantine period when some sites show a mortality ratio of 8.3–17.2%. Then in many late Byzantine and Medieval sites the situation worsens once more with mortality ratio as high as 22.4–50%. As the infant mortality ratio is accepted as one of the main and the most sensitive indicators of socio-cultural conditions and life quality in present and past populations (Bourbou 2004; Koç *et al.* 2009), this trend may indicate favourable living conditions in Imperial Roman times, conditions which worsen in the chaotic environment of early Byzantine times, until the situation improves in the mid-Byzantine period. In the later periods life changes again to poorer quality (cf. Demirel (in press); Günay *et al.* (2010); Ingvarsson-Sundström (2014); and Kiesewetter and Teegen, both in this volume).

Perinatal and neonatal mortality ratios are noted particularly as the indicator of maternal health (Pakiş and Koç 2009; Tezcan *et al.* 2009). It has also been reported that neonatal and earlier deaths are mainly connected to endogenous factors such as birth complications, genetic factors, and insufficient diet of the mother; while the post-neonatal deaths are connected to exogenous factors such as infection, respiratory disorders, and diarrhoea (Eryurt and Koç 2009; Lewis 2007a; 2007b). Recent research has shown that infections constitute the most frequent reason for mortality in the perinatal and neonatal period (Pakiş and Koç 2009; Kaya *et al.* 2010; Yılmaz *et al.* 2010), and stillbirths are firmly connected to premature births, foetal hypoxia, infection, birth trauma, and congenital anomalies (Pakiş and Koç 2009). Lewis (2007a, 81) states that the transition of a child from a stable uterine environment to the external environment with its variety of pathogens and other stimuli is the first crisis in a human's life and, if the child fails to adapt to its new environment, it could easily lose its life. Thus, it is evident that the quality of the physical environment substantially determines the newborn's survival through to later age. On the other hand, Bourbou (2004) reports that, if a newborn is sufficiently breast-fed during their first year of life, its survival is more probable as the immune system develops as long as it takes breast milk. Also Byzantine literature indicates that the Byzantines were aware of the importance of breast feeding and even hired wet nurses when the mother's breast milk was not enough for the baby (Bourbou and Garvie-Lok 2009; Yurdakök 2005; Tritsaroli and Valentin 2008). In this population, almost half (49.2%) of the babies died around birth. This ratio probably means that the newborns here were not lucky enough to live to develop their immune system. Therefore, it can be argued that complications listed above were the most probable reasons of perinatal and neonatal deaths in Amorium.

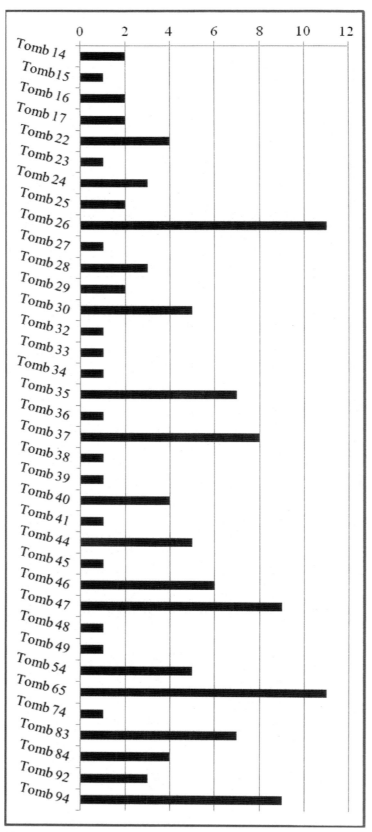

Fig. 19.4. Amorium. Distribution of individuals according to the graves in A20 area.

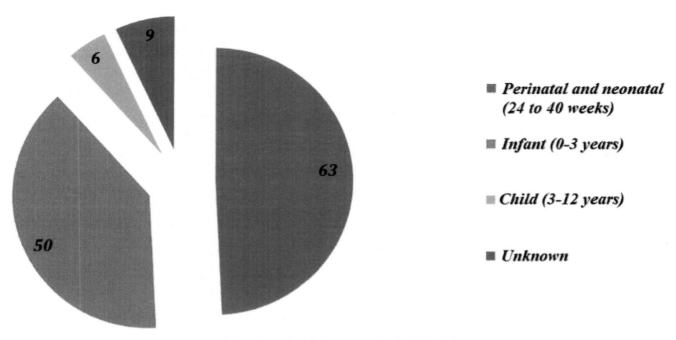

Fig. 19.5. Amorium. Age distribution of the individuals (n=128).

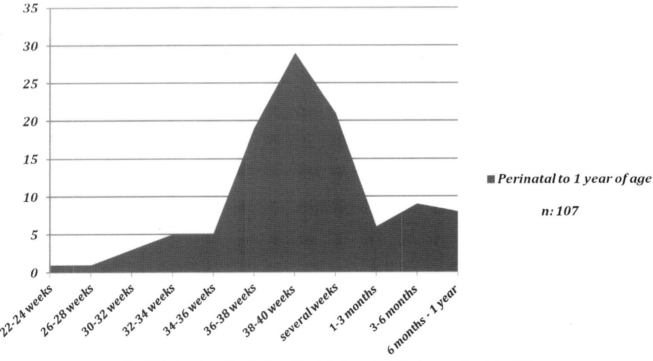

Fig. 19.6. Amorium. Distribution of the infants (perinatal to 1 years of age) (n=107).

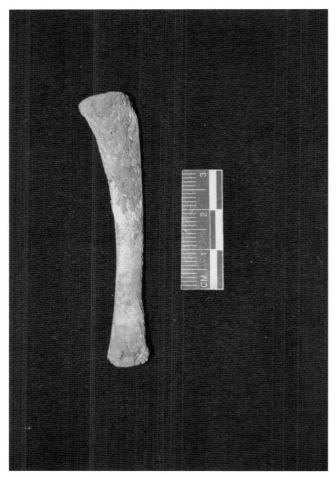

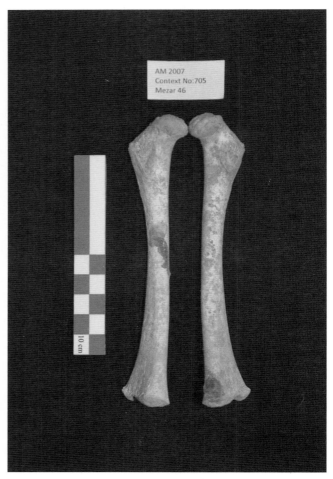

Fig. 19.7. Amorium. Infectional lesion on a neonatal left tibia (Grave 14) (photo by the author).

Fig. 19.8. Amorium. Possible case of rickets on infant femur (Grave 46) (photo by the author).

One possible reason why babies were buried separately from their mothers could be the mother's survival during the delivery of the baby (Lightfoot *et al.* 2009). Byzantines were a people of strong faith, who believed that death represented a passage to a better world and afterlife (Talbot 2009, 291; Dennis 2002). Talbot (2009, 301) notes that burial of infants and children in a segregated area is only occasionally found in the Byzantine world. In this respect, the cemetery in Amorium is extremely important since it gives a special insight into mid-Byzantine burial practices.

According to Byzantine law, the subadult life was divided into three age groups: infants up to seven years, children or juveniles up to 12 years for girls and to 14 for boys, and adolescents or older subadults up to 25 years old (Tritsaroli and Valentin (2008), quoting Kiousopoulou (1997)). In this cemetery all but one of the tombs contain no burial over 8 years (±24 months) of age, which means that this area was specially designed for infants. The only tomb that belongs to adults in this area is Tomb 43 and its anthropological investigation is also in progress.

In the Byzantine world, members of society with higher social status and clerics were buried within the church (Ivison 1993). The cemetery area A20 is located immediately outside the north wall of the church and to the east of the baptistery. Tritsaroli and Valentin (2008) record that in general the church was the main element influencing cemetery organization, and burial location in the church has significant social implication which means that the social status of the individuals was protected after death. Talbot (2009, 301) also points out the possibility that this cemetery in Amorium was reserved for unbaptized children. Until the 6th century, only adults were baptized (Lightfoot and Lightfoot 2007), and the usual ritual intervals to be observed between birth and baptism varied in Byzantium (Baun 1994, 117). By contrast, in the mid-Byzantine period babies were also baptized 40 days after birth, but in cases where there was the risk of death, baptism was performed on the eighth day or just after birth (Tritsaroli and Valentin (2008), quoting Koukoules (1951), and Congourdeau (1993). At Amorium it was impossible to apply the skeletal criteria showing the

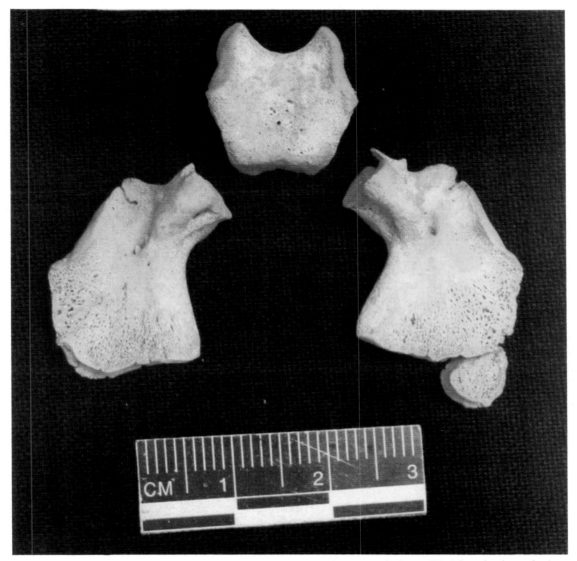

Fig. 19.9. Amorium. Congenital anomaly at the base of the skull (perinatal: Grave 27) (photo by the author).

viability of the foetus (Kósa 1989) since the skeletal remains, especially the skulls, from area A20 are very fragmentary. Therefore, it was not possible to determine whether the babies lived long enough to be baptized. On the other hand, Baun (1994, 122) indicates that the soul is only a formless matter before baptism and baptism is literally the beginning of existence for the Christian soul. Baun (1994, 118) also points out that the high infant mortality ratios show that death before baptism was an ever-present threat in Byzantine world, but the issue arises most often with reference to the moral and theological predicament of the parents alone and not in relation to the fate of the babies themselves. In this case, Christian beliefs in Byzantine times should not affect the baby, as it is very likely that these babies were accepted as having *pure souls*. At Assos, for example, infant

and child burials were placed over the early Byzantine baptistery and east of the apse of the Ayazma church (Böhlendorf-Arslan 2012, 50–1, pls. 10–11; 2013, 236–7). Likewise, at Amorium it may be suggested that the placement of the cemetery next to the baptistery was chosen deliberately, since it would have been an appropriate area in which to bury such *pure souls*. The arrangement of many of the tombs next to the east wall and small apse of the baptistery would have meant that they would have been washed by 'holy' rainwater pouring off the baptistery roof. This symbolic baptism may have also offered consolation to the grieving parents. In relation to the burial of the deceased infants it should be noted that none were buried with their mothers, who also may have died during labour, childbirth, or from subsequent complications.

Table 19.1. Mortality rates of infants in some Roman, Byzantine, and Medieval settlements in Anatolia.

Site	Publication	Date	No. individuals	Age range	Infants	% infants
Bôrûkçû	Sağir *et al.* 2004	Geometric – Roman	54	0–3	6	11.1
Adramytteion	Atamtürk-Duyar 2008	5th – 6th centuries	28	foetal–3	4	14.3
Elaiussa Sebaste	Paine *et al.* 2007	Mid-6th to mid-7th centuries	116	foetal–1.5	25	21.5
Iasos	Yilmaz Usta 2013	6th century	230	foetal–2.9	34	14.7
Tepecik/Çiftlik	Büyükkarakaya *et al.* 2009	Late Roman – early Byzantine	71	0–2.5	17	23.9
Topakli	Güeç 1988	6th – 7th centuries	187	0–5	38	20.3
Alanya	Üstündağ-Demirel 2008	10th century	27	foetal–2.5	3	11.1
Amorium*	Demirel (this volume)	10th – 11th centuries	128	0–1	44	34.4
Church of St Nicholas	Erdal Ö. D. 2009	11th – 13th centuries	28	0–2.5	7	8.6
Herakleia Perinthos	Demirel 2016	9th – 13th centuries	109	foetal–3	14	12.8
İznik	Erdal Y.S. 1993	13th century	170	0–1	38	22.4
İznik*	Özbek 1991	Late Byzantine	22	0–1	11	50
Kadikalesi / Anaia	Üstandağ 2008	12th – 13th centuries	58	0–2.5	10	17.2
Kovuklukaya	Erdal Y. S. 2004	10th century	36	0–2.5	3	8.3
Kyme	Teegen (this volume)	Late Byzantine	49	foetal–1	23	48
Phokaia	Üstündağ 2009	8th, mid-10th – mid-11th centuries	26	foetal–3	3	11.5
Yortanli	Nalbantoğlu *et al.* 2000	Late Byzantine	107	0–2.5	2	7.4
Değirmentepe	Özbek 1986	Medieval	49	foetal–1	11	22.4
Değirmentepe*	Erdal, Ö. D.-Özbek 2009	Medieval	22	0–1	10	45.6
Karagündüz	Gözlük 2004a	Medieval	890	foetal–2.5	246	27.6
Minnetpinari	Yiğit *et al.*	Medieval	86	0–2.5	8	9.3

* The marked sites provide data solely for the subadults.

Results

Amorium was attacked by the Arabs during the Dark Ages (7th to the mid-9th century AD), besieged and destroyed by the Abbasid caliph al-Mu'tasim in 838, and good evidence of this catastrophe has been collected during the excavations. The excavations also revealed that peace and prosperity returned to the city in the 10th–11th centuries. This was the time when new buildings were constructed, large amounts of money were spent on the refurbishment of the Lower City (and possibly other churches in the city), and local industries were re-established (Lightfoot and Lightfoot 2007). Despite the city's apparent wealth, the unfavourable geographic conditions of Amorium's location might have produced inadequate hygienic conditions and malnourishment among mothers, which in turn caused infections and complications during birth, leading to premature and stillbirths. There is another factor that should not be neglected, namely, 'Sudden Infant Death Syndrome', which is defined as 'the sudden and unexpected death of an infant younger than one year old and usually beyond the immediate perinatal period, which remains unexplained after a thorough case investigation including performance of complete autopsy and review of circumstances of death and clinical history' (Beckwith 2003, 289). This syndrome is also counted as one of the main reasons of infant death in archaeological populations (Bourbou 2004, 66).

Interviews with the modern-day villagers at Amorium made it clear that they were still suffering high infant mortality rates. Other evidence for this phenomenon comes from Turkish Statistical Institute's official data on the mortality rates for 2009, 2010, and 2011.[1] According to these statistics, the one year of age period still has the highest mortality ratios compared to other age groups in the city of Afyonkarahisar (75%, 76.6% and 73.3% for the years 2009, 2010 and 2011 respectively). The unfavourable geographical conditions surrounding Amorium clearly still have negative effects on the infant survival rates in the present day. In providing protection for their social status after life, it can be argued that those infants buried within the church complex were likely to be children of families with a higher status in society. These unfortunate infants nevertheless had no chance to develop a proper immune system, mostly owing to the disadvantages of the physical environment, which had many negative effects on both the mothers and the infants. But if they managed to survive this critical time of life, they were largely able to advance to older ages.

Further studies on pathological conditions and stable isotope and trace element analysis on the nutritional status of skeletal remains will hopefully provide more information about the health status and also about the factors leading to death in the infant and child population in mid-Byzantine Amorium.

Acknowledgements

I would like to express my sincere thanks to Dr Christopher Lightfoot for inviting me to study the material, his continuous support and very helpful comments on the text; and to Professor Eric Ivison for his contributions. I would also like to express my sincere thanks to my students for their help throughout the challenging laboratory work of such delicate material.

Note

1   The data is obtained from the Turkish Statistical Institute's official website, www.turkstat.gov.tr, read on 21.02.2013.

Bibliography

Atamtürk, D. and Duyar, İ. (2008) Adramytteion (Örentepe) iskeletlerinde ağız ve diş sağlığı. *Hacettepe Üniversitesi Dergisi* 25 (1), 1–15.
Aufderheide, A. C. and Rodriguez-Martin, C. (2008) *The Cambridge encyclopaedia of human paleopathology.* Cambridge, Cambridge University Press.
Baun, J. (1994) The fate of babies dying before baptism in Byzantium. In D. Wood (ed.) *The church and childhood. Papers read at the summer meeting and the winter meeting of the Ecclesiastical History Society,* 115–25. Oxford, Blackwell Publishers.
Beckwith, B. (2003) Defining the sudden infant death syndrome. *Archives of Pediatrics and Adolescent Medicine* 157, 286–90.
Böhlendorf-Arslan, B. (2012) Ayazma Kilisesi. *Kazı Sonuçları Toplantısı* 33.3, 50–5.
Böhlendorf-Arslan, B. (2013) Ayazmakirche. *Archäologischer Anzeiger* 2013.1, 230–8.
Bourbou, C. (2004) *The people of Early Byzantine Eleutherna and Messene (6th – 7th centuries A.D.). A bioarchaeological approach.* Athens, University of Crete.
Bourbou, C. (2010) *Health and disease in Byzantine Crete (7th – 12th centuries AD).* Farnham Surrey, Ashgate Publishing Limited.
Bourbou, C. and Garvie-Lok, J. S. (2009) Breastfeeding and weaning patterns in Byzantine times: Evidence from human remains and written sources. In Papaconstantinou and Talbot (eds.), 65–83.
Buikstra, J. E. and Ubelaker D. H. (1994) *Standards for data collection from human skeletal remains.* Fayetteville, Arkansas Archaeological Survey Research Series No. 44.
Büyükkarakaya, A. M., Erdal, Y. S., and Özbek M. (2009) Tepecik/Çiftlik insanlarının antropolojik açıdan değerlendirilmesi. *Arkeometri Sonuçları Toplantısı* 24, 119–38.
Congourdeau, M. H. (1993) Regards sur l'enfant nouveau-né á Byzance. *Revue des études Byzantines* 51, 161–76.
Demirel, F. A. (in press) Anthropology. In P. Niewöhner (ed.) *The archaeology of Byzantine Anatolia.*
Demirel, F. A. (2016) Analysis of human remains from Herakleia Perinthos. In S. Westphalen (ed.) *Die Basilika am Kalekapı in Herakleia Perinthos: Bericht über die Ausgrabungen von 1992–2010 in Marmara Ereğlisi* (Istanbuler Forschungen 55), 130–44. Tubingen, Ernst Wasmuth Verlag.
Dennis, G. T. (2002) Death in Byzantium. *Dumbarton Oaks Papers* 55, 1–7.
El-Nofely, A. A. and İşcan, M. Y. (1989) Assessment of age from the dentition of children. In M. Y. İşcan (ed.) *Age markers in the human skeleton,* 237–54. Springfield, Charles C. Thomas Publisher.
Erdal, Ö. D. (2009) Demre Aziz Nikolaos Kilisesi geç Bizans ve Yakınçağ insanlarının yaşam biçimleri. *Adalya* 12, 361–88.
Erdal, Ö. D. and Özbek, M. (2009) Değirmentepe (Malatya) çocuk iskeletlerinin antropolojik analizi. *Arkeometri Sonuçları Toplantısı* 25, 279–96.
Erdal, Y. S. (1993) İznik geç Bizans topluluğunun demografik analizi. *Arkeometri Sonuçları Toplantısı* 8, 243–57.
Erdal, Y. S. (2004) Kovuklukaya (Boyabat, Sinop) insanlarının sağlık yapısı ve yaşam biçimleriyle ilişkisi. *Anadolu Araştırmaları,* 17.2, 169–96.

Eryurt, M. A. and Koç, İ. (2009) Yoksulluk ve çocuk ölümlülüğü: hanehalkı refah düzeyinin çocuk ölümlülüğü üzerindeki etkisi. *Çocuk Sağlığı ve Hastalıkları Dergisi* 52, 113–21.

Gözlük, P. (2005) Karagündüz toplumunun paleodemografik açıdan incelenmesi. *Antropoloji Dergisi* 20, 75–106.

Güleç, E. (1988) Topaklı populasyonunun demografik ve paleoantropolojik analizi. *Araştırma Sonuçları* 5, 347–57.

Günay, I., Satar, S. and Şimşek, N. (2010) Laodikeia iskeletlerinin osteolojik analizi. *Arkeometri Sonuçları Toplantısı* 25: 329–42.

Ingvarsson-Sundström, A. (2014). Human remains from Tegea. In E. Østby (ed.) *Tegea II. Investigations in the sanctuary of Athena Alea 1990–94 and 2004.* Papers and Monographs from the Norwegian Institute at Athens. Volume 4, 427–42. Athens, The Norwegian Institute at Athens.

Ivison, E. A. (1993) Mortuary practices in Byzantium (c. 950–1453): An archaeological contribution. Unpublished thesis, University of Birmingham.

Ivison, E. A. (2010) Kirche und religiöses leben im Byzantinischen Amorium. In F. Daim and J. Drauschke (eds.) *Byzanz – das Römerreich im Mittelalter, Teil 2,1 Schauplätze,* 84/2, 1, 309–43. Mainz, Monographien des Römisch-Germanischen Zentralmuseums.

Ivison, E. A. (2012) Excavations at the Lower City enclosure, 1996–2008. In C. S. Lightfoot and E. A. Ivison (eds.) *Amorium reports 3: The Lower City enclosure, finds reports and technical studies,* 5–151. Istanbul, Ege Yayınları.

Kaya, A., Bilgin, U. E., Şenol, E., Koçak, A., Aktaş, E. Ö., and Şen F. (2010) İzmir'de yapılan bebeklik dönemi adli otopsiler: 1999–2007. *Ege Tıp Dergisi* 49.3, 177–84.

Kiesewetter, H. (this volume) Toothache, back pain, and fatal injuries: What skeletons reveal about life and death at Roman and Byzantine Hierapolis, 268–85.

Kiousopoulou, A. (1997) Hronos kai êlikies stê Byzantinê koinônia: ê klimaka tôn êlikion apo ta agiologika keimena tês mesês epohês. *Istoriko arheo ellênikês neolalias* 30. Kentro Neoellênikôn Ereunôn. Athens, E.I. E.

Koç, İ., Yüksel, İ., and Eryurt, M. A. (2009) Bebek ve çocuk ölümlülüğü. *Türkiye nüfus ve sağlık araştırması, TNSA-2008,* 131–41. Ankara, Hacettepe Üniversitesi, Nüfus Etütleri İdaresi.

Kósa, F. (1989) Age estimation from the fetal skeleton. In M. Y. İşcan (ed.) *Age markers in the human skeleton,* 21–54. Springfield, Charles C. Thomas.

Koukoules, P. (1951), Byzantinôn bios kai politismos (Vie et civilisation Byzantine), vol. 3. Athens, Institut Français d'Athénes.

Lewis, M. E. (2007a) *The bioarchaeology of children: Perspectives from biological and forensic anthropology.* Cambridge, Cambridge University Press.

Lewis, M. E. (2007b) Brief and precarious lives: Infant mortality in contrasting sites from Medieval and Post-Medieval England (AD 850–1859). *American Journal of Physical Anthropology* 134, 117–29.

Lightfoot, C. and Lightfoot, M. (2007) *Anadolu'da bir Bizans kenti: Amorium.* İstanbul, Homer Kitapevi.

Lightfoot, C., Ivison, E., Şen, M., and Yaman, H. (2009) Amorium kazısı-2007. *30. Kazı Sonuçları Toplantısı* 30.1, 201–26.

Nalbantoğlu, E., Türk, H., and Nalbantoğlu, C. (2000) 1996 yılı Yortanlı nekropolis kazısı iskelet populasyonu üzerinde paleoantropolojik çalışmalar, *Türk Arkeoloji ve Etnografya Dergisi* 1, 27–36.

Ortner, D. J. (2003) *Identification of pathological conditions in human skeletal remains* (second edition). London, Academic Press.

Özbek, M. (1986) Değirmentepe eski insan topluluklarının demografik ve antropolojik açıdan analizi. *Arkeometri Sonuçları Toplantısı* 1, 107–130.

Özbek, M. (1991) İznik Açıkhava Tiyatrosu'ndaki kilisede bulunan bebek iskeletleri. *Belleten,* 55 (213), 315–31.

Paine, R. R., Vargiu, R., Coppa, A., Morselli, C., and Schneider, E. E. (2007) A health assessment of high status Christian burials recovered from the Roman-Byzantine archaeological site of Elaiussa Sebaste, Turkey. *Homo. Journal of Comparative Human Biology* 58, 173–90.

Pakiş, I. and Koç, S. (2009) Perinatal ve neonatal dönem bebek ölümleri. *Klinik Gelişim, Adli Tıp Özel Sayısı* 22, 60–3.

Papaconstantinou, A. and Talbot, A. M. (eds.) (2009) *Becoming Byzantine: Children and childhood in Byzantium,* 283–310. Dumbarton Oaks Research Library Collection. Cambridge (MA), Harvard University Press.

Roberts, C. and Manchester, K. (1995) *The archaeology of disease.* New York, Alan Sutton Publishing Ltd.

Sağır, M., Özer, İ., Satar, Z. and Güleç, E. (2004) Börükçü iskeletlerinin paleoantropolojik açıdan değerlendirilmesi. *Arkeometri Sonuçları* 19, 27–40.

Scheuer, L. and Black, S. (2000) *Developmental juvenile osteology.* London, Elsevier Academic Press.

Scheuer, L. and Black, S. (2004) *Juvenile skeleton.* London, Elsevier Academic Press.

Smith, B. H. (1991) Standards of human tooth formation and dental age. In M. A. Kelley and C. S. Larsen (eds.) *Advances in dental anthropology,* 143–68. New York, Wiley-Liss.

Talbot, A. M. (2009) The death and commemoration of Byzantine children. In Papaconstantinou and Talbot (eds.), 283–310.

Teegen, W-R. (this volume) Pergamon – Kyme – Priene: Health and disease from the Roman to the late Byzantine period in different locations of Asia Minor, 250–67.

Tezcan, S., Ergöçmen, B., and Tunçkanat, F. H. (2009) Düşükler ve ölü doğumlar. *Türkiye nüfus ve sağlık araştırması, TNSA-2008,* 97–108. Ankara, Hacettepe Üniversitesi, Nüfus Etütleri İdaresi.

Tritsaroli, P. and Valentin, F. (2008) Byzantine burial practices for children: Case studies based on a bioarchaeological approach to cemeteries from Greece. In F. G. Jener, S. Muriel and C. Olária (eds.) *Nasciturus, infans, puerulus vobis Mater Terra: La muerte en la infancia,* 93–113. Diputació de Castelló, Servei d'Investigacions Arqueològiques i Prehistòriques.

Ubelaker, D. H. (1989) The age estimation of age at death from immature human bone. In M. Y. İşcan (ed.) *Age markers in the human skeleton,* 55–70. Springfield, Charles C. Thomas Publisher.

Üstündağ, H. (2008) Kuşadası Kadıkalesi/Anaia kazısında bulunan insan iskelet kalıntıları. *Arkeometri Sonuçları Toplantısı* 24, 209–28.

Üstündağ, H. (2009) Phokaia (Turkey), Season 2007. *Bioarchaeology of the Near East* 3, 27–31.

Üstündağ, H and Demirel, F. A. (2008) Alanya Kalesi kazılarında bulunan insan iskelet kalıntılarının osteolojik analizi. *Türk Arkeoloji ve Etnografya Dergisi* 8, 79–90.

White, T. D. and Folkens, P. A. (2005) *The human bone manual.* San Diego, Elsevier Academic Press.

Yılmaz, R., Pakiş, I., Turan, N., Can, M., Kabakuş, Y., and Gürpınar, S. S. (2010) Adli Tıp Kurumu Birinci Adli Tıp İhtisas Kurulu'nca ölüm sebebi verilen 0-1 yaş grubu bebeklerin ölüm sebebi açısından değerlendirilmesi. *Türk Pediatri Arşivi Dergisi* 45, 31–6.

Yılmaz Usta N. D. (2013) Iasos (Bizans) toplumunda ağız ve diş sağlığı. *Antropoloji Dergisi* 25: 117–24.

Yiğit, A., Gözlük Kırmızıoğlu, P. Durgunlu, Ö., Özdemir, S. and Sevim Erol, A. (2008) Kahramanmaraş/Minnetpınarı iskeletlerinin paleoantropolojik açıdan değerlendirilmesi. *Arkeometri Sonuçları Toplantısı* 23, 91–110.

Yurdakök, M. (2005) Bizanslılarda pediatri. *Çocuk Sağlığı ve Hastalıkları Dergisi* 48, 93–9.

The wrestler from Ephesus: Osteobiography of a man from the Roman period based on his anthropological and palaeopathological record

Jan Nováček, Kristina Scheelen, and Michael Schultz

Abstract

An osteobiography is based on the attempt to reconstruct the medical history of an individual from his skeletal remains and may serve to improve our insight into the living conditions of past populations. This paper presents the osteobiography of a man from Roman Ephesus. His skeleton was found in a large limestone burial chamber in grave house 1/08 from the Ephesian Harbour Necropolis, together with the remains of more than 60 other individuals. The man was sturdily built and muscular, and died between the ages of 35 and 45 years. His bones bear vestiges of various pathological conditions. During his life, he must have incurred many injuries; including muscular traumata, fractures of ribs and nose, as well as injuries of the spine, several joints, mandible and neck. While each injury alone could have happened due to various activities, their accumulation and distribution tend to indicate interpersonal, unarmed violence, as, for example, wrestling, pankration, or similar freestyle fighting. The high frequency of abscesses and gingival pockets, missing teeth and gum disease suggest poor oral hygiene. Furthermore, the skeleton shows vestiges of physical strain beginning in childhood, resulting, for example, in the formation of an os acromiale. During his youth or early adulthood, the man must have survived severe pleurisy. His external cranial surface bears vestiges of long-term scalp inflammation. In both external auditory canals exostoses were observed. Due to the severely narrowed auditory canal, he probably was almost deaf in his right ear.

Keywords: auditory exostosis, close combat, Ephesus, multiple traumata, *os acromiale*, osteobiography, palaeopathology, pleurisy, wrestling

Introduction

With a total of at least 169 buried individuals from five different burial chambers, grave house 1/08 (Fig. 20.1) from the Harbour Necropolis in Ephesus (3rd to 5th century AD) provides important information about late Roman life and lifestyle in the city (Fig. 20.2). One of these skeletons, individual I from burial chamber 3, will be introduced in this paper. The focus is placed on another aspect of palaeopathology than statistical analysis of the frequencies of various pathological processes in a population which, though of major importance, will be presented within another context (Nováček *et al.* forthcoming). The present case allows reconstruction of an osteo- or palaeobiography (cf. Schultz 2011). Based on the data of the skeletal human

remains investigated, this 'patient's record' is best compared to the work of modern forensic sciences (Blau and Ubelaker 2011; Christensen *et al.* 2014; İşcan and Steyn 2013). Whereas the frequencies of illnesses or injuries provide essential data on the understanding of life conditions of the population, the fate of one single individual should also not be forgotten. Osteobiographical reconstruction allows an insight into the daily life of this one Ephesian. As is often the case in legal medicine, the patient's record is not complete. Depending on the state of preservation of a body under investigation, some facts can no longer be detected and others can only be determined within a rather wide scope (for example, the famous 'blunt trauma' caused by an unknown weapon). Primarily, the patient can no longer

Fig. 20.1. Ephesus. An overview of grave house 1/08 after the excavation in 2008. The grave house is situated directly next to the harbour channel of Ephesus. Inside, two of the emptied burial chambers are visible. Burial chamber 3 is on the right, below a collapsed arch (Courtesy of Österreichisches Archäologisches Institut).

detected in the bone if its course is rapid and aggressive, and the patient dies within a short time. A persistent pneumonia, however, may result in pleurisy, which can leave visible traces on the surface of the bone (Schultz et al. 2001; 2003; Schultz 2010). Many pathological processes affecting the skeleton do not leave any distinct changes which would allow a clear diagnosis. For example, traces of physical strain or overload are manifested in the bone and can be properly diagnosed, also including the muscles involved. However, correlation with a defined activity is only possible in very few cases. An overexertion of the upper arm muscles could result from swinging a sword or scythe, as well as throwing a spear or fishing net. Without claiming to be necessarily true, however, in consideration of the requirements, an interpretation of the evidence may have a good chance of being correct. In this way, many aspects, especially from the patient history of the individual investigated, can be illuminated and provide information about the individual's lifestyle.

be questioned. Therefore, the completeness of insights into his problems cannot bear comparison with the patient's record from a family physician. However, in many cases, long-term diseases may be detected, even if they primarily did not affect the bone. Pneumonia, for example, cannot be

Materials and methods

The skeletal remains from individual I were found on the upper level of those entombed in burial chamber 3 from grave house 1/08 (Fig. 20.3) of the Ephesian Harbour

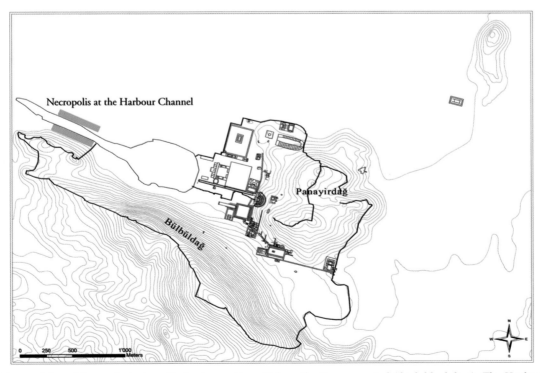

Fig. 20.2. Ephesus. Map of the ancient city with the large city wall from the Roman period (thick black line). The Harbour Necropolis was situated outside the city wall, along the channel (red). The position of grave house 1/08 was on the northern bank, close to the earlier city wall (Courtesy of Österreichisches Archäologisches Institut).

Fig. 20.3. Ephesus. Overview of the in situ interments in burial chamber 3. The skeletons from the upper layers were mostly found and recovered in their anatomical context. The skeletons from the lower layers were more mixed and disturbed. Commonly, bone elements were missing or severely fragmented (Courtesy of Österreichisches Archäologisches Institut).

Necropolis. Thus, they must have belonged to one of the latest burials in this chamber, which contained the burials of at least 61 individuals (cf. Nováček *et al.*, forthcoming). In contrast to many of the other skeletons from the lower layers of the grave, which were mixed up and could not be recovered in their anatomical context, individual I was recovered almost complete and placed in one single skeleton box at the excavation (Fig. 20.4).

The anthropological investigation was conducted using the methodological standards of Ferembach *et al.* (1980), İşcan and Steyn (2013), and Rösing *et al.* (2007). For the morphological age estimation, the degree of tooth abrasion according to Perizonius and Pot (1981), Brothwell (1981), and Lovejoy (1985) were also considered.

For the light microscopic age estimation, the inner bone structure was estimated by qualitative histomorphological (Nováček 2012) as well as quantitative histomorphometrical means (Kerley 1965; Kerley and Ubelaker 1978). For this purpose, one piece of each left or right femur was embedded in resin and a 60 μm thick thin section was cut (method by Schultz 1988b; 2003). For the body height estimation, the equation of Trotter and Gleser (1952; 1977) was employed.

The palaeopathological investigation was performed using the standard methods according to Schultz (1988a), particularly the classification of degenerative joint disease and dental pathologies. The degree of joint degeneration according to the morphological condition is evaluated on a scale from 0, or I (a healthy joint) to VI (a completely destroyed joint; cf. Schultz 1988a). The edges and articular surfaces are evaluated separately. To avoid undervaluation, in the combined assessment of the severity of the changes in the joint units, only the more highly degenerative degrees from the edge and articular surface are selected (cf. Schultz 1988a). The degree of periodontal disease is evaluated on a five-stage scale, from I (light) to V (severe; cf. Schultz 1988a). Furthermore, both literature on palaeopathology (e.g. Aufderheide and Rodríguez-Martín 1998; Mann and Hunt 2013; Ortner 2003) and bone pathology (Adler 2005; Niethard *et al.* 2009), as well as forensics (Blau and Ubelaker 2011; Christensen *et al.* 2014; İşcan and Steyn 2013; Kreutz and Verhoff 2002), were, in our case, consulted.

Results and discussion

Preservation and estimation of sex and age-at-death

The state of preservation of the skeleton is very good. It was recovered almost complete (cf. Fig. 20.4) and the surfaces are only minimally weathered (<20%). The bone tissue is solid. Only some of the spongy parts of the bones (e.g. some of the vertebral bodies, ribs and metaphysis of the long bones) are fragile. The irregular and blotched discolouration

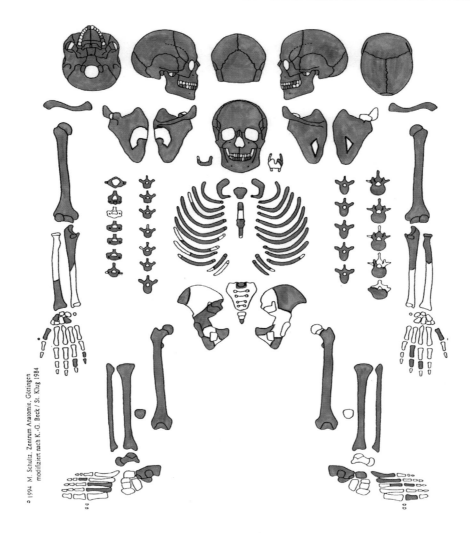

Fig. 20.4. Ephesus. The skeletal scheme of individual I. The red coloured parts of the skeleton are preserved, the white ones are missing (Drawing J. Nováček).

of the bones, often with bleached-out water spots, indicate that the burial chamber has not been filled with earth for a prolonged period of time and was probably alternately flooded by water from the harbour channel followed by dry intervals.

Morphological traits both of the pelvic bone, as well as of the skull, indicate that the individual was most probably male (Ferembach *et al.* 1980; İşcan and Steyn 2013; Rösing *et al.* 2007).

The biological age-at-death was estimated to be between 35 and 45 years. By means of the macroscopic investigation, an age of 35–50 years was determined. With the light microscopic investigation, the result could be stated more precisely as being 35–45 years. The biological age of an individual may differ from the actual, chronological one. The macroscopically detectable features on the skeleton are under a strong environmental influence. The condition

of teeth and periodontium is not only determined by age, but also nurture and dental hygiene are significantly involved. The sutural closure may be accelerated by individual predispositions or pathological processes of the skull vault. The condition of the spongy and compact bone of the long bones mainly depends on physical strain, hormonal balance and health condition, as well as the age of the individual. Therefore, the macroscopical age determination should be determined as a rather broad category. Also, the microscopical structure of bone tissue is influenced by strain, hormonal balance and health. However, there are some determining characteristics, whose presence or absence represent distinct limits for the biological age of bone tissue (Hummel and Schutkowski 1993; Nováček 2012). Thus, the detectable remains of an external circumferential lamella on the surface of the bone can exclude an age determination of more than 50 years and set the presumable upper limit to 45 years. A physical-strain-induced, higher turn-over of the bone tissue may accelerate a little the disappearance (reduction) of this lamella, but not slow it down. Hence, a false higher age classification can be excluded (Nováček 2012).

The average life expectancy of adult men from late Roman Ephesus, as represented by the population buried in grave house 1/08, was about 37 years (cf. Nováček *et al.* forthcoming). At 35 to 45 years, individual I at least had reached an average or even slightly above average age. Therefore, the man presumably was considered to be one of the more experienced members of the Ephesian society.

Body height, body type and musculature

From a present-day perspective, the man seems to have been rather short with a body height between 158 and 166 cm (161.9±4 cm; Trotter and Gleser 1952; 1977). Still among his fellow citizens, at least compared to the other investigated individuals from grave house 1/08 (average body height of men 165.6±4 cm), he was neither remarkably tall nor short (cf. Nováček *et al.* forthcoming).

However, remarkably outstanding for any society must have been his muscle strength. All bones of the upper limbs show strong or extremely strong muscle marks. The muscle marks from the lower limbs are distinct but, in comparison, rather less pronounced than the ones on his arms. Beyond doubt, it can be presumed that the man practised activities which generally resulted in a strong strain on the muscoskeletal system, though with a higher load on the upper than on the lower limbs. His daily, professional occupation is not able to be securely ascertained; it can only be determined as some manual activity with a high physical effort (cf. Villotte *et al.* 2010). The asymmetric development of muscle marks from the left and right arm, as well as further skeletal traits (for example the length of the clavicles and arm bones) imply that he was probably a right hander (cf. Kreutz and Verhoff 2002; Schultz 2011; Schultz and Schmidt-Schultz 2004; Sládek *et al.* 2007). The muscle marks on the man's postcranial skeleton indicate that he must have possessed generally strongly developed muscles. Although muscle marks can partly still enlarge after body growth is completed, the strong degree present rather points to the active growth of a juvenile or young adult person. Unlike musculature, which is reduced without strain, the muscle marks, at least macroscopically, do not remodel again, once they have been developed (cf. Maggiano *et al.* 2008). Thus, it cannot be determined with certainty whether the musculature of the man was still strong when he died. In the case of long-term underloading or even atrophy, lasting for years, the inner bone structure of the muscular marks would have been reduced (cf. Schlecht *et al.* 2012), which it was not. However, with respect to the last six months before death, no statement can be hazarded (cf. Schultz 1997; 2003).

The man's stature seems to have been stocky and robust. The morphological traits of his shoulders, with rather long, but robust clavicles, represent an athletic (Kretschmer 1977) or mesomorphic (Kreutz and Verhoff 2002; cf. Sheldon *et al.* 1940) body type. On the other hand, the comparatively short, broad manubrium and broad body of the sternum, as well as the width of the long bone epiphyses and rather short, broad shafts, might rather hint at a pyknic (Kretschmer 1977) or endomorphic (Kreutz and Verhoff 2002; cf. Sheldon *et al.* 1940) body type. Therefore, a mix of an athletic and a pyknic body type may be assumed, with a broad and robust trunk and short, robust, muscular limbs.

Individual traits and traces of pathological changes
Individual traits

The styloid process of the temporal bone, usually a pointed, 1.5–3 cm long process on the lower surface of the skull base, where some muscles of the pharynx, the hyoid and the tongue (stylopharyngeus, styloglossus, and stylohyoid muscle) are attached, is bilaterally shortened and rudimentarily atrophied (Fig. 20.5). With a length of only 5 mm, its morphology is atypical (Williams *et al.* 1995). A similar, rudimentary structure of this process, probably related to an atypical course of some muscles, was also observed in several other skeletons from burial chamber 3. Because of this observation, it seems likely that this morphological trait is epigenetic. It might indicate that the man was related to other buried individuals from grave chamber 3. It is unlikely that the atypical morphology of the origin of the muscles caused a noticeable anomaly.

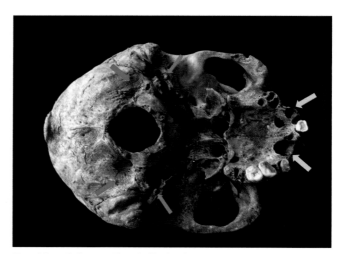

Fig. 20.5. Ephesus. The skull of individual I in basal view. Newly built bone formations behind and next to the occipital condyles (red arrows) are asymmetric in morphology, possibly indicating two injuries from different time points. Bilateral very short, rudimentary styloid processes (green arrows) are an individual, possibly familial trait. Large abscesses in the maxilla (yellow arrows) destroyed the bone tissue and perforated the bony surface of the hard palate (Photo K. Scheelen).

Diseases during childhood and youth

Vestiges of pathological processes on this skeleton document an arduous life from childhood. On the left shoulder blade, an atypical variation of the acromial process was detected. The process was not fused with the scapular spine, as it normally should be, but separated by a persisting epiphyseal notch, probably filled with hyaline cartilage during lifetime (Fig. 20.6). The unfused notch shows vestiges of physical overload. There are two possible reasons for the development of such an *os acromiale*. On the one hand, these are epigenetically (familial) predispositions (Angel *et al.* 1987), which occur, on average, in between 1.4% (Sterling *et al.* 1995) and 7% (Yammine 2014) of the population and mostly do not cause any symptoms (Sterling *et al.* 1995). On the other hand, it could result from a childhood trauma, because before the age of about 16 years, the epiphyseal notch on this part of the shoulder blade is not closed (Scheuer and Black 2000). In this case, presumably the morphological variation was acquired due to an injury of the epiphyseal notch. A later trauma during adulthood does not seem likely, as the notch runs through the widest part of the acromial process, not at the least-resistant spot. A unilateral development of such a separately ossified nucleus due to genetic predisposition cannot be excluded, but the most probable reason seems to be a trauma. Such injuries of epiphyseal notches often result from physical overload during childhood or youth (Schultz and Schmidt-Schultz 2004). If the injury was caused by a non-recurring stress, like a fall or child abuse, it probably would have ossified anyway and more or less vanished over the years. However, the morphology of the acromial metaphysical surface corresponds to changes which may be observed in a so-called 'false joint' or pseudarthrosis, a syndesmotic connection of both surfaces. Its development was

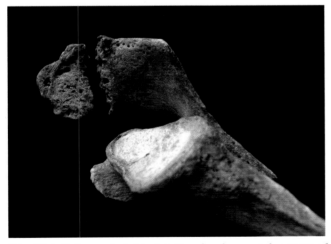

Fig. 20.6. Ephesus. Individual I. Non-fused acromial process of the left shoulder blade with irregular, porous edges, which indicate physical overload. The size of the acromial bone corresponds to the position of the epiphyseal notch at subadult age. Therefore, the probable reason is an injury of the growth cartilage in childhood or youth (Photo K. Scheelen).

probably caused by a permanent or at least long-term strain. Such strain-related pseudarthrosis can often be observed in people who have trained in archery since childhood (Stirland 1984; 2001; Schultz and Schmidt-Schultz 2004). This variation has also been observed in modern bowmen (Hershkovitz *et al.* 1996). Hence, a possible explanation is that the man, already during his youth, was a practised archer. However, also a rather non-specific kind of strain could have been the reason for the development of the *os acromiale*, for example, overloading of the shoulder girdle because of extensive gymnastic exercise or athletic sports. In modern medicine, more cases than normal of unfused epiphyseal notches of the acromial processes can be found in athletes who practise weight lifting or exercises with the arms lifted above their heads (e.g. rings, parallel bars, high bar, cf. Pagnani *et al.* 2006). These cases are of medical interest, as they can result in painful lesions. All of the athletes examined for this study started training in their youth, before the fusion of the acromial process (Pagnani *et al.* 2006). Therefore, especially regarding the extremely strong muscle attachments of the upper limbs and shoulder girdle, the *os acromiale* of individual I could have been the result of physical overload due to extensive training during his younger years.

Three preserved teeth (27, 33, and 43) show weak transversal enamel hypoplasia (stage 1 according to Schultz 1988a), which developed between the ages of three and five years (Ubelaker 1989). Transversal enamel hypoplasia are lines in the dental enamel. For example, due to malnutrition or severe diseases (cf. Skinner and Goodman 1992; Schultz *et al.* 1998), they emerge during childhood as disturbances of the growing and developing enamel. The formation of transversal enamel hypoplasia does not necessarily correlate with other skeletal stress markers, such as, for example, Harris lines. The origin of different skeletal traits partly seems to be triggered by various influences (cf. Asuming and Schultz 2000; Alfonso-Durruty 2011). As not all teeth grow and develop contemporaneously, only some are affected together at a certain age. Therefore, the enamel hypoplasia were found on the only preserved second molar as well as both of the preserved canines. Incisors and first molars are not affected. The initiating event, presumably a severe disease or malnutrition during childhood, must have occurred quite precisely between the ages of three and five years, an age, at which otherwise many children from the investigated population from grave house 1/08 died (cf. Nováček *et al.* forthcoming).

Diseases during adulthood

Muscle and ligament injuries

The bones from the shoulder girdle of the man's right arm show vestiges of several healed pulled muscles and ligaments. The muscle marks of the rotator cuff, especially on the humeral points of insertion, are strongly developed on both arms, but more pronounced on the right. Both humeri (greater and minor humeral tubercles as well as greater and minor tubercle crests) show traces of long-term healed injuries, accompanied by inflammable changes. On the right arm, these changes are more pronounced. These originally frayed depressions with rounded edges lead back to a bilateral overload, which are stronger on the right, and muscular injury of the supraspinatus and subscapular muscles. Also, an injury of the sub-spinal muscle cannot be excluded, as it inserts into the greater humeral tubercle and, in this case, cannot clearly be distinguished from the supraspinatus muscle, which is located in front (cf. Williams *et al.* 1995). All of these muscles are involved in the inner or outer rotation of the arm and stabilize the shoulder joint. Because of several pit-shaped depressions, apparently not connected to each other, it is probable that these injuries did not occur at the same time. As differential diagnosis, subsequent to the pulled muscle, also an inflammation induced osteoclastic process may have led to the described substantial loss of bone tissue and the obviously healed depressions (cf. Aufderheide and Rodríguez-Martín 1998; Ortner 2003; Mann and Hunt 2013). A probable reason for such distinct, perhaps recurrent bilateral injuries of several of the rotator cuff muscles, is the excessive external rotation of the arm, for example, simultaneously with a retroversion. Such movement might well occur in wrestling matches or similar kinds of fighting. However, an injury caused by an accident cannot be excluded.

Within the joint capsule of the right acromioclavicular joint, an irregular, pit-shaped depression with originally frayed, but now rounded (healed) edges, can be interpreted as vestiges of pulled ligaments. The location directly below the capsule of the right acromioclavicular joint, which connects the clavicle with the shoulder blade, points to a strained capsule of this joint (cf. Williams *et al.* 1995). On the same arm, traces of a pulled deltoid muscle origin, manifested as a flat, necrotic depression on the frontal margin of the acromial extremity of the right clavicle, were observed. The deltoid particularly is responsible for the abduction of the arm. Its frontal part, which left the injury mark on the clavicle, additionally is responsible for the anteversion of the arm, as well as the shoulder joint adduction. Presumably, the strain of the clavicular part of the deltoid muscle was caused by an excessive retroversion of the right arm against an opposition, for example by fierce dragging of the arm to distal and dorsal. Whether the cause was an accident or, for example, close combat, cannot be determined.

Also the lower limbs show vestiges of physical overload, although less distinctly than the upper ones. Irregular, necrotic depressions and neoplasms on the upper margins of the acetabula can be interpreted as a bilateral strain of the iliofemoral ligament, the strongest ligament between pelvis and the neck of the femur (Fig. 20.7). This could have been caused, for example, by a retroversal overexpansion

Fig. 20.7. Ephesus. Individual I. Newly built bone formations, irregular depressions and deep lesions directly on the upper edge of the former cartilaginous rim of the acetabular fossa on the right pelvic bone, corresponds to the attachment place of the iliofemoral ligament, which was probably injured (Photo K. Scheelen).

of the joint capsule, combined with excessive adduction. A further pulled muscle on the thigh could have been related to a similar injury. Probably this small, compact neoplasm on the medial surface of the right femur, about 10 cm below the minor trochanter, results from a subperiostal, long-term healed and integrated bone layer, probably caused by a minor muscle trauma (Schultz *et al.* 2001; Schultz and Walker 2013). A possible cause could be a strain of the quadriceps femoris medial head (medial vastus), accompanied by a haemorrhage of the supplying vessels. However, there are various potential reasons for such strain.

Nearly the whole surface of the right fibula shaft is rough and covered with many well-integrated, subperiostal bony neoplasms, presumably induced by muscle traumata and the thereby induced haemorrhages (cf. Schultz *et al.* 2001). It cannot be confirmed with any certainty whether these multiple traumata developed at the same time. However, their similar morphology points to a rather narrow time frame. Since all of these changes occurred long before death, neither a reliable chronological order nor concurrence of these earlier haemorrhages can be verified, even by light microscopic means. If they developed at the same time, a connection with the following injury (see below) would be likely. In the lateral malleolar fossa and the malleolar sulcus of the right fibula, on the rearmost margin of the right ankle, an enlarged, rounded depression with a smooth, newly built surface, which shows rough, porous lesions, is located. On the opposing, lateral side of the right anklebone, a matching, nearly symmetric, pit-like depression can be found. The two depressions fit together and, therefore, form an almost round cavity (Fig. 20.8). Probably, these are vestiges of an old injury of the joint capsule or the ligamentous apparatus (talofibular and/or posterior tibiofibular ligament). Due to

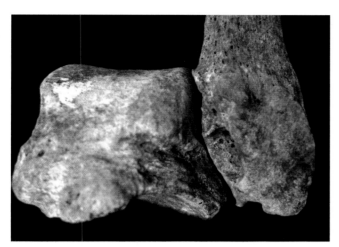

Fig. 20.8. Ephesus. Individual I. Detail of the right ankle bone and fibula from dorsal, focused on the roundish pit-like depression between the two bones. This structure is probably a result of a badly healed injury of the upper ankle joint or/and the ligamentous apparatus of this joint (Photo K. Scheelen).

this injury, a secluded, aseptic structure formed at the dorsal margin of the right fibulas lateral malleolus. Presumably, the soft tissue, which originally filled this cavity, consisted of scar-like connective tissue. A possible cause could have been a severe strain of the ankle joint, which did not properly heal due to continuous, physical load. Taking this into consideration, a connection with the described, and various muscle traumata seems possible, although not obligatory. There are many potential causes for such severe strain. A working accident, during which the man was dragged along by his leg from a bolting beast of burden; a jammed foot in the cordages of a crane in the harbour or being run over in a tilted position by a cartwheel, for example after a fall from a carriage? Imaginable, though, would also be an injury during a wrestling match, such as forceful turning and dragging of the foot of an immobilized opponent, lying face down, whose knee was held bent and whose right foot was dislocated.

PROLIFERATIVE AND DEGENERATIVE CHANGES OF THE LIMB JOINTS AND THE SPINE

Primarily and most commonly, arthrosis of the joints develops because of mechanical damage of the layers of articular cartilage. Secondarily, it may develop because of deposits due to different metabolic processes (e.g. in the case of gout), in the wake of arthritis or aseptic bone necrosis (for example, *osteochondrosis dissecans*). Mostly at the beginning, arthrosis can proceed almost painlessly, but at an advanced stage of disease, it often causes severe joint pain (Adler 2005; Niethard *et al.* 2009). As the layer of articular cartilage is directly adjacent to the articular surface, changes of this cartilage are soon followed by

changes of the subjacent bone, although changes of the cartilage itself are not visible on macerated bone (Ortner 2003). Usually, inflammatory changes of joints originate from a synovitis (Adler 2005). In the case of arthritis, the inflammation encroaches on the articular cartilage. In the case of osteoarthritis, it disperses to the bones (Aufderheide and Rodríguez-Martín 1998; Ortner 2003). Generally, there are three main groups of arthritis (Adler 2005). Unspecific arthritis does not give any precise information concerning pathogens of the disease. Amongst others, directly or haematogenously introduced, microbial pathogens (e.g. viruses, bacteria, fungi, purulent arthritis) may be a cause. Specific arthritis is caused by a particular pathogen (e.g. tuberculous arthritis). The third common origin lies in various immunopathological processes. Often, their causes are multifactorial or even unknown, i.e. idiopathic. In summary, these diseases are referred to as rheumatoid arthritis, including, for example, ankylosing spondylitis (Adler 2005; Aufderheide and Rodríguez-Martín 1998; Ortner 2003).

The joints of the man's postcranial skeleton confirm the indication of severe physical overload. Many joints show vestiges of distinct degenerative changes which indicate physical overload or incorrect weight-bearing. An inflammatory component was only rarely detected. Both shoulder joints, as well as the right elbow joint, show arthritic changes (degree 4; Schultz 1988a), while the other elbow, the wrists and the knuckles are without pathological findings. The right sacroiliac joint (degree 4/5; Schultz 1988a) and the right hip joint (degree 4; Schultz 1988a) are strongly arthrotic, with a distinct inflammatory component. Likewise, both knee joints show clear vestiges of degeneration (degree 4; Schultz 1988a), while the retropatellar joints only show minor changes (degree 2; Schultz 1988a). Seemingly, the knee joint muscle conduction was very good, because usually, the retropatellar joints would be sooner and more severely affected than the actual knee joints between thighbone and shinbone (Niethard *et al.* 2009). Due to the old injury of the right upper ankle joint described (see above), it secondarily became arthrotic (degree 4; Schultz 1988a). Both lower ankle joints show distinct vestiges of overload or incorrect weight-bearing, and are bilaterally severely degenerated (degree 4; Schultz 1988a).

The joints of the spine also demonstrate severe arthrosis, as well as trauma-induced changes. Most striking is a pathological change of the median atlantoaxial joint, which is described in connection with changes of the occiput (see below). Further vertebral bodies of the cervical spine are arthrotically or arthritically deformed (mostly degree 4 or 5; Schultz 1988a, Fig. 20.9). Some of the spinal vertebral bodies show osteophyte development and lipping, which can be interpreted as a rather weak spondylosis deformans. Along the front surface of the cervical vertebral bodies, an ossified anterior longitudinal ligament was detected from C4 to the upper edge of C7. Vertebrae C5 and C6 are ankylosed

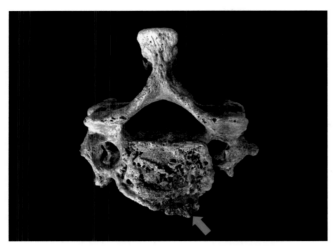

Fig. 20.9. Ephesus. Individual I. Caudal view of the sixth cervical vertebra. The ground plate of the vertebral body shows vestiges of osteophytic proliferation on the ventral edge of the plate (red arrow) and very strong (degree 5 according to Schultz 1988a) arthritic changes of the plate surface; the intervertebral discus was probably severely damaged or even no longer present. Above the sixth cervical vertebra, the ankylosed fifth vertebra can be identified (Photo K. Scheelen).

(cf. Fig. 20.9). The intervertebral disc obviously was missing and underwent bony transformation, merely with a small dorsal opening. Primarily, the ankylosis was caused by a severe arthritis and an inflammation of the intervertebral disc, presumably combined with the ossification of the anterior longitudinal ligament, as the vertebral bodies laterally do not show vestiges of injuries (e.g. compression fractures). Nevertheless, a trauma could also have been the reason, although this cannot be verified. Except for vertebra Th4 (degree 4; Schultz 1988a), the vertebral bodies of the upper part of the thoracic spine are without pathological findings and do not exceed the age-appropriate, minor changes of the articular surfaces (degree 2–3; Schultz 1988a). Downwards from thoracic vertebral body Th7 to the promontory, all vertebrae show at least slight degenerative changes (degree 4; Schultz 1988a). Vertebrae Th4, Th7, Th9–Th12, as well as most of the vertebral bodies of the lumbar spine, demonstrate the characteristic lipping of a spondylosis deformans. Pathological changes of the facets of the superior and inferior articular processes are comparatively less well developed than those of the vertebral bodies. Still they bear witness to physical strain. The facets of the articular processes of the cervical spine, the upper thoracic and the lower lumbar vertebrae at the transition to the sacral bone do not exceed the age-appropriate, probably symptom- and pain-free arthrosis (degree 2–3; Schultz 1988a). Some of the articular processes in the middle of the thoracic spine (Th4–Th6), as well as at the transition of thoracic and lumbar spine (Th12–L4) are arthrotically changed (degree 4; Schultz 1988a). The facets of both the left and the right

articular processes are equally affected. Furthermore, the vertebral arches show vestiges of physical overload, which manifest as ossified spinal ligaments. The elastic ligamenta flava connect the edges of the stacked vertebral arches and from behind, serve as protection for the spinal canal (Williams *et al.* 1995; Niethard *et al.* 2009). A draught-induced overload leads to the formation of bony neoplasms on their attachments; shaped like drawn-out, often sharp-edged spikes on the superior edges of the lower as well as the inferior edges of the upper vertebral arches. Directly at the edges of the facets of the articular processes, ligament-strengthened joint capsules consisting of connective tissue are attached. Determined by an unduly heavy overload through overextension of the articular processes, also these ligament attachments can be changed by spiky or bulging bony neoplasms. Ossifications of the ligamenta flava were detected on most of the man's thoracic vertebrae (Th4–L1). Ossified joint capsules of the articular processes even affected a longer part of the spine: from the third thoracic to the third lumbar vertebra (Th3–L3).

Based on the pattern of arthrotic and arthritic changes of vertebral and other joints, no distinct load profile of the body may be drawn. According to the rather moderate degree of pathological changes of the lower spine, the man hardly worked as a carrier of heavy loads. Apart from that, no kind of physical load can be excluded.

TRAUMATA OF THE SPINE

The man's spine shows two different kinds of trauma-induced changes. On the one hand, the facet of the left inferior articular process of the ninth thoracic vertebra (Th9) was affected by a small, well-healed fatigue fracture, which is only visible dorsomedially as a fine, curved line. Obviously, this small fracture healed a long time before death, without further consequences for the mobility of the spine. As the adjacent vertebra is without pathological findings, it apparently was a singular, overload-induced injury of the articular process.

The aetiology of the frequent Schmorl's nodes, affecting several thoracic and lumbar ground and deck plates of the vertebrae (Th6–Th12 as well as ground plate L2 and deck plate L3), is quite different. Schmorl's nodes originate in a prolapse of the colloidal centres of the vertebral discs (nucleus pulposus; Williams *et al.* 1995). In the case of such a prolapse, parts of the nucleus are compressed into the osseous endplate of the vertebral body. If the impressions remain defined and with a smooth surface, they probably emerged individually (genetically) and without discomfort, at least not primarily due to a traumatic incident (cf. Dar *et al.* 2009; 2010). A possible reason, for example, might be Scheuermann's disease, which typically occurs during youth (Aufderheide and Rodríguez-Martín 1998; Ortner 2003). However, the Schmorl's nodes present on the man's lumbar spine, penetrated into the vertebral canal medially or laterally. Such Schmorl's nodes might originate from a trauma, caused by a rip of the ring-shaped, fibrous cartilage arranged around the colloidal centre (*anulus fibrosus*). As the fibres of the fibrous cartilage are tightly attached to the surface of the vertebral plates, such injuries can also be observed on macerated bone. Depending on the localization, the penetrated liquid may either cause restriction, disconnect one of the spinal nerves (mediolateral prolapses), or constrict the dural sac (median prolapses). A mediolaterally herniated disc affects one, at the most a few, spinal nerves. A median prolapse, in contrast, may cause multifaceted pain (Niethard *et al.* 2009; Trepel 2012). However, the course may also proceed completely free of pain, restrictions, or any other symptoms. In the present case, it cannot be determined whether the man suffered from neurological complaints or from pain.

RIB FRACTURES

The man's thorax shows vestiges of multiple, long-term healed rib fractures. During at least two different time events, several ribs were simultaneously fractured. On the left, the ventral third of seventh, eighth and ninth ribs was affected. The oblique fracture line runs in a cranial to caudal direction, towards the sternal end. The callus is healed very well, with a smooth surface. It is only visible as a flat, unremarkable scar on the surface. On the left, also the short twelfth rib presents a callus in the middle of the bone. The fracture healed almost in a straight position, but left a distinct, less well-integrated callus than the one observed in the upper ribs. The injury must have happened later than the other rib fractures, at the latest a year or two before the man's death (cf. Blau and Ubelaker 2011; Christensen *et al.* 2014). On the right, the fifth rib was centrically broken, but healed similarly well to the left ribs 7–9. Only a flat, unremarkable callus is visible. Likewise, the fracture must have happened several years before death, maybe at the same time. Furthermore, the right seventh and eighth ribs were fractured at the costal angle (Fig. 20.10), but healed without dislocation. However, both ribs show a solid, irregular callus with pointed neoplasms on the surfaces, corresponding to the course of the intercostal muscles. Probably, both of these fractures correlated with a trauma of the intercostal muscles. Judging from the stage of healing, the injury could have happened at the same time as those of the left ribs 7–9 and the right fifth rib. Otherwise, a third traumatic event would also be possible. For such traumata, many probable reasons are imaginable. Rib fractures may result from accidents (for example, during work with animals, a fall from a height) or interpersonal, violent conflict (cf. Matos 2009; Simon and Sherman 2011). Especially the location of the partly healed fracture of the left twelfth rib indicates that it might have occurred in a fight. The fracture exactly corresponds to the

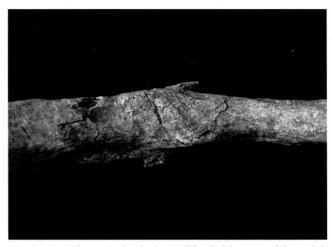

Fig. 20.10. Ephesus. Individual I. Well healed fracture of the eighth right rib close to the costal angle. Irregular bone formations on both cranial and caudal edge of the rib indicate a trauma of the intercostal muscles. The callus, however, is well healed, the fracture apparently occurred at least several years before death (Photo K. Scheelen).

position of the left kidney, adjacent to the trunk wall and close to the spine (cf. Williams *et al.* 1995). In humans, the short, twelfth rib is located directly above the superior half of the kidney. Because of their algesic capsules, kidneys represent a weak point within close-combat fights ('kidney punch'). Such a precisely placed injury, for example, caused by a punch or kick, could well have happened during a fight. In any case, after such an injury in the kidney region, including a rib fracture, the man must have suffered pain and maybe even a kidney trauma (cf. Schmid 1970, Hoppes *et al.* 1991).

INJURIES OF THE HEAD JOINTS AND THE NECK MUSCLES

Of particular interest is the pathological evidence from the head joints. Pathological changes were observed on the skull base, in the insertion area of the short neck muscles on the occipital bone, as well as laterally of the foramen magnum, on the dorsal edge of the occipital bones' condyles and the back part of the atlanto-occipital joints capsule. On the left, the surface is merely slightly rough and apparently scarred. It shows several atypical, radiating, vascular impressions around a depression on the posterior edge of the occipital condyle, adjacent to the foramen magnum (Fig. 20.11). On the right, the neoplasm is less well integrated and obviously more massive (cf. Fig. 20.5). Close to the dorsal edge of the articular surface, a deep depression is located, enclosed by porous, newly built bone tissue from both sides. Medially, the depression adjoins the edge of the foramen magnum. Laterally and ventrally, a thick, bubonic structure of porous bone tissue is located, which extends to the jugular foramen. Dorsally, behind the depression, a porous, newly built bone

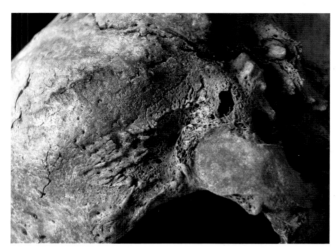

Fig. 20.11. Ephesus. Individual I. Detail of the skull basis and the right occipital condyle. On the dorsal edge of the joint itself, there is an arthrotic rim, which could have developed secondarily due to a traumatic damage of the joint capsule, but not primarily due to the trauma. Behind and lateral to the joint, separated from it by an apparent, deep, curved groove (which could indicate the location of the joint capsule), the newly built bone structure is situated, with porous tissue and various openings, probably vascular canals. This porous structure filled the space between the joint, the neck muscles dorsal and dorsolateral as well as the large blood vessels and nerves lateral and ventrolateral from the joint. The most probable reason is an old trauma due to an overload of the occipital joint. (Photo K. Scheelen).

formation is located, which contains several integrated, radiating, vascular impressions. It ends in front of the inferior nuchal line. In both cases, the extensively healed, newly built bone formations most likely resulted from an injury of the short neck muscles (probably *rectus capitis lateralis, rectus capitis posterior major* and/or *obliquus capitis superior* muscles) and a collateral haemorrhage in the capsule of the atlanto-occipital joint. Due to the location of the atlanto-occipital joint's capsule, the position of the internal jugular vein and the insertions of further neck muscles (e.g. *semispinalis capitis* and *rectus capitis posterior minor* muscle), the haemorrhage obviously was restricted to the frontal area of the inferior nuchal line. The degenerative changes of the right occipital condyle are not connected to the described injuries. These processes did not occur at the same time. While the one on the left probably happened several years before the man's death, the one on the right must have been more recent. When he died, it doubtlessly was healed, but the bony neoplasms were not yet fully integrated into the surface of the bone. Obviously, it occurred at most a few years before his death, although at least a few months before.

A further, probably coherent injury was found on the lower head joint. The edge of the dental fovea on the anterior arch of the atlas (Fig. 20.12) is extremely arthrotic (degree 5; Schultz 1988a). The lipping of the edge of the facet elongates especially cranially, towards the anterior edge of the foramen magnum. The arthrotic lipping of the

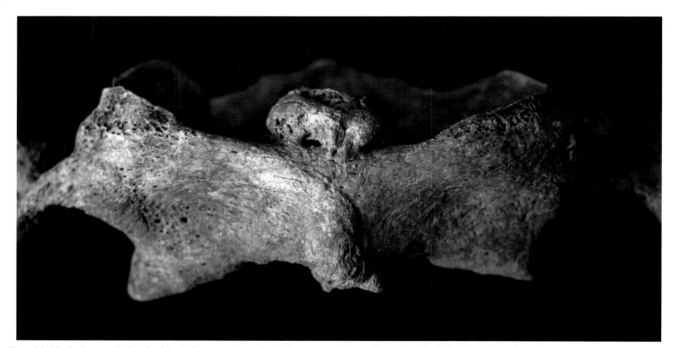

Fig. 20.12. Ephesus. Individual I. A frontal view of the first cervical vertebra, detail of the frontal arch with a large newly built bone structure on the edge of the joint socket of the atlantodental joint. This extremely large rim probably originated in an old trauma, possibly even a partial dislocation of the joint and damage to the joint capsule, long time before death (Photo K. Scheelen).

edge of the dental fovea is irregular, but well-integrated and must have developed a long time before death. On the cranial edge of the dental fovea, a probably partly integrated impression of the dental apical ligament is visible between the dens axis and the edge of the foramen magnum. The dens axis only shows lipping on the superior edge of the anterior articular facet. A probable reason for this complex evidence might be a repeated, ventral overexpansion of the head. Presumably, an extreme dorsal pressure towards the occipital bone led to an overexpansion and strain of the dorsal part of the joint capsule, as well as the short neck muscles, which could not resist such a sudden, jerky movement. Apparently, also an overextensive ventral flexion of the lower head joint occurred (cf. Babar 1999; Schünke *et al.* 2009). Thereby, the connection between the frontal atlas arch and the dens axis was overstretched and damaged. Also, such an injury might originate from, for example, a strong overstretching of the head backwards, for example by pulling the opponent's jaw upwards from behind. Supposedly, from this injury, a subluxation of the atlantoaxial joint resulted. A sudden, jerky movement, as often seen nowadays in car accidents (cf. Rubinstein *et al.* 1996; Myung-Sang *et al.* 2006), would presumably have led to a broken neck, once the resistance of the head-joints was overcome and even the capsule lacerated (Richards 2005). A slow, increasing, continuous pressure on the occipital bone seems rather more likely. Such a situation may be imaginable apart from a bodily confrontation. However, the explanation of an action in the sense of a nelson hold (for example, half or full nelson) during a wrestling match perfectly meets all requirements. This grappling hold is executed from behind the opponent and used to control him. One or both arms are used to encircle the opponent's arm under the armpit, and secured at the opponent's neck. By cranking the hands forwards, in a full nelson, pressure is applied on the neck of the defenceless opponent. In this way, he can be forced to give up. Nowadays, the usage of a full nelson in amateur wrestling is banned, as it can lead to severe or even life-threatening injuries (cf. https://unitedworldwrestling.org/sites/default/files/1-wrestling_rules_july_2014_eng.pdf). The effectiveness of this or similar grappling holds was apparently well known during Antiquity, as they are frequently portrayed in statues or images of pankratiasts or wrestlers (Poliakoff 2004; Werner 1965). In case of individual I, repeated, severe injuries during wrestling matches or other ('freestyle'?) fights seem to be the most plausible explanation for the extraordinary diagnostic findings on the occipital bone and upper cervical spine.

NOSE FRACTURE

The man's nasal bones are completely preserved. They were once broken and healed in a slightly left-shifted position (Fig. 20.13). The right side of the nose was broken more caudally (towards the tip of the nose) and was crushed into

Fig. 20.13. Ephesus. Individual I. The nasal bones were broken, slightly dislocated to the left, crushed into the nasal cavity and healed well with hardly visible callus. The broken nose was apparently corrected and treated in a professional manner. However, the crooked nose was visible on the man's face (Photo K. Scheelen).

the nasal cavity. The left side was broken more cranially (towards the front) and laterally displaced. The fracture healed well. There are only minimal vestiges of callus, which is limited to the original fracture lines. Partly, the fracture lines are rounded, but open (for example, the left nasomaxillary suture is busted open). The nasal septum is deformed in its frontal, upper part (perpendicular plate of the ethmoid bone) and shifted to the left. Yet, the actual fracture line is well healed. The fracture might well have been caused by an accident (as a fall), or, for example, by a punch on the nose. Both nasal bones, as well as the nasal septum, broke and were shifted to the left. Presumably, the injury occurred several years before death (cf. Niethard *et al.* 2009; Ortner 2003). The nasal bones were corrected professionally. They healed in a position only slightly shifted to the side. Therefore, the 'saddle nose' often formed after such injuries did not develop. It can be assumed, that the fracture was treated by a specialist, like a physician (for example, *medicus ordinarius* in the Roman army) or a surgeon (*capsarius*).

Injury of the temporomandibular joint

The medial corner of the left mandibular condyle was flattened intravitally and damaged. The almost oval lesion has a diffuse margin and a smooth and only slightly bulging surface (Fig. 20.14). At this point, the lateral pterygoid muscle inserts. The purpose of the muscle is to draw the mandible forwards and inwards, to the opposite side (cf. Schünke *et al.* 2009). Therefore, the mandible is shifted to the left and slightly forward, if the right muscle is shortened. A congenital anomaly cannot be excluded as a possible reason for such a peculiar morphology of the jaw joint, although it seems rather unlikely that such an anomaly would be precisely restricted to the insertion of the lateral pterygoid muscle. A more probable differential diagnosis is a trauma-induced strain of the muscle. It could result from a strong impact from front and side, probably from the opposite right side. Such an impact would shift the mandible in the opposite direction and could cause a strain of the muscle tendon. Apart from dental and periodontal diseases, the mandible still does not show any further pathological findings. If the anomaly of the jaw joint really resulted from an injury, it neither damaged the bone on the point of impact, nor caused a fracture of the mandible. Recent studies revealed that about one third of the impact power in direct mandible strikes is usually transferred to the skull base and occiput, and more than half of the impact power to the condyles (Tuchtan *et al.* 2015). Therefore, it is not necessary to observe any bone changes at the point of impact. A fall on the jaw (most common in bicycle accidents, cf. Hausamen *et al.* 2012; Niethard *et al.* 2009), or sporting accidents (e.g. football or boxing, cf. Fonseca *et al.* 2013) nowadays frequently leads to fractures of the mandible. An injury due to a punch seems possible. In this case, however, it did not leave recognisable traces on the point of impact, but only on the opposite side. A possible explanation could be that the man's mouth was open when he was hit. So the mandible itself could have been shifted to the side (cf. Fonseca *et al.* 2013). Therefore, the bone did not break, but only the muscle was damaged.

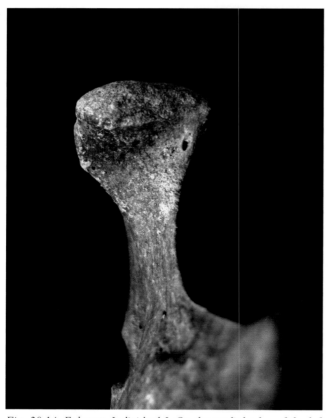

Fig. 20.14. Ephesus. Individual I. On the medial edge of the left mandibular condylar process, close to, but apparently aside and separated from the edge of the jaw joint, a thickening of the bone with an irregular, originally lacerated surface is visible. The probable reason for this morphology is a trauma of the lateral pterygoid muscle. It indicates a strain of this muscle (Photo K. Scheelen).

Inflammation of the scalp

The external lamina of the man's skull is diffuse-porous on both parietal bones (Fig. 20.15), as well as the occipital plate. The surface is slightly scarred and bulging. The porous areas show different stages of healing of an inflammatory process. Different openings of varying size and with rounded edges, represent blood vessel canals. They lend the surface its porous appearance. At the time of death, many of these vessel openings were obviously about to close, while others were still completely open. Also the smaller, bulging bumps, which developed during the scarring process, are partially already elapsed. These peculiar, porous and bulging parts are restricted to areas framed by lines of muscle insertions (e.g. temporal line, nuchal line, cf. Fig. 20.15). The described restriction of these changes proves that the structures did not grow out of the spongy bone, as in a weak, anaemic reaction (*cribra cranii externa* or 'hair-on-end' phenomenon, cf. Ortner 2003; Hollar 2001). They are rather vestiges of an external infection, which spread from the scalp and the adjacent epicraneal aponeurosis *(galea aponeurotica)* to the bone surface of the skull (cf. Schünke *et al.* 2009; Williams *et al.* 1995). In those parts additionally covered by other muscles (e.g. origin of the temporal muscle on the temporal line; origin of the neck muscles on the superior nuchal line), the surface is without pathological findings. An affliction of the scalp may first spread on the *galea aponeurotica* and, from there, on to the whole adjacent surface of the skull (cf. Schultz *et al.* 2001; 2003). Those parts of the skull without direct connection to the inflamed scalp stay unaffected. Different pathological processes might cause such severe, presumably chronic inflammations of the scalp. Any long-lasting skin affliction connected with intensive itching may be the reason. For example, it could have been caused

Fig. 20.15. Ephesus. Individual I. The irregularly porous, scarred, rough surface of the right parietal bone emerges from the sagittal suture to the temporal line (visible in the lower right corner of the photo). Therefore, the process, which caused the porous structure of the skull vault, must have originated from the scalp (inflammatory infection of the skin). The part of the skull vault below the temporal line, which was covered by the temporal muscle, was not affected (Photo K. Scheelen).

by parasites in the person's hair (as, for example, head lice) or directly on, in or under the skin (for example, dermatomycosis or parasites, such as *tinea capitis*; cf. Abeck 2014; Ely *et al.* 2014). Also, skin diseases (as, for example, psoriasis; cf. Korman *et al.* 2015) may lead to intensive itching of the scalp. Scratching the itching spots leads to open wounds, which may become inflamed in the case of poor hygiene (bacteria or other pathogens from dirt on skin and hair or under the nails). A similar result might be caused by long-term wearing of an air-impermeable headdress, like a helmet or thick cap (Schultz *et al.* 2001; 2003). Due to insufficient ventilation, they lead to increased sweating and, therefore, drying-out of the scalp, which becomes cracked and itches. In the present case, the different stages of healing of the pathological changes indicate a long-term inflammation of the *galea aponeurotica* and scalp. It must have persisted for a long time, until death. The precise cause of the inflammation cannot to be determined from the skeletal remains, although, certainly, the man suffered from a permanent itching, maybe painful scalp, which he often had to scratch.

PLEURISY

On the inner surface of the man's right ribs, from the 6th rib downwards in the region of the costal angle, vestiges of a healed pleurisy can be observed. The puckered, scarred surfaces with some partly integrated blood vessel impressions, indicate periostal inflammation of the intrathoracal, bony surfaces of the ribs healed a long time

before death (Fig. 20.16). This kind of inflammation is typically caused by pleurisies (cf. Schultz 2010; Schultz *et al.* 2001; 2003). The pleura consists of two layers (visceral pleura, adjacent to the lungs; parietal pleura, lines the thorax); separated only by a narrow, liquid-filled gap (cf. Williams *et al.* 1995). Pleurisy may develop from pneumonia. In that case, the infection first affects the inner and then the outer layer of the pleura (cf. Gerok *et al.* 2007). The result is an infection of the periosteum of the rib and the development of bony neoplasms on the inner surface of the rib. Typically, these neoplasms are restricted to the area of the rib surface, directly adjacent to the pleura. Those parts of the ribs covered by muscle insertions, as well as the costal grooves, mostly remain without changes (cf. Schultz 2010; Schultz *et al.* 2001; 2003). Pleurisy can be severe and also life-threatening for a patient, even nowadays, with modern medical care (cf. Fauci *et al.* 2008). In spite of the fact that pleurisies usually heal without consequences, they may lead to agglutinations of the pleural layers and thus, a reduction of the patient's lung volume (cf. Fauci *et al.* 2008). Certainly, the man suffered from severe pneumonia for at least several weeks. During that time, he must have been cared for and probably had some sort of medical treatment, particularly since antibiotics were not yet available in Roman times. Nevertheless, he must have had a strong physical constitution as well as a good immune system to sustain such a disease (cf. Thomas 2010). It is not determinable when exactly he suffered from the illness. The pathological changes are thoroughly integrated, and

Fig. 20.16. Ephesus. Individual I. The ninth right rib, detail of the middle of the internal (intrathoracal) surface. The surface above the costal groove is irregularly scarred, with various well integrated newly built bone structures, which indicate an inflammatory process of the adjacent external layer of the pleura. The costal groove is not affected, as it is not directly adjacent to the pleura. This inflammation must have originated from severe pleurisy, which had long healed when the man died (Photo K. Scheelen).

only small, compact bumps and a scarred surface remained from the originally porous neoplasms. Probably, the pleurisy healed many years before death. Certainly, it must have occurred after the age of about 20, because his thorax must have been fully grown. Otherwise, the growth would have led to a reduction of the pathological changes. As the man died in his late 30s or early 40s, he probably sustained pleurisy during his early adulthood, between the ages of about 20 and 30 years.

MENINGEAL REACTION

On the internal lamina of both parietal bones, a large area with atypically arranged blood vessel impressions bears witness to pathological process between the dura mater and inner surface of the skullcap, healed a long time before death. Aside from those blood vessel impressions, no further characteristics allow a more precise diagnosis. Therefore, only a haemorrhagic-inflammatory reaction of the dura mater, healed a long time before death, could be diagnosed. The most probable interpretation for the development of such atypical vessels is a haemorrhage (hematoma) of unknown origin. The affected area is large and diffuse. The changes rather indicate a widely spread, slow, long-term haemorrhage, such as, for example, oozing bleedings due to a small defect in the vascular wall of the middle meningeal artery (for example, in scurvy, Ortner 2003). Also, it could be interpreted as a result of inflammatory damage to the dura mater (for example, pachymeningitis, cf. Jacobson *et al.* 1996; Lieb *et al.* 1996), although trauma-induced bleeding cannot be excluded. Aside from the primary cause, an epidural bleeding in the area of the cerebrum may become life-threatening. Rising intracranial pressure may displace the brain. However, in spite of the extensive expansion of the haematoma, in the present case, the bleeding was survived with at worst temporary disorders, as the internal lamina only shows vestiges of the well-healed haematoma. It is not possible to determine whether the process had caused remote damages. For example, large and thick scars of the connective tissue could have caused reoccurring headaches (Matta *et al.* 2007) or convulsions (cf. Skierczynski *et al.* 2007). It cannot precisely be determined when the haemorrhagic process occurred; certainly not until the skull was fully grown. It probably happened during youth or young adult age, as later, the remodelling ability of the outer layer of the dura mater decreases (Williams *et al.* 1995). Although the ability never completely disappears, bone remodelling is increasingly slowed down in older people. Presumably, a minor bleeding would not have caused such wide-ranging changes. Furthermore, the haemorrhage must have occurred and healed at least some years before the man's death. Thus, it can be stated that it probably happened before the age of about 30 years.

EXOSTOSES OF THE EXTERNAL AUDITORY CANAL

During the investigation, two rounded, bony neoplasms were detected in the man's right external acoustic meatus (Fig. 20.17). The larger, regularly rounded structure is located on the anterior wall of the auditory canal, probably originally close to the front of the ear drum. On the posterior wall, closer to the exit of the bony canal, a second, smaller and rather cone-shaped structure is located. These exostoses are benign, bony neoplasms. With respect to appearance and structure, they resemble button osteoma. Clinical studies have revealed that, in recent cases, the differentiation from the latter is difficult (cf. Carbone and Nelson 2012) even by histopathological means. The two exostoses nearly completely obstructed the lumen of the external auditory canal with only a small passage left open. If the soft tissue of the auditory canal is added, it probably was completely closed. The probable result was a substantial loss of hearing. Due to insufficient aeration of the auditory canal, the exostoses frequently may have induced inflammations of the external ear (cf. Carbone and Nelson 2012; Goto *et al.* 2013). On the posterior wall of the left auditory canal, a smaller exostosis is located. As a slightly convex bump, it protrudes into the lumen of the auditory canal. In this case, the exostosis seems to be at an early stage. The left auditory canal was slightly narrowed, but at most connected with only minor hearing loss. Most commonly, such exostoses of the external auditory canal originate from long-term influence of cold water exposure. Depending on the study, their appearance correlates with the amount of years (ten years or more), but not with the daily frequency. Also

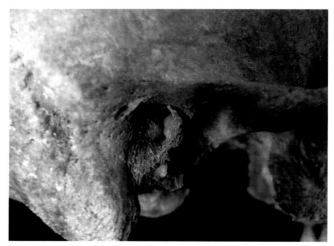

Fig. 20.17. Ephesus. Individual I. The right external auditory canal is almost completely obstructed due to two roundish newly built bone formations on the ventral and the dorsal side of the canal, about 6–8 mm deep. These structures are of compact composition, similar to small osteomas. Considering the soft tissue covering them, the canal must have been completely closed, causing a substantial loss of hearing in the right ear (Photo K. Scheelen).

genetic predispositions play a minor role in their genesis (cf. Manzi *et al.* 1991; Capasso *et al.* 1999). In modern medicine, these neoplasms are called 'surfer's ear', because they are particularly often found in surfers (Chaplin and Stewart 1998; Wong *et al.* 1999) and divers (Sheard and Doherty 2008). As part of two different studies, in more than 70% of the surfers examined, who practised surfing also at cold temperatures, these changes were observed (for example, in 73% of 62 surfers from New Zealand examined, Chaplin and Steward 1998; 73.5% of 307 Californian surfers, Wong *et al.* 1999). A similar prevalence was observed in free divers (87.7% of 101 free divers, Sheard and Doherty 2008). The observed severity of changes was comparable in all studies. More than half of the persons examined suffered from exostoses narrowing 50% or more of the external auditory canal (Chaplin and Steward 1998; Wong *et al.* 1999). As coastal winds mostly drift from one direction, the circumstances of evaporation in left and right ears are different. Many of the Californian surfers examined show an asymmetry of exostoses between the left and the right ear (King *et al.* 2010). Probably, the exostoses in the Ephesian man's ears should not be interpreted as evidence for the first surfing activities in the Mediterranean. Nevertheless, a profession as diver, for example, on the docks or the coast, connected with fishing or gathering sponges, seems possible (cf. Crowe *et al.* 2010). Differentially diagnostic, such exostoses frequently occur in long-term, habitual sauna users (Goto *et al.* 2013). In an archaeological context, external auditory exostoses were often observed in coastal habitants (c.f. Capasso 1987; Crowe *et al.* 2010; Özbek 2012). For the Imperial Roman coastal communities of Velia and Portus (1st – 3rd century AD), for example, isotope analysis revealed a correlation of the prevalence of external auditory exostoses with high nitrogen isotopic values. These circumstances point to marine food consumption and, therefore, probably fishing as responsible reason for the development of the exostoses (cf. Crowe *et al.* 2010). In case of the middle class Roman males from Portus (Isola Sacra), also the known, frequent use of baths and *thermae* was supposed to might have caused the high prevalence of external auditory exostoses (cf. Capasso *et al.* 1999; Manzi *et al.* 1991). Whether or not the man's auditory exostoses were caused by cold water diving and/or fishing, frequent use of the *thermae*, or another, unknown reason, it is certain that his right ear was almost deaf.

DENTAL PATHOLOGIES

The man's ante- and perimortal tooth loss might either have resulted from a trauma, or from other dental pathologies. The second and third left (37, 38) and the first right (46) mandibular molars were already missing long before his death. Only shortly before he died, he lost the first premolar and second and third molar from the right maxilla (14, 17, 18), as well as

the second left maxillary premolar (25). None of these teeth commonly tend to be lost due to trauma. In such cases, it is more likely that the front teeth, especially the incisors, would be involved (Hausamen *et al.* 2012). The man's entire dental status is extremely bad and indicates poor dental hygiene (see below). All tooth loss would probably have resulted from caries and periodontal diseases (cf. Alt *et al.* 1998). A definite diagnosis is not possible in this case. With regard to the condition of the teeth of the adult population from grave house 1/08 (cf. Nováček *et al.* in print), the man's abysmal dental status is neither surprising nor much different, also considering his comparatively advanced age. Besides the ante- and perimortal tooth loss (see above), some of the man's teeth were lost postmortally and, therefore, could not be investigated. From the 15 preserved teeth, 11 were located in the mandible and four in the maxilla. All these teeth were still good and not devitalized, but show vestiges of advanced abrasion (anterior teeth between degree 4 and 5, posterior teeth degree 5 to 5+; Brothwell 1981). In some cases, even the pulp was laid open due to abrasion and represented a potential target surface for caries. As a reaction, on some of the teeth, adventitious dentine had developed and sealed the pulp, so that no caries infection had occurred. However, such severe dental abrasion corresponds well to the usually observed wearing of the teeth of (pre-) historic humans of an advanced age (cf. Perizonius and Pot 1981; Brothwell 1981; Lovejoy 1985). Teeth 26, 27, 36, 47 and 48 show irregular tooth wear (cf. Fig. 20.18). Among them, teeth 26 and 47 are oblique abraded; mesial being much more pronounced than distal. Tooth 36 corresponds well to 26, as it

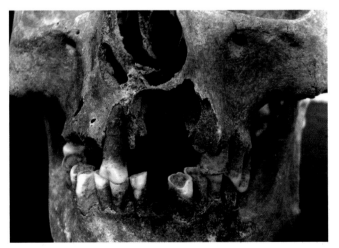

Fig. 20.18. Ephesus. Individual I. A very large abscess (degree V, 'cherry size'; Schultz 1988a) destroyed the alveoli of teeth 21 and 22 and was about to reach the alveolus of tooth 23 by the time of the man's death. Originally, it developed at tooth 21 or 22 and spread to tooth 23, as well as through the hard palate into the oral cavity (cf. Fig. 20.5). The abscess almost reached the nasal cavity, but was still separated by a thin layer of bone tissue, still intact. On the left side of the dental arch, the irregular wear of the molars is clearly visible (Photo K. Scheelen).

is mesial and buccal strongly abraded, but distal and lingual only to a minor degree. Teeth 27 and 48 are hardly abraded and clearly elongated. They protrude out of the dental alignment. This anomaly is based on the biomechanics of the denture (cf. Alt *et al.* 1998). Obviously, the corresponding teeth from the opposite side (17, 18, 37, and 38) were already missing long before death. Therefore, those teeth still present at the edge of the dental arch (26, 36 and 47) were irregularly strained. Both non-abraded teeth (27 and 48) were elongated due to the lack of counter pressure from the lost antagonists. Obviously, the Ephesian man had already lost several teeth during young adulthood, as the abrasion of teeth 27 and 48 correlates with an age of at the most 25–30 years (Perizonius and Pot 1981; Brothwell 1981; Lovejoy 1985). Since then, the antagonists must have been missing. The loss of teeth 17, 18, 37 and 38 could have been due to caries or abscess. Regarding the man's rather young age at that time, also another possible reason (for example, trauma) cannot be excluded.

All alveolar edges, which are not atrophied or closed due to tooth loss, show vestiges of periodontal disease. In most cases, the degree is weak or moderate (degree 1–2; Schultz 1988a), although some alveoli present an advanced degree (teeth 26, 27, 33 and 45 degree 3; Schultz 1988a) which might have threatened the dental stability. Periodontitis is an inflammation of the gum. It often develops due to poor oral hygiene and spreads to the adjacent bone tissue. The bone tissue pathologically changes, in the context of a periostitis, and is typically reduced irregularly and is jaggy. This means that, after some time, the teeth are no longer properly attached to the alveoli and can be lost (Schultz 1988a). Also, such inflammations may be connected to the development of dental calculus, gingival pockets, or dental abscesses.

Dental calculus develops from accretion, food particles and bacteria in the mouth cavity, influenced by saliva. Therefore, dental calculus is frequently observed near the openings of the great salivary gland, for example, on the inner surface of the lower incisors (sublingual and submandibular gland) or the outer surface of the maxillary molars (parotid gland; cf. Pfeifer and Müller-Hermelink 2003). If dental calculus is not removed for a longer period of time, it may form on any teeth, depending on the individual saliva composition and concentration, the food, the oral hygiene and further endo- and exogenous influences. Thus, it is not surprising that all the teeth of the man present show at least weak traces of dental calculus (mostly degree 1; Schultz 1988a). Considering the evidence of a rather rudimentary oral hygiene, which supposedly did not include frequent removal of dental calculus, he obviously naturally did not tend to strongly develop it.

Also the development of gingival pockets is closely connected to oral hygiene, condition of the gum and dental calculus formation. Gingival pockets can be observed as bulges between the root of the tooth and the alveolus. In these pockets, bacteria and food particles, especially from meat, are deposited (Pfeifer and Müller-Hermelink 2003;

Staufenbiel *et al.* 2013). Gingival pockets may also develop due to extreme formation of dental calculus, which shifts the gum aside and may cause a pressure atrophy-induced reduction of the alveoli. Especially large gingival pockets, which expand to the root apex, may not only lead to the loss of the tooth concerned, but may also cause an inflammation or apical abscess. In the Ephesian man, gingival pockets were particularly frequently observed. Teeth 13, 24, 25, 28, 36, 45 and 47 present gingival pockets of different sizes. Most of them led to the formation of an abscess. As the teeth are only affected from dental calculus to a minor degree, presumably the gingival pockets are connected with the periodontitis. Both pieces of evidence may indicate frequent meat consumption (cf. Staufenbiel *et al.* 2013).

The man's jaws demonstrated many, in some cases very large abscesses. Particularly large are the abscesses of teeth 12–13 (degree III 'pea-size'; Schultz 1988a) and teeth 21–22 (degree V 'cherry-size'; Schultz 1988a, Fig. 20.18), in the frontal part of the maxilla. The abscess of teeth 21–22 is so large that it spread to several alveoli. It cannot be determined from which tooth it originated. As a possible differential diagnosis, a large nasopalatinal duct cyst, cannot be excluded (cf. Bains *et al.* 2016), although, in a typical case, the nasopalatine duct should be involved (Wu *et al.* 2015). The alveolar walls are completely reduced and atrophied, and the lesion connects the oral cavity to the buccal cavity. Certainly, it caused severe symptoms and pain. Purulent exudates oozed into the oral cavity (cf. Fig. 20.5), accompanied by a purulent taste and smell and ultimate loss of the affected teeth. The inflammation-induced reduction of the maxillary alveolar process almost extends to the apertura piriformis and the nasal cavity. However, the persisting bone layer had not been perforated before death. Nevertheless, the mucous membranes of the oral and possibly also the nasal cavity must have been severely affected, as the inflammation could have spread into these structures through the already laterally perforated incisor canal. Supposedly, the man also suffered from a dentally induced stomatitis and perhaps also a purulent rhinitis. Obviously, the abscess perforated the mucous membranes of the oral cavity and the incisor canal. The presumably connected spread of inflammation must have happened only shortly before death, because the bony surfaces do not show detectable changes which would indicate a long-term, chronic inflammation. The perforation of the hard palate must have occurred earlier, as the porous and slightly bulging surfaces of the edges of the defect and the adjacent hard palate indicate a periostitis. The abscess from alveolus 13 obviously developed from a gingival pocket and spread to alveolus 12. Through the hard palate, it perforated into the oral cavity, probably at about the same time as the abscess of teeth 21–22, because the edges of the defect show similar changes. Furthermore, another three teeth (28, 36 and 45) were affected by abscesses (degree I 'sesame seed-size'; Schultz 1988a) which probably all developed from gingival pockets.

Only the first left lower molar (36) was affected by a small, carious lesion, detectable as a punctiform defect of the enamel (degree 1; Schultz 1988a). Probably, it had not yet caused any pain. Regarding the generally bad state of dentition, this is rather surprising. Differential diagnostic, such little prevalence of caries could imply that the man's nutrition only contained few carbohydrates (e.g. flour, sugar, starches; cf. Staufenbiel *et al.* 2013), or that his enamel was genetically determined strong and resistant. Otherwise, all carious teeth could have, by chance, been lost postmortally, so that a precise diagnosis cannot be established. The devitalized premolar 24 could also have resulted from caries. Due to carious lesions or traumatic damage, devitalized teeth lose their crowns. Only a dead snag of a tooth root remains, at best with parts of the hollowed neck of the tooth. In a premolar, a severe caries infection would not be surprising. In contrast, the trauma-induced loss of a premolar would be no common diagnosis, as the necessary impact power usually exceeds the power of a punch or even a kick. As both neighbouring teeth were lost postmortally, the diagnosis cannot be specified more precisely.

Concluding summary

This paper presents the osteobiography of individual I from burial chamber 3 in grave house 1/08 of the Roman Harbour Necropolis from Ephesus. The skeletal evidence shows that this man died in his late thirties or early forties which means that he reached a slightly above-average age compared to other men buried in the same grave house. Certain morphological traits might indicate a genetic relationship with other individuals from the same burial chamber. Judging from his skeleton, he was seemingly rather short (at most slightly more than 1.6 m tall) and sturdily built, with short, muscular limbs and a wide trunk and chest. From his childhood on, he apparently endured intense physical strain, resulting in generally extremely strong muscle marks on his bones. Moreover, probably due to physical overload during childhood and youth, his left acromial process did not fuse with the shoulder blade. This morphological trait often occurs in relation to intense physical training during youth, for example, connected to archery or athletics. Transversal enamel hypoplasia on several teeth bear witness to health disturbances or strain, such as, for example, severe infections or malnutrition, during his early childhood. In late youth or early adulthood, he survived potentially life-threatening intracranial bleeding, as well as pleurisy, presumably connected to severe pneumonia.

The man's teeth and gums were in an abysmal state. Aside from severe gum disease, gingival pockets and missing teeth, he suffered from several huge abscesses. Astonishingly, only one molar had a small carious lesion. The pathological profile of his dental status could indicate that the man's nutrition did not contain a high share of carbohydrates, but rather more often meat and/or fibrous vegetables or fruits.

On the outer surface of the skull, traces of a chronic inflammation were detected. Such inflammation might either be related to parasitic infestation of the hair and scalp, or to habitual wearing of airtight headgear.

Especially on his right side, the man's external auditory canals were narrowed by bony neoplasms. On his right side, these auditory canal exostosis were so pronounced that the man probably was almost deaf on his right ear. In modern medicine, these neoplasms are referred to as 'surfer's ear', as they most commonly occur in cold-water surfers or divers. Hypothetically, an occupation connected to fishing or diving seems possible. In such case, the particularly strongly developed muscle marks of his upper limbs might be connected to rowing or throwing nets. Nevertheless, also a connection of auditory exostoses to cold-water bathing, for example after the sauna, is known, and seems well possible in late Roman Ephesus.

Both joints of the extremities, as well as parts of the vertebral column, are arthrotically and arthritically changed. Moreover, the man suffered a variety of injuries during his life. Many bones show traces of pulled muscles or ligaments, as well as bleeding or more minor injuries. With regard to muscle injuries, the upper limbs were rather more often affected than the lower ones.

The skeleton also shows a number of more severe injuries. The right ankle was traumatically damaged and disarticulated, possibly resulting in lowered resilience and resistance of the foot. In addition to the observed pathologies connected to arthrosis, the lumbar vertebra show vestiges of multiple Schmorl's nodes. Two or three times, the man sustained fractures of one or several ribs. At least in one case, the fracture could have been caused by a punch to the kidney. The head joints were also repeatedly injured. The morphology of these injuries allows an interpretation that they were a result from wrestling or pankration matches or similar close-combat situation. The injury to the mastication muscles on the mandible might have been caused by a blow to the jaw. Also the broken nose could have resulted from a punch. Subsequently, it was straightened and healed in only a slightly crooked position. Many of the observed vestiges of injuries probably result from interpersonal violence, possibly some kind of wrestling-like fights.

Although the osteobiography does not cover the whole 'patient record' of the man presented here, it illuminates many aspects of his life, and his medical history. The interpretation of these diagnostic findings does not claim to be of ultimate accuracy. Still, the methodological comparison with modern forensics allows a choice of the most probable explanation. In this way, the man, originally only known as individual I from grave chamber 3, steps forward from the anonymous crowd of late Roman Ephesus. The analysis of his palaeopathological data lets him tell us, in his own way, something about his personal story and medical history.

Acknowledgements

First of all, we would like to thank the Head Office for Antiquities and Museums of the Republic of Turkey in Ankara for the research permit. Many thanks to Sabine Ladstätter, director of the Austrian Archaeological Institute and head of the excavation at Ephesus, as well as Martin Steskal, head of the Harbour Necropolis research project, for the possibility to investigate the human skeletal remains from grave house 1/08 in Ephesus. Furthermore, we would like to thank Cyrilla Maelicke for her help with the English manuscript. Last, but not least, we thank the editors for the possibility to publish our results in this volume, as well as the anonymous peer reviewer for the helpful, constructive comments on our text.

Bibliography

Abeck, D. (2014) Frequent pathogen-induced diseases of the scalp. *Hautarzt* 65 (12), 1050–5.

Adler, C. P. (2005) *Knochenkrankheiten. Diagnostik makroskopischer, histologischer und radiologischer Strukturveränderungen des Skeletts* (3rd ed.). Berlin, Heidelberg, and New York, Springer.

Alfonso-Durruty, M. P. (2011) Experimental assessment of nutrition and bone growth's velocity effects on Harris-lines formation. *American Journal of Physical Anthropology* 145, 169–80.

Alt, K. W., Rösing, F. W., and Teschler-Nicola, M. (eds.) (1998) *Dental anthropology: Fundamentals, limits, and prospects.* Vienna and New York, Springer.

Aufderheide, A. C. and Rodríguez-Martín, C. (1998) *The Cambridge encyclopedia of human paleopathology.* Cambridge, Cambridge University Press.

Asuming, R. and Schultz, M. (2000) Häufigkeit und Intensität transversaler Schmelzhypoplasien und ihre Korrelation mit den HARRIS-Linien dargestellt an der frühmittelalterlichen Population von Barbing-Kreuzhof. In *Schnittstelle Mensch – Umwelt in Vergangenheit, Gegenwart und Zukunft, Proceedings, 3. Kongress der Gesellschaft für Anthropologie (GfA).* Göttingen, Cuvillier, 263–7.

Babar, S. M. A. (1999) *Neck injuries.* London, Berlin, Heidelberg, New York, Barcelona, Hong Kong, Milan, Paris, Santa Clara, Singapore, and Tokyo, Springer.

Bains, R., Verma, P., Chandra, A., Tikku, A. P., and Singh, N. (2016) Nasopalatine duct cyst mimicking an endodontic periapical lesion: A case report. *General Dentistry* 64.1, 63–6.

Blau, S. and Ubelaker, D. H. (2011) *Handbook of forensic anthropology and archaeology* (World Archaeological Congress Research Handbooks in Archaeology). Walnut Creek, Left Coast Press.

Brothwell, D. R. (1981) *Digging up bones. The excavation, treatment and study of human skeletal remains* (3rd ed.). Ithaca (NY), British Museum (Natural History), Cornell University Press.

Capasso, L. (1987) Exostoses of the auditory meatus in pre-Columbian Peruvians. *Journal of Palaeopathology* 1.3, 113–6.

Capasso, L., Kennedy, K. A. R., and Wilzcak, C. A. (1999) *Atlas of occupational markers on human remains.* Teramo, Edigrafital.

Carbone, P. N. and Nelson, B. L. (2012) External auditory osteoma. *Head and Neck Pathology* 6.2, 244–6.

Chaplin, J. M. and Stewart, I. A. (1998) The prevalence of exostoses in the external auditory meatus of surfers. *Clinical Otolaryngology and Allied Sciences* 23.4, 326–30.

Christensen, A. M., Passalacqua, N. V., and Bartelink, E. J. (2014) *Forensic anthropology. Current methods and practice.* Amsterdam, Boston, Heidelberg, New York, Oxford, Paris, San Diego, San Francisco, Singapore, Sydney, and Tokyo, Academic Press, Elsevier.

Crowe, F., Sperduti, A., O'Connell, T. C., Craig, O. E., Kirsanow, K., Germoni, P., Macchiarelli, R., Garnsey, P., and Bondioli, L. (2010) Water-related occupations and diet in two Roman coastal communities (Italy, 1st century AD): Correlation between stable carbon and nitrogen isotope values and auricular exostosis prevalence. *American Journal of Physical Anthropology* 142, 355–66.

Dar, G., Peleg, S., Masharawi, Y., Steinberg, N., May, H., and Hershkovitz, I. (2009) Demographic aspects of Schmorl nodes: A skeletal study. *Spine* 34.9, 312–5.

Dar, G., Masharawi, Y., Peleg, S., Steinberg, N., May, H., Medlej, B., Peled, N., and Hershkovitz, I. (2010) Schmorl's nodes distribution in the human spine and its possible etiology. *European Spine Journal* 19.4, 670–5.

Ely, J. W., Rosenfeld, S., and Seabury Stone, M. (2014) Diagnosis and management of tinea infections. *American Family Physician* 90.10, 702–10.

Fauci, A. S., Braunwald, E., Kasper, D. L., Hauser, S. L., Longo, D. L., Jameson, J. L., and Loscalzo, J. (2008) *Harrison's principles of internal medicine* (17th ed.). New York, McGraw-Hill.

Ferembach, D., Schwidetzky, I., and Stloukal, M. (1980) Recommendations for age and sex diagnoses of skeletons. *Journal of Human Evolution* 9, 517–49.

Fonseca, R. J., Walker, R. V., Barber, H. D., Powers, M. P., and Frost, D. E. (2013) *Oral and maxillofacial trauma* (4th ed.). Saint Louis (MI), Sounders, Elsevier.

Gerok, W., Huber, C., Meinertz, T., Zeidler, H. (eds.) (2007) *Die innere Medizin – Referenzwerk für den Facharzt.* Stuttgart, Schattauer.

Goto, T., Tono, T., Nakanishi, H., Matsuda, K., Ganaha, A., and Suzuki, M. (2013) Three cases of external auditory exostoses in a habitual sauna user. *Nihon Jibiinkoka Gakkai Kaiho* 116.11, 1214–9.

Hausamen, J.-E., Machtens, E., Reuther, J. F., Eufinger, H., Kübler, A., and Schliephake, H. (2012) *Mund-, Kiefer- und Gesichtschirurgie* (4th ed.). Berlin, Heidelberg, and New York, Springer.

Hershkovitz, J., Bedford, L., Jellema, L., and Lattimer, B. (1996) Injuries to the skeleton due to prolonged activity in 'hand-to-hand combat'. *International Journal of Osteoarchaeology* 6, 167–78.

Hollar, M. A. (2001) Hair on end sign. *Radiology* 221.2, 347–8.

Hoppes, D. A., Olson, K., and Jenkinson, S. A. (1991) Renal response to boxing: An investigation of changes in the urine in amateur boxers. *Journal of the American Osteopathic Association* 91.5, 461–4.

Hummel, S. and Schutkowski, H. (1993) Approaches to the histological age determination of cremated human remains. In G. Grupe and N. Garland (eds.) *Histology of ancient human bone: Methods and diagnosis. Proceedings of the paleohistology workshop from 3–5 October 1990 at Goettingen,* 111–23. Berlin, New York, and Tokyo, Springer.

İşcan, M. Y. and Steyn, M. (2013) *The human skeleton in forensic medicine* (3rd ed.). Springfield (IL), Charles C. Thomas Publisher Ltd.

Jacobson, D. M., Anderson, D. R., Rupp, G. M., and Warner, J. J. (1996) Idiopathic, hypertrophic, cranial pachymeningitis: Clinical-radiological-pathological correlation of bone involvement. *Journal of Neuro-Ophthalmology* 1996.4, 264–8.

Kerley, E. R. (1965) The microscopic determination of age in human bone. *American Journal of Physical Anthropology* 23, 149–63.

Kerley, E. R. and Ubelaker, D. H. (1978) Revisions in the microscopic method of estimation age at death in human cortical bone. *American Journal of Physical Anthropology* 49, 545–6.

King, J. F., Kinney, A. C., Iacobellis, S. F. 2nd, Alexander, T. H., Harris, J. P., Torre, P. 3rd, Doherty, J. K., and Nguyen, Q. T. (2010) Laterality of exostosis in surfers due to evaporative cooling effect. *Otology and Neurotology* 31.2, 345–51.

Korman, N. J., Zhao, Y., Li, Y., Liao, M., and Tran, M. H. (2015) Clinical symptoms and self-reported disease severity among patients with psoriasis – Implications for psoriasis management. *Journal of Dermatological Treatment* April 2015, 1–6.

Kretschmer, E. (1977) *Körperbau und Charakter: Untersuchungen zum Konstitutionsproblem und zur Lehre von den Temperamenten* (26th ed.). Berlin, Springer.

Kreutz, K. and Verhoff, M. (2002) *Forensische Anthropologie.* Berlin, Lehmanns LOB.

Lieb, G., Krauss, J., Kollmenn, H., Schrod, L., and Sörensen, N. (1996) Recurrent bacterial meningitis. *European Journal of Pediatrics* 155.1, 26–30.

Lovejoy, C. O. (1985) Dental wear in the Libben population: Its pattern and role in the determination of adult skeletal age at death. *American Journal of Physical Anthropology* 68, 1, 47–56.

Maggiano, I. S., Schultz, M., Kierdorf, H., Sierra Sosa, T., Maggiano, C. M., and Tiesler Blos, V. (2008) Cross-sectional analysis of long bones, occupational activities and long-distance trade of the Classic Maya from Xcambó – archaeological and osteological evidence. *American Journal of Physical Anthropology* 136, 470–7.

Mann, R. W. and Hunt, D. R. (2013) Photographic regional atlas of bone disease: A guide to pathologic and normal variation in the human skeleton (3rd ed.). Springfield (IL), Charles C. Thomas Publisher LTD.

Manzi, G., Sperduti, A., and Passarello, P. (1991) Behavior-induced auditory exostosis in imperial Roman society: Evidence from coeval urban and rural communities near Rome. *American Journal of Physical Anthropology* 85, 253–60.

Matos, V. (2009) Broken ribs: Paleopathological analysis of costal fractures in the human identified skeletal collection from the Museu Bocage, Lisbon, Portugal (late 19th to middle 20th centuries). *American Journal of Physical Anthropology* 140, 25–38.

Matta, A. P., Ribas, M. C., Morreira-Filho, P. F. (2007) Postmeningitis headache: Case report. *Arquivos de Neuro-Psiquiatria* 65.2B, 521–3.

Myung-Sang, M., Jeong-Lim, M., Doo-Hoon, S., and Young-Wan, M. (2006) Treatment of dens fracture in adults. A report of thirty-two cases. *Bulletin of the Hospital for Joint Disease* 63, 108–12.

Niethard, F. U., Pfeil, J., and Biberthaler, P. (2009) *Orthopädie und Unfallchirurgie. Duale Reihe* (6th ed.). Stuttgart, Thieme.

Nováček, J. (2012) *Möglichkeiten und Grenzen der mikroskopischen Leichenbranduntersuchung.* Diss. rer. nat. University of Hildesheim.

Nováček, J., Scheelen, K., and Schultz, M. (forthcoming) *Cives metropolis asiae. Die anthropologische und paläopathologische Untersuchung spätantiker Kollektivgräber aus Grabhaus 1/08 in der Hafennekropole von Ephesos.*

Ortner, D. J. (ed.) (2003) *Identification of pathological conditions in human skeletal remains.* Amsterdam, Boston, London, New York, Oxford, Paris, San Diego, San Francisco, Singapore, Sydney, and Tokyo, Academic Press, Elsevier.

Özbek, M. (2012) Auditory exostoses among the Prepottery Neolithic inhabitants of Çayönü and Aşıklı, Anatolia; its relation to aquatic activities. *International Journal of Paleopathology* 2, 181–6.

Pagnani, M. J., Mathis, C. E., and Solman, C. G. (2006) Painful os acromiale (or unfused acromialapophysis) in athletes. *Journal of Shoulder and Elbow Surgery* 2006.4, 432–5.

Perizonius, W. R. K. and Pot, T. (1981) Diachronic dental research on human skeletal remains excavated in the Netherlands in Dorestad's cemetery on 'the Heul'. *Berichten van de Rijksdienst voor het Oudheidkundig Bodermonderzoek* 31, 369–413.

Pfeifer, U. and Müller-Hermelink, H. K. (eds.) (2003) *Grundmann Pathologie für Zahnmediziner* (2nd ed.), München, Stuttgart, and Jena, Urban & Fischer.

Poliakoff, M. B. (2004) *Kampfsport in der Antike: Das Spiel um Leben und Tod.* Düsseldorf, Verlag Patmos.

Richards, P. J. (2005) Cervical spine clearance: A review. *Injury – International Journal of the Care of the Injured* 36, 248–69.

Rösing, F., Graw, M., Marre, M., Ritz-Timme, S., Rothschild, M. A., Roetzscher, K., Schmeling, A., Schroeder, I., and Geserick, G. (2007) Recommendations for the forensic diagnosis from sex and age from skeletons. *Homo Journal of Comparative Human Biology* 58, 75–89.

Rubinstein, D., Escott, E. J., and Mestek, M. (1996) Computed tomographic scans of minimally displaced type II odontoid fractures. *Journal of Trauma and Acute Care Surgery* 40, 204–10.

Rudolph, W. (1965) *Olympischer Kampfsport in der Antike: Faustkampf, Ringkampf und Pankration in den Griechischen Nationalfestspielen.* Berlin, Akademie-Verlag.

Scheuer, J. L. and Black, S. (2000) *Developmental juvenile osteology.* London, Academic Press.

Schlecht, S. H., Pinto, D. C., Agnew, A. M., Stout, S. D. (2012) Brief communication: The effects of disuse on the mechanical properties of bone: What unloading tells us about the adaptive nature of skeletal tissue. *American Journal of Physical Anthropology* 149, 599–605.

Schmid, L. (1970) Kidney lesions in boxing. *Journal of Sports Medicine and Physical Fitness* 10.4, 265–8.

Schultz, M. (1988a) Paläopathologische Diagnostik. In R. Knussmann (ed.) *Anthropologie. Handbuch der vergleichenden Biologie des Menschen 1, 1,* 480–96. Stuttgart, G. Fischer.

Schultz, M. (1988b) Methoden der Licht- und Elektronenmikroskopie. In R. Knussmann (ed.) *Anthropologie. Handbuch der vergleichenden Biologie des Menschen 1, 1,* 698–730. Stuttgart, G. Fischer.

Schultz, M. (1997) Microscopic investigation of excavated skeletal remains: A contribution to paleopathology and forensic medicine. In W. D. Haglund and M. H. Sorg (eds.) *Forensic taphonomy: The post-mortem fate of human remains,* 201–26. Danvers, CRC Press.

Schultz, M. (2001) Paleohistopathology of bone: A new approach to the study of ancient diseases. *Yearbook of Physical Anthropology* 44, 106–47.

Schultz, M. (2003) Light microscopic analysis in skeletal paleopathology. In Ortner (ed.), 73–108.

Schultz, M. (2010) The biography of the wife of Kahai: A biological reconstruction. In A. Woods, A. McFarlane, and S. Binder (eds.) *Egyptian culture and society. Studies in honour of Naguib Kanawati* (Supplément aux Annales du Service des Antiquités de l'Égypte 38, Vol. 2), 163–79. Cairo, Publications du Conseil Suprême des Antiquités de l'Égypte.

Schultz, M. (2011) Paläobiographik. In G. Jüttemann (ed.) *Biographische Diagnostik*, 222–36. Lengerich, Berlin, and Bremen, Pabst Science Publishers.

Schultz, M., Carli-Thiele, P., Schmidt-Schultz, T. H., Kierdorf, U., Kierdorf, H., Teegen, W. R, and Kreutz, K. (1998) Enamel hypoplasias in archaeological skeletal remains. In Alt, Rösing, and Teschler-Nicola (eds.), 293–311.

Schultz, M. and Schmidt-Schultz, T. H. (2004) 'Der Bogenschütze von Pergamon' – Die paläopathologisch-biographische Rekonstruktion einer interessanten spätbyzantinischen Bestattung. *Istanbuler Mitteilungen* 54, 243–56.

Schultz, M. and Walker, R. (2013) Report on the mummy of Djau, governor of Upper Egyptian Provinces 8 and 12 (6th Dynasty). In N. Kanawati (ed.) *Deir-el Gebrawi. Volume III. The southern cliff. The tomb of Djau/Shemai and Djau* (The Australian Center for Egyptology: Reports 32), 64–78. Oxford, Aris and Phillips Ltd.

Schultz, M., Walker, R., Strouhal, E., and Schmidt-Schultz, T. H. (2001) Skeletal remains. II Merinebti, Hefi and Iries. In N. Kanawati and M. Abder-Raziq (eds.), *The Teti cemetery at Saqqara, Volume VII. The tombs of Shepsipuptah, Meriri (Merinebti), Hefi and others* (The Australian Centre for Egyptology: Reports 17), 65–74. Warminster, Aris and Phillips Ltd.

Schultz, M., Walker, R., Strouhal, E., and Schmidt-Schultz, T. H. (2003) Report on the skeleton of *Jj-nfrt* from his mastaba in the North Cemetery of Unis's Pyramid (5th Dynasty). In N. Kanawati and M. Abder-Raziq (eds.) *The Unis cemetery at Saqqara, Volume II. The tombs of Iynefert and Ihy (reused by Idut)* (The Australian Centre for Egyptology: Reports 19), 755–86. Oxford, Aris and Phillips Ltd.

Schünke, M., Schulte, E., and Schumacher, U. (2009) Prometheus. *LernAtlas der Anatomie. Kopf, Hals und Neuroanatomie* (2nd ed.). Stuttgart and New York, Thieme Verlag.

Sheard, P. W. and Doherty, M. (2008) Prevalence and severity of external auditory exostoses in breath-hold divers. *Journal of Laryngology and Otology* 122.11, 1162–7.

Sheldon, W. H., Smith Stevens, S., and Tucker, W. B. (1940) *The varieties of human physique – an introduction to constitutional psychology.* New York, Harper.

Simon, R. R. and Sherman, S. C. (2011) *Emergency orthopedics* (6th ed.) New York, Chicago, San Francisco, Lisbon, London, Madrid, Mexico-City, Milan, New Deli, San Juan, Seoul, Singapore, Sydney, and Toronto, McGraw-Hill.

Skierczynski, P. A., Goodman, J. M., Signal, P., Payner, T. D., and Bonnin, J. M. (2007) Idiopathic hypertrophic pachymeningitis resulting in delayed panhypopituitarism. *Endocrine Practice* 13.5, 481–6.

Skinner, M. F. and Goodman, A. H. (1992) Anthropological uses of developmental defects of enamel. In S. R. Saunders and A. Katzenberg (eds.) *Skeletal biology of past peoples: Research methods*, 153–74. New York, Wiley-Liss.

Sládek, V., Berner, M., Sosna, D, and Sailer, R. (2007) Human manipulative behavior in the Central European Late Neolithic and Early Bronze Age: Humeral bilateral asymmetry. *American Journal of Physical Anthropology* 133, 669–81.

Staufenbiel, I., Weinspach, K., Förster, G., Geurtsen, W., and Günay, H. (2013) Periodontal conditions in vegetarians: A clinical study. *European Journal of Clinical Nutrition* 67.8, 836–40.

Sterling, J. C., Meyers, M. C., Chesshir, W., and Calvo, R. D. (1995) Os acromiale in a baseball catcher. *Medicine and Science in Sports and Exercise* 1995.6, 795–9.

Stirland, A. (1984) A possible correlation between os acromiale and occupation in the burials of the Mary Rose. In V. Capecchi and E. Rabino Massa (eds.) *Proceedings of the 5th European meeting of the paleopathology association in Siena.* Siena, Tipografia Sienese, 327–34.

Stirland, A. J. (2001) *Raising the dead: The skeleton crew of King Henry VII's great ship the Mary Rose.* Chichester, John Wiley and sons Ltd.

Thomas, C. (ed.) (2010) *Atlas der Infektionskrankheiten.* Stuttgart, Schattauer.

Trepel, M. (2012) *Neuroanatomie. Struktur und Funktion* (5th ed.). Munich, Urban & Fischer.

Trotter, M. and Gleser, G. C. (1952) Estimation of stature from long bones of American whites and negroes. *American Journal of Physical Anthropology* 10, 463–514.

Trotter, M. and Gleser, G. C. (1977) Corrigenda to 'Estimation of stature from long bones of American whites and negroes'. *American Journal of Physical Anthropology* 47, 355–6.

Tuchtan, L., Piercecchi-Marti, M.-D., Bartoli, C., Boisclair, D., Adalian, P., Léonetti, G., Behr, M., and Thollon, L. (2015) Forces transmission to the skull in case of mandibular impact. *Forensic Science International* 252, 22–8.

Ubelaker, D. A. (1989) *Human skeletal remains. Excavation, analysis, interpretation* (2nd edition). Washington D.C., Taraxacum.

Villotte, S., Castex, D., Couallier, V., Dutour, O., Knüsel, C. J., and Henry-Gambier, D. (2010) Enthesopathies as occupational stress markers: Evidence from the upper limb. *American Journal of Physical Anthropology* 142, 224–34.

Williams, P. L., Lawrence, H., and Bennister, P. L. 1995 *Gray's anatomy* (38th ed.). Edinburgh, Churchill Livingstone, Elsevier.

Wong, B. J., Cervantes, W., Doyle, K. J., Karamzadeh, A. M., Boys, P., Brauel, G., and Mushtaq, E. (1999) Prevalence of external auditory canal exostoses in surfers. *Archives of Otolaryngology – Head and Neck Surgery* 125.9, 969–72.

Wu, Y. H., Wang, Y. P., Kok, S. H., and Chang, J. Y. (2015) Unilateral nasopalatine duct cyst. *Journal of the Formosan Medical Association* 114.11, 1142–4.

Yammine, K. (2014) The prevalence of os acromiale: A systematic review and meta-analysis. *Clinical Anatomy* 2014.4, 610–21.

https://unitedworldwrestling.org/sites/default/files/1-wrestling_rules_july_2014_eng.pdf (consulted 22.07.2016).

Index

Index words in captions are included, but not words in the Figures themselves.

Many entries are collected under a common denomination, as, for example, age groups, animals, artefacts, family, names, palaeoanthropology, palaeopathology, pottery, professions, stone, tomb, etc.

See also refers to all the main entries; *see also here* refers to sub-entries under the actual main entry.

constitution 121 (family), 209 (doctrine), 331 (physical)

construction technique xix, 12, 39, 48–54, 153, 171

consumer 240, 245, 264

consumption, consume xx–xxi, 230, 234–5, 237–8, 241, 243–4, 246–7, 264, 333–4

container for the dead; *see also* funerary architecture; tomb xix, 69–71, 77, 79–81, 157, 204, 238

contamination (of DNA) 220–3, 230 (contaminants)

continuity/continuum xix, 12, 16 n. 17, 86 (cultural), 154, 204, 212

conversion 207, 228

cooking 238

Corinth xvii, 205, 282

Corpus Iuris Civilis. Digesta 205–6

correspondence factor analysis 77

corridor; *see also* access; anteroom; *anticella*; door; *dromos*; entrance 28

Corsica 286

Çorum 288

cosmological 207

cost; *see also* price xix, 51, 133–5, 140–1, 145 nn. 13+16,

costume 280

council 14, 87, 88, 139

countryside xix–xx, 151, 198, 211

crafts/craftsmanship xviii–xix, 103, 105, 111, 116, 160

craftsman; *see* professions

cremation; *see also* inhumation xix, 70–1, 77, 82, 154–5, 157, 159–61, 167, 169, 173, 196, 206–7, 209, 212, 212–3 nn. 4–5, 238, 250, 252, 254, 256, 268, 271–2, 281

crepidoma; *see also* tomb: tumulus 21–2, 25–9, 31–3, 36, 62–3, 66–7 nn. 2+23, 74

Crete 173, 279, 282

crime/criminal act 81, 87–8

crisis 205–6, 309

crops xx

cross 170 (pendant), 194, 208, 219, 229, 269 (pendant)

crypt 223

Çukurbag 21–23, 36

cult 6–9, 11–12, 16, 87–9, 106, 139, 145 n. 23, 207
  Christian 14, 16 n. 14
  of martyrs 3, 7, 9, 16
  of saints 3, 16

cultivation/cultivate xix, 56, 134, 150, 171, 235

cura aquarum 8

curse 88, 151, 173

Çürüksu 19

custom; *see also* burial; funerary; practice xvi, 64, 66, 74, 79–80, 85–6, 106, 121, 144, 167, 205–9, 212

Cyaneae 129 nn. 15+16

Dadastana 168, 170

Dağmarmara 27

Dalawa, *see* Tlawa/Tlos

damage 12, 27–30, 34–37, 44, 59, 79, 87, 110, 192, 220, 325–6, 328–30, 332, 335

Danishmend 287

Danubian region 167

dead/the dead xviii–xix, 70–1, 76, 80, 88, 131, 149, 157, 172–3, 176, 179, 197, 205, 207–8, 210, 212, 271, 281, 288, 299–300, 335

deaf; *see* palaeopathology

death xviii–xx, xxii, 14, 64, 70, 74, 81, 110, 121–2, 125, 127, 129 n. 7, 131–2, 144 n. 6, 170, 176, 179, 181, 183, 185–6, 190, 196, 204–5, 207 (myth), 207–10, 228, 230, 235, 257 (risk), 261, 268, 272–4, 276, 281–3, 290–1, 296, 302, 309, 312–3, 315, 320–2, 324, 326–34

decline 167, 283, 286–7, 291, 300, 303

decomposition/decompose 70–1, 76–7, 80–1

decoration/decorate xxi, 5, 12–3, 16 n. 12, 25–6, 31, 41, 48, 53, 59, 64, 66, 66–7 nn. 10+19, 70, 73, 89, 103, 116, 122, 135, 145 n. 12, 160, 178, 185, 191–2, 197, 281

decree; *see also* doctrine; dogma; law; norm 205, 209

Dede Mezarı 172

dedication/dedicate 3, 5, 9, 13–4, 106, 110, 152–3, 212

deficiency xviii, xxii, 151, 268, 272, 274, 282, 287, 290–1, 301, 303

deforestation 283

deformation xxii, 258, 260

Değirmentepe 314

degradation (of DNA) 220, 225, 229

deities, Greco-Roman and Lycian 27, 66 n. 7, 87
  Aphrodite 121–2
  Apollo 8–9, 12, 27 (Karios), 66 n. 7, 203, 207, 269
  Artemis 3, 6, 224–5
  Asclepius/Asklepios xxii, 14
  Cronos 106 n. 12
  Cybele 207
  Dionysos 89
  Hades 8–9, 14
  Heracles 66 n. 7, 114, 116, 118
  Huwedri 87
  Kore 9
  Maliya 87
  Nemesis 145 n. 24
  Pluto 203, 207, 269
  Serapis 9, 14
  Trqqas (storm god) 87–8, 106, 106 n. 12

delivery; *see* birth

Delphi 118, 205

demand 16, 150

demarcation/demarcate 153, 160

demography xvi, xviii, xx, 131–2, 136, 138–9, 144, 150, 153, 209, 224, 250, 252, 268, 272–3, 281, 286, 288, 303, 306, 308–9

demolition/demolish 8–9, 12, 97, 192, 194, 203–4, 207

demos 88, 99, 102

Denizli xvii, 19, 26, 168–9, 228

depiction/depict 86–7, 91, 95, 105–6, 114, 118, 208

deposition 3, 52, 54, 59, 64, 66 n. 2, 69–1, 75–8, 80–2, 152, 154, 159, 167, 172, 185, 190, 199, 202–3, 206, 208, 211–2, 272, 306

descendant; *see* heir

destruction/destruct 8–9, 14, 36, 110, 190, 203–4, 269, 283

Develi 21

diachronic xvii–xviii, 71, 126, 149–51, 184, 196, 268–9

diagenesis 230, 300

diagnosis/diagnostic 113–4, 262, 272, 274, 276, 279–80, 282–3, 289, 291–2, 294–5, 300–3, 309, 319, 324, 329–30, 332–5

diaspora 69–71, 78, 81,

diet/dietary xvi, xviii, xxi–xxii, 134, 191, 228–35, 237–47, 250, 264, 273–4, 309

digestion/digest 231, 238

dilek agacı (tree of desires) 7

disability 286

disaster (natural) xx

disease; *see* palaeopathology

dishonour; *see also* honour 70

dislocation/dislocate 71, 77, 81, 140, 204, 325, 327–9

disturbance/disturb xx (climatic), xxii (mental), 22, 73 (non-), 76, 78, 80 (un-), 82, 89, 118, 157, 160, 167–8, 190 (un-), 194, 199, 202, 212 n. 4, 255, 260–1, 270, 282, 306 (un-), 320, 323, 335

diversity 66, 121–4, 144, 151, 220–1, 223–5, 232

divination xix

divine/godly 90–1, 105

divine retribution 88